Electrochemical Sensors Based on Carbon Composite Materials

Fabrication, properties and applications

IOP Series in Sensors and Sensor Systems

The IOP Series in Sensors and Sensor Systems includes books on all aspects of the science and technology of sensors and sensor systems. Spanning fundamentals, fabrication, applications and processing, the series aims to provide a library for instrument and measurement scientists, engineers and technologists in universities and industry.

The series seeks (but is not restricted to) publications in the following topics:

- Advanced materials for sensing
- Biosensors
- Chemical sensors
- Industrial applications
- The Internet of Things
- Lab-on-a-chip
- Localization and object tracking
- Manufacturing and packaging
- Mechanisms, modelling and simulations
- MEMS/NEMS
- Micro- and nanosensors
- Non-destructive testing
- Optoelectronic and photonic sensors
- Optomechanical sensors
- Physical sensors
- Remote sensors
- Sensing for health, safety and security
- Sensing principles
- Sensing systems
- Sensor arrays
- Sensor devices
- Sensor networks
- Sensor technology and applications
- Signal processing and data analysis
- Smart sensors and monitoring
- Telemetry

Authors are encouraged to take advantage of electronic publication through the use of colour, animations, video and interactive elements to enhance the reader experience.

A full list of titles published in this series can be found here: https://iopscience.iop.org/bookListInfo/iop-series-in-sensors-and-sensor-systems.

Electrochemical Sensors Based on Carbon Composite Materials

Fabrication, properties and applications

Edited by
Jamballi G Manjunatha
Department of Chemistry, Field Marshal KM Cariappa College, Constituent College of Mangalore University, Madikeri, Karnataka, India

IOP Publishing, Bristol, UK

ISBN 978-0-7503-5127-0 (ebook)
ISBN 978-0-7503-5125-6 (print)
ISBN 978-0-7503-5126-3 (myPrint)
ISBN 978-0-7503-5129-4 (mobi)

DOI 10.1088/978-0-7503-5127-0

Version: 20220901

IOP ebooks

British Library Cataloguing-in-Publication Data: A catalogue record for this book is available from the British Library.

Published by IOP Publishing, wholly owned by The Institute of Physics, London

IOP Publishing, Temple Circus, Temple Way, Bristol, BS1 6HG, UK

US Office: IOP Publishing, Inc., 190 North Independence Mall West, Suite 601, Philadelphia, PA 19106, USA

Dedicated to my mother, J Sharadamma Jamballi.

Contents

Preface

The developments in *Electrochemical sensors based on carbon composite materials: fabrication, properties, and applications* have been the considerable methodology in the improvement of electrochemical detection in several research areas such as soil analysis, metal ion detection, forensic sample analysis, environmental monitoring, food quality assessment, biomedical applications, biosensors, antibiotic drug testing, organic molecular analysis, carbon nanomaterial-based sensors, herbicides detection, disposable sensors, detection of environmental pollutants, industrial analysis, hormones detection, energy storage devices, and so on. In fact, innovative approaches in the fabrication of electrochemical sensors based on carbon composite sensors and their properties and applications for the above-mentioned research themes have been established. Carbon composite materials are the best sensing tools and their surface properties are enhanced through surface activation using chemical and material-based surface-active agents which provide the simplest, most affordable, sensitive, eco-friendly, and toxic-free electrochemical sensors. Carbon composite materials have a prodigious influence on the construction of electrochemical sensors and their real-time appliances due to their exceptional features such as higher mechanical, thermal, and electronic properties with improved biological, physical and chemical compatibility, larger electrochemical active surface area, and so on. Electrochemical sensors are well known for their many useful properties which include: simple to handle, rapid response, low consumption of sample, high sensitivity, selectivity, reproducibility, and so on. This book contributes a prospect for a new carbon-based composite sensor materials in the electrochemical analysis with advanced features and applications.

Acknowledgement

I want to extend my sincere thanks to John Navas (Senior Commissioning Manager, IOP Publishing, UK) and Phoebe Hooper (ebooks Editorial Assistant, IOP Publishing, UK) for providing necessary information to complete the book. Also, I extend my sincere thanks to all the contributors for contributing most important chapters to complete the book.

Editor biography

Dr Jamballi G Manjunatha

Dr J G Manjunatha is an Assistant Professor in the Department of Chemistry, Field Marshal KM Cariappa College, Constituent College of Mangalore University, Madikeri, Karnataka, India. He received his PhD in Chemistry from Kuvemmpu University, India and a Postdoc from the University of Kebangsaan, Malaysia. His research interests focus on the fabrication of electrochemical sensors for the detection of biologically active molecules and the fabrication of binderless supercapacitors. He published more than 142 research articles in various reputed international Journals and he is an editor for many books (Elsevier IOP publishing, and Bentham science publishers) and special issues (IOP Science Publisher, Frontiers in Sensors, and MDPI). Also, he is an editorial member for various reputed peer-reviewed journals and he has been ranked as the World's Top 2% Scientist in the Field of Energy and Analytical Chemistry in 2019 and 2020 (Listed in the published articles by Stanford University Scientists, USA).

List of contributors

Roberto Acevedo
Facultad de Ingeniería y Tecnología, Universidad San Sebastián, Chile

Fatma Ağın
Karadeniz Teknik University, Faculty of Pharmacy, Department of Analytical Chemistry, Turkey

S Ananda
Department of Studies in Chemistry, Manasagangotri, University of Mysore, India

Nida Aydogdu
Afyonkarahisar Health Sciences University, Faculty of Pharmacy, Department of Analytical Chemistry, Turkey

Hadi Beitollahi
Environment Department, Institute of Science and High Technology and Environmental Sciences, Graduate University of Advanced Technology, Iran

Amoolya Chandran
Department of Applied Chemistry, Cochin University of Science and Technology, India

H C Charan Kumar
Department of Chemistry, PES College of Engineering, India

M M Charithra
Department of Chemistry, FMKMC College, Constituent College of Mangalore University, India

B Chethan
Department of Physics, Indian Institute of Science, India

Nilakantha Dash
Department of Physics, School of Applied Sciences, KIIT, India

Anerise de Barros
Institute of Chemistry, University of Campinas, Brazil

Amanda de S M de Freitas
Institute of Science and Technology (ICT), Federal University of São Paulo, Brazil

Ersin Demir
Afyonkarahisar Health Sciences University, Faculty of Pharmacy, Department of Analytical Chemistry, Turkey

Zahra Dourandisha
Environment Department, Institute of Science and High Technology and Environmental Sciences, Graduate University of Advanced Technology, Iran

Cem Erkmen
Ankara University, Faculty of Pharmacy, Department of Analytical Chemistry, Turkey

Marystela Ferreira
Center of Science and Technology for Sustainability, CCTS, Federal University of São Carlos, Brazil

Mohammad Abu Haija
Department of Chemistry, Khalifa University of Science and Technology; Center for Catalysis and Separations, Khalifa University of Science and Technology, United Arab Emirates

K Girish Kumar
Department of Applied Chemistry Cochin University of Science and Technology, India

Çiğdem Kanbeş Dindar
Ankara University, Faculty of Pharmacy, Department of Analytical Chemistry, Turkey

Keerthi Kodakat
Department of Applied Chemistry Cochin University of Science and Technology, India

K Girish Kumar
Department of Applied Chemistry, Cochin University of Science and Technology, India

Sevinc Kurbanoglu
Ankara University, Faculty of Pharmacy, Department of Analytical Chemistry, Turkey

Cristiane Carla Maciel
Science and Technology Institute of Sorocaba (ICTS), Brazil

Renan S Mariano
Universidade Estadual de Londrina, Centro de Ciências Exatas, Departamento de Química, Brazil

Gabriel J Mattos
Universidade Estadual de Londrina, Centro de Ciências Exatas, Departamento de Química, Brazil

Milan Z Momčilović
Department of Chemical Dynamics and Permanent Education, 'Vinča' Institute of Nuclear Sciences—National Institute of the Republic of Serbia, University of Belgrade, Serbia

Fariba Garkani Nejad
Environment Department, Institute of Science and High Technology and Environmental Sciences, Graduate University of Advanced Technology, Iran

Srikantamurthy Ningaiah
Department of Chemistry, Vidyavardhaka College of Engineering, Visvesvaraya Technological University, India

Goldamol S Pallam
Department of Applied Chemistry, Cochin University of Science and Technology, India

Sulagna Parhi
Department of Physics, School of Applied Sciences, KIIT, India

Swetapadma Praharaj
Department of Physics, School of Applied Sciences, KIIT, India

V Prasad
Department of Physics, Indian Institute of Science, India

R Ramesh
Department of Physics, Periyar University, India

Varsha Rani
Department of Biotechnology, Shoolini University, India

Dibyaranjan Rout
Department of Physics, School of Applied Sciences, KIIT, India

S Swathy
Department of Applied Chemistry, Cochin University of Science and Technology, India

Sonia Sam
Department of Applied Chemistry Cochin University of Science and Technology, India

Monima Sarma
Department of Chemistry, KL Deemed to be University (KLEF), India

Elen R Satori
Universidade Estadual de Londrina, Centro de Ciências Exatas, Departamento de Química, Brazil

Ozge Selcuk
Ankara University, Faculty of Pharmacy, Department of Analytical Chemistry, Turkey

Shajahan Shanavas
Department of Chemistry, Khalifa University of Science and Technology, United Arab Emirates

R Shashanka
Department of Metallurgical and Materials Engineering, Bartin University, Turkey

R Shilpa
Department of Chemistry Yuvaraja's College, University of Mysore, India

Pallavi Singhal
Environmental Monitoring and Assessment Division, Bhabha Atomic Research Centre, India

Devika Sudha Ravindran
Department of Applied Chemistry Cochin University of Science and Technology, India

Somayeh Tajikb
Research Center of Tropical and Infectious Diseases, Kerman University of Medical Sciences, Iran

Arun Thirumurugan
Laboratorio de Medio Ambiente y Energía, Universidad de Atacama Sede Vallenar, Chile

S Thomas
International and Inter University Centre for Nanoscience and Nanotechnology, MahatmaGandhi University, India

Didem Nur Unal
Ankara University, Faculty of Pharmacy, Department of Analytical Chemistry, Turkey

Bengi Uslu
Ankara University, Faculty of Pharmacy, Department of Analytical Chemistry, Turkey

Madan L Verma
Department of Biotechnology, School of Basic Sciences, Indian Institute of Information Technology Una, India

Chapter 1

An overview of voltammetric techniques to the present era

Sulagna Parhi, Nilakantha Dash, Swetapadma Praharaj and Dibyaranjan Rout

The last few decades have evidenced the growth and maturation of electroanalytical techniques with voltammetry at the center of the field. The early stages of this method encountered many problems making it less significant as compared to spectroscopy, chromatography, and electrophoresis. However, advancement in the late 20th century was mostly driven by sophisticated and inexpensive instrumentation including operational amplifiers, Polaroid land cameras, etc. The voltammetry techniques exhibit merits of superior sensitivity with a wide range of inorganic and organic species (10^{-12}–10^{-1} M), a large temperature span, simultaneous measurement of several analytes, easy and rapid analysis, capability to govern mechanistic and kinetic parameters, generation of different types of waveforms with simplicity, and capability to measure small currents. Researchers can usually explore these advantages to quantify the electrochemical properties of a diverse range of compounds. It has been proved to be useful to conduct fundamental studies on redox reactions in different media, electron transfer processes, adsorption mechanisms, transport properties, and thermodynamics of solvated species. Applications also extend to study compounds of pharmaceutical interest and can prove to be effective in the analysis of complex structures when combined with HPLC. This chapter will cover the development of the major types of voltammetric techniques and their applications to different fields of study.

1.1 Introduction

1.1.1 Voltammetry

The Czech scientist Jaroslav Heyrovsky was the first to discover the technique 'voltammetry' in the 1920s. The term voltammetry is taken from 'voltamperometry'. At first, the word 'polarography' was suggested for this technique which is today used widely for 'voltammetry with a dropping mercury electrode'. This technique is considered to be a form of 'electrochemistry'. As these techniques outnumber all the

doi:10.1088/978-0-7503-5127-0ch1

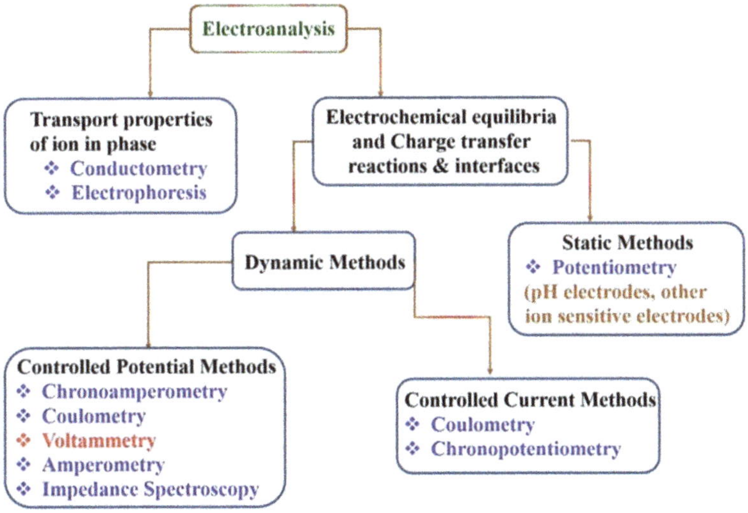

Figure 1.1. List of electroanalysis techniques.

other electroanalytical techniques it is often identified by the term electroanalysis (figure 1.1). This chapter provides a brief discussion on one of the important techniques in electroanalysis i.e., voltammetry. The technique mainly includes measurement of current (I) flow via working electrodes submerged in a solution having an electroactive sample to be scrutinized by varying potential (v), following some systematic manner. Today this technique is becoming important due to its simple, cheap, and rapid way of determining biologically and environmentally vital genera. It offers supreme selectivity and sensitivity, as the analyte can be quickly recognized by its voltammetric peak potential.

In voltammetric experiments, the electrochemical cell, as shown in figure 1.2, has a working electrode (also called an indicator), a reference electrode, and a counter electrode (also called auxiliary). The effect of charge can be realized across the interface of the electrode. When the reaction/transfer of interest is taking place, it is considered to be the working electrode. With the application of the required potential, a redox reaction takes place at the surface of the working electrode resulting from the flow of current. The reference electrode has a known and fixed electrode potential. The reference electrode controls the working electrode potential by measuring the voltage among the two electrodes. At the surface of the electrode, the potential difference of the working electrode is varied with the help of the potentiostat. Generally, the working electrode is grounded to control the solution potential.

1.1.2 General theory

There are several well-known laws to describe voltammetry, i.e., the effects of applied potential (E) and behavior of the redox current [1]. According to Nerst or Butler–Volmer equations, at the electrode surface ($C_O{}^o$: concentration oxidation species; and $C_R{}^o$: concentration redox species) the redox species concentrations are controlled by the potential applied and the rate of reaction (k^o) respectively.

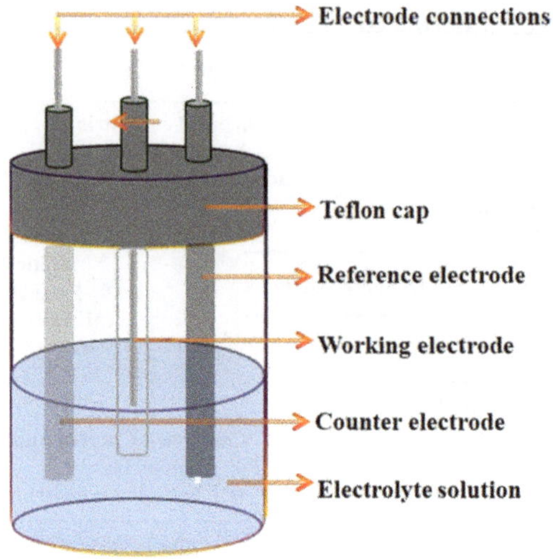

Figure 1.2. A three-electrode electrochemical cell.

Fick's law describes the relationship between the redox current and material flux at the solution–electrode interface where diffusion plays an important role (this current is also called the Faradaic current). The applied potential (*E*), in a reversible electrochemical reaction is given as $O + ne^- \Leftrightarrow R$, forces the concentrations of O and R at the electrode surface (i.e., $C_O{}^o$ and $C_R{}^o$) to a ratio following the Nernst equation:

$$E = E^\circ - \frac{RT}{nF}\ln C_R^\circ / C_O^\circ$$

where R = molar gas constant (8.3144 J mol^{-1} K^{-1}), E^o = standard reduction potential of the redox couple, F = Faraday constant (96,485 C/equiv.), T = absolute temperature (K) and n is the number of electrons transferred. As the applied potential changes, the ratio ($C_R{}^o/C_O{}^o$) also changes. For a more negative potential the ratio becomes larger (O is reduced) and for a more positive potential the ratio becomes smaller (R is oxidized). The Butler–Volmer equation links potential, concentration, and current and is given by,

$$\frac{i}{nFA} = k^\circ\left\{C_O^\circ \exp[-\alpha\theta] - C_R^\circ \exp[(1-\alpha)\theta]\right\}$$

where $\theta = nF\frac{(E - E^\circ)}{RT}$, k^o = heterogeneous rate constant, α = transfer coefficient and A = area of the electrode. Again, the flow of current is directly dependent on the flux of material and the electrode surface. Fick's law describes the concentration gradient and mass transport due to the formation of new O and R and the destruction of O and R. According to this law, the flux of matter (φ) which is proportional to the concentration gradient is given by

$$\varphi = -AD_o(\partial C_o / \partial x)$$

where x = distance from the electrode surface and D_o is the diffusion coefficient of O. A similar expression can be written for R. The flux of O and R at the electrode surface control the rate of reaction and the Faradaic current in the cell. A quantitative analysis of current suggests the rate at which a species can be oxidized or reduced at the electrode surface. Factors that affect the value of current are: the cell volume, the number of electrons transferred, the solution resistance, the concentration of redox species, and the size, shape, and material of the electrode. Mass transfer can occur by migration (i.e., the movement of a charged ion in an electric field) or convection (movement of the electroactive species by thermal current or density gradients or stirring the solution or rotating the electrode) [1].

1.1.3 Voltammetric techniques and applications

The widely used voltammetric techniques are generally classified into sweep type and polarographic type according to their applications. The basic difference in both voltammetric techniques is that the solution is stirred (polarographic) or not stirred (sweep) after each set of applied potential. There are also some additional methods. A schematic diagram listing all the voltammetric techniques is given in figure 1.3. It was observed that the voltammetric technique has become a promising analytical tool in the study of electrochemical reactions, industrial quality control, environmental monitoring, solar energy conservation and many more. This technique is used to measure kinetic rates and constants, determine thermodynamic properties of solvated species, determine organic/inorganic compounds in aqueous/nonaqueous solutions, determine electron transfer, and reaction mechanisms, determine adsorption processes on surfaces, basic studies of oxidation and reduction processes in various media etc.

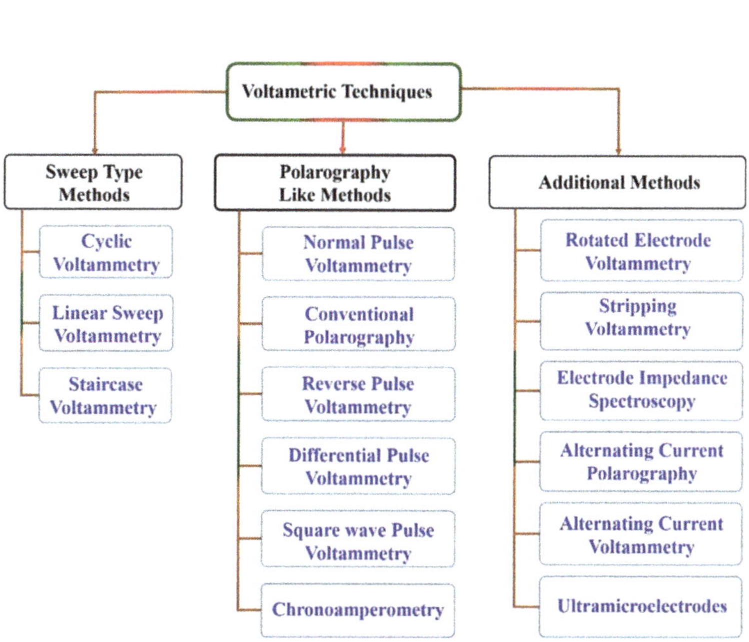

Figure 1.3. Types of voltammetric techniques.

The common applications of this technique are to determine the redox potential, the metal ion concentrations in H_2O to sub-parts-per-billion levels, pharmaceutical compounds, eluted analytes in high-performance liquid chromatography (HPLC), and flow injection analysis, number of electrons in redox reactions, and study the kinetics of the reaction. Recently, using carbon-based electrodes, the electrochemical properties of many pharmaceutical samples and biological fluids were analyzed through voltammetric techniques particularly cyclic voltammetry and differential pulse voltammetry [2–5]. Some of the important techniques and their applications in recent years are discussed in the following sections.

1.2 Voltammetric techniques

1.2.1 Sweep voltammetric techniques

1.2.1.1 Cyclic voltammetry

In recent years, cyclic voltammetry (CV) has proved to be one of the most versatile and frequently used electroanalytical techniques as it provides a lot of experimental information and its insight into kinetic as well as thermodynamic details of several chemical systems. It gives exact mathematical analysis and has simple diagnostic criteria as well. It is used in many fields, e.g.: electrochemistry, organic chemistry, inorganic chemistry, and biochemistry, due to its versatility and simple measurement. In this technique, the potential difference between two potentials varies at a constant rate and when either of the potentials reached a limiting value, the direction of the potential change is swapped. Any of the potentials can be the starting potential, however is often 0 V. Between the limiting values of the potentials, each continuous change is referred to as a 'linear sweep'. Among voltammetric techniques, cyclic voltammetry has two main advantages: (i) ease in the interpretation of results, and (ii) large potential ranges can be rapidly examined. Generally, a cyclic voltammogram as shown in figure 1.4 can be used to extract information like the reversibility of oxidation and reduction, and the approximate formal potential readily.

1.2.1.1.1 Determining the reversibility of a reaction

The reversibility of a reaction can be determined by using the CV technique. There are two types of reversibility reaction: 'chemically reversible' and 'thermodynamically reversible'. The Faradaic electrode reaction is a heterogeneous charge transfer between the electroactive species A and the electrode. It depends upon the electrode potential; A is either oxidized or reduced to B. The shape of the voltammetric current–voltage curve depends on the above two processes: (i) the heterogeneous charge transfer, and (ii) diffusional mass transport. The Butler–Volmer equation describes the heterogeneous charge transfer at the electrode surface and is given by,

$$j_A(0, t) = \frac{i}{nFA} = C_A(0, t)k^\circ \exp[-\alpha nF/RT(E - E^\circ)] - C_B(0, t)k^\circ$$
$$\exp[(1 - \alpha)nF/RT(E - E^\circ)]$$

where, k° = heterogeneous standard rate constant, α = transfer coefficient, E° = standard potential, A = electrode area, j = flux at the electrode interface and F = Faraday constant. The current density (heterogeneous flux) depends upon

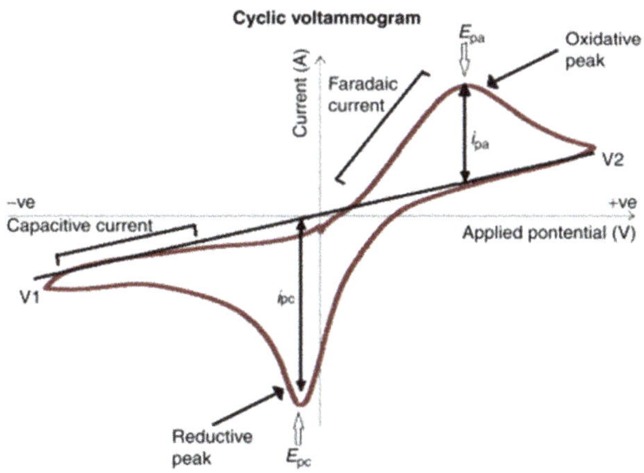

Figure 1.4. A schematic of cyclic voltammogram, reprinted from [6], copyright (2016) with permission from Elsevier.

the surface concentration of the redox partner, actual electrode potential, and upon $k°$. In the thermodynamic equilibrium condition, $E°$ experience surface concentration of the redox reaction to be equal. According to Fick's law, the time-dependent concentration distribution in the expanding diffusion layer is,

$$\partial c_i/\partial t = D_i(\partial^2 c_i/\partial x^2), \quad \text{where, i} = A, B$$

This equation holds for purely linear semi-infinite diffusion. However, for analytical purposes, the electrode area to volume ratio is chosen to be extremely large to increase the sensitivity in reverse voltammetry. Here the diffusion is infinite instead of semi-infinite. The concentration gradient at the electrode surface is directly proportional to the charge flux i.e.,

$$j_A(0, t) = -D_A(\partial C_A/\partial x)_{x=0} = D_B(\partial C_B/\partial x)_{x=0}$$

 # Case 1: reversible process, $k°> 10^{-1}$ cm s^{-1}

A dynamic equilibrium is established at the phase boundary due to the high rate of heterogeneous charge transfer. The Butler–Volmer equation is reduced to the Nerst equation. This indicates that the surface concentration does not depend upon heterogeneous kinetic effects, it only depends upon the actual electrode potential. This case is known as reversible as thermodynamic equilibrium exists at the phase boundary. The current is mainly influenced by mass transport. With time-dependent potential $E(t)$ the surface concentration of reactant A decreases and that of B increases. Hence, the gradient and the charge flux increase until $E°$ exceeds and the surface concentration of A becomes negligible. Thereby the current becomes maximum and the diffusion layer expands, decreasing the concentration gradient at the electrode surface. The current curve decays. For the opposite potential direction, similar phenomena occur. A positve current is produced by the potential-dependent conversion of B into A. The current gradually decreases and reaches a minimum value and then slowly decays as the species B in the diffusion layer is used up. In a reversible reaction, $\frac{i_{pa}}{i_{pc}} = 1$ and $E_{pa} \bullet E_{pc} = 0.52/nV$. Where i_{pa} and i_{pc} are the peak currents, E_{pa} and E_{pc} are the peak potentials.

 # Case 2: irreversible process, $k° < 10^{-5}$ cm s^{-1}

At the electrode, the charge transfer reaction is very slow, and the cathodic or anodic heterogeneous reaction rate depends on the potential. In an irreversible process, the surface concentration depends upon heterogeneous reaction and is far removed from thermodynamic equilibrium. Here, the Nerst equation is not applied. Hence, we cannot compare the measured potential values with the thermodynamic equilibrium potential.

 # Case 3: quasi-reversible: $10^{-1} > k° > 10^{-5}$ cm s^{-1}

Using sweep rate, the time scale of the CV experiment can be varied up to a wide range, as a result the electrode reactions can be steered out of charge transfer control and that of diffusion control and vice versa. The two components dominate the electrode process and the characteristics are reflected in the cyclic voltammetric curve. Reversible and quasi-reversible cases occur frequently in practice under standard measuring conditions. The current is given by,

$$i = nFAC_A^*(D_A a)^{1/2}\pi^{1/2} \times (at)$$

where, $at = nFvt/RT = nF/RT(E(t) - E(i))$ and C_A^* = substrate concentration of the electrolyte. Quantitative and qualitative information about thermodynamic and kinetic properties of the system can be obtained from the electron–transfer reactions without analyzing the whole cyclic voltammogram. Two peak potentials, i.e., E_{pa}, and E_{pc} and the two peak currents i_{pa}, and i_{pc} are the important factors. Irrespective of sweep rate charge transfer is reversible and quasi-reversible ($\alpha = 0.5$) process without coupled chemical reactions holds good

$$\frac{i_{pa}}{i_{pc}} = 1$$

To determine the current base sweep in the reverse sweep Nicholson derived an empirical formula based on the switching potential E_λ lies ca. 60 mV n^{-1} after the peak potential.

$$\frac{i_{pa}}{i_{pc}} = \frac{i_{ao}}{i_{pc}} + 0.485 i_{\lambda 0}/i_{pc} + 0.086$$

A voltammogram is called quasi-reversible when it has both cathodic and anodic peaks but does not have $E_{pa} \cdot E_{pc} = 0.52$ nV^{-1}. The reaction is not chemically reversible when $i_{pa}/i_{pc} \neq 1$ indicates other reactions except the redox reaction are taking place in the compound. When additional reactions occur that is called a coupled reaction [7].

1.2.1.1.2 Determination of formal reduction potential of a species

The formal potential is the measurement of the potential of a cell with both oxidized and reduced species having equal concentration. It involves the measurement of potential in a galvanic cell. There are two solutions of different redox pairs in a galvanic cell having two electrodes and a connection between the two solutions for charge transfer. A positive sign of potential refers to reduction at the right and oxidation at the left electrodes and vice-versa. The difference in potential is related to Gibb's free energy and is given by $\Delta G = -nFE$, n is the number of molecules. The standard reduction potential is the standard potential where the reference electrode is the normal hydrogen electrode. Under the standard conditions, the potential is given by

$$E^{\circ}_{\text{cell}} = E^{\circ}_{\text{right}} - E^{\circ}_{\text{left}}$$

Now pure E processes can be interpreted as, $\Delta E_p = E_{pa} - E_{pc}$.

Using the polarographic method the half-wave potential can be calculated easily. The half-wave potential of the redox reaction is given by,

$$E_{1/2} = E_{pa} - \Delta E_p/2 = \frac{(E_{pa} + E_{pc})}{2},\ 0.3 \leqslant \alpha \leqslant 0.7 \text{ and } E_{\lambda} \leqslant E_{pc} - 100\text{ mV}$$

$E_{1/2}$ is associated with standard redox potential E^0 as, $E_{1/2} = E^{\circ} + (RT/nF)\ln(D_r/D_0)^{1/2}$ $E_{1/2}$ corresponds to the thermodynamic redox potential. For very fast charge transfer ΔE_p under a Nerst condition is given by $\Delta E_p = 2.3RT/nF = \left(\frac{56.5}{n}\right)$ mV (25 °C) [7].

1.2.1.1.3 Measurement of electron transfer kinetics

Nicholson introduced the kinetic parameter or the standard rate constant k° to explain the kinetics of electron transfer. When $E = E^{\circ}$ (formal potential). Nicholson introduced the kinetic parameter for various values of $\Psi = \left(\frac{D_0^{\alpha/2}}{D_R^{\frac{\alpha}{2}}} k^{\circ}\right)\left(\frac{D_0 \pi n F v}{RT}\right)^{\frac{1}{2}}$ where, D_o and D_R are the diffusion constant of oxidized species and reduced species, v = scan rate, T = temperature, F = Faraday constant, n = number of electrons in the redox reaction, R = universal gas constant. The major drawback of this method is the uncompensated resistance (i.e., the resistance between the working electrode and the entire equipotential surface that traverses the tip of the reference electrode) may dominate the cyclic voltammogram [7].

1.2.1.1.4 Determining the energy levels of semiconducting polymers

CV is one of the most useful techniques for the characterization of organic materials and the energy band diagram. There are two main technical terms used for organic semiconductors, i.e., HOMO and LUMO. HOMO is the energy required to extract an electron from a molecule which represents an oxidation process. LUMO is the process of introducing an electron to a molecule which is a reduction process. In CV, by the measurement of redox potentials i.e., E_{red} and E_{ox} one can measure HOMO and LUMO. To calculate the energy of HOMO and LUMO, ferrocene with a ferrocene value of –4.4 eV is used.

The equations to calculate energy levels are,

$$E(\text{HOMO}) = -e[E_{\text{ox}}^{\text{onset}} + 4.4]$$

$$E(\text{LUMO}) = -e[E_{\text{red}}^{\text{onset}} + 4.4]$$

Energy levels corresponding to different energy values of the electronic states correspond to UV or visible radiation in the case of organic molecules. The energy emitted by the electromagnetic radiation can be absorbed by the molecule at resonance, as a result, an electron can jump from a lower to a higher energy level. These transitions can be measured by a UV–vis spectrophotometer. The optical gap energy can be calculated as,

$$E_g = 1242/\lambda_{\text{onset}}$$

where E_g= optical gap energy and λ_{onset}= longest absorption wavelength. One can estimate if two materials can work together in the doner–accepter system by combining these two systems [8].

1.2.1.1.5 Assignment and characterization of coupled reaction

To allocate coupled reactions to the electrochemically formed species, CV is used. The characterization of processes can be done by estimating the shape, plotting dependence on peak position and $i_{pa}i_{pc}$ on scan rate.

1.2.1.1.6 Recent advances

In a study by Thomas M Asrat *et al* the sub-second co-detection of guanine, adenine, and cytosine, first as free nucleosides then within plasmid DNA, plasmid RNA, and custom synthesized oligos are done by fast-scan cyclic voltammetry (FSCV). They observed FSCV have higher sensitivity and produce higher peak oxidation currents when detecting ssDNA samples and shorter oligonucleotides at an equivalent concentration of nucleobase (figure 1.5(a)). FSCV with CFMEs (carbon fiber microelectrode) can co-detect nucleobases when it is polymerized into RNA and DNA [9]. In another study made by Rocha *et al* image-based analysis of FSCV color

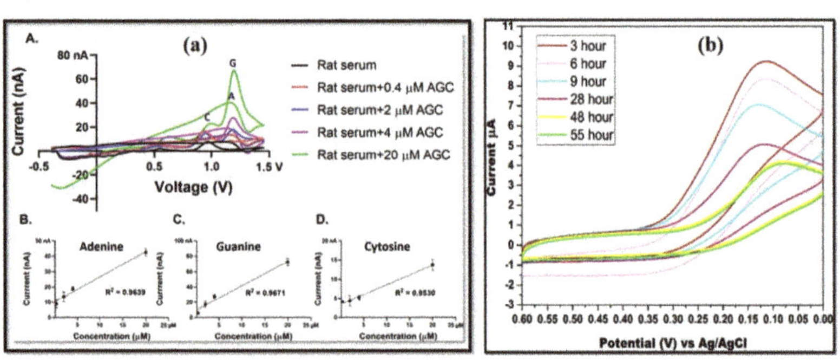

Figure 1.5. (a) Cyclic voltammograms of the colligative spike-in test with adenine, guanine, and cytosine. A linear variation of the concentration of adenine, guanine, cytosine, and peak oxidation current with increasing spike-in amount, reprinted from [9], copyright (2021) ACS publication, (b) cyclic voltammogram at different time intervals of 60 mg L^{-1} Cr(VI) in C + G beads system (potential range: 0.0–0.6 V; scan rate: 0.1 V s^{-1} and sensitivity: 0.001 A V^{-1}), reprinted from [11], copyright (2021), the Author(s), with permission of Springer Nature.

plots especially the structural similarity index (SSIM) to detect rapid neurochemical events is done. First, they focus on the identification of spontaneous adenosine events and then the oxidation peaks that grow with time. They conclude for FSCV data analysis, SSIM is a general strategy that uses 3-D data to identify multiple analytes in an automated, efficient analysis [10].

Watwe and Kulkarni reported cyclic voltammetry with the gold electrode as an effective electrochemical technique for the detection of Cr(VI) in wastewater. The experimental data of Cr(VI) adsorption on C+G beads are used for thermodynamic and kinetic studies (figure 1.5(b)) [11]. Pushpanjali *et al* prepared a poly(L-ApA) layered graphene nanoplatelet paste electrode (PApA/GNpPE) for the selective determination of riboflavin and ascorbic acid. They studied the electrochemical behavior of riboflavin (RF) through cyclic voltammetry [12]. Hareesha and Manjunatha studied the electro-oxidation nature of formoterol fumarate in the presence of uric acid on the surface of poly thiazole yellow-G (TY-G) layered multi-walled carbon nanotube paste electrode (MWCNTPE). The good electrocatalytic nature of the electrode was confirmed by CV technique [13].

1.2.1.2 Linear sweep voltammetry

Linear sweek voltammetry (LSV) is considered to be one of the simple electrochemical techniques. Although LSV is similar to CV, it involves only one linear sweep from the lower to the higher potential limit. LSV uses a three-electrode system like that of CV. In this voltammetric technique, potential is swept linearly with time between the reference electrode and the working electrode, and the current is measured at the working electrode. The potential with which the species starts to reduce and oxidize is described by a trough or a peak in the current signal (figure 1.6). The concentration can be found from the limiting current's height and the identification of the unknown species is given by $E_{1/2}$. With the increase of scan rate the current sensitivity versus voltage also increases. The redox reaction at the WE increases with a higher potential per second.

The LSV set up has three electrodes (reference electrode RE, working electrode WE, and auxiliary electrode AE) and a potentiostat. The potentiostat provides potential to the three-electrode setup. An electrochemical cell having these three electrodes and the electrolytic solution of interest is connected to the potentiostat.

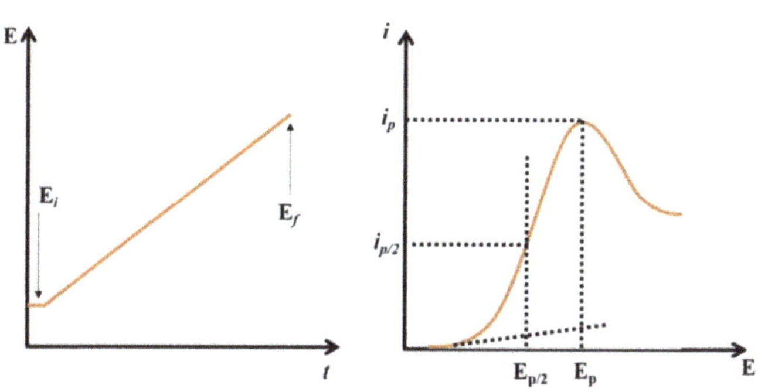

Figure 1.6. Linear potential sweep staring at E_i and resulting i–E curve.

The WE delivers the potential and the redox reaction occurs at this electrode. The scan rate is the slope of potential versus time and ranges from mV s^{-1} to 10^9 mV s^{-1}. At AE, also known as the counter electrode, the opposite mechanism to that in WE occurs. The amount of charge removed or added by the WE is observed by the reference and auxiliary electrode as a union. The AE keeps the RE at its known reduction potential by balancing the WE. At WE, the current is measured to plot the electrochemical response of a particular material. The interactions of cytochrome c (cyt c) with the calixarene carboxylic acid derivative (CX) incorporated to the lipid membrane sBLM consisting of dimyristoyl phosphatidylcholine (DMPC) were studied in a metal arrangement of isothermal charge transient spectroscopy (IQTS) with voltammetry. Hysteresis is observed in the double layer capacitance for the mixed DMPC+CX layer and is interpreted by the reduction of the charge involved in the ongoing redox reaction. Figure 1.7 shows a schematic experimental arrangement and the time domain capacitance vs. voltage along with the voltammetric characteristics on sBLM in an aqueous environment in the electrochemical cell configuration [14].

One can observe many key parameters using LSV to find the thermodynamic reversibility of a material such as the peak current (i_p) and the corresponding potential (E_p), and the potential at half the peak current ($E_{\mathrm{p}/2}$). For the system, $O + e^- \rightleftharpoons R$, with R absent initially (no coupled reaction) these parameters can be interpreted. The electrochemical parameters can be determined using these values at quasi-reversible and reversible limits.

1.2.1.2.1 The peak current

The peak current i_p in amperes, with bulk concentration $C_O{}^*$ in mol cm^{-3}, at 25 °C, the diffusion constant D_o in cm^2 s^{-1}, electrode area A in cm^2, and the scan rate v in Vs^{-1} is given as,

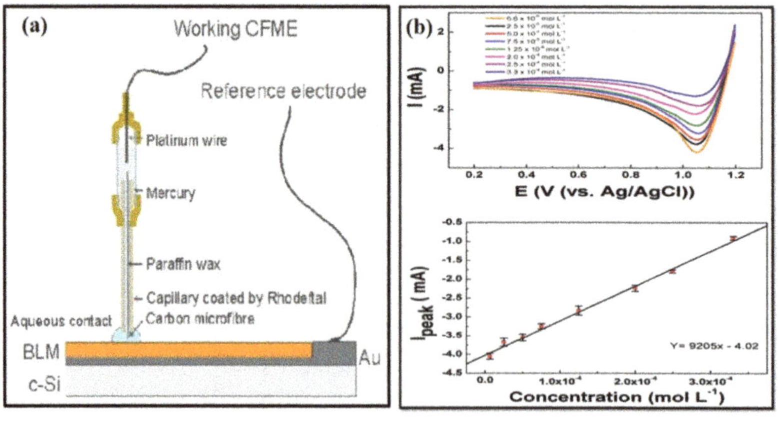

Figure 1.7. (a) Experimental set up of LSV (CFME = Continuous friction measuring instrument, BLM = bilayer lipid membrane), reprinted from [14], copyright (2012) with permission @Elsevier, (b) LSV curves registered for increasing concentrations of urea and linear relationship between the registered I_peak versus the concentration of urea. Reprinted with permission from [15].

$$i_p = (2.69 \times 10^5) A D_0^{1/2} C_0^* v^{1/2} k, \quad [k = 1, \text{ for reversible system}]$$

Value of k for the irreversible system is given as, $k = \alpha^{1/2}\left(\frac{2.99 \times 10^5}{2.69 \times 10^5}\right)$ [15, 16].

1.2.1.2.2 The peak current potential

The value of peak current is given as,

$$E_p - E_{\frac{1}{2}} = -\Xi\left(\frac{RT}{F}\right)$$

For reversible system, $\Xi = 1.109$.

$$\Xi = \frac{1}{\alpha}\left[0.780 + \ln\left(\frac{D_O^{1/2}}{k^0}\right) + \ln\left(\frac{\alpha F v}{RT}\right)^{1/2} + \alpha \ln\left(\frac{D_R^{1/2}}{D_O^{1/2}}\right)\right] \text{ [16, 17].}$$

1.2.1.2.3 The half-peak current potential

The formula for half-peak current potential is given as,

$$| E_{\frac{p}{2}} - E_p | = \Delta\left(\frac{RT}{F}\right)$$

In the case of reversible system, $\Delta = 2.20$ and in the case of irreversible system, $\Delta = \left(\frac{1.857}{\alpha}\right)$ [16, 17].

1.2.1.2.4 Recent advances

Lee *et al* reports an indirect electroanalytical qualification method (linear sweep voltammetry) for the sensitive and accurate qualification of urea in an aqueous sample using dimensional stable anode (DSA) of $Ti/RuO_2–TiO_2–SnO_2$. This method also allows urea quantification with a competitive LOD of 1.83×10^{-6} mol L^{-1} in a linear range of 6.66×10^{-6} to 3.33×10^{-4} mol L^{-1} of urea concentration [15]. Figure 1.7(b) shows the LSV curves registered for increasing concentrations of urea in KCI 0.10 mol L^{-1} and a linear change between the registered I_{peak} and urea concentration. LSV curves were recorded on the working electrode $Ti/RuO_2–TiO_2–SnO_2$ (50:40:10 atom. %) at $U = 50$ mV s^{-1} scan rates after holding 1.2 V versus Ag/AgCI for 60 s as analytical preconditioning. Mazzara *et al* developed an electrochemical sensor for the primary detection of uric acid and its validation in urine and milk samples. They used indium tin oxide coated on flexible polyethylene by co-electrodeposition at –0.8 V versus SCE for 200 s. Characterization of samples was done by electrode diffusion spectroscopy and electron scan microscopy. Detection of uric acid was done by the LSV method and it was observed that with increasing uric acid concentration from 10 to 1000 μM, the I_{peak} value increases with a limit of detection about 7.1 μM. The sensor has high selectivity for various interferents found in milk and urine matrix, as calcium, chloride ammonium ions and sodium [18].

1.2.1.3 Staircase voltammetry

Staircase voltammetry (SCV) is considered to be a derivative of LSV. Here in the case of SCV, the current is noted after a time delay with an electrode at a fixed potential. Then by changing the potential, the corresponding value of current is recorded and the process is repeated again and again over a range of potentials. The change in potential is linear, in direction and size, making a staircase as shown in the inset of figure 1.8. Although this technique is designed in such a way that it can be used in polarographic electrodes, still we can use it in the case of more standard, non-polarographic electrodes. In the case of a static mercury drop electrode, first of all, mercury drop is to be produced, then the motion of mercury is stopped, finally the mercury drop is kept for the whole process of applied potential steps. After the applied potential steps finish its running process, the mercury is finally dropped by restoring the solution for the other scan. To regain the initial concentration, in the case of non-polarographic electrodes, the solution can be stirred or left for some time. This technique results in the same way as that of linear sweep voltammetry. A similar cyclic voltammetry can be obtained by the combination of various staircases. The intention of this technique includes the performance of linear sweep voltammetry or cyclic voltammetry with suppressed charging currents. The sampling rate and scan rate is made lower to reduce the charging currents, which subsequently prevents fast scans. This method offers poor signal-to-noise. Recently, Schilder *et al* found the staircase voltammetry technique promising for the determination of the antioxidant nature of beverages such as wines, teas, and juices. For the characterization of the antioxidant nature, with its antioxidant activity and capacity, a new parameter is introduced called 'the antioxidant index (AI)'. As a result, they observed green tea, red wine, and non-superfoods have better antioxidant behavior than other teas, white wine, and superfoods respectively [19]. Ppetkovska and Gulaboski have reported a theoretical work on SCV of a surface catalytic mechanism that leads to reversible chemical step (surface catalytic EC_{rev}). They have considered the specific surface regenerative mechanism and their results can be

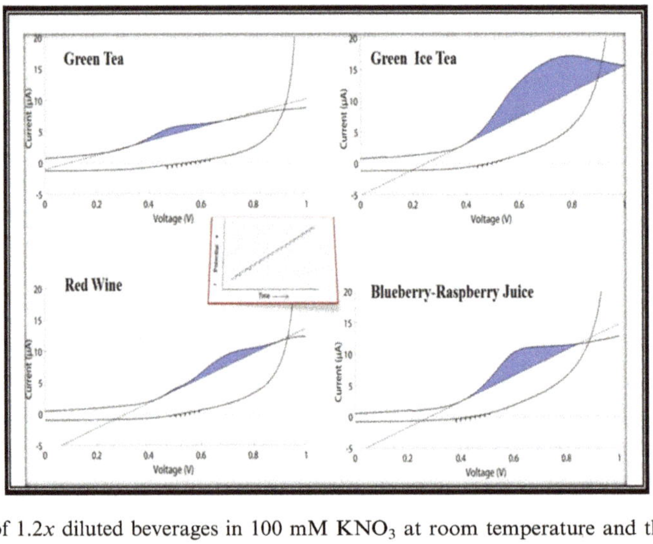

Figure 1.8. SCV of 1.2*x* diluted beverages in 100 mM KNO_3 at room temperature and their baselines. The blue represents the calculated area under the curve. SCV was measured clockwise on a 7.07 mm^2 glassy carbon working electrode at v_{scan} 50 mV s^{-1}. Adapted from [19], copyright (2020), with permission from Elsevier.

utilized in the analysis of thermodynamics and kinetics of drug-DNA and drug–drug interactions [20].

1.2.2 Polarography-like methods

1.2.2.1 Normal pulse voltammetry

This is based upon the electrochemical technique which undergoes measurement of cell current as a function of time and potential between the working and reference electrode. Initially, it has been observed that for a particular value of the voltage applied to the working electrode, there could not be any flow of current. But by applying a single pulse at the working electrode, the current starts flowing. Generally, the NPV experiment continues several times in which a series of pulses having certain amplitudes have been applied and each time magnitude of current flow has been recorded. NPV gives larger sensitivity when compared with the non-pulse method. Moreover, this method can measure diffusion coefficients and other physical quantities with more accuracy. Using this technique, the diffusion constant, diffusion concentration, electrode area, polarographic half-wave potential, and kinetic parameters can be calculated. Zhang *et al* correlate the diffusional current with the concentration of UCI_3 and $MgCI_2$ in molten eutectic LiCI-KCI using NPV. They found that NPV is a promising method (figure 1.9) for simultaneous analysis of UCI_3 and $PuCI_3$ in molten salt electrolytes for the use of spent nuclear fuel electrorefining [21]. Takayuki *et al* was able to differentiate norepinephrine (NE) from dopamine (DA) using NPV with 0.1–0.3 V rectangular pulse. They found with

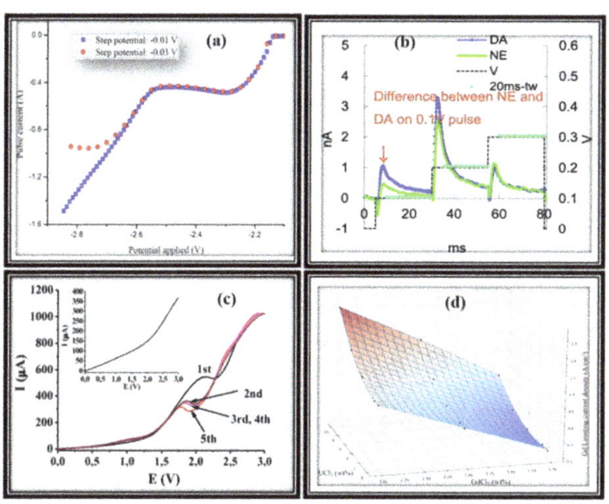

Figure 1.9. (a) NPV movements with 5 wt% UCl_3 and 1.5 wt% $MgCl_2$ in eutectic LiCl-KCl at different step potentials with pulse time 0.3 s and interval time 15 s [21] copyright IOP Publishing Ltd, CC By 4.0, (b) Current-time curve of the differential current (nA) of NPV. Light-blue line indicates the 20 ms time windows, reprinted from [22], copyright (2017) with permission Elsevier, (c) NPV curves for 50 mM phenol in different acetonitrile solvents with a scan rate of 0.1 V s^{-1} and pulse width of 5 ms in the presence of 0.1 M $TBuClO_4$ [23], copyright 2019, the Authors, with permission of Springer, (d) Gd NPV plateau current densities versus respective $GdCl_3$ and UCl_3 concentrations in LiCl-KCl eutectic at 773 k with plasma time 0.01 s and interval time 15 s, reprinted from [18], copyright (2018) with permission Elsevier.

a 0.2 V rectangular pulse, the major electrochemical current of NE and DA is obtained, but for 0.1 V the relative current is higher in the case of DA than NE (figure 1.9(b)). This results in the estimation of sequential equations which helps to differentiate between NE and DA [22]. Moreover, Laszlo Kiss *et al* studied the electrooxidation of phenol on the platinum electrodes in five aprotic non-aqueous solvents (acetonitrile, dimethyl sulfoxide, dimethylformamide, acetone, and tetrahydrofuran) (figure 1.9(c)). This NPV investigation indicates that the extent of electrode passivation is insignificant in three of the solvents used due to the application of the short anodic potential pulses [23]. In another work by Chao Zhang *et al* the analytes GdCI3 and UCI3 were studied in LiCI—KCI eutectic solvent at 773 K. A mixture matrix of varying concentrations of GdCI3 and UCI3 from 1 to 3 wt% and 1 to 10 wt % respectively was tested (figure 1.9(d)). The NPV measurements suggesting the presence of GdCI3 does not affect the U3+ NPV reduction current, whereas, for high U concentration, Gd3+ NPV reduction current was affected [24].

1.2.2.2 Chronoamperometry

Chronoamperometry is a unique electrochemical and time-dependent approach that utilizes a square wave potential to generate information in terms of electrode current. A link between current response and time is derived after employing a square step potential function to the working electrode of the system. Hence, this technique is used to estimate the time-dependent diffusion-controlled processes at the electrode surface. The current at the electrode swing is as per the diffusion of the analyte from the bulk solution. Experiments can be performed using either a single step (application of forward potential and then registering the current) or a double step (the second step comes as a reverse potential in addition to the forward step). It is different from both chronocoulometry and chronopotentiometry. In chronocoulometry method, the charge is measured with variation of time due to the applied potential step perturbation while in chronopotentiometry, change of potential is measured with time. Moreover, large current is produced in chronoamperometry which decays exponentially with respect to time as seen in RC circuits following the Cottrell equation. This basic equation records the current obtained every time after a high potential step in a reversible redox reaction as, $I(t) = \dfrac{nFAC_0 {D_0}^{1/2}}{\pi^{1/2} t^{1/2}}$, where n represents the stoichiometric electro number involved in redox reaction; F denotes Faraday's constant; A = surface area of the electrode and; C and D represent the concentration and diffusion constant respectively of the electroactive species in the electrolyte. A three-electrode system is used most commonly and large currents integrated over long periods give a good signal-to-noise ratio which is an advantage as compared to other amperometry methods. Chronoamperometry, being a very simple process, is utilized in multifarious applications. Ensafi *et al* employed a modified electrode by decorating Ag nanoparticles on functionalized MWCNTs as a fast, stable, and efficient non-enzymatic glucose sensor based on the hydrodynamic chronopotentiometry method [25]. Interestingly, very low levels of detection limit (0.03 μM), linear dynamic range (1.3–1000 μM) high sensitivity (1.06×10^3 μA mM^{-1}) were obtained for this noble biosensor. Apart from that, this technique can be employed for the determination of ascorbic acid on paper-based devices, chemical sensing of liquids, determination of glucose with enzymatic paper-based sensors, etc.

1.2.2.2.1 Advances

In recent years, some developments have been made in electrode systems to improve their overall performance for the chronoamperometry technique. Nanotechnology has proved to be a useful tool in this regard. Huges *et al* in 2020 designed counter and reference electrodes with deposition of cross-linked nanostructures using newly developed confined atmospheric pulsed laser deposition (CAP) and laser ablation synthesis in solution (LASiS). The three-electrode system constructed using this nanofabrication method yielded good results in chronoamperometry experiments (figure 1.10(a) and (b)) [26]. In another report, Ezenarro *et al* used carbon screen-printed electrodes in chronoamperometry to detect Legionella, a pathogenic bacterium usually found in the freshwater environments. The system lessened the time of detection from 10 days to 2–3 h which is very important during outbreaks. It also exhibited a satisfactory linear response with $R^2 \sim 0.99$ and a detection limit of 4 cfu mL^{-1} for Legionella concentrations of 10^1–10^4 cfu mL^{-1} [27]. A group of Russian researchers headed by Mikhelson measured the K+ ion concentration in complicated blood serum matrix using chronoamperometry, chronopotentiometry, and coulometry techniques and found that the results acquired in later methodologies were at par with the well-established ion-selective electrode (ISE) technique figure 1.10(c) and (d) [28].

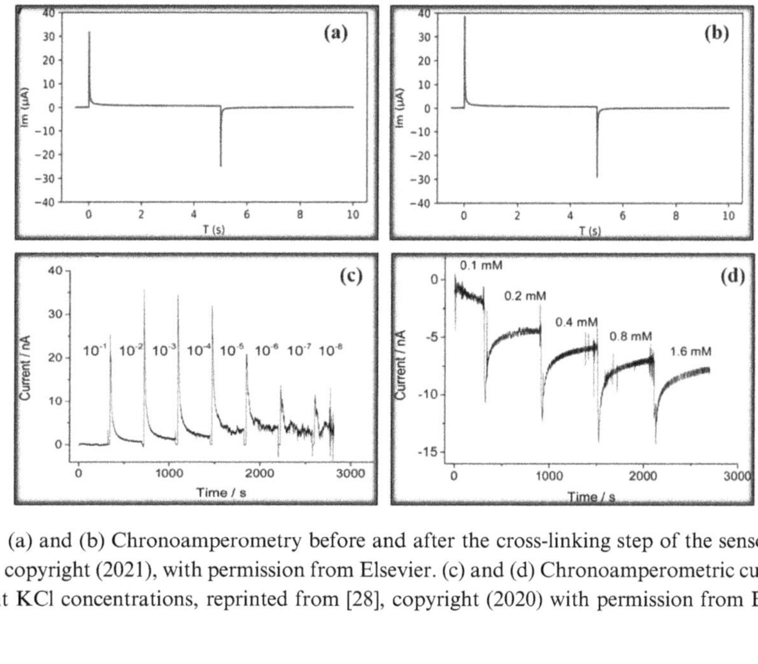

Figure 1.10. (a) and (b) Chronoamperometry before and after the cross-linking step of the sensor fabrication process [26], copyright (2021), with permission from Elsevier. (c) and (d) Chronoamperometric curves with ISE 1 at different KCl concentrations, reprinted from [28], copyright (2020) with permission from Elsevier.

1.2.2.3 Differential pulse voltammetry

Differential pulse voltammetry (DPV) is one of the earliest analytical techniques to detect catecholamines. In this method, short pulses of small amplitudes are superimposed on a linear ramp and a base potential having no Faradaic reaction is chosen and given to the electrode. Then it is increased in a step-wise manner with equal increments. A difference in the measured currents just before and after the pulse application is considered which effectively lowers the background current due to dc ramp. Such a process leads to a Faradaic current which is almost free of capacitive components and may be considered to be one of the advantages leading to high sensitivity. Apart from that, small step potential results in narrow voltammetric peaks, and DPV can thus be utilized to segregate analytes having similar oxidation potentials. One of the important areas of interest with regard to DPV is probing the intercellular processes. Earlier, Meulemans *et al* employed DPV to estimate the number of electro-active drugs in the cytoplasm of cholinergic neurons of Aplysia California and find out the cellular uptake rate and clearance [29]. The same group used a needle-tipped glass insulated platinum electrode for identifying intracellular serotonin in metacerebral neurons from Aplysia [30]. Similarly, multipulse DPV with platinum electrodes could measure dopamine levels in the giant dopamine neuron of Planorbiscorneus [31]. Prinith *et al* fabricated electro-polymerized L-Ser modified graphite carbon nanotube composite paste electrode for the analysis of Riboflavin (RF) in both commercial and real samples like natural food supplement, beverage and pharmaceutical supplement. The electrochemical redox activity of the PSR/CNTPE to RF was investigated through CV and DPV [32]. Hareesha *et al* functionalized the electrochemically polymerized glutamic acid layered multi-walled carbon nanotube paste electrode for the sensing of indigo carmine (IC) through DPV [33].

1.2.2.3.1 Advances

In the recent past, extremely sensitive differential pulse voltammetry measurements along with CV proved to be beneficial to investigate the electrochemical reduction of nitrate by hemoglobin. Experiments were carried out over a large potential range of +0.3 V to −1.3 V (versus Ag/AgCl) by immobilizing various hemoglobin species over DDAB liquid crystal film (figure 1.11). It is noteworthy to state that the nitrate reductase activity (NRA) measurements by this approach ensure less quantifying time at significantly lower levels of reagents at the electrode surface [34]. Besides, differential pulse voltammetry acts as a trump card in the case of detecting a very low concentration of the analytes. Hence DPV controlled smartphones (with electronic display) have been employed to explore very low levels of cortisol by Liu *et al* [35]. Cortisol is often considered to be a remarkable biomarker of psychological or physical stress. So, detection of cortisol levels becomes essential for personal health monitoring (PHM). This research group fabricated an ultra-sensitive immunosensor based transducer utilizing Au nanoparticles/MoS2/Au nanoparticles and could successfully examine non-invasive salivary cortisol at point of care (POC) with a miniaturized DPV system based on a smartphone. Not only a high detection limit of 0.11 nM could be achieved but also cortisol could be distinguished from other substances in saliva. Differential pulse voltammetry also

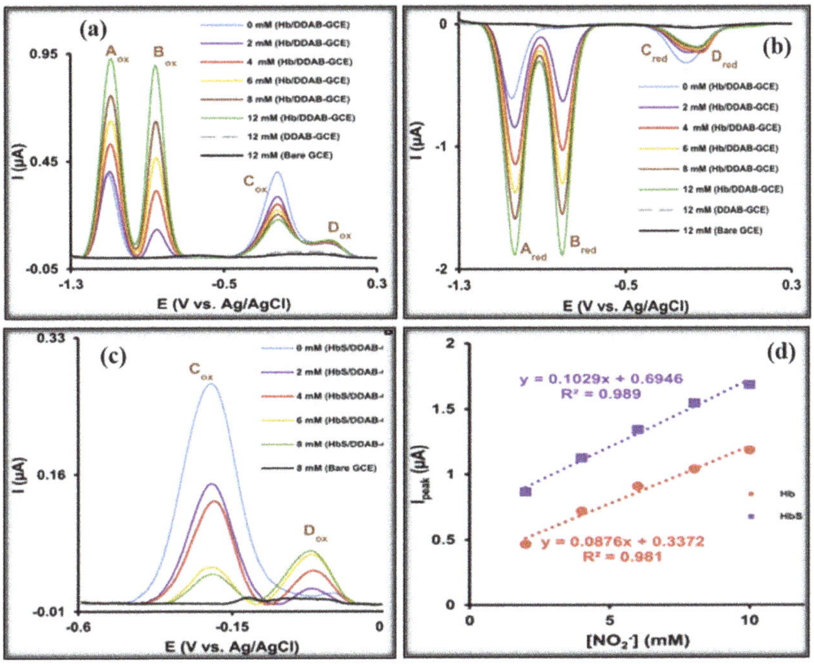

Figure 1.11. (a) and (b) Cathodic and anodic DPV for Hb/DDAB-modified GCE in 100 mM PBS (pH 7.4) with 50 mM NaBr at various concentrations of sodium nitrite, (c) DPV of Hb-S/DDAB-modified electrode in 100 mM PBS (pH 7.4) with 50 mM NaBr, and (d) peak current versus concentration of sodium nitrite, adapted from [34], copyright (2020) with permission from Elsevier.

proved to be useful in detecting small concentrations of Hg (II) ions during biosorption of Hg ions by garlic adsorbent. The main benefit of this technique is that it prevents the release of carcinogenic mercury vapor.

1.2.2.4 Square wave voltammetry

The square wave voltammetry (SWV) technique discovered by Barker and Jankins in 1957, is a type of linear sweep voltammetry which utilizes a blended square wave and staircase potential applied to a stationary electrode. It is a relatively rapid and very sensitive pulse voltammetry technique. Moreover, its detection limit is perhaps compared with certain chromatographic and spectroscopic techniques. In this measurement, the stationary current is usually recorded while the voltage applied between reference and the working electrode is pulsed forward as well as backward keeping the frequency constant. By using the square wave voltammetry, we can plot the differential current curve instead of using individual forward and reverse current plots which will lead to highly effective screening. The peak magnitudes in this plot are proportional to the concentrations of the various redox-active species as stated by,

$$\Delta i_p = \frac{nFAD_0^{1/2}C_0^*}{(t_p)^{1/2}}\Psi_p$$

where, $\Delta i_p \rightarrow$ value of the differential current peak, $A \rightarrow$ electrode surface area, $C_0^* \rightarrow$ species concentration, $D_0 \rightarrow$ diffusivity of the species, $t_p \rightarrow$ pulse width and $\Delta \psi_p \rightarrow$ dimensionless factor. This is one of the widely used techniques in different fields that includes the sensor and medical applications owing primarily to its ability to measure the Faradaic current at a time when the double layer current is negligible and presents output current in the form of a symmetrical peak. Rattanarat *et al* fabricated a three-layered selective electrochemical paper-based analytical device to determine dopamine in the presence of ascorbic acid and uric acid in human model serum by SWV detection (linear range: 1–100 μM, detection limit: 0.37 μM) [36]. Teegam *et al* used SWV as a detection method for quantification of human papillomavirus by a paper-based electrochemical DNA biosensor [37]. Daizong *et al* developed a smartphone-based SCV system with screen-printed graphene electrodes in order to overcome some issues like sensor performance, system sensitivity, and sensor stability. The system consists of a coin-sized detector, sensor, and smartphone. To construct flexible screen-printed graphene electrodes, the graphene ink was printed on a soft substrate, and these electrodes were found to have great potential in the field of POC testing and wearable detection [38]. In another work by Zhang and Yang, SWV is used for the detection of phenol in industrial wastewater. In this method, phenol is first oxidized to phenoxyl radicals, then becomes hydroquinone and catechol for detection. This method provides the detection of phenol in the sub-ppm range over a wide pH range with a detection limit of 0.1 ppb [39].

1.2.3 Additional methods

1.2.3.1 Stripping voltammetry

The analytical applications of electroanalytical techniques which required high sensitivity are described using the stripping voltammetry (SV) technique. Among all voltammetric methods, this technique is the most highly sensitive due to the preconcentration step which gathers the wished analyte on the electrode surface. Depending on the stripping method used, the analyte is preconcentrated either by physical adsorption or by electrodeposition. After achieving sufficient preconcentration the working electrode's potential is swept to strip off the analyte of the electrode surface and the faradaic current is calculated to determine the concentration of the involved analyte. The most used techniques are anodic stripping voltammetry (ASV) and cathodic stripping voltammetry (CSV). In this method, at the electrode surface, the concentration of the electroactive species increases and the electroactive species in case of highly dilute solution (10^{-10} to 10^{-11} M) can be studied. The compound may easily deposit on the electrode. Certain potential is given to the electrolyte so that electrolysis can start, and the redox species deposits on the electrode. Then the species is stripped or, redissolved and analyzed. If the species is first reduced then oxidized then it is called ASV and if the species undergo oxidation first then reduction then it is called CSV. If the species are adsorbed by the electrode surface without electrolysis then it is called adsorptive SV. This technique is widely used in the electrochemical method to detect heavy metals in water and soil samples because it can measure metal ions at trace concentrations. ASV is used to test surface water, quality of drinking water and wastewater. CSV is used to analyze trace metals and organic compounds containing sulfur in aqueous solutions.

1.2.3.1.1 Recent advances

Marzieh Nodehi *et al* modified a glassy-carbo electrode with a new nanocomposite of polymethyldopa and gold nanoparticles on the surface of magnetic graphene oxide (GCE/GO/Fe_3O_4@PMDA/AuNPs) and determined As^{3+} and Cu^{2+} with great sensitivity (figure 1.12). By square wave SV, the oxidation peaks of As^{3+} and Cu^{2+} are clearly distinguished which leads to good recovery values for the analysis of As^{3+} and Cu^{2+} in pool and drinking water samples [40]. Hossein *et al* prepared a new adsorbent of magnetite–carbon/dithizone nanocomposite and they found that the magnetic sorbent is a viable material for the preconcentration of mercury (II) which is quantified by differential pulse anodic SV technique. The method was validated by the analysis of the certified reference material NIST SRM $_{15}{}^{66b}$, and applied to the preconcentration and qualification of mercury (II) in industrial wastewater and spiked water samples [41]. In a study, Suji Park *et al* demonstrated the unique capability of a carbon-based ion-selective electrode (ISE) to perform highly sensitive square wave anodic stripping voltammetry, while maintaining all the properties of an ISE, in terms of sensitivity, detection limit, response time, and selectivity. Square wave anodic stripping voltammetry involves deposition and dissolution steps of metal ions, which means adsorption and desorption of metal ions on the conductive ion-selective membrane without losing its ion-sensing property. They concluded that

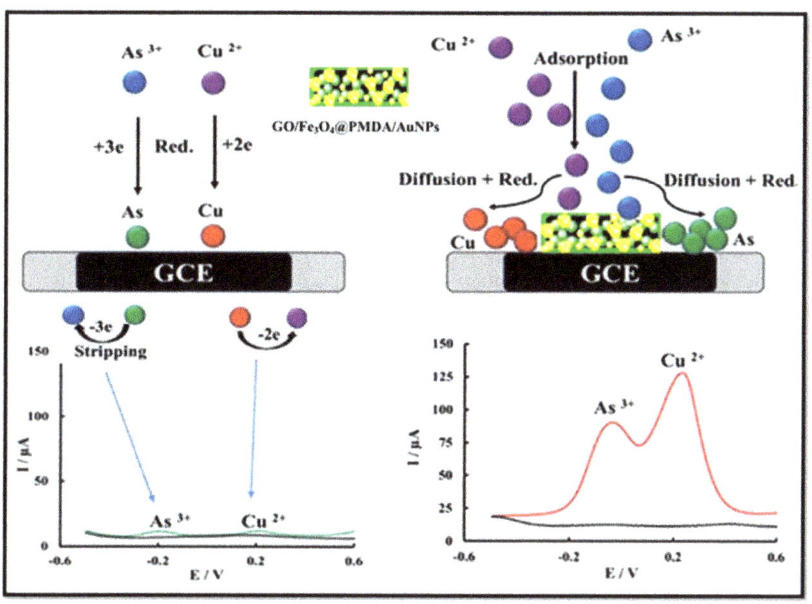

Figure 1.12. Schematic for simultaneous determination of As^{3+} and Cu^{2+} on the surface of bare GCE and GCE/GO/Fe_3O_4@PMDA/AuNPs, reprinted from [40], copyright (2021) with permission from Elsevier.

it was an opportunity to use the low detection limit of anodic stripping voltammetry and the high selectivity of ISE-based potentiometry [42].

1.2.3.2 Ultramicroelectrodes

Ultramicroelectrodes can be used in experiments which do not need mercury dropping. In the case of ultramicroelectrodes, the electrode area is small (electrode length <20 μm) as compared to the diffusion layer formed during the experiment. It provides decent information which a conventional electrode is unable to provide, when in at least 1-D. As one of the electrodes is minimum in size, we can observe phenomena such as electroactive species, mass transport varies from linear diffusion to electrode surface to 2- or 3-D diffusion, the current is small and is independent of the electrode area and high current density. Earlier these properties were assessed negatively because of the complicated mass transport, involvement of kinetic complications in current, small S/N ratio, technical limitations to build the electrode. Due to the small electrode size, the diffusion at the electrode surface is independent of time [43]. The small characteristic length of ultramicroelectrode tends to be either a plane or a point or a line. On this basis, ultramicroelectrodes are of three types and those are plane electrode, point electrode, and line electrode. This classification is used to understand basic behavior. Realization of mass transport at the electrode surface is necessary for the characterization of ultramicroelectrodes [43]. Electric migration, convection of electroactive species, diffusion causes mass transport at ultramicroelectrodes. Mass transport by migration and diffusion is due to the local gradient of the electrochemical potential of electroactive species. Let, C_i = concentration of electroactive species i, D_i = diffusion coefficient, Ψ = inner potential of the solution, Z_i = no. of charge. Then from the Nernst–Plank equation the transport rate, J_i is,

$$J_i = -D_i \mathrm{grad} C_i - (Z_i F/RT) D_i C_i \mathrm{grad}\Psi \tag{1.1}$$

This equation does not depend upon electrode geometry. Applying equation of continuum, i.e., $\partial C_i/\partial t = -\mathrm{div} J_i$ to the above equation it becomes,

$$\partial C_i/\partial t = D_i \Delta C_i + (Z_i F/RT) D_i (\mathrm{grad} C_i)\bullet(\mathrm{grad}\Psi) + (Z_i F/RT) D_i C_i \Delta\Psi \tag{1.2}$$

where "•" is the inner product. This is the basic time-independent equation of mass transport. Depending upon the magnitude of grad Ψ, the above equation can be of three categories; (i) when supporting electrolytes have high concentration (grad $\Psi = 0$), equation (1.2) takes the form of Fick's second law, (ii) when grad Ψ varies with supporting electrolyte concentration, the Laplace equation ($\Delta\Psi = 0$) exist under the boundary conditions of the electrode and cell geometry, and grad Ψ in equation (1.2) can be evaluated by the Laplace equation, (iii) for small supporting electrolyte concentration, Laplace equation is replaced by the Poisson equation, i.e.,

$$\Delta\Psi = -(1/\epsilon_o \epsilon_s) \sum Z_i F C_i$$

where, ϵ_o = permitivity of vacuum and ϵ_s = relative permitivity of solution [43]. These ultramicroelectrodes can operate with much smaller timescales. This is because when the applied potential is changed the electrodes take some time to respond to it. By the decay of time constant the electrode's potential achieves the applied potential called the 'cell time constant'. Smaller electrode size leads to a small time constant. Ultramicroelectrodes are capable of minimizing the potential expansion on the conducting charges in the solution. Hence errors can be minimized even in the solutions having poor conductivity. Application of ultramicroelectrodes include voltammetry under steady-state conditions voltammetry with high scan rates, electrochemistry in poorly conducting electrolytic media, exploiting the smallness of microelectrodes, and in case of several chemical analyzes [44].

1.2.3.2.1 Recent advances

In a study by David T. Boyle *et al* investigation of the electron transfer kinetics of lithium electrodeposition is done by using ultramicroelectrodes with transient voltammetry, however, the obtained results deviate from the Butler–Volmer model of electrode kinetics. Further, the role of the interplay of electron transfer and mass transport resistances on the morphology of lithium is understood by comparing the results to electrochemical impedance spectra [45]. In another study, Naiyun Liu *et al* review the applications and research methods of *in situ* electrochemical Raman spectroscopy. It summarizes the methods of electrode preparation and demonstrates the use of the correlation between Raman spectroscopy and electrochemistry at the ultramicroelectrode. As a result, they observed ultramicroelectrode as a working electrode in *in situ* electrochemical Raman spectroscopy expands the research of Raman spectroscopy. It also leads to a promising tool in the study of photo-electrochemical and electrochemical reactions [46].

1.3 Summary

Electrochemistry has moved a long way in terms of speed, accuracy, and size reduction; and hence voltammetric techniques have also evolved along the same lines. The use of spatial resolution in such techniques for analyzing materials up to smaller domains on surfaces and membranes seems to have had a remarkable impact on analytical and physical chemistry, bio-chemistry related studies. For instance, the use of micro/ultramicroelectrodes has enabled measurement of currents scaled down to nanoampere, picoampere, and femtoampere. Integration of nanotechnology in voltammeric methods has also been instrumental in improving the sensitivity which is essential for designing electrochemical sensors. Thus, it is noteworthy to say that advancements in the field of voltammetric techniques will add new dimension to the exploration of different material properties. These discussions on the recent advancements will definitely provide impetus to the chemists/bio-chemists to investigate more and come up with new strategies to integrate chemistry on a chip (a far-reaching dream!).

References

[1] Kounaves S P 1997 Voltammetric techniques *Handbook of Instrumental Techniques for Analytical Chemistry* ed F A Settle (Englewood Cliffs, NJ: Prentice Hall) pp 709–26

[2] Tigari G, Manjunatha J G, D'Souza E S and Sreeharsha N 2021 *ChemistrySelect* **6** 2700–8
[3] Pushpanjali P A, Manjunatha J G, Hareesha N, D'Souza E S, Charithra M M and Prinith N S 2021 *Surf. Interfaces* **24** 101154
[4] Charithra M M and Manjunatha J G 2021 *Mater. Chem. Phys.* **262** 124293
[5] Hareesha N, Manjunatha J G, Amrutha B M, Sreeharsha N, Asdaq S B and Anwer M K 2021 *Colloids Surf. A Physicochem. Eng. Asp.* **23** 127042
[6] Guy O J and Walker K A D 2016 Graphene functionalization for biosensor applications *Silicon Carbide Biotechnology* 2nd edn, ed S E Saddow (Amsterdam: Elsevier) pp 85–141
[7] Heinze J 1984 *Angew. Chem. Int. Ed. Engl.* **23** 831–47
[8] Leonat L, Sbarcea G and Branzoi I V 2013 *UPB Sci. Bull. Ser. B.* **75** 111–8
[9] Asrat T M, Cho W, Liu F A, Shapiro S M, Bracht J R and Zestos A G 2021 *ACS Omega* **6** 6571–81
[10] Puthongkham P, Rocha J, Borgus J R, Ganesana M, Wang Y, Chang Y, Gahlmann A and Venton B J 2020 *Anal. Chem.* **92** 10485–94
[11] Watwe V and Kulkarni P 2021 *J. Anal. Sci. Technol.* **12** 37
[12] Pushpanjali P A, Manjunatha J G, Hareesha N, Amrutha B M, Raril C, ALOthman Z A, Alanazi A M and Pandith A 2022 *Mater. Chem. Phys.* **276** 125392
[13] Hareesha N and Manjunatha J G 2021 *Sci. Rep.* **11** 12797
[14] Nádaždy V, Gmucová K, Poturnayová A, Šnejdárková M, Karpišová I, Lányi Š and Hianik T 2012 *Procedia Chem.* **6** 60–8
[15] Vasconcellos M D, Silva L R, Lee C S, Fajardo A S, Garcia-Segura S and Ribeiro J 2021 *Sensors* **21** 3450
[16] Bard A J and Faulkner L R 2001 *Electrochem. Methods.* **2** 482
[17] Matsuda H and Ayabe Y 1955 *Z. Elektrochem.* **59** 494–503
[18] Mazzara F, Patella B, Ganci F, O'Riordan A, Aiello G, Torino C, Vilasi A, Sunseri C and Inguanta R 2021 *Chem. Eng. Trans.* **87** 421–6
[19] Schilder W H, Tanumihardja E, Leferink A M, van den Berg A and Olthuis W 2020 *Heliyon* **6** e04210
[20] Petkovska S and Gulaboski R 2020 *Electroanalysis.* **32** 992–1004
[21] Zhang C, Rappleye D and Simpson M 2017 *J. Electrochem. Soc.* **164** H5218
[22] Jo T, Yoshimi K, Takahashi T, Oyama G and Hattori N 2017 *J. Electroanal. Chem.* **802** 1–7
[23] Kiss L, Bősz D, Kovács F, Li H, Nagy G and Kunsági-Máté S 2019 *Polym. Bull.* **76** 5849–64
[24] Zhang C, Wallace J and Simpson M F 2018 *Electrochim. Acta* **290** 429–39
[25] Ensafi A A, Zandi-Atashbar N, Rezaei B, Ghiaci M, Chermahini M E and Moshiri P 2016 *RSC Adv.* **6** 60926–32
[26] Hughes C, McCann R M, Freeland B and Brabazon D 2021 *J. Electroanal. Chem.* **880** 114813
[27] Ezenarro J J, Párraga-Niño N, Sabrià M, Del Campo F J, Muñoz-Pascual F X, Mas J and Uria N 2020 *Biosensors* **10** 102
[28] Kondratyeva Y O, Tolstopjatova E G, Kirsanov D O and Mikhelson K N 2020 *Sens. Actuators B Chem.* **310** 127894
[29] Meulemans A, Poulain B, Baux G, Tauc L and Henzel D 1986 *Anal. Chem.* **58** 2088–91
[30] Meulemans A, Poulain B, Baux G and Tauc L 1987 *Brain Res.* **414** 158–62
[31] Chen T K, Lau Y Y, Wong D K and Ewing A G 1992 *Anal. Chem.* **64** 1264–8
[32] Prinith N S, Manjunatha J G, Tigari G, ALOthman Z A, Alanazi A M and Pandith A 2021 *ChemistrySelect* **6** 10746–57

[33] Hareesha N, Manjunatha J G, Amrutha B M, Pushpanjali P A, Charithra M M and Subbaiah N P 2021 *J. Electron. Mater.* **50** 1230–8
[34] Fini H and Kerman K 2020 *Anal. Chim. Acta* **1104** 38–46
[35] Liu J *et al* 2020 *Sensors* **20** 1422
[36] Rattanarat P, Dungchai W, Siangproh W, Chailapakul O and Henry C S 2012 *Anal. Chim. Acta* **744** 1–7
[37] Teengam P, Siangproh W, Tuantranont A, Henry C S, Vilaivan T and Chailapakul O 2017 *Anal. Chimica Acta* **952** 32–40
[38] Ji D *et al* 2020 *Smart Mater. Med.* **1** 1–9
[39] Yi Z and Kun-Lin Y 2021 *Anal. Chim. Acta* **1178** 338788
[40] Nodehi M, Baghayeri M and Veisi H 2021 *Talanta* **230** 122288
[41] Abdolmohammad-Zadeh H, Mohammad-Rezaei R and Salimi A 2020 *Microchim. Acta* **187** 2
[42] Park S, Maier C S and Koley D 2021 *Electrochim. Acta* **390** 138855
[43] Aoki K 1993 *Electroanalysis* **5** 627–39
[44] Heinze J 1993 Ultramicroelectrodes in electrochemistry *Angew. Chem. Int. Ed. Engl.* **32** 1268–88
[45] Boyle D T, Kong X, Pei A, Rudnicki P E, Shi F, Huang W, Bao Z, Qin J and Cui Y 2020 *ACS Energy Lett.* **5** 701–9
[46] Liu N, Wu L and Huang Y F 2021 *Scientia Sinica Chim.* **51** 256–63

Chapter 2

Development of electrochemical sensors for toxic metal detection

Devika Sudha Ravindran, Sonia Sam, Keerthi Kodakat and K Girish Kumar

A substantial development has been made in the synthesis, characterization, processing, and applications of carbon composite materials during the past few decades because of their remarkable physical and chemical properties. Carbon composite materials such as graphite, graphene, carbon nanotubes, carbon fibers, carbon nanoribbons, graphene oxide, glassy carbon, and graphite reinforced polymers offer numerous advantages like high chemical stability, mechanical strength, electrical conductivity, surface-to-volume ratio, and good biocompatibility. These notable properties make them trustworthy candidates in chemical sensing applications. Today, electrochemical techniques have gained enormous attention over other traditional analytical techniques owing to their merits such as simplicity, cost effectiveness, rapid response, and ease of miniaturization. In the current world scenario, where environmental pollution and related health hazards due to toxic metal ions have been increasing continuously, electrochemical sensing strategy can provide a significant breakthrough in their determination. The contamination by toxic metals can occur due to various agricultural, mining, and industrial activities, and their drastic increase in the environment may lead to bioaccumulation, which is a serious threat to the whole living world.

In this regard, this chapter comprises the advancements in the application of various carbon composites in the development of electrochemical sensors for toxic metals.

2.1 Introduction

A substance is categorized as 'toxic' when it becomes hazardous to nature or living organisms at a certain concentration. Metals, which are found everywhere, also exhibit such toxicity and may even cause death. Heavy metals having a density and atomic weight about five times more than that of water are predominant in this

doi:10.1088/978-0-7503-5127-0ch2

category. According to Duffus 'the term "heavy metal" refers to a group name for metals and metalloids that are associated with contamination and potential eco toxicity' [1]. Lead, cadmium, zinc, mercury, silver, chromium, copper, etc, come under this class of elements and, in total there are around 40 of them [2]. Volcanic eruptions and the weathering process of metal-containing rocks are the important natural sources of these heavy metals. These metals are also released into the environment from industrial waste, pesticides, phosphate fertilizers, combustion of fossil fuels, and also during mining and smelting operations [3]. In fact some of these heavy metals like Cu and Zn are essential to maintain various biological functions in living organisms, but when their concentration exceeds a certain limit they become virulent. Hence toxic metals are one of the prominent environmental pollutants with varying adverse effects on living organisms in which Cd, Hg, Pb, and As are the most common and dangerous. The bioaccumulation of toxic metals can damage living cells since they are potent to denature proteins and can even cause oxidative DNA damage. Oxidative stress has been found to be the major toxicity mechanisms by these deadly metals in living organisms since they can reduce the level of some antioxidants in the body [4]. Modern industrialization has increased the exposure of humans to these toxic metals which results in serious health issues like Parkinson's disease, Alzheimer's disease, heart disorders, kidney, lung, and stomach cancers, etc [4]. Similarly, accumulation of these toxic metals has also caused some serious imbalances in other components of the biosphere like soil, plants, water, etc. When the concentration of heavy metals in soil exceeds a threshold value, it may ruin its microbiological balance and reduce the fertility. Long term intake of these heavy toxic metals by plants from contaminated soil may retard their growth due to oxidative damage in cells which may even result in severe food chain contamination, since plants constitute the first trophic levels in the chain [5]. Similarly, concentration of toxic metals in water bodies will affect the aquatic life and thereby slowly destroy aquatic ecosystems [2]. Severe mercury contamination in the coastal waters of Minamata in Japan and a serious neurological disorder (Minamata disease) that followed it remains in history as the most striking example of heavy metal toxicity [6]. In order to avoid such disasters in future, authorities are very anxious to limit such heavy metal pollution in the environment. Water quality standards which limit the concentration of heavy metals in water resources to an optimum value is one such initiative. In order to follow these standards, assays which detect even trace amounts of such toxic heavy metals in different samples is very important [7].

The literature clearly reveals that the scientific community has been carrying out very robust research for many years in developing efficient analysis platforms for toxic metal determination. Laboratory based techniques which require separation and tedious pre-concentration of the samples such as mass spectrometry, atomic absorption spectrometry, x-ray fluorescence, etc are in use for this purpose [7]. Today, chemical sensors which enable the detection as well as determination of a wide range of analytes have quickly become the main preferences for heavy metal detection. Both optical as well as electrochemical techniques have been used in the development of such sensors. Of the available techiques, electrochemical stripping techniques offer great promise in the real sample detection of toxic metal ions.

Square wave voltammetry (SWV) and differential pulse voltammetry (DPV) are the techniques commonly incorporated with the stripping method for quantification of the metals. This can be attributed to the need for low cost real time analysis of samples within minimal time and with high sensitivity. Electrochemical techniques with high sensitivity, multicomponent analysis possibilities, low cost, and simple instrumentation have thus become the most suitable for heavy metal sensor development [8–10]. In addition potentiostatic stripping techniques are preferred as they involve a simple pre-concentration step followed by a stripping step which enables highly sensitive determination of these trace toxic metals *in situ* [11]. Conventionally such anodic stripping techniques were carried out with the aid of mercury based electrodes such as dropping mercury electrode (DME) and hanging mercury dropping electrode (HMDE) for toxic metal analysis. But now studies are focussed on the use of mercury-free electrodes in toxic metal detection owing to the high cost and toxicity of mercury based electrodes [12]. Hence, chemically modified carbon-based electrodes and gold electrodes are functioning as efficient alternatives to the conventional mercury based electrodes [13, 14]. Moreover, electrochemical sensors which function on simple cyclic voltammetry (CV), SWV, and DPV techniques without any accumulation steps have been developed for this purpose. Here, suitable chemical modifications of the electrodes are adopted to achieve sensitivity on par with the stripping methods [15].

Since the introduction of carbon nanomaterials (CNMs) as novel nanosized materials, they have been utilizing in multiple disciplines for diverse functions. In particular, zero dimensional fullerenes, one-dimensional carbon nanotubes (CNTs) and two dimensional graphene have brought remarkable revolution in the field of material science [16]. The low toxicity and cost-effective nature of these materials have found applications in vast fields like electronics, gas storage, production of batteries, paints, textiles, etc [17]. The fascinating properties exhibited by CNMs such as high electrical conductivity, high surface-to-volume ratio, biocompatibility, and chemical stability have also made them very favored materials in electro-analytical studies [17, 18]. Graphene, reduced graphene oxide (RGO) and its derivatives, carbon nanotubes (CNTs), graphitic carbon nitride (g-C_3N_4), carbon nanofibres (CNFs), carbon nanoribbons (CNRs), carbon nano onions (CNOs), carbon nanohorns (CNHs), etc are the common carbon-based nanomaterials employed in the fabrication of electrochemical sensors as electrode functional materials [19]. At the same time materials like glassy carbon, graphite, etc, are used as efficient electrode materials in electrochemical sensor fabrication. On analyzing the trends that have been followed in the electrochemical sensing of toxic metal ions, the weighty role played by CNMs in creating significant developments in the field is explicitly revealed. It is evident that the strategies of incorporation of CNMs on to electrochemical sensing platforms have paved the way for many significant advancements in the analytical performance of these assays in an economical manner. The interesting inbuilt porous structure of CNMs like CNTs, can smoothly entrap toxic metal ions and their excellent electrocatalytic ability enables the highly sensitive determination of the latter [20]. When CNMs like nitrogen and sulfur doped graphene, RGO, and CNTs alone are reported to catalyze

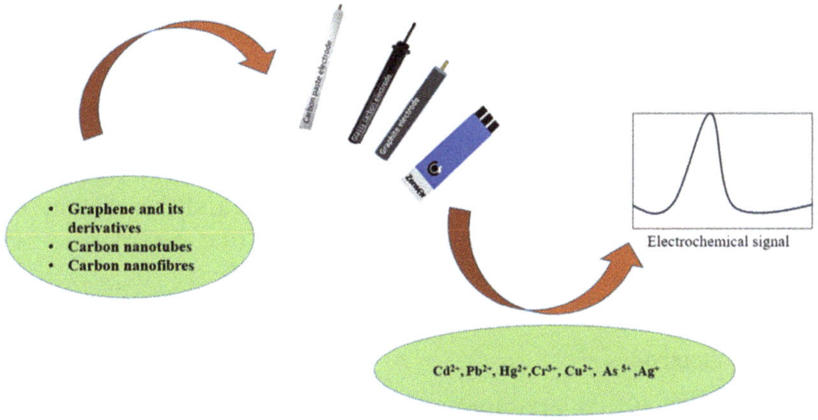

Figure 2.1. Schematic representation of different carbon-based electrode materials and modifiers used for the determination of heavy metals. Reprinted from [43], Copyright (2019), with permission from Elsevier.

toxic metal ion electrochemical sensing, their composites with other CNMs, metal and metal oxides nanoparticles, conducting polymers are also proposed for the same [20–22]. Even though the majority of the toxic metal ion sensors have made use of glassy carbon, a non-graphitizable carbon as the transduction element, other electrochemical signal transducers such as graphite and carbon paste electrodes (CPEs) have been also utilized for the same (figure 2.1) [23–25].

This chapter will see the ways of utilization of CNMs as electrode modifiers and electrode material in some recent electrochemical sensors for toxic metal ions and discuss their advancements in comparison with sensors based on other materials.

2.2 Carbon-based electrode materials for the electrochemical sensing of toxic metals

The geometry and composition of the working electrode plays a key role in the performance of an electrochemical sensor [26]. For so long, the determination of heavy metals was preferably done using HMDE, since Hg can form amalgam with many metals and also it exhibits excellent sensitivity [27]. Apart from its advantages, Hg is extremely toxic, so researchers were always in search of more eco-friendly electrode materials [27]. Carbon materials have gained the interest of electrochemists due to their fascinating properties including high surface area and electrical conductivity [28]. As the field of electrochemical sensors became more advanced, new electrode materials have been encountered to replace conventional ones and now, carbon-based electrode materials are at the forefront for electrochemical sensing applications. Carbon exhibits numerous advantages to be explored as an electrode material. Being an abundant element, it is less expensive and is easily available in different forms [26]. Because of their slow oxidation kinetics, carbon electrodes exhibit a broad potential window particularly in the anodic direction (up to +2.0 V), in contrast to a Hg electrode where the anodic range is only up to +0.4 V due to its low oxidation over potential [29, 30]. Also, the electrochemically inert nature of carbon compared to other metals and its ability to adsorb heavy metals make it an excellent electrode material [31]. Currently, various types of carbon-based electrodes such as graphite electrodes (GEs), CPEs, glassy carbon electrodes (GCEs), screen printed carbon electrodes (SPCEs), and boron doped diamond electrodes (BDDEs) are used for electrochemical sensing applications [32–36]. This classification of electrodes is not strictly based on the carbon material used in the electrode but based on the form of carbon material used and this section focuses on the most widely used carbon-based electrodes, GEs, CPEs, GCEs, and SPCEs for the electrochemical determination of toxic metals.

2.2.1 Graphite electrodes

Graphite electrodes (GEs) are among the most commonly used electrodes in electrochemical sensors and are considered as ideal for sensing heavy metals owing to their low cost (in comparison to glassy carbon, gold, indium tin oxide, and platinum electrodes), wide potential window and low background current [37]. Graphite is mainly employed in electrochemical sensors as graphite rod and pencil graphite electrodes (PGEs). The PGE is regarded as one of the most significant inventions in the field of electrochemical sensors as they are inexpensive, easily available, less toxic and most importantly, they are disposable [38]. There has always been an increasing interest in the application of disposable electrodes for toxic metal analysis as they are ideal for on-site determination of samples. Also, the use of PGE does not require a polishing step prior to analysis as is required in case of other commercial electrodes such as GCEs and they can be easily renewed by removing the projected part, which further makes them a simple and easy to use electrode material [39]. The increasing number of studies involving PGEs for the electrochemical determination of toxic metals proves that their impressive properties make them suitable candidates for environmental analysis.

Graphite pencils, which have long been used for writing purposes were identified as the low cost, easily available electrode material only in late 1997 when Bond *et al* used them for an anodic stripping voltammetric study of the heavy metals, cadmium (Cd), and lead (Pb) [40]. Graphite pencil leads are mixtures of approximately 65% graphite, 30% clay together with wax, resin or polymers as binders [41]. A high percentage of graphite which has metallic and non-metallic properties that are favorable to be used as electrode material, combined with clay particles providing proper hardness to the electrode makes PGE a suitable working electrode in electrochemical sensing applications [41].

One of the widespread applications of PGEs in electrochemical sensing is in the determination of heavy metals. Numerous reports are available for the quantification of toxic metals in various environmental samples based on PGEs. Apart from the increase in concern over the environmental pollution resulting from the toxic metals, the need for their efficient, on-field, rapid and cost-effective analysis from environmental samples has always encouraged scientists in electrochemical fields to come up with new sensing technologies [42]. The surge in studies for toxic metal determination in the past years using PGEs can be attributed to their greater sensitivity, reproducibility, ease of use, and renewability compared to other conventional electrodes such as GCEs [43]. In addition to PGEs, other GEs such as exfoliated graphite (EG) and graphite rods are also used for sensing toxic metals [44]. EG is a low density material which has good mechanical stability and electrical conductivity. One of the advantages in preparing EG electrodes compared to PGE is that they can be easily prepared by compression at high pressure without the use of any binders which tend to degrade on use [45]. Also, graphite rods sourced from waste batteries have been employed as electrodes as they are abundantly available and there are reports where such electrodes are used for the sensitive and selective quantification of Cd and Pb from herbal food supplements [32].

Putting aside all the merits offered by GEs over other commercial electrodes, unmodified GEs suffer from lower sensitivity when it comes to trace level determination [46]. Pre-treatment methods and modification of PGE surfaces are found to be very effective for increasing the sensitivity and selectivity towards target metals, and to date several studies have been reported exploring different modifications for the quantification of toxic metal ions [43]. Various electroactive materials such as nanoparticles and polymers are employed for modifying GEs [47]. Metal nanoparticles (MNPs) on account of their high electroactive surface area, fast electron transfer rate and high signal to noise ratio find wide applications in the determination of toxic metals [47]. One of the metals most frequently used as a modifier for the determination of heavy metals is bismuth (Bi). Apart from its lower toxicity, Bi can act as a replacement for Hg based electrodes due to its ability to form alloys with heavy metals such as Pb and Cd [42]. Bi film modified GE has been widely explored for the single as well as simultaneous determination of heavy metals such as As, Cd, Pb, Zn, Cu, etc, and most of the studies have been carried out using anodic stripping voltammetry in combination with square wave or differential pulse voltammetric techniques as a pre-concentration step is involved that helps to accumulate the heavy metals on the electrode surface [39, 42, 44, 48, 49]. Besides Bi, other MNPs, metallic films, conducting polymers, CNMs, macrocyclic molecules such as crown ethers and composites of MNPs with different polymers have also been incorporated into GEs as modifiers [49–54]. Composites of polymers along with MNPs or CNMs have also proved to be an effective chemical modification for the more selective and sensitive determination of heavy metals, since they can provide synergistic effects of the high electrical conductivity as well as the large surface area offered by the nanoparticle and the polymer [55, 56].

Incorporating bio elements such as DNA, DNAzymes, etc, onto the GE surface is found to enable more selective and sensitive detection of some heavy metal ions such as Hg, Pb, and Ag [57, 58]. The selective binding of metal ions with DNA is attributed to their strong affinity with the DNA bases that results in the formation of base pairs that are even stronger than Watson–Crick base pairs [59]. Although there are many reports of the electrochemical biosensing of heavy metals using enzymes, DNA, aptamers etc, studies carried out on GEs are limited.

GEs exhibit promising characteristics for the electrochemical determination of heavy metals and are a good alternative to toxic mercury electrodes as well as expensive carbon-based electrodes. GEs have been successfully used for the determination of a wide range of toxic metals including As, Pb, Hg, Cr, Zn, and Cu in environmental samples. Among GEs, PGEs have been explored to a greater extent for the determination of toxic metals owing to their disposability, renewability and ease of miniaturization. Also, the electrocatalytic activity and thereby sensitivity and selectivity towards the metals can be increased by suitable modification of the electrode surface. Although they offer excellent opportunities in the rapid and on-site determination of toxic metals, they are still in the infant stage of commercialization. The challenge to be faced lies in their miniaturization without losing the reliability and accuracy of determination.

2.2.2 Carbon paste electrodes

Carbon paste electrodes (CPEs) have gained the interest of scientists due to their low ohmic resistance, renewability, non-toxicity, chemical inertness, and ease of preparation [60]. They are basically graphite powder mixed with a non-electrolytic binder such as paraffin oil or silicon oil tightly packed into a Teflon body [61]. It was Adams in 1958 who first reported the paste of carbon with an organic liquid during the study regarding a dropping carbon electrode [62]. Now, looking back, CPEs have evolved to be one of the important electrode material for electrochemical sensors owing to their interesting electrochemical and mechanical properties. CPEs are widely employed for the determination of toxic metals as they are considered to be an eco-friendly alternative to toxic mercury electrodes [63]. As pointed out earlier, their ease of renewability, facile preparation and possibility of miniaturization makes them suitable for on-site determination of environmental samples [63].

The applicability of CPEs for the determination of toxic metals is limited due to their lack of selectivity and sensitivity. It was in this context that scientists came up with different chemical modifications of CPEs in an attempt to improve the properties and to achieve the desired selectivity and sensitivity towards target metal ions [61]. A vast variety of chemical modifications so far done for the determination of toxic metals proves the ease of modification of CPEs [64–67]. They can be easily modified by dissolving the modifier in the binder or by just mixing the modifier into the paste itself. Also, the already prepared carbon paste can be modified *in situ* using drop casting or electrochemical techniques [61].

By exploiting the ease of modification of CPEs, diverse compounds including MNPs, CNMs, crown ethers, conducting polymers, ionic liquids, metal organic frameworks (MOFs), etc, have been used as modifiers in the determination of single as well as simultaneous determination of heavy metals such as Pb, Cd, Hg, Cr, Cu, Co, and Tl, [65, 68, 69]. The unique properties of MNPs such as high surface area, high signal to noise ratio, and electrocatalytic activity make them suitable modifiers in electroanalytical applications. MNPs and metal oxides of Bi, Sb, Zn, Au, Ti, Ni, and Cr have been incorporated in CPEs enabling selective and sensitive determination of Hg, Pb, Zn, Cu, and Cd in environmental, food and biological samples [69–74]. Another important modification which gained much interest among researchers is MOFs. These highly ordered crystalline compounds with high specific surface area, tuneable porosity and active electrocatalytic sites find wide applicability in electrochemical sensors [75]. CPEs incorporated with MOFs prepared from metals such as Zn, V, and Ca with various organic linkers have been successfully used for the determination of heavy metals like Pb, Cu, and Hg [63, 76, 77]. Despite their unique properties, one of the main difficulties faced in the development of MOFs based electrochemical sensors include their lower conductivity which may cause lower sensitivity. Advances in studies to overcome these problems are progressing and incorporating other conducting materials such as MNPs and polymers into MOFs is found to be an attractive alternative to improve the sensing performance [75].

CPEs modified with CNMs have also been extensively used for the determination of heavy metals. CNMs have, obviously, evolved to be an integral part of electrochemical sensors due to their high electrical conductivity, large surface area, biocompatibility, and chemical versatility [78]. CNMs such as MWCNT alone or in combination with MNPs incorporated in CPEs are found to greatly improve the sensitivity and selectivity for the determination of metals like Cr, Pb, Cd, etc, and the better sensing performance can be attributed to the synergistic effect arising from the faster electron transfer rate and high electroactive surface area of MNPs and CNMs [33, 79, 80]. The good and reliable performance of a developed sensor greatly depends on the electrode modification and CNMs shows excellent chemical and mechanical properties to be incorporated in CPEs for the selective and sensitive determination of heavy metals.

CPEs modified with biomolecules such as DNA or microbes are also used for the determination of heavy metals. As mentioned earlier some heavy metals can interact selectively with nucleo bases and if the CPE can be effectively modified with the proper nucleic acid sequences, then the specific coordinating ability of the target metal with the nucleo bases can be used for its quantification [59, 81, 82]. Besides nucleic acids, microbes such as *Rhizopus arrhizus* and *Pseudomonas aeruginosa* show strong biosorption capacities towards certain metals like Pb and have been used for modifying CPEs for the selective and sensitive detection of Pb [83, 84]. Although not as much literature has been found for biosensors for heavy metals based on CPEs, integrating these less expensive bio elements into CPEs can pave way for more specific, sensitive, and cost-effective determination of heavy metals.

CPEs are low cost electrodes that can be easily adapted to versatile chemical modifications and hence find wide application in the effective determination of heavy metals. They are good alternatives to toxic Hg based electrodes and further research on the possibility of miniaturization may shed light on their faster on-site determination in environmental and biological samples.

2.2.3 Glassy carbon electrodes

Glassy carbon electrodes (GCEs) are among the most commonly used electrodes for sensing heavy metals. Glassy carbon is a non-graphitized carbon produced by the pyrolytic degradation of polymer resins [85]. It has excellent temperature resistance, chemical and mechanical stability, and impassibility to gases and liquids and hence finds wide applications as an electrode material [86]. Bare GCE shows very slow electron transfer kinetics and lower selectivity and sensitivity for the determination of target metals. Numerous studies have been reported for the determination of various heavy metals using GCEs employing diverse chemical modifications using a variety of materials such as conducting polymers, MNPs, MOFs, macromolecules like calixarenes, etc [87–89]. GCEs can be easily modified using electrochemical deposition, electropolymerization, drop casting or by self-assembled monolayer methods. Applicability of wide potential windows and possibility of versatile modifications have attracted the interest of researchers to experiment with new modifications for the determination of toxic metals using GCEs.

As explained in the case of other carbon-based electrodes, MNPs have always proved to be among the best possible modification for enhanced sensing performance due to their high surface area, conductivity, and electrocatalytic activity [90, 91]. Metal and metal oxide nanoparticles of Au, Ag, Bi, Sb, Fe, Mn, and Ti combined with other modifications such as CNMs or conducting polymers, are widely explored for the determination of heavy metals in environmental and food samples [92–96]. More selectivity and sensitivity for the determination of heavy metals such as Fe, Cd, Pb, and Hg have been acquired by modifying GCEs with macromolecules such as crown ethers, calixarenes or cyclodextrins by exploiting their ability to form host–guest complexes with the target metals enabling ion selective recognition [97–100]. MOFs also find wide applications as GCE modifiers and a large number of studies are available for the electrochemical determination of environmental pollutants including heavy metals [34, 101–104]. In most cases, one of these modifications are combined with others and these composite modifications have proved to provide enhanced sensing ability due to the collective effect of the characteristic properties of the modifications involved.

Apart from the modifications mentioned above, great advances in electrochemical sensing of heavy metals are advancing by incorporating biological elements like DNA, aptamers, etc. Since the toxic heavy metals are present in complex matrices in the environment, one of the main challenges to overcome is their specific and sensitive detection from the test samples [105]. Certain heavy metals such as Hg, Pb, Ni, and As can bind specifically to biomolecules such as aptamers and DNA. This may lead to changes in conformation of the biomolecule based on which the metal can be quantitatively determined [106]. Hence, these less expensive bio elements can enable fast and selective detection of toxic metal ions from real samples and several studies have been reported where the effective immobilization of strands of DNAs or aptamers on GCE were achieved by modifying the electrode with suitable MNPs (Au, Ag, Pt, etc) in combination with MOFs and CNMs [107–110]. In some cases the changes in their conformation upon interaction with the metals ion have been transduced by using redox labels such as methylene blue enabling more sensitive detection [105]. GCE, however, involves a difficult surface renewal method, and is one of the commonly used electrode materials for sensing heavy metals. Ease of surface modification, wide potential window and low limit of detection (LOD) values make them suitable for sensing applications.

2.2.4 Screen printed carbon electrodes

The increase in environmental pollution and the continuing need for selective, reliable, and cost-effective sensing of heavy metals has always encouraged scientists across the world to aim for advanced technologies. Screen printed electrode (SPCE) related sensors have gained much attention owing to their possibility of miniaturization and lower complexity of preparation in comparison to conventional electrodes, and their disposability [111, 112]. In the production of screen printed electrodes, the electrode material is prepared in the form of a paste or ink and is then printed on to the surface of a ceramic or plastic substrate [113]. A non-conducting binder and a volatile solvent is also added while preparing the ink to provide proper adhesion onto the substrate [112]. Different forms of carbon including graphite, carbon black, and activated carbon can be printed to make SPCEs [112]. Because of their small size and reliability, SPCEs can pave way for rapid field analysis of heavy metals.

Only a limited number of studies are available for the determination of heavy metals using a bare SPCE [114–116]. The surface of SPCEs are coated with thin metallic films of Hg, Bi, or Au such that it then behaves like a conventional Hg, Bi, or Au electrode. Also, the electrode can be bulk modified while preparing the ink, prior to screen printing [35, 117–120]. In addition to this, different chemical modifications are also introduced on the SPCE to increase sensitivity and selectivity in the determination of heavy metals [121]. Also, ASV is the predominant technique used for their determination as it involves a pre-concentration step that can increase the sensitivity of detection [122]. CNMs, nafion, chitosan, macromolecules like crown ethers, various MNPs and their composites have been incorporated into SPCEs for efficient sensing performance [35, 123–130].

SPCEs are also used for the development of biosensors for heavy metals by serving as a platform for immobilizing aptamer, DNA, bacteria, and enzymes [131–135]. The specificity and sensitivity provided by these bio elements combined with the advantages of SPCEs such as disposability and their applicability for *in situ* determination can provide more opportunities for the real time determination of heavy metals.

Despite the advantages, SPCE based sensing of heavy metals is still in its infant stage. Studies for more effective modifications employing less toxic and environmentally friendly materials are progressing and new inventions in this area coulc open the possibilities of cost-effective and portable heavy metal sensors.

2.3 Carbon composite materials for the electrochemical sensing of toxic metals

Since, conventional working electrodes in their bare form are less effective in meeting the requirements of a better electrochemical sensor, electrochemists devoted much attention to modifying the surface of electrode surfaces [136]. Electrode modifications by the purposeful attachment of multifarious materials to alter the structural and electronic properties of an electrode which eventually led to better performance of the electrochemical sensor [137]. Carbon-based materials as electrode modifiers is a hot area of research and many advances happened during the past few decades. The key for successful modification depends on the selection of a suitable composite that can effectively interact with the target molecule. For instance, the affinity of mercuric ion for heteroatoms such as N and S has paved the way to use different carbon-based materials like graphene and multiwalled carbon nanotubes doped with the elements for the successful determination of the metal ion [138]. Similarly integrating other materials with CNMs for the synergistic effects of their properties in electrochemical performance has resulted in highly sensitive sensing platforms for toxic metal ions. The following is a brief discussion of the application of CNMs like graphene, RGO, CNFs, and CNTs as electrode modifiers in the development of electrochemical sensors for toxic metal ions

2.3.1 Graphene and its derivatives

Graphene, the thinnest and strongest known material, is nothing but a flat honeycomb monolayer of tightly packed carbon atoms. This versatile material can be modified into materials like 0-D fullerene, 1D CNTs, and 3D graphite by simply wrapping, rolling, and stacking respectively. Owing to the high electrocatalytic activity, thermal and electrical conductivity, atomic thickness, surface area and electron mobility, graphene-based electrochemical sensors are miles ahead [139–141]. But the properties like irreversible agglomeration due to weak Van der Waals forces and p–p interactions, less sensitivity and interference from other metal ions induce challenges while developing high performance sensors for toxic metal ions [141]. The agglomeration of graphene into graphitic forms can badly affect the surface area characteristics and the charge transfer ability of graphene-based materials which in turn leads to a decrease in efficiency of the sensors based on such classes of materials. However, these issues can be easily overcome by incorporating advanced materials on graphene or its counterparts, since such modified electrodes can enhance the surface characteristics of an electrode which subsequently increase the sensitivity and reproducibility of that particular sensor [142]. Most importantly, many modifications can provide a wide potential window for the detection and thus open a way to improve the simultaneous determination of numerous metal ions. Recently, there have been big reports on trace level quantification of toxic metal ions using graphene and its modified forms [143, 144]. A single layer of graphene sheet generally exhibits two structural regions such as basal plane with sp2 carbon atoms and one atom thick graphitic line of carbon called edge having dangling bonds and several capping moieties like carbonyl, carboxy, and hydroxyl groups. But the generation of a single layer of graphene with only carbon atoms has become a major challenge in its electrochemical applications. Returning to history, it can be seen that first monolayer of graphene was obtained from graphite by its mechanical exfoliation. Later, other methods like atomic layer deposition, chemical vapor deposition and thermal oxidation of graphite have resulted in the production of pristine graphene on large scales [145]. Exploitation of the intrinsic properties of graphene can be enabled by making it into highly porous structures, where these porous graphene networks provide extremely conductive multiplexed pathways as well as improved diffusion of the electrolytes and analytes. Moreover, those pores offer numerous active sites for interaction with wide variety of analyte molecules [141]. Depending on the porosity, graphene-based materials can be classified into micro (<2 nm), meso (2–50 nm) and macro (>50 nm) porous where the distribution of sizes and shapes of void spaces crucially determine the functions and applications of such materials. Hence, by modifying their exceptional porous and textural properties one can expect better performance from devices made from those materials [145].

Graphene related materials are generally obtained from graphene oxide (GO). Chemically, GOs are the thin graphene layers having several oxygen-bearing functional groups. Those functional groups can be hydroxyl, epoxy, carboxyl, and carbonyl, where they decorate edges and basal planes of GO [146]. Rather than GO, its reduced form (reduced graphene oxide RGO) is widely utilized due to various reasons because, on reducing GO to RGO its conducting property significantly changes. The removal of oxygen moieties from GO helps to obtain properties close to those of graphene. Currently, there are reports that show the relationship between degrees of reduction of RGO and its sensing efficiency on detecting various heavy metal ions. RGO of various reduction degrees were generated using the cyclic voltammetric technique by varying the number of cycles and that platform was successfully utilized for the simultaneous determination of Cd^{2+} and Pb^{2} in the presence of bismuth ions using a differential anodic stripping voltametric technique [147]. Bismuth plays an important part in improving the properties of graphene-based electrodes for the electrochemical sensing of heavy metal ions where advantages such as resolution of nearby peaks, alloy formation with the metal ions, and lower toxicity of Bi plays a crucial role. Rather than Bi films, nanoparticles made from Bi have been widely explored owing to the profuse surface area and abundant number of active sites on the surface [148]. Incorporating the advantages of both graphene and Bi yielded a highly sensitive sensing platform for the determination of toxic metal ions [144]. *In situ* deposition of Bi on RGO and gold electrodes has been explored as an integrated electrochemical microsensor for toxic metal ions on a silicon substrate where the micro patterned film enhanced the properties without affecting electrochemical characteristics [149].

The process of doping using heteroatoms like N, S, B, F, and Cl can intrinsically modify the electrochemical and physicochemical properties of graphene and its derivatives [150]. Among the heteroatom dopants, nitrogen gained wide attention and there are extensive reports on nitrogen doped graphene-based electrochemical sensors for toxic metal ions where N plays a crucial role in tuning the mechanical properties and modifying the electronic as well as the electrocatalytic activity of graphene materials [151, 152]. Similarly, doping with nitrogen can also improve the sensitivity as well as biocompatibility which in turn boosts the biosensing applications of these materials [153]. The combined effects of nitrogen doping as well as the advantages of graphene and its derivatives, have paved the way to enhance the stripping peak currents of toxic metal ions like Cd^{2+}, Pb^{2+}, Cu^{2+}, Hg^{2+} which significantly enable the simultaneous determination of these metal ions [154]. Likewise, N doped carbon dots and GO hybrids have provided a sensing platform for $Cd^{2+}$and Pb^{2+}. In this hybrid format, the presence of oxygen containing functionalities and enormous electroactive surface area could help in the adsorption of metal ions through electrostatic interactions which ultimately improve the sensitivity of the electrochemical sensor. In addition to that, homogeneously distributed nitrogen doped carbon dots can also impart a crucial role in re-oxidation of metal ions for stripping analysis [155]. The high affinity of mercuric ions towards sulfur and nitrogen atoms has paved the way in the fabrication of sensors for the highly toxic Hg^{2+} using sulfur or nitrogen containing functionalities on graphene. Thiourea can act as a source of both nitrogen and sulfur in S/N doping, which enables interaction between these heteroatoms with toxic metal ions like Hg^{2+} where these heteroatoms containing functional groups act as hooks for mercuric ions [138]. Similarly, fluorine doped graphene derivatives can also yield extraordinary pathways for toxic metal ions. The electronegativity difference between fluorine and carbon opens a way to ionic, semi-ionic, and covalent bonding characteristics. They come from the electron withdrawing nature of F atoms as well as the electron donating characteristics of F lone pair electrons [156]. Fluorination on graphene can be generally done using the method of plasma fluorination and direct gas fluorination [157, 158]. This pragmatic material can also act as a suitable candidate for various other applications like catalyst support, capacitors, batteries, and biomedical devices [156]. Since, fluorinated graphene has better electrochemical performance than pristine GO, it turned into a simultaneous detection platform for several toxic metal ions like Cd^{2+}, Pb^{2+}, Cu^{2+}, and Hg^{2+} using the anodic stripping voltammetric technique [159]. Melioration in properties due to the doping of heteroatoms gave an insight into the use of conductive polymers having heteroatoms for electrode modification. Polymeric materials like polyaniline (PA), polypyrrole, etc, with N and poly 3,4-ethylenedioxithiophene with both O and S have been widely recognized for the sensitive determination of mercuric ions. In particular, polypyrrole with relatively easy substitution and polymerization, prominent electrical conductivity, and exceptional chemical stability has resulted in their exceeding the utility of their nanocomposites for sensing purposes [160]. Polysaccharides have been incorporated to improve the hydrophilic and amorphous nature of a conducting polymer. Pectin, a polysaccharide with both N and O, helps to increase the ease of amide bond formation and diffusion of particles in the polymer matrix [161]. Owing to these characteristics, interaction between metal ions and conducting polymer–graphene composites could improve and thus an innumerable number of sensing strategies for toxic metal ions have been evolved. In many of these cases, simultaneous determination of toxic metals with a very low limit of detection was also achieved [162]. Most importantly, porous carbon materials and their derivatives can combine with polypyrrole to boost adsorption of heavy metal ions. In addition to that, affinity of conducting polymers towards functionalized nanomaterials also help to improve the accretion of various metal ions on the surface of electrodes. This in turn act as an unswerving approach in their electrochemical sensing [163]. Likewise, polyaniline and graphene-based materials have also been employed for the same purpose. Conventionally the composites of graphene and PA can be made using *in situ* electrochemical synthesis and *in situ* polymerization [164, 165]. Owing to the advantages of polymer hydrogels such as biocompatibility and stimuli responsive properties, the PA–GO composites have been introduced for the ultrasensitive determination and separation of highly hazardous lead ions [166]. Similarly, graphene and several poly amino acids, their analogous and derivative based composites have also been extensively aided for the same purpose. Since poly (aminoacid)s mainly contain several amino, carboxyl groups on them and hence improve the properties of the sensor by efficient binding with metal ions [167]. Moreover, in order to improve the polymer–graphene interactions, porphyrins were introduced to the systems as terminal groups of poly(aminoacid)s where the conjugate porphyrin interact easily by p–p stacking with conductive graphene [168].

Chitosan (poly D-glucosamine), a natural polysaccharide having primary amino groups on the polymer chain has been widely utilized for several decades in water treatment owing to the formation of metal chelates. Relatively low cost, non-toxicity, hydrophilicity, and biocompatibility make chitosan a suitable candidate in the fabrication of a variety of materials [169]. In addition to that, its numerous metal chelating sites and excellent film forming ability, is considered to be suitable as a substrate for ion imprinting polymer films. Owing to the synergistic effect, its combination with MNPs like silver, gold, and platinum enhance the electron transfer as well as electrochemical properties. A combination of GO or RGO, chitosan and MNPs are widely recognized for trace determination of toxic metal ions [170]. Similarly, L-cysteine as well as poly L-cysteine and chitosan composites on GO have been explored for the simultaneous determination of toxic metals like Cd, Pb, and Cu [171, 172]. An anionic polysaccharide produced from cellulose—carboxy methyl cellulose has been explored for the functionalization of porous RGO which helps to improve the mechanical properties and enhance dispersibility in physiological solvents. Incorporating another pentasaccharide, Fondaparinux with the highest negative charge density also helps to increase the affinity towards metal ions. This nanocomposite on GCE yielded an efficient platform for sensitive determination of cadmium and lead ions [173].

By hybridizing different MNPs on GO sheets, the performance of electrochemical detection can be enhanced. Among the different MNPs, gold (Au) and silver (Ag) nanoparticles show versatile features like quantum characteristics and hence they have been expansively utilized for the detection of metal ions incorporating the advantages of RGOs [174]. In the case of Au-graphene composites, high catalytic activity and electron transport kinetics play a crucial role in electrochemical sensing of various species. Owing to the higher affinity of mercuric ion towards AuNPs, the utilization of AuNPs as one component in the electrode modification for Hg^{2+} sensing also attracted much attention. But the homogenous distribution of AuNPs on the electrode surface is literally challenging and need much attention during sensor fabrication. But there are reports on simple, one step solvothermal approach for the synthesis of AuNPs using ethylene glycol as both reducing agent as well as solvent during the synthesis and which reduces the use of any additional surfactants and reducing agents during the synthesis [175]. Nanoparticles of iron oxide have also attracted much attention over the past few years. The super paramagnetic nanoparticle Fe_2O_3, and graphene composite modified on GCE worked as an efficient electrochemical sensor for the simultaneous determination of metals such as zinc, lead, and cadmium [148]. But, as in the case of Fe_3O_4 alone, there are more chances for aggregation and undesirable oxidation and they are not selective towards complex matrices [176]. Moreover, they were commonly used as a core in core–shell systems. So, the incorporation of other nanomaterials with more suitable features like durability, stability and cost effectiveness can provide a better response. TiO_2 is one of them which also helps to protect the core. Apart from this, Fe_3O_4@TiO_2 exhibits high adsorption capability towards lead ions. Also, introducing AuNPs and making an Au–S bond with an organic compound ETBD (2,2-((1E)-((4-((2-mercaptoethyl) thio)-1,2-phenylene) bis (azanylylidene)) bis (methanylylidene)) diphenol) along with Fe_3O_4@TiO_2 on N doped graphene enhances the properties [177]. Among the organic modifiers, dendritic macromolecules (dendrimer) have gained world-wide attention. Polyamidoamine (PAMAM) dendrimers are the most widely recognized. They show promising extraction quality towards various metal ions from water samples. Even though they have great selectivity towards various toxic metal ions, their uses have been limited in electrochemical techniques, owing to their poor conductivity. But their combination with graphene and magnetic nanoparticles like Fe_3O_4 has resulted in a highly sensitive sensor for the simultaneous detection of lead and cadmium ions. Similarly, the nanocomposite of nickel tungstanate and graphene has been widely recognized as an efficient material for the individual as well as simultaneous electrochemical detection of toxic metal ions. This binary metal oxide NP acts both as an electrochemical catalyst as well as a spacer to reduce restacking of individual graphene sheets which enhances overall efficiency of the sensor [178]. Graphene-based metal oxide hybrids are also an important candidate in toxic metal detection where the rigid structure of metal oxide helps to avoid the graphene restacking. The synergistic effect of nanosized cerium dioxide and graphene on GCE yields a better sensing platform for simultaneous determination of metal ions like Cd^{2+}, Pb^{2+}, Cu^{2+}, and Hg^{2+} [179]. Another important class of materials which helps to avoid the aggregation and restacking of graphene are CNTs which act as spacers between the graphene sheets. Combining the effects of RGO and CNTs composite on Au substrate along with *in situ* plated Bi film, simultaneous determination of Cd and Pb were possible [180]. Likewise, GCE modified with a composite of electrochemically reduced GO multi-walled carbon nanotubes (MWCNTs)-L-cysteine with the help of the drop casting method was exclusively chosen for the selectivity improvement in the effective determination of lead ion [181].

MOFs are also among the well explored modifiers owing to advantages like porosity, high surface area, and pore volume, open metal sites and tunable structures. They are nothing but an assembly of metal clusters and ligands. They are extensively employed in numerous applications such as separation, sensing, storage, catalysis, and drug delivery. Due to the poor conductivity of this material, it is usually incorporated with MNPs, graphene and its derivatives as well as conducting polymers [182]. So, MOFs along with graphene and its derivatives can enhance the properties of electrochemical sensors and ultimately enable simultaneous determination of various species. Another important class of materials combined with graphene for enhancing the electrode properties is calixarene. The three-dimensional cup or basket shaped material has the ability to form suitable ionophores for metal ions. The complexes formed between calixarenes and metals enhance the sensitivity of the respective sensor for toxic metal detection [97].

Construction of sensors based on molecular/ion imprinted polymers have resulted in the selective recognition of structurally similar species. There are several sensors for toxic metal ions based on ion imprinted polymers and graphene. In the determination of Pd^{2+}, a palladium ion imprinted membrane was made on a GCE using allyluread as the monomer and ethylene glycol dimethacrylate (EGDMA) as the cross linker. Azobisisobutylnitrile was taken as the initiator in the *in situ* polymerization in acetonitrile where allyluread effectively coordinated to Pd^{2+} through two nitrogen atoms according to the HSAB concept [183]. Similarly, the role of ionic liquids in extracting metal ions and as pasting binders for CPEs also helped to develop electrochemical sensors for several metal ions. A composite electrode made from a new synthetic phosphorus ylide (1-*n*-octylpyridinum hexafluorophosphate (OPFP)), and (2,4-Cl2C6H3C(O)CHPPh3) and graphene have been utilized for toxic ions such as Tl^{+}, Pb^{2+} and Hg^{2+} [184]. Likewise, a combination of graphene, ionic liquid (1-butyl-2,3-dimethylimidazolium tetrafluoroborate) and nafion composite along with Bi film modified SPCE was fabricated for the determination of Zn^{2+}, Cd^{2+}, and Pb^{2}. Herein, the combined advantages of nafion with excellent antifouling capacity, high permeability to cations and chemical inertness and tetrafluroborate with high electrochemical stability and electrical conductivity resulted in the generation of an efficient electrode modifier [185]. Introducing single stranded DNA probes on graphene and nano Au electrodeposited GCE has increased the sensitivity to attomolar level in the electrochemical sensing of mercuric ion. Functionalization with a 10-mer thymine-rich DNA probe and a 29-mer guanine-rich DNA probe labeled with methyl blue helped to strengthen the response of the signals. The sensing strategy works on the thymine–Hg^{2+}–thymine coordination chemistry [108]. The quantification of toxic metal ions with high selectivity and sensitivity can be achieved through different forms of pragmatic graphene and its composites. Combining the advantages of graphene and other important classes of materials enhances the sensing parameters. Although there are a large number of advantages, the randomly distributed pores and presence of defects reduces the overall merits. There is much scope for this particular area of research since there is more chance for miniaturization and micro fabrication which in turn allows the real time monitoring of environmental samples more prominently.

2.3.2 Carbon nanotubes

Since their discovery in 1991, carbon nanotubes (CNTs), which are graphene sheets rolled to a cylindrical structure, have attracted the attention of many scientists around the world. The literature clearly reveals the remarkable applications of CNTs in various fields of science such as transparent electrodes in light emitting diodes and lithium ion batteries, super capacitors, transistors, catalysts, in chemical sensors, etc [186]. Without doubt the unique chemical, mechanical and electronic properties possessed by these CNTs make it the most favored nanomaterial among researchers these days. CNTs because of their excellent electrical conductivity, high surface area, and adsorptive properties enable them to function as very effective electrode modifiers in electrochemical sensors to improve sensitivity and as a matrix to immobilize biorecognition elements. Since the surface and inner layers of CNTs possess small defects, it enables them to interact with heavy metals ions very easily [22]. Evidently this interaction has been effectively utilized in many of the electrochemical sensing strategies for heavy metal detection. Both single-walled, double-walled, and MWCNTs can be seen in these sensors with remarkable performance.

Vertically aligned single-walled carbon nanotubes (vSWCNTs) with superior binding ability with substrates and high electrical conductivity holds great promise for advanced applications in electrochemistry. These nanostructures, which can also increase the number of binding sites and thereby enhance the access of the analyte molecules to the electrode surface, is very much favored in electro analysis. It has also been proven to facilitate electron transfer in a faster manner. These vSWCNTs with desired orientation and homogeneity can be fabricated with the aid of self-assembled monolayers (SAM). Thus, fabricated vSWCNTs with excellent substrate binding strength can successfully incorporate biorecognition elements like exonuclease-III which have target recycling ability for the determination of even trace levels of heavy metal ions [187]. It can be seen in the literature that nanocomposites of CNTs with other CNMs such as graphene, graphitic carbon nitride, carbon quantum dots, etc, when used as electrode modifiers can be a promising approach for monitoring heavy metal ions. Synergistic effects of the high electrocatalytic activity of graphitic carbon nitride and high electrical conductivity of oxidized MWCNTs makes simultaneous determination of even four heavy metals such as Cd^{2+}, Hg^{2+}, Pb^{2+}, and Zn^{2+} [12]. The strategy of anchoring carbon quantum dots on MWCNTs has also been adopted for the fabrication of electrochemical sensors for heavy metal sensing with improved stability and sensitivity [15]. The strong chelation ability of amino functionalized MOFs and the strong electrostatic interaction of carboxyl functionalized MWCNTs to heavy metal ions have been also combined in an electrochemical sensing platform for the detection of the latter [15]. Reports also show that metal and metal oxide nanoparticles when encapsulated on CNMs like CNTs exhibits high sensitivity towards heavy metal ions like Hg^{2+}, Pb^{2+}, Cd^{2+}, and Zn^{2+} in electrochemical determination. It is noteworthy that the excellent adsorption ability of fluorinated MWCNTs for heavy metal ions when integrated with the promising electrocatalyst magnetite nanoparticles, has exhibited remarkable performance in multiple (five) ions sensing. Such types of CNT based sensors with proven applicability in environment and food analysis, are cost-effective and eco-friendly alternatives for the reliable simultaneous determination of multiple heavy metal ions [188].

2.3.3 Carbon nanofibers

Similar to CNTs, one-dimensional carbon nanofibers (CNFs) are also a great contribution of material science to other fields of science for several remarkable advancements. The electrochemical sensing field of analytical science has also achieved a significant leap with the aid of CNFs. It has been proven that CNFs are excellent electrode modifiers as it is a part of many of the recent efficient electrochemical sensors for the determination of several biomolecules, drugs, pollutants, toxic metals, etc. Recent reports show that CNFs and thei composites with excellent structural stiffness and affinity for heavy metals can be utilized for the effective determination of a number of heavy toxic metal ions like Cd^{2+}, Pb^{2+}, Hg^{2+}, and Cu^{2+}.

Nano electrode arrays fabricated through vertically aligned CNFs (VACNFs) on suitable electrode surfaces have gained considerable attention in various fields [189]. Interestingly, VACNFs arranged on a GCE have been proven to be an efficient alternative to the conventional mercury electrodes for Pb^{2+} determination. It can be seen that the employment of relatively cheaper hole mask colloidal lithography in the above patterning can offer a cost-effective nature to the assay along with a sensitivity as low as nano molar range [189]. Nanocomposites obtained through a suitable blend of CNMs and conducting polymers with exceptional properties are very favored candidates for electrode modification for sensor development. A number of electrochemical sensors have been reported so far for the individual and simultaneous determination of heavy metal ions using CNMs like graphene, RGO, MWCNTs, carbonaceous nanospheres, etc, with different conducting polymers [189]. In addition to this CNFs are also used in combination with a conducting polymer for the real sample analysis of heavy metals like Pb^{2+}. The promising nanomaterial CNFs with remarkable electrical conductivity, chemical and mechanical stability, good surface area when combined with the conductivity and stability offered by polymer like polypyrrole, could achieve a sensitivity up to angstrom molar range for the determination of the metal ions [190]. It is demonstrated that the incorporation of N and S into CNMs can impart good sensitivity and selectivity to sensors for the determination of heavy metals. This is attributed to the ability of N and S to improve the surface chemistry of CNMs by breaking the electro neutrality and thereby increasing the surface sites. Also the built-in affinity of CNFs for heavy metals becomes enhanced upon an increase in its porous structure on doping with N [191]. Wide linear dependence and low detection limits exhibited by the as prepared sensors clearly shows the excellent electrocatalytic ability of N and S doped CNFs towards heavy metal ions. When a single electrode modification can act both as an electrocatalyst as well as an immobilizer for a biorecognition element, it definitely simplifies the sensor [192]. CNFs can be functionalized with oxygen moieties using acid treatment and those hydrophilic CNFs can act as good immobilizers for biorecognition elements. Such incorporation of biorecognition elements could overcome many of the drawbacks of electrochemical techniques such as anodic stripping voltammetry which is a better approach in heavy metal detection. Further, the use of screen printed electrodes as electrochemical signal transducers along with the single modification using CNFs can provide a cost-effective disposable platform in heavy metal detection [193]. Among the different CNMs, CNFs are very suitable in supporting metal nanostructures due to their high length to diameter ratio. A nanohybrid formed by the successful deposition of MNPs and CNMs on highly conducting CNFs allows efficient adsorption and charge transfer of several heavy metal ions simultaneously [194, 195]. The synergistic effect of the specific complexing ability of synthetic receptors like bis(indolyl)methane (BIM) with heavy metal ions and the ability of mesoporous CNFs to enhance the matrix conductivity have found to enhance the electrochemical response in heavy metal detection [196].

2.4 Conclusion

In the current world scenario, toxic metals are contaminating the water we drink, the food we eat, thereby reaching our body and badly affecting our internal organs. The cataclysmic effects of toxic metals on the lithosphere during the past few decade demands simple, sensitive, and cost-effective sensing strategies for their effective determination. Owing to the fascinating advantages of carbon-based materials that meets the above demands, electrochemical sensors based on the latter have been well explored. This chapter discussed the use of such materials as electrode materials as well as modifiers in the electrochemical determination of toxic metal ions in the past few years. Carbon-based electrodes are found to effectively replace conventional toxic mercury based electrodes for the determination of toxic metals and GEs, CPEs, GCEs, and SPCEs stand out different in their unique properties. Advances are made to realize portable devices for the effective on-site determination of toxic metals and invention of SPCEs are considered to be a milestone in this area. Also, carbon materials are widely explored as electrode modifiers for the determination of toxic metals. The successful utilization of CNMs such as graphene, RGO and its derivatives, CNTs and CNFs can be seen in electrochemical sensors for toxic metal determination. Rather than its pristine form, different modifications of graphene and RGO can be seen in these sensors and are discussed in detail here. In order to enhance their properties, strategies such as doping with heteroatoms and introducing materials such as conducting polymers, metal organic frameworks, biomolecules, as well as various metal nanoparticles have been used. Similarly, CNTs and CNFs in different alignments and functionalizations also have been integrated in many electrochemical sensors for toxic metals to achieve advanced sensing performance. Also the combination of CNFs and CNTs with other CNMs, polymers, metal organic frameworks, and metal nanoparticles have been adopted for better adsorption of toxic metal ions onto the transducer to enhance the sensitivity of various voltammetric sensors. On the whole, this chapter discussed how the versatile properties of carbon-based materials are meeting the global demand for the sensitive determination of hazardous metal ions. Incorporating the advantages of carbon-based materials along with other multifaceted materials have resulted in the development of extremely fruitful electrochemical sensing platforms for toxic metals. The most important advantage of electrochemical sensing is the possibility of miniaturization and thus the real time analysis. But, the on-site determination of toxic metal ions using simple devices are still in a nascent stage and much effort has to be devoted to the development of facile and reliable technologies for such a relevant objective.

Owing to the innumerable hazards caused by toxic metals, completely banning them may seem crucial. But in this era of modernization, completely banning these toxic metals is an Utopian idea, proper disposal of them can avoid their massive discharge to the environment.

References

[1] Duffus J H 2002 *Pure Appl. Chem.* **74** 793–807

[2] Length F 2002 *Int. J. Phys. Sci.* **2007** 112–8

[3] Ali H, Khan E and Ilahi I 2019 *J. Chem.* **2019** 6730305

[4] Turkez H, Geyikoglu F, Tatar A, Keles M S and Kaplan I 2012 *Exp. Toxicol. Pathol.* **64** 93–101

[5] Ovečka M and Takáč T 2014 *Biotechnol. Adv.* **32** 73–86

[6] Förstner U and Wittmann G T W 1981 *Metal Pollution in the Aquatic Environment* (Berlin: Springer)

[7] Garciá-Miranda Ferrari A, Carrington P, Rowley-Neale S J and Banks C E 2020 *Environ. Sci. Water Res. Technol.* **6** 2676–90

[8] Mathew M R and Girish Kumar K 2020 *J. Electrochem. Soc.* **167** 047519

[9] Anand S K, Mathew M R, Radecki J, Radecka H and Girish Kumar K 2020 *J. Electroanal. Chem.* **878** 114531

[10] Anand S K, Mathew M R and Girish Kumar K 2021 *J. Electrochem. Soc.* **168** 047507

[11] Wang J, Tian B, Wang J, Lu J, Olsen C and Yarnitzky C 1999 *Anal. Chim. Acta* **385** 429–35

[12] Ramalingam M, Ponnusamy V K and Sangilimuthu S N 2019 *Microchim. Acta* **186** 69

[13] Manjunatha J G 2020 *Anal. Bioanal. ELectrochem.* **12** 893–903

[14] Broli N, Vallja L, Shehu A and Vasjari M 2019 *J. Food Process. Preserv.* **43** e13838

[15] RasulKhan B B, Periakaruppan P, Ponnaiah S K, Venkatachalam G and Jeyaprabha B 2021 *J. Clust. Sci.* **32** 135–44

[16] Liu Z, Robinson J T, Tabakman S M, Yang K and Dai H 2011 *Mater. Today* **14** 316–23

[17] Kour R, Arya S, Young S-J, Gupta V, Bandhoria P and Khosla A 2020 *J. Electrochem. Soc.* **167** 037555

[18] Raril C, Manjunatha J G and Tigari G 2020 *Instrum Sci. Technol.* **48** 561–82

[19] Hareesha N, Manjunatha J G, Amrutha B M, Sreeharsha N, Basheeruddin Asdaq S M and Anwer M K 2021 *Colloids Surf. A Physicochem. Eng. Asp.* **626** 127042

[20] Liu F, Cheng Y, Tan J, Li J, Cheng H and Hu H 2021 *Front. Chem.* **9** 1–7

[21] Duoc P N D, Binh N H, Van Hau T, Thanh C T, Van Trinh P and Tuyen N V 2020 *J. Hazard. Mater.* **400** 123185

[22] Deshmukh M A, Shirsat M D, Ramanaviciene A and Ramanavicius A 2018 *Crit. Rev. Anal. Chem.* **48** 293–304

[23] Hareesha N and Manjunatha J G 2020 *J. Sci. Adv. Mater. Devices* **5** 502–11

[24] Manjunatha J G 2020 *Chem. Data Collect.* **25** 100331

[25] Hareesha N, Manjunatha J G, Raril C and Tigari G 2019 *ChemistrySelect* **44** 559–67

[26] Uslu B and Ozkan S A *Anal. Lett.* **40** 817–53

[27] Locatelli C and Torsi G 2003 *Microchem. J.* **75** 233–40

[28] Kumar S and Srivastva A 2021 *Application of Carbon Nanomaterial Decorated Electrochemical Sensor for Analysis of Environmental Pollutants* (London: Intechopen)

[29] Zhang W, Zhu S, Luque R, Han S, Hu L and Xu G 2016 *Chem. Soc. Rev.* **45** 715–52

[30] Vyskočil V and Barek J 2009 *Crit. Rev. Anal. Chem.* **39** 173–88

[31] Karnib M, Kabbani A, Holail H and Olama Z 2014 *Energy Procedia.* **50** 13–20

[32] Palisoc S, Vitto R I M and Natividad M 2019 *Sci. Rep.* **9** 18491

[33] Le Hai T, Hung L C, Phuong T T B, Ha B T T, Nguyen B S and Hai T D 2020 *Microchem. J.* **153** 104456
[34] Ru J, Wang X, Cui X, Wang F, Ji H and Du X 2021 *Talanta* **234** 122679
[35] Mandil A, Idrissi L and Amine A 2010 *Microchim. Acta* **170** 299–305
[36] Wei M, Zeng G Y, Liu Y and Lu Q 2013 *Asian J. Chem.* **25** 861–3
[37] Palisoc S, Espique E M, Ribay A M, Sy S, Tan K A and De Jesus N 2018 *Int. J. Electrochem. Sci.* **13** 10355–67
[38] Annu S S, Jain R and Raja A N 2020 *J. Electrochem. Soc.* **167** 037501
[39] Ustabasi G S, Ozcan M and Yilmaz I 2021 *J. Electrochem. Soc.* **168** 097508
[40] Bond A M, Mahon P J, Schiewe J and Vicente-Beckett V 1997 *Anal. Chim. Acta* **345** 67–74
[41] David I G, Popa D E and Buleandra M 2017 *J. Anal. Methods Chem.* **2017** 1905968
[42] Palisoc S, Gonzales A J, Pardilla A, Racines L and Natividad M 2019 *Sens. Bio-Sensing Res.* **23** 100268
[43] Kawde A N, Baig N and Sajid M 2016 *RSC Adv.* **6** 91325–40
[44] Ndlovu T, Mamba B B, Sampath S, Krause R W and Arotiba O A 2014 *Electrochim. Acta* **128** 48–53
[45] Ntsendwana B, Peleyeju M G and Arotiba O A 2016 *J. Environ. Sci. Heal—Part A Toxic/ Hazardous Subst. Environ. Eng.* **51** 571–8
[46] Tasić Ž Z, Petrović Mihajlović M B, Simonović A T, Radovanović M B and Antonijević M M 2021 *Results Phys.* **22** 103911
[47] Akanda M R, Sohail M, Aziz M A and Kawde A N 2016 *Electroanalysis* **28** 408–24
[48] Bedin K C, Mitsuyasu E Y, Ronix A, Cazetta A L, Pezoti O and Almeida V C 2018 *Int. J. Anal. Chem.* **2018** 1473706
[49] Pierini G D, Pistonesi M F, Di Nezio M and Centurión M E 2016 *Microchem. J.* **125** 266–72
[50] Giao N Q, Dang V H, Yen P T H, Phong P H, Ha V T T and Duy P K 2019 *J. Appl. Electrochem.* **49** 839–46
[51] Lima T M, Soares P I, do Nascimento L A, Franco D L, Pereira A C and Ferreira L F 2021 *Chem. J.* **168** 106406
[52] Lahrich S, Maanoun B E L and Mhammedi M A 2019 *Int. J. Environ. Anal. Chem.* **101** 140–52
[53] Elobeid W H and Elbashir A A 2019 *Prog. Chem. Biochem. Res.* **2** 24–33
[54] El Badry Mohamed M, Yossri Frag E and Marzouk H A 2021 *Food Chem.* **363** 130324
[55] Liao Y, Li Q, Yue Y and Shao S 2015 *RSC Adv.* **5** 232–8
[56] Ardalani M, Shamsipur M and Besharati-Seidani A 2020 *J. Electroanal. Chem.* **879** 114788
[57] Akbari Hasanjani H R and Zarei K 2019 *Biosens. Bioelectron.* **128** 1–8
[58] Kokkinos C, Prodromidis M, Economou A, Petrou P and Kakabakos S 2015 *Electrochem. Commun.* **60** 47–51
[59] Saidur M R, Aziz A R A and Basirun W J 2017 *Biosens. Bioelectron.* **90** 125–39
[60] Tajik S, Beitollahi H, Garkani Nejad F, Sheikhshoaie I, Nugraha A S and Jang H W J 2021 *Mater. Chem.* A **9** 8195–220
[61] Taylor P, Švancara I, Vyt K and Svancara I 2010 *Crit. Rev. Anal. Chem.* **31** 37–41
[62] Adams R N 1963 *Rev. Polarogr.* **11** 71–8
[63] Wang Y, Wu Y, Xie J and Hu X 2013 *Sens. Actuators B Chem.* **177** 1161–6
[64] Tonle I K, Ngameni E and Walcarius A 2005 *Sens. Actuators B Chem.* **110** 195–203
[65] Ping J, Wu J, Ying Y, Wang M, Liu G and Zhang M 2011 *J. Agric. Food Chem.* **59** 4418–23

[66] Prabhakaran D C, Ramamurthy P C, Sivry Y and Subramanian S 2020 *Int. J. Environ. Anal. Chem.* **102** 2053–73
[67] Zhou W, Chai Y, Yuan R, Guo J and Wu X 2009 *Anal. Chim. Acta* **647** 210–4
[68] Laghlimi C, Ziat Y, Moutcine A, Hammi M, Zarhri Z and Maallah R 2020 *Chem. Data. Collect.* **29** 100496
[69] Niu P, Fernández-Sánchez C, Gich M, Ayora C and Roig A 2015 *Electrochim. Acta* **165** 155–61
[70] Devnani H and Satsangee 2015 *Int. J. Environ. Sci. Technol.* **12** 1269–82
[71] Tesarova E, Baldrianova L, Hocevar S B, Svancara I, Vytras K and Ogorevc B 2009 *Electrochim. Acta* **54** 1506–10
[72] Ashrafi A M, Cerovac S, Mudrić S, Guzsvány V, Husáková L and Urbanová I 2014 *Sens. Actuators, B Chem.* **191** 320–5
[73] Moutcine A, Laghlimi C, Ziat Y, Smaini M A, El Qouatli S E and Hammi M 2020 *Inorg. Chem. Commun.* **116** 107911
[74] Koudelkova Z, Syrovy T, Ambrozova P, Moravec Z, Kubac L and Hynek D 2017 *Sensors* **17** 1832
[75] Liu L, Zhou Y, Liu S and Xu M 2018 *Chem. Electro. Chem.* **5** 6–19
[76] Niu B, Yao B, Zhu M, Guo H, Ying S and Chen Z 2021 *J. Electroanal. Chem.* **886** 115121
[77] Vlachou E, Margariti A, Papaefstathiou G S and Kokkinos C 2020 *Sensors* **20** 4442
[78] Baptista R, Belhout S A, Giordani S and Quinn S J 2015 *Chem. Soc. Rev.* **44** 4433–53
[79] Hilali N, Ghanam A, Mohammadi H, Amine A, García-Guzmán J J and Cubillana-Aguilera L 2015 *Electroanalysis* **30** 2750–9
[80] Guo J, Chai Y, Yuan R, Song Z and Zou Z 2011 *Sens. Actuators, B Chem.* **155** 639–45
[81] Ruhan Q, Yinyin L, Dawei C, Yushi G, Zongping L and Haidong L 2020 *Int. J. Electrochem. Sci.* **15** 7347–58
[82] Ebrahimi N, Raoof J B, Ojani R and Ebrahimi M 2021 *J. Iran. Chem. Soc.* **18** 407–13
[83] Yüce M, Nazir H and Dönmez G 2010 *Bioresour. Technol.* **101** 7551–5
[84] Mohanraj P, Bhuvaneswary S, Sreelakshmi M S, Chandrashekhar V, Narasimhulu K, Kiran B and Kumar S 2019 *Water Sci. Technol.* **80** 2058–66
[85] Serp P 2013 *Carbon* (Amsterdam: Elsivier)
[86] Hassler M 2012 *Coatings for Biomedical Applications* (Cambridge: Woodhead Publishing)
[87] Kokab T, Shah A, Iftikhar F J, Nisar J, Akhter M S and Khan S B 2019 *ACS Omega* **4** 22057–68
[88] Hassan K M, Gaber S E, Altahan M F and Azzem M 2020 *Sens. Bio-Sensing Res.* **29** 100369
[89] Lee S, Park S K, Choi E and Piao Y 2016 *J. Electroanal. Chem.* **766** 120–7
[90] Buledi J A, Amin S, Haider S I, Bhanger M I and Solangi A R 2021 *Environ. Sci. Pollut. Res.* **28** 58994–9002
[91] He Y, Wang Z, Ma L, Zhou L, Jiang Y and Gao 2020 *RSC Adv.* **10** 27697–705
[92] Hao C, Shen Y, Shen J, Xu K, Wang X and Zhao Y A 2016 *Microchim. Acta* **183** 1823–30
[93] Palisoc S T, Natividad M T, De Jesus N and Carlos J 2018 *Sci. Rep.* **8** 17445
[94] Ding L, Zhai J, Bond A M and Zhang J 2013 *J. Electroanal. Chem.* **704** 96–101
[95] Mnyipika S H, Munonde T S and Nomngongo P N 2021 *Membranes* **11** 517
[96] Sun F, Ge D and Fan G 2017 *Int. J. Electrochem. Sci.* **12** 7377–85
[97] Göde C, Yola M L, Yılmaz A, Atar N and Wang S 2017 *J. Colloid Interface Sci.* **508** 525–31
[98] Zheng H, Yan Z, Dong H and Ye B 2007 *Sens. Actuators, B Chem.* **120** 603–9

[99] Dehdashtian S and Shamsipur M 2018 *Colloids Surf. B. Biointerfaces* **171** 494–500
[100] Hu Y, Liu Z, Zhan H and Shen Z 2018 *Anal. Methods* **10** 767–74
[101] Tajik S, Beitollahi H, Nejad F G, Safaei M, Zhang K and Van Le Q 2020 *RSC Adv.* **10** 21561–81
[102] Lu Z, Zhao W, Wu L, He J, Dai W and Zhou C 2021 *Sens. Actuators, B Chem.* **326** 128957
[103] Wang X, Xu Y, Li Y, Li Y, Li Z and Zhang W 2021 *Food Chem.* **357** 129762
[104] Kwon D and Kim 2021 *J. Appl. Electrochem.* **51** 1207–16
[105] Zhang X, Huang C, Jiang Y, Jiang Y, Shen J and Han E 2018 *J. Agric. Food Chem.* **66** 10106–12
[106] Kanellis V G and dos Remedios C G 2018 *Biophys. Rev.* **10** 1401–14
[107] Xu W, Zhou X, Gao J, Xue S and Zhao J 2017 *Electrochim. Acta* **2251** 25–31
[108] Zhang Y, Zeng G M, Tang L, Chen J, Zhu Y and He X X 2015 *Anal. Chem.* **87** 989–96
[109] Wang Y, Wang Y, Wang F, Chi H, Zhao G and Zhang Y 2022 *J. Colloid Interface Sci.* **606** 510–7
[110] Yadav R, Kushwah V, Gaur M S, Bhadauria S, Berlina A N and Zherdev A V 2020 *Int. J. Environ. Anal. Chem.* **100** 623–34
[111] Menon S, Mathew M R, Sam S, Keerthi K and Girish Kumar K 2020 *J. Electroanal. Chem.* **878** 114596
[112] Fletcher S 2016 *Adv. Electrochem. Sci. Eng.* **16** 425–43
[113] https://pineresearch.com/shop/wp-content/uploads/sites/2/2016/10/DRP10036-Screen-Printed-Electrodes-Overview-REV004.pdf
[114] Padilla V, Serrano N and Díaz-Cruz J M 2021 *Chemosensors* **9** 94
[115] Aragay G, Pons J and Merkoi A 2011 *J. Mater. Chem.* **21** 4326–31
[116] Núñez C, José Triviño J, Segura R and Arancibia V 2020 *Microchem. J.* **159** 105393
[117] Palchetti I, Mascini M, Minunni M, Bilia A R and Vincieri F F 2003 *J. Pharm. Biomed. Anal.* **32** 251–6
[118] Zaouak O, Authier L, Cugnet C, Castetbon A and Potin-Gautier M 2010 *Electroanalysis* **22** 1151–8
[119] Meucci V, Intorre L, Pretti C, Laschi S, Minunni M and Mascini M 2009 *Vet. Res. Commun.* **33** 249–52
[120] Chen C, Niu X, Chai Y, Zhao H and Lan M 2013 *Sens. Actuators, B Chem.* **178** 339–42
[121] Sullivan C, Lu D, Senecal A and Kurup P 2021 *Food Chem.* **352** 129327
[122] Barton J, García M B G, Santos D H, Fanjul-Bolado P, Ribotti A and McCaul M 2016 *Microchim. Acta* **183** 503–17
[123] Zhang L, Li D W, Song W, Shi L, Li Y and Long Y T 2010 *IEEE Sens. J.* **10** 1583–8
[124] Avuthu S G R, Wabeke J T, Narakathu B B, Maddipatla D, Eshkeiti A and Emamian S 2015 *IEEE Sensors—Proc.* pp 3–6
[125] Yao Y, Wu H and Ping 2019 *J. Food Chem.* **274** 8–15
[126] Keawkim K, Chuanuwatanakul S, Chailapakul O and Motomizu S 2013 *Food Control* **31** 14–21
[127] Khaled E, Hassan H N A, Habib I H I and Metelka R *Int. J. Electrochem. Sci.* **5** 158–67
[128] Pauliukaite R, Metelka R, Svancara I, Krolicka A, Bobrowski A and Norkus E 2004 *Sci. Pap. Univ. Pardubice Ser.* A **10** 47–58
[129] Nellaiappan S, Pillai K C and Kumar A S 2018 *Anal. Methods* **10** 799–808
[130] Hwang J H, Fox D, Stanberry J, Anagnostopoulos V, Zhai L and Lee W H 2021 *Micromachines* **12** 649

[131] Ding J, Liu Y, Zhang D, Yu M, Zhan X and Zhang D 2018 *Microchim. Acta* **185** 545
[132] Irvine G W, Tan S N and Stillman M J 2017 *Biosensors* **7** 14
[133] Ferancová A, Adamovski M, Gründler P, Zima J, Barek J and Mattusch J 2007 *Bioelectrochemistry* **71** 33–7
[134] Florescu M, Badea M, Coman G and Mitrica M 2009 *Chemistry* **2**
[135] Sanllorente-Méndez S, Domínguez-Renedo O and Julia Arcos-Martínez M I 2010 *Sensors* **10** 2119–28
[136] Zen J M, Kumar A S and Tsai D M 2003 *Electroanalysis* **15** 1073–87
[137] Alkire R C, Kolb D M, Lipkowski J and Ross P N 2011 *Chem. Modif. Electrodes* **11** 1–267
[138] Bi C C, Ke X X, Chen X, Weerasooriya R, Hong Z Y and Wang L C 2020 *Anal. Chim. Acta* **1100** 31–9
[139] Geim A K 2009 *Science* **324** 1530–4
[140] Geim A K and Novoselov K S 2007 *Nat. Mater.* **6** 183–91
[141] Baig N, Waheed A, Sajid M, Khan I, Kawde A N and Sohail M 2021 *Trends Environ. Anal. Chem.* **30** e00120
[142] Justino C I L, Gomes A R, Freitas A C, Duarte A C and Rocha-Santos T A P 2017 *TrAC —Trends Anal. Chem.* **91** 53–66
[143] Coroş M, Pruneanu S and Staden R-I S 2019 *J. Electrochem. Soc.* **167** 037528
[144] Zou J, Zhong W, Gao F, Tu X, Chen S and Huang X 2020 *Microchim. Acta* **187** 433
[145] Yang T, Lin H, Zheng X, Loh K P and Jia B 2017 *J. Mater. Chem.* A **5** 16537–58
[146] Pan X, Ji J, Zhang N and Xing M 2020 *Chin. Chem. Lett.* **31** 1462–73
[147] AL-Gahouari T, Sayyad P, Bodkhe G, Ingle N, Mahadik M and Shirsat S 2021 *Appl. Phys. A Mater. Sci. Process* **127** 1
[148] Lee S, Oh J, Kim D and Piao Y 2016 *Talanta* **1605** 28–36
[149] Xuan X, Hossain M F and Park J Y 2016 *Sci. Rep.* **6** 1
[150] Yuan W, Zhou Y, Li Y, Li C, Peng H and Zhang J 2013 *Sci. Rep.* **3** 2248
[151] Muralikrishna S, Sureshkumar K, Varley T S, Nagaraju D H and Ramakrishnappa T 2014 *Anal. Methods* **6** 8698–705
[152] Mei C Y, Bao F H, Yin W, Jun H C, Qun H D and Mei L F 2016 *J. Solid State Electrochem.* **20** 327–35
[153] Wang Y, Shao Y, Matson D W, Li J and Lin Y 2010 *ACS Nano* **4** 1790–8
[154] Xing H, Xu J, Zhu X, Duan X, Lu L and Wang W J 2016 *Electroanal. Chem.* **760** 52–8
[155] Li L, Liu D, Shi A and You T 2018 *Sens. Actuators, B Chem.* **255** 1762–70
[156] Gusmão R, Sofer Z, Šembera F, Janoušek Z and Pumera M 2015 *Chem—A Eur. J.* **21** 16474–8
[157] Lataste E, Legein C, Body M, Buzaré J Y, Tressaud A and Demourgues A 2009 *J. Phys. Chem.* C **113** 18652–60
[158] Chen M, Zhou H, Qiu C, Yang H, Yu F and Sun L 2012 *Nanotechnology* **23** 115706
[159] Thiruppathi A R, Sidhureddy B, Keeler W and Chen A 2017 *Electrochem. Commun.* **76** 42–6
[160] Alekseeva E, Bober P, Trchová M, Šeděnková I, Prokeš J and Stejskal J 2015 *Synth. Met.* **209** 105–11
[161] BeMiller J N 1986 *An Introduction to Pectins: Structure and Properties* (Washington, DC: ACS Publication)
[162] Arulraj A D, Devasenathipathy R, Chen S M, Vasantha V S and Wang S F 2016 *J. Colloid Interface Sci.* **483** 268–74
[163] Seenivasan R, Chang W J and Gunasekaran S 2015 *ACS Appl. Mater. Interfaces* **7** 15935–43

[164] Ameen S, Akhtar M S and Shin H S H 2012 *Sens. Actuators, B Chem.* **173** 177–83
[165] Yang Y, Kang M, Fang S, Wang M, He L and Zhao J 2015 *Sens. Actuators, B Chem.* **214** 63–9
[166] Muralikrishna S, Nagaraju D H, Balakrishna R G, Surareungchai W, Ramakrishnappa T and Shivanandareddy A B 2017 *Anal. Chim. Acta* **990** 67–77
[167] Promphet N, Rattanarat P, Rangkupan R, Chailapakul O and Rodthongkum N 2015 *Sens. Actuators, B Chem.* **207** 526–34
[168] Yi W, He Z, Fei J and He X 2019 *RSC Adv.* **9** 17325–34
[169] Sakthivel B and Dhakshinamoorthy A 2017 *J. Colloid Interface Sci.* **485** 75–80
[170] Wu S, Li K, Dai X, Zhang Z, Ding F and Li S 2020 *Microchem. J.* **155** 104710
[171] Zhou W, Li C, Sun C and Yang X 2016 *Food Chem.* **192** 351–7
[172] Guo Z, Di L D, Ke L X, Hui L Y, Zhao Q N, Li M and Meng S 2017 *J. Colloid Interface Sci.* **490** 11–22
[173] Priya T, Dhanalakshmi N, Karthikeyan V and Thinakaran N 2019 *J. Electroanal. Chem.* **833** 543–51
[174] Sang S, Li D, Zhang H, Sun Y, Jian A and Zhang Q 2017 *RSC Adv.* **7** 21618–24
[175] Sahoo P K, Sahoo S, Satpati A K and Bahadur D 2015 *Electrochim. Acta* **180** 1023–32
[176] Abolhasani J, Khanmiri R H, Ghorbani-Kalhor E, Hassanpour A, Asgharinezhad A A and Shekari N 2015 *Anal. Methods* **7** 313–20
[177] Mei L F, Zhang Y, Yin W, Jun H C, Qun H D and He B 2017 *Sensors Actuators, B Chem.* **242** 889–96
[178] Kumar R, Bhuvana T and Sharma A 2017 *RSC Adv.* **7** 42146–58
[179] Xie Y L, Zhao S Q, Ye H L, Yuan J, Song P and Hu S Q 2015 *J. Electroanal. Chem.* **757** 235–42
[180] Xuan X and Park J Y 2018 *Sensors Actuators, B Chem.* **255** 1220–7
[181] AL-Gahouari T, Bodkhe G, Sayyad P, Ingle N, Mahadik M and Shirsat S M 2020 *Front Mater.* **7** https://doi.org/10.3389/fmats.2020.00068
[182] Lu M, Deng Y, Luo Y, Lv J, Li T and Xu 2019 *J. Anal. Chem.* **91** 888–95
[183] Bai H, Wang S, Liu P, Xiong C, Zhang K and Cao Q 2016 *J. Electroanal. Chem.* **771** 29–36
[184] Bagheri H, Afkhami A, Khoshsafar H, Rezaei M, Sabounchei S J and Sarlakifar M 2015 *Anal. Chim. Acta* **870** 56–66
[185] Chaiyo S, Mehmeti E, Žagar K, Siangproh W, Chailapakul O and Kalcher K 2016 *Anal. Chim. Acta* **918** 26–34
[186] Schnorr J M and Swager T M 2011 *Chem. Mater.* **23** 646–57
[187] Shi L, Wang Y, Chu Z, Yin Y, Jiang D and Luo J A 2017 *J. Mater. Chem.* B **5** 1073–80
[188] Mao A, Li H, Cai Z and Hu X 2015 *J. Electroanal. Chem.* **751** 23–9
[189] Gupta R K, Meyyappan M and Koehne J E 2014 *RSC Adv.* **42** 2642–50
[190] Oularbi L, Turmine M and El Rhazi M 2017 *J. Solid State Electrochem.* **21** 3289–300
[191] Liu S and Zhang X 2020 *Int. J. Electrochem. Sci.* **15** 9838–48
[192] Gao S, Liu J, Luo J, Mamat X, Sambasivam S and Li Y 2018 *Microchim. Acta* **185** 282
[193] Fakude C T, Arotiba O A, Arduini F and Mabuba N 2020 *Electroanalysis* **32** 2650–8
[194] Zhang S, Zhu H, Ma P, Duan F, Dong W and Du M 2017 *Anal. Methods* **9** 6801–7
[195] Singh S, Pankaj A, Mishra S, Tewari K and Pratap Singh S 2019 *J. Environ. Chem. Eng.* **7** 103250
[196] Liao Y, Li Q, Wang N and Shao S 2015 *Sensors Actuators, B Chem.* **215** 592–7

Chapter 3

Voltammetric sensors for environmental monitoring

Hadi Beitollahi, Zahra Dourandish, Fariba Garkani Nejad and Somayeh Tajik

Rising environmental pollution concerns emphasize the need to develop environmental pollution measuring devices with multiplexed, rapid, cost-effective, and sensitive approaches. Carbon composite sensors are efficient and easy to manufacture, capable of providing data used skillfully to monitor environment and health. Porous carbons, hollow sphere, fullerene (C_{60}), graphite carbon nitride (g-C_3N_4), carbon nanotubes (CNTs) and graphene (Gr) have attracted further attention because of their structurally specific properties for the fabrication of sensitive sensors. This chapter discusses new advances in the construction of electrochemical sensors using different carbon composites along with voltammetric detection approaches to control environmental contaminants.

3.1 Introduction

Increasing urbanization and industrialization due to the emission of numerous organic and inorganic contaminants had severely influenced the environment and humans, and has thus posed dangerous challenges to the survival of humans and all living things [1, 2]. The environmental contaminants have been reported to be organic ones such as pesticides, nitroaromatic explosives, aromatic amine, and phenolic substances; inorganic ones such as nitrite and heavy metals; and endocrine disruptors such as nonylphenol and bisphenol; as well as biomaterials such as pathogenic microorganisms. Persistent inhalation of polluted air and long-term consumption of contaminated food and drinking water are extremely threatening to human life [3, 4]. Consequently, permanent monitoring of the concentrations of such pollutants is always necessary, which can be achieved through the development of sensitive, fast, selectable, reliable, and cost-effective analytical instruments and methods. Some of these tools and techniques include fluorescence [5, 6], chemiluminescence [7], surface-enhanced Raman spectroscopy [8, 9], mass spectrometry [10]

doi:10.1088/978-0-7503-5127-0ch3

and high-performance liquid chromatography [11, 12]. Despite the significant advantages of these approaches, they face serious challenges and disadvantages such as sophisticated pre-treatment, the need for expensive instruments, length of time, and the need for skillful operators. These techniques are restricted to on-site, in-site, and instantaneous analytical protocols. In contrast, electrochemical procedures have been reported to have admirable benefits in detecting a variety of natural and man-made contaminants such as pesticides and heavy metal ions [13–15]. Some of these attractive features include portability, miniaturization probability, unparalleled selectivity, high sensitivity, minimum need for power and space, wide linear range, and cost-effectiveness [16, 17]. Bare electrodes have defects in the process of contaminant analysis, including electrode inactivation, slow direct electron transfer, and high potential for analyte responses. Accordingly, it is essential to select and design the appropriate electrode material taking into account the sensor function. Electrode surface modification can help reduce the overpotential required for electrochemical responses and pre-concentration capacity for certain analytes, resulting in the production of a modifying electrode interface that ensures the generation of bridges to allow electron displacement to eventually promote the amplification of signals [18–20].

Carbon nanomaterials have exceptional properties, which have turned them into strong supporting materials for the construction of various devices such as catalysts, sensors, and energy instruments. The carbon used in the structure of electrochemical electrodes has astonishing traits such as ultra-narrow background current, cost-effectiveness, and high-potential window. They have a biocompatible nature that is considered to be an influential advantage in the field of electrochemical detection of diverse analytes [21–23]. The concentrations of electroactive analytes can be quantitatively and qualitatively measured by electrochemical analysis.

This chapter scrutinizes recent advances in environmental monitoring using carbon composites including porous carbons, hollow spheres, fullerene, graphite carbon nitride, carbon nanotubes, and graphene. Our main focus has been on introducing electrochemical sensors constructed by diverse carbon composites along with voltammetric detection approaches to control the environmental contaminants that are heavy ions, organic compounds, and pesticides.

3.2 Monitoring the environment contaminants by voltammetric sensors constructed by carbon-based nanocomposites

The mechanism of action of an electrochemical sensor is based on the oxidation and reduction of analytes during electrochemical reactions. The electrochemical determinations are on the basis of a three-electrode cell containing a reference electrode, a counter electrode, and a working electrode. Electroanalytical sensing refers to an interaction of chemistry with electricity, meaning the determination of electrical quantities that is current, potential, or charge and their correlation with chemical indices. Measurements of electrical potential, current, and other electrical signals are required to determine the concentration of analytes implicated in the reaction. Electrochemical techniques are divided based on the electrical signal required to be measured by voltammetric, potentiometric, amperometric, and conductometric methods [24, 25]. The voltammetric sensors work on the basis of electroanalytical approaches, so that the concentration of the analyte is measured quantitatively by changing the potential and measuring the current resulting from the electrochemical reaction of the analyte with the surface of the working electrode. Some of the common voltammetric sensing methods are cyclic voltammetry (CV), square wave voltammetry (SWV), differential pulse voltammetry (DPV), and linear sweep voltammetry (LSV).

LSV is more efficient for irreversible reactions, such as using a biocathode to investigate direct production of methane. The derivative LSV leads to staircase voltammetry, so that chemical redox reaction is determined as a trough or crust in the current signal. SWV refers to a large-amplitude differential approach, so that a waveform is formed from the symmetrical square waves. The exceptional sensitivity of the SWV can be attributed to the high net current when compared with the backward or forward current, with efficient discrimination versus the charge current. CV includes tests in which the potential of the working electrode increases linearly with time. The working electrode potential in the CV is scanned linearly from an initial amount to a final amount and back, which records simultaneously the intensity of the corresponding current. The CV is the most extensively employed approach to achieve qualitative data on the features of electrochemical responses [26, 27], and on the redox thermodynamics and the heterogeneous electron transfer kinetics, as well as on the paired chemical or adsorption behaviors. Differential pulse voltammetry, cathodic stripping voltammetry (CSV), anodic stripping voltammetry (ASV), and mormal pulse voltammetry (NPV) are voltammetric strategies to reinforce the sensing sensitivity [28, 29], which can be helpful for the development of environmental monitoring.

3.2.1 Voltammetric sensors constructed by graphene-anchored composites

Semi-metallic graphene is an allotrope of carbon consisting of a hexagonal covalent bond of a single layer of atoms, which are together placed in a honeycomb lattice [30, 31]. Three of the four carbon atoms show sp^2 hybridization. Following hybridization, elevated bond energy (approximately 5.9 eV) is revealed from the triangular system of carbon atoms, which are isolated via sigma bonds possessing 1.42 Å bond length. The p-orbital alone, half of which is filled, generates the pi bonds with neighboring carbon atoms [32, 33]. The graphene has slippery monolayer carbon sheets that are stacked to form the three-dimensional structure of graphite. The connection of these sheets to each other is mediated by the weak forces of van der Waals. These sheets are slippery due to the vibrational phonons that also exist in three-dimensional solids. The graphene is an optimal electrode material due to its impressive electrical conductivity (64 mS cm^{-1}) and large surface area (2630 m^2 g^{-1}). The graphene also has admirable electrochemical traits, including rapid electron transfer, a big potential window, appreciable electrochemical performance, and small charge transfer resistance. Such unique physico-electrochemical functions of graphene make it a suitable candidate for electrochemical detection [34–36].

In a study by Radhakrishnan *et al* a nitrite sensor was constructed on the basis of Fe_2O_3/rGO. They investigated this composite to electrochemically detect nitrite based on DPV and CV methods. According to electrochemical determinations, this composite linearly determined the nitrite at different concentrations (5.0×10^{-8}–7.8×10^{-4} M) and exhibited a low limit of detection (LOD=1.5×10^{-8} M). The modified Fe_2O_3/rGO electrode displayed an exceptional anti-interference activity versus metal ions and electroactive substances [37].

In a study by Bai *et al* an electrochemical sensor was constructed on the basis of a TiO_2-modified graphene nanoparticle (NP) anchored on a screen-printed electrode (TiO_2–graphene oxide (GO)/SPE) to sensitively detect 4-chlorophenol (4-Cp). The electrochemical response of TiO_2–GO/SPE to 4-Cp was determined via the CV method. The TiO_2–GO film showed greater performance for detection of 4-Cp with bimodal response when compared with the GO film. Under optimized circumstances, the peak currents of oxidation were proportional to the concentrations of 4-Cp over the ranges of 0.05–20.0 mM and 20.0–200.0 mM, with a low LOD value of 0.02 mM [38].

A new protocol was reported by Yu *et al* to electrochemically detect the hexachlorobenzene (HCB). According to the DPV method, the HCB reduction was sensitive on the surface of a glassy carbon electrode (GCE) modified with nitrogen-doped graphene (NG) and chitosan in deaerated solutions of organic electrolyte. The modified electrode in the best circumstances exhibited impressive sensing activities to detect HCB, which had a broad linear dynamic range (3 μg L^{-1}–10 mg^{-1} L^{-1}) and a low LOD (1.72 μg L^{-1}) [39].

The graphene nanosheet functionalized with cyclodextrin and molecularly imprinted poly-vinylferrocene (GN/CD-MIPVF) composite was developed by Tang *et al* with vinylferrocene as the functional monomer. Then, they produced an electrochemical sensor (GN/CD-MIPVF/GCE) through the modification of the GCE surface with as-fabricated composite for 1-naphthylamine (1-NA) determination. Vinylferrocene acted as an electron mediator and anchored on the surface imprinted composite via a covalent bond, thereby improving GN/CD-MIPVF sensor sensitivity. The DPV method was applied to characterize the GN/CD-MIPVF sensor in terms of electrochemical responses, selectivity, and adsorption dynamics. The GN/CD-MIPVF sensor performance to detect the 1-NA levels (0.3–100.0 μM) in the optimized conditions displayed a broad linear range and a low LOD value of 0.1 μM [40].

In a study by Sha *et al* a new procedure was introduced to determine phenol using GCE modified with a rGO-zinc oxide (ZnO) composite (rGO-ZnO/GCE), which removed the surface fouling effect by accurately selecting the sensing peak. The DPV method was applied to characterize the phenol determination, the results of which revealed peaks at 0.35 and 0.94 V. The phenol was detected at a lower potential (0.35 V) because it removes the need to renew the electrode surface before each scan owing to surface fouling, thereby providing stable and reproducible determination. Two linear response ranges were found for the sensor toward phenol, including 15–40 μM (sensitivity: 0.389 μA μM^{-1} cm^{-2}) and 2–15 μM (sensitivity: 1.79 μA μM^{-1} cm^{-2}), with impressive reproducibility, selectivity, and stability, as well as a lower LOD value of 1.94 μM [41].

Manasa *et al* modified a carbon paste electrode (CPE) with cetyltrimethylammonium bromide (CTAB) and ZnO/G (ZnO/G/CTAB/MCPE) to electrochemically detect Bisphenol F (BPF). This economical, biocompatible, and conductive sensor in optimized circumstances was ultra-sensitive towards BPF detection exploiting the DPV method, the result of which showed a low LOD value of 0.06 μM [42].

Wei *et al* introduced a dip coating approach to sensitively and selectively detect Cu (II) using novel electrochemical sensors of ion imprinting polymer (IIP) by the modification of GCE with chitosan–graphene oxide composites (CS/GO-IIP). The fabrication of polymers imprinted with Cu (II) ions was performed via chemical cross linking with epichlorohydrin after anchoring the CS/GO/Cu (II) composite on the GCE surface. Figure 3.1 shows the fabrication of a CS/GO-IIP sensor and corresponding chemical proceedings. A differential pulse anodic stripping voltammetry method (DPASV) in the optimized circumstances provided the linear dynamic range of 0.5–100 μM and the LOD value of 0.15 μM [43].

Cd(II) and Pb(II) were determined simultaneously by a new electrochemical sensor on the basis of GCE modification with poly(amidoamine) dendrimer functionalized magnetic graphene oxide (GO-Fe_3O_4-PAMAM) nanocomposite introduced by Baghayeri *et al* who used the square wave anodic stripping voltammetry (SWASV) technique to detect the desired metal ions following closed-circuit mode. The sensor in optimized circumstances was able to present a linear voltammetric behavior for Pb (0.4–120.0 μg L^{-1}) and Cd (0.2–140.0 μg L^{-1}) showing detection coefficients of ˃0.99, with a narrow LOD for Pb (130 ng L^{-1}) and Cd (70 ng L^{-1}) [44].

Migliorini *et al* introduced a new electrochemical sensing nano-platform to determine malathion (man-made organophosphate insecticide) in real water resources based on polymeric electrospun nanofibers of polypyrrole (PPy) and polyamide 6 (PA6) anchored on fluorine tin oxide (FTO), modified with graphene species. Thus, electrochemically reduced graphene oxide (ERGO) and chemically reduced graphene oxide (CRGO) were utilized for the investigation into how the amount of graphene oxide reduction affected electrical conductivity and sensing profiles.

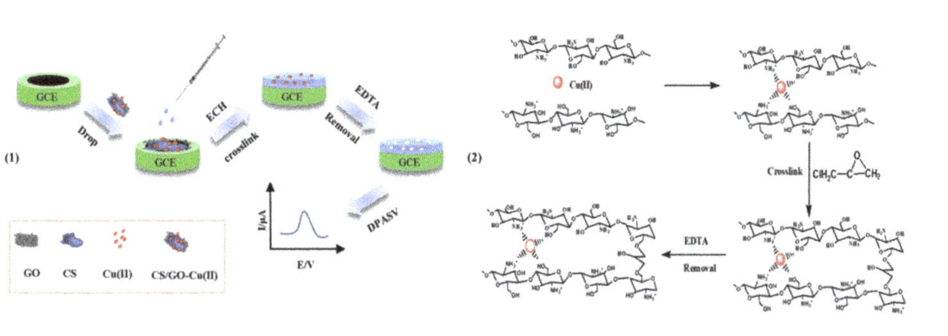

Figure 3.1. Diagram of the construction process for CS/GO-IIP sensor (1) and the corresponding chemical proceedings involved (2). Note: the outer green layer is not GCE but the electrode insulation; GCE is the black inner disk. Reprinted from [43], copyright 2019, with permission from Elsevier.

CRGO-modified platform, with higher electrical conductivity than ERGO, was able to successfully detect malathion electrochemically, with a low detection limit (0.8 ng mL^{-1}) [45].

Al-Hamry *et al* applied poly (diallydimethylammonium chloride) (PDAC)/rGO-modified Ag electrodes to determine dimethoate (DMT), the results of which demonstrated the role of temperature in the GO reduction in PDAC/GO bi-layer on the Ag electrode. The voltammetric measurements verified the direct electrochemistry of DMT on as-fabricated electrodes, and the temperature of 150 °C provided the best electrode to achieve the optimal cathodic and anodic current responses versus the DMT levels. Further analysis revealed that the linear voltammetric responses of the electrode at different analyte concentrations ranged from 3×10^{-4} to 9×10^{-4} mol dm^{-3}, with R^2 value more than 0.94 [46].

Wang *et al* tried to detect BPA concentration using a new electrochemical sensor on the basis of gold NPs-MoS_2 nanoflowers/ionic liquid supported graphene (AuNPs/MoS_2-NFs/IL-graphene). The DPV and CV techniques were employed to assess the sensor electrochemical response, the findings of which confirmed its impressive electrochemistry towards the BPA. The sensor assessed in the optimized circumstances displayed a broad linear dynamic range (0.05–4.0 μM) for the relationship of peak current with BPA concentration, with a small LOD (0.028 Mm) [47].

Mnyipika *et al* detected Cu, Cd, and Zn simultaneously and sensitively in the real surface water specimens exploiting the DPASV method based on a MnO_2@RGO-modified GCE (MnO_2@RGO@GCE). According to the electrochemical responses, the as-produced sensor improved the current response on the determination of Cu, Cd, and Zn ions because of the interaction between RGO nano-sheets and MnO_2 nanomaterials. Appreciable reproducibility, sensitivity, and stability were observed for the modified electrode in the simultaneous detection of Cu, Cd, and Zn ions. The LOD values were calculated to be between 0.002 and 0.015 μg L^{-1} for detection of Cu, Cd, and Zn ions simultaneously [48].

Graphene-based voltammetric sensors reported for environmental analysis are listed in table 3.1.

3.2.2 Voltammetric sensors constructed by carbon nanotubes composites

Carbon nanotubes, long and hollow cylindrical tubule molecules, contain rolled-up sheets of one or more graphite-like layers with diameters typically measured in the range between 0.4 and tens of nanometers. Two classes of CNTs are categorized based on their count of carbon layers: single-walled carbon nanotubes (SWCNTs) and multi-walled carbon nanotubes (MWCNTs). The SWCNTs are graphene monolayers with a diameter of 0.4 to 2 nm and commonly appear as hexagonal packages. The MWCNTs have two or more cylindrical layers, each of which is composed of graphene sheets and their diameter is variable between 1 and 3 nm [49, 50]. The CNTs have spurred extensive interest in the area of healthcare systems as an alternative to environmental equipment due to their specific traits such as admirable mechanical strength, impressive electrical conductivity, electrochemical stability in

Table 3.1. Different analytical indices of graphene-based voltammetric sensors reported for environmental analysis.

Electrochemical sensor	Analyte	Analytical method	LOD	Linear range	References
Fe_2O_3/rGO-modified GC electrode	Nitrite	DPV	1.5×10^{-8}M	5.0×10^{-8} to 7.8×10^{-4}M	[37]
TiO_2–GO/SPE	4-Chlorophenol	CV	0.02 mM	0.05 mM–200.0 mM	[38]
$(NG/CS)_{3.5}$/GCE	Hexachlorobenzene	DPV	1.72 μg L^{-1}	3 μg L^{-1}–10 mg^{-1} L^{-1}	[39]
GN/CD-MIPVF/GCE	1-Naphthylamine	DPV	0.1 μM	0.3–100.0 μM	[40]
rGO-ZnO/GCE	Phenol	DPV	1.94 μM	2.0–40.0 μM	[41]
ZnO/G/CTAB/MCPE	Bisphenol F	DPV	0.06 μM	0.5–10.0 μM	[42]
CS/GO/GCE-IIP	Cu (II)	DPASV	0.15 μM	0.5–100.0 μM	[43]
GO-Fe_3O_4-PAMAM/GCE	Pb(II)	SWASV	130 ng L^{-1}	0.4–120.0 μg L^{-1}	[44]
	Cd(II)		70 ng L^{-1}	0.2–140.0 μg L^{-1}	
FTO/PA6/PPy/CRGO	Malathion	DPV	0.8 ng mL^{-1}	500.0–2.0×10^4 ng ml^{-1}	[45]
PDAC/rGO-Ag electrode	Dimethoate	CV	–	3×10^{-4}–9×10^{-4} mol dm^{-3}	[46]
AuNPs/MoS_2-NFs/IL-graphene/GCE	Bisphenol A	DPV	0.028 μM	0.05–4.0 μM	[47]
MnO_2@RGO/GCE	Cd (II), Zn (II) and Cu (II)	DPV	0.002–0.015 μgL^{-1}	0.05–700.0 μgL^{-1}	[48]

aqueous and non-aqueous media, and unparalleled thermal conductivity. Moreover, the CNTs have exceptionally large surface area, thereby providing immobilization of practical units on the CNT surface such as receptor moieties for sensing purposes [51, 52].

In a study by Wan *et al* the voltammetric and electrochemical determination of nano-molar 4-tert-octylphenol was evaluated in water. They employed an electrode coated with polymer/CNT film. The electrode was fabricated first by coating a polished GCE with a CNT film and subsequently by coating the electrode covered with the CNT film with a film conductive of L-lysine (polymer/CNT/GCE). The anodic peak current was proportional to 4-*tert*-octlyphenol levels (6.5–20 nM), with the LOD value of 0.5 nM [53].

Zhang *et al* selectively and sensitively detected 2-NP at a molecularly imprinted film on a GCE using the DPV method. The immobilization of imprinted film on GCE was carried out via electrodeposition of a molecularly imprinted polymer (MIP) sol–gel containing MWNT-Nafion (NF) composite (MIP/sol–gel/MWNT-NF/GCE). The MIP sol–gel film was differentiated from the target molecule through the measurement of alterations in the DPV response of the redox probe, $K_3Fe(CN)_6$ on GCE. A linear response was achieved in optimized circumstances, covering different concentrations (0.2–360.0 μM) with the LOD value of 0.06 μM [54].

Hassan *et al* fabricated, characterized, and investigated the electrocatalytic performance of crown-ether/MWCNTs/CPEs to voltammetrically detect mercury (Hg^{2+}) present in various real biological and environmental specimens. Thus, the selectivity and sensitivity were improved by integrating 12-crwon-4-ether with MWCNTs (10%w/w). The as-produced 12-Crown-4/MWCNTs/CPE sensor in the optimized circumstances displayed an admirable linear response to Hg^{2+} at different concentrations (5.0–110.0 μg mL^{-1}) with the LOD value of 0.25 μg mL^{-1} [55].

In a study carried out by Messaoud *et al* a new electrochemical sensor was constructed to determine BPA based on GCE modified with MWCNT and AuNP (AuNP/MWCNT/GCE). The sensor in the optimized circumstances portrayed a linear response to BPA (between 0.01 and 0.7 μM), with the LOD value of 4.3 nM [56].

Rajkumar *et al* utilized a thermal reduction procedure beginning with urea and nickel (Ni) salt under an inert atmosphere to fabricate N-doped MWCNTs modified with Ni NPs (Ni/N-MWCNT). The Ni/N-MWCNT nanocomposite was anchored on the SPCE to produce an electrochemical sensor (Ni/NMWCNT@ SPCE). The DPV and CV techniques were utilized to evaluate the electrocatalytic performance of this sensor towards to simultaneously determine hydroquinone (HQ) and catechol (CC). The LOD value was 11 nM for HQ and 9 nM for CC, with the linear dynamic range of 0.3 to 300.0 μM for HQ and 0.1 to 300.0 μM for CC [57].

Sipa *et al* modified the GCE surface with β-cyclodextrins and MWCNTs (β-CDs/MWCNTs/GCE) to electroanalytically detect dichlorophen (Dcp) pesticide using square wave adsorptive stripping voltammetry (SWAdSV). The as-produced sensor in the optimized circumstances was able to detect broad Dcp concentration (between 5.0×10^{-8} M and 2.9×10^{-6} M), with the thin LOD of 1.4×10^{-8} M and appreciable repeatability, selectivity, and stability [58].

Balram *et al* detected the HQ as environmental pollutant by constructing an electrochemical sensor using functionalized MWCNT (f-MWCNT) anchored ruskin michelle-like zinc oxide (ZnO) nanocomposite, which was then used to modify GCE (f-MWCNT/ZnO/GCE) and characterized by DPV, CV, and amperometric methods. The results exhibited a low LOD (0.02 μM), a broad linear range (0.1–190 μM) and a great sensitivity (6.37 $\mu A\ \mu M^{-1}\ cm^{-2}$) [59].

Koyun *et al* constructed a pencil graphite electrode (PGE) modified with MWCNTs (MWCNT-COOH) and copper(II) phthalocyanine (CuPc) to electrochemically detect the BPA with this new sensor (CuPc/MWCNT-COOH/PGE), followed by optimization of parameters like adsorption time and pH. The results indicated a broad linear range (between 2.75×10^{-5} and 1.0×10^{-7} M) and low LOD (0.0189 mM). The sensor was examined for practical applicability by detecting BPA in the real specimens, the results of which confirmed a successful detection with a great recovery rate of more than 98% [60].

Manasa *et al* introduced a facile and affordable electrochemical sensor produced on the basis of 77Maghemite/MWCNT (M/MWCNT) modified CPE (M/MWCNT/MCPE) to detect and quantify the resorcinol (RS), which was examined for voltammetric response of analyte using DPV and subsequently the findings presented the smallest LOD at 0.02 μM [61].

Wu *et al* reported the production of Fe_3O_4 NPs, Fe_3O_4/MWCNTs and Fe_3O_4/fluorinated MWCNTs (FMWCNTs) nanocomposites and then their application for the GCE surface modification to simultaneously determine the heavy metals of Pb^{2+}, Cd^{2+}, Hg^{2+}, and Cu^{2+}. The three sensors produced were compared in terms of simultaneously determination of the heavy metals, the findings of which revealed preponderant performance of Fe_3O_4/F-MWCNTs sensor with higher sensitivity towards detection of Pb^{2+}, Cd^{2+}, Hg^{2+}, and Cu^{2+} (125.91, 108.79, 312.65, and 160.85 $\mu A\ mM^{-1}\ cm^{-2}$, sequentially). Furthermore, a broader linear dynamic range was reported for the Fe^3O^4/F-MWCNTs sensor, namely 0.5 to 30, 0.5 to 30, 0.5 to 20, and 0.5 to 30 μM as well as the LOD (S/N=3) value of 0.08, 0.05, 0.05, and 0.02 nM for Pb^{2+}, Cd^{2+}, Hg^{2+} and Cu^{2+}, individually [62].

Le Hai *et al* produced a series of MWCNTs modified with antimony oxide (Sb_2O_3/MWCNTs) paste electrodes to simultaneously determine Pb^{2+} and Cd^{2+} exploiting a sensitive anodic stripping voltammetry (ASV). First, the MWCNT composite was employed to construct a paste electrode. Secondly, the modification of MWCNT bulk electrode was performed by different Sb_2O_3 concentrations (1, 3, and 4 wt.%, sequentially) to enhance Pb^{2+} and Cd^{2+} detection capacity. From Sb_2O_3/MWCNTs sensors, 3 wt.% Sb_2O_3-modified electrode displayed a significant analytical determination for Pb^{2+} and Cd^{2+} through linear anodic stripping voltammetry (LASV). There were appropriate linear curves of current reaction against Pb^{2+} (5–35 ppb) and Cd^{2+} (80–150 ppb) concentrations. It should be mentioned that the sensitivity of the Pb^{2+} and Cd^{2+} mixture was 2.694 and 1.932 $\mu AL\ \mu g^{-1}$, sequentially [63].

Bashir *et al* reported the production of $CoMn_2O_4$ nanocomposite functionalized with carbon nanotubes using the co-precipitation approach to detect the trace concentration of Pb^{2+} based on SWASV. According to CV and EIS analyses, the

CNT modification enhanced the electrochemical performance of GCE modified with $CoMn_2O_4$ NPs ($CoMn_2O_4$/CNT/GCE). There was a linear relationship between Pb^{2+} concentrations (0.01–0.85 μM) and current response, with the LOD value of 0.004 μM [64].

Ghodsi *et al* introduced a voltammetric sensor to detect the pesticide diazinon on the basis of diazinon reduction on GCE modified with MWCNTs covered by TiO_2 NPs (GCE/MWCNTs/TiO_2NPs), tested then by LSV, CV, SWV, and DPV. Appropriate electrocatalytically synergic effects on the reduction of diazinon were reported for TiO_2NPs and MWCNTs, thereby leading to high sensitivity in detection of diazinon present in the real specimens (agricultural well water and urban piped water). The as-fabricated sensor successfully exhibited the linear range of 11–8360 nM, the LOD of 3 nM and the limit of quantitation (LOQ) of 10 nM [65].

The CNTs-based voltammetric sensors reported for environmental analysis are listed in table 3.2.

3.2.3 Voltammetric sensors constructed by g-C_3N_4 composites

Graphite carbon nitride has a graphite-like planar system in which the N atoms have three-fold coordination and two-fold coordination atoms, with resemblance and proximity of 2D structure to graphene and even superiority over graphene structure. The nitrogen-substituted graphite in a conventional method provides more active edge locations and cavities when compared with graphene in such a structure [66, 67]. Metal ions are small enough to penetrate g-C_3N_4 layers and holes through amino condensation or electrostatic bonding. The g-C_3N_4 has a two-dimensional stacked structure and has been introduced as a suitable option with a variety of applications owing to its appreciable properties, including high accessibility, affordability, thermal and chemical stability, pure composition, structural regulations, environmental safety, and easy production [68, 69].

Gao *et al* modified the GCE surface with g-C_3N_4/Nafion (g-C_3N_4/Nafion/GCE) via a self-assembling method for DPASV to detect Cd ions. Their sensor determined Cd^{2+} in aqueous solution in addition to electrostatic interaction by creating holes during the DPASV measurement. The Cd^{2+} standard curves exhibited two ranges (1.0 nM–1.0 μM and 1.0–100.0 μM), with the LOD of 0.5 nM [70].

Ponnaiah *et al* employed the ultra-sonication at 40 ± 3 kHz and 200 W to construct a new nanocomposite containing ruthenium NPs, polyaniline, and graphitic carbon nitride (Ru^0/PANI/g-C_3N_4). The GCE surface modification with Ru^0/PANI/g-C_3N_4 led to BPA detection in the real urine specimens of humans and animals, showing a broad linear range between 0.01 and 1.1 μM and a pico-molar-level LOD. Figure 3.2 illustrates the synthesis and BPA sensing of Ru0/PANI/g-C3N4/GCE [71].

A facile phenol sensor on the basis of Au/CeO_2/g-C_3N_4 was produced by Yin *et al* through gold–cerium oxide (Au/CeO_2) nanoparticle loading on g-C_3N_4 using precipitation–reduction protocols. According to their observations, Au/CeO_2 plus g-C_3N_4 enhanced sensing activity for the detection of phenol. The linear dynamic range was reported to be between 10.0 and 90.0 μM for the sensor in optimized

Table 3.2. Different analytical indices of CNTs-based voltammetric sensors reported for environmental analysis.

Electrochemical sensor	Analyte	Analytical method	LOD	Linear range	References
Polymer/CNT/GCE	4-tert-Octylphenol	CV	0.5 nM	6.5–20 nM	[53]
MIP/sol–gel/MWNT-NF/GCE	2-Nonylphenol	DPV	0.06 μM	0.2–360.0 μM	[54]
12-Crown-4/MWCNTs/CPE	Hg^{2+}	DPV	0.25 μg mL^{-1}	5.0–110.0 μg mL^{-1}	[55]
AuNP/MWCNT/GCE	Bisphenol A	DPV	4.3 nM	0.01–0.7 μM	[56]
Ni/NMWCNT-modified SPCE	Catechol	DPV	9 nM	0.1–300.0 μM	[57]
	Hydroquinone		11 nM	0.3–300.0 μM	
β–CDs/MWCNTs/GCE	Dichlorophen	SWAdSV	1.4×10^{-8} M	5.0×10^{-8}–2.9×10^{-6} M	[58]
f-MWCNT/ZnO/GCE	Hydroquinone	DPV	0.02 μM	0.1–190 μM	[59]
CuPc/MWCNT-COOH/PGE	Bisphenol A	DPV	0.0189 mM	2.75×10^{-5}–1.0×10^{-7} M	[60]
M/MWCNT/MCPE	Resorcinol	DPV	0.02 μM	0.5–10.0 μM	[61]
Fe_3O_4/F-MWCNTs/GCE	Cd^{2+}	SWASV	0.05 nM	0.5–30.0 μM	[62]
	Pb^{2+}		0.08 nM	0.5–30.0 μM	
	Cu^{2+}		0.02 nM	0.5–30.0 μM	
	Hg^{2+}		0.05 nM	0.5–20.0 μM	
Sb_2O_3/MWCNTs-electrode	Cd^{2+}	LASV	16.77 ppb	80.0–150.0 ppb	[63]
	Pb^{2+}		6.12 ppb	5.0–35.0 ppb	
$CoMn_2O_4$/CNT/GCE	Pb^{2+}	SWASV	0.004 μM	0.01–0.85 μM	[64]
GCE/MWCNTs/TiO_2NPs	Diazinon	CV and SWV	3.0 nM	11.0–8360.0 nM	[65]

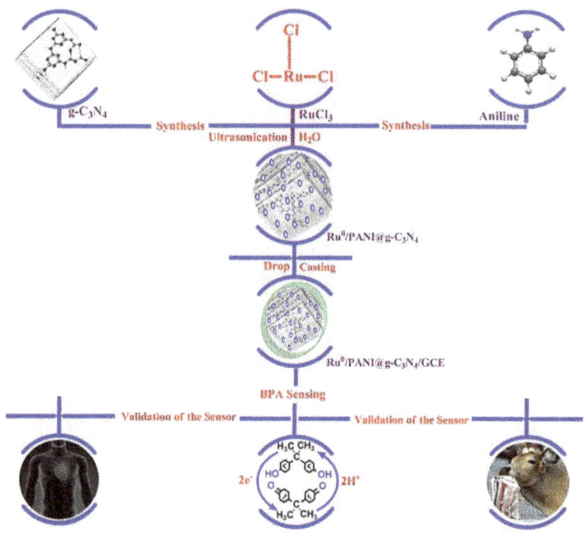

Figure 3.2. Production processes and BPA sensing of Ru_0/PANI/g-C_3N_4/GCE. Reprinted from [71], copyright 2019, with permission from Elsevier.

circumstances, and the LOD value was very low (2.33 μM) and the sensitivity was very high (0.1080 mA μM^{-1}) for detection of phenol [72].

In the study of Rajaji *et al* the Bi_2Te_3 binary nano-sheets (BNs) were fabricated according a facile hydrothermal and solvothermal protocols. Then, the product was anchored with g-C_3N_4 using hydrothermal energy, resulting in Bi_2Te_3/g-C_3N_4 BNs that were subsequently used to modify the GCEs. The electrochemical performance was EIS, DPV, and CV techniques, the results of which demonstrated appreciable electrochemical response towards ractopamine (RAC) with high sensitivity and a broad linear range between 0.015 and 456.4 μM, as well as nano-molar LOD of 1.77 nM [73].

Mahmoudian *et al* directly exerted Pt (II) reduction in exposure to the reducing agent of L-cysteine to construct Pt/g-C_3N_4/polyaniline nanocomposites (Pt/g-C_3N_4/PAn NCs), which was then applied for the GCE surface modification to selectively and sensitively detect Hg^{2+}. The Hg^{2+} adsorption was increased due to the active sites highly available in the PAn and g-C_3N_4. There was linear voltammetric response at different Hg^{2+} concentrations (1.0 to 500.0 nM), the LOD of 0.014 Nm, and the sensitivity of 0.6 Ma Nm^{-1} cm^{-2} [74].

Han *et al* anchored g-C_3N_4 with PdNP (PdNPs/g-C_3N_4) through hydrazine hydrate-mediated reduction, which was then employed for the modification of GCE (PdNPs/g-C_3N_4/GCE) to detect Hg^{2+} using DPV analysis. Their results revealed exceptional detection selectivity and sensitivity, with unparalleled repeatability, anti-interference, and stability in identical optimized circumstances. The findings showed the LOD value of 0.009 μg L^{-1}, lower than that published by the World Health Organization (1.0 μg L^{-1}). The as-produced sensor could successfully detect Hg^{2+} in the real tap, river, and lake water samples [75].

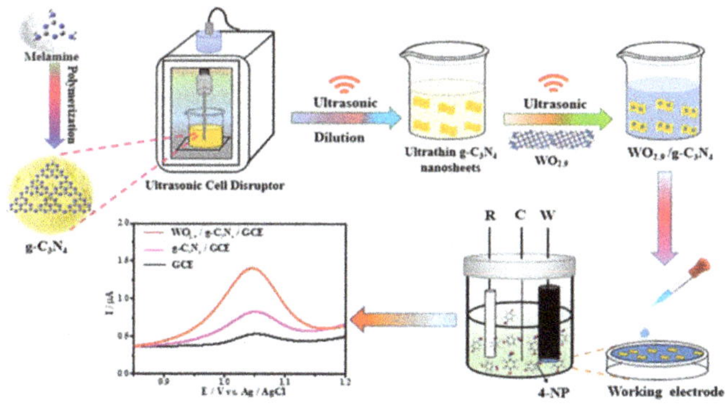

Figure 3.3. Schematic view of the construction process electrochemical sensor for 4-NP based on Vo-S-scheme $WO_{2.9}$/g-C_3N_4 heterojunctions. Reprinted from [76], copyright 2021, with permission from Elsevier.

He *et al* reported the production of step-scheme (S-scheme) heterojunction of $WO_{2.9}$/g-C_3N_4 mediated by oxygen vacancy (Vo) employing an electrostatic self-assembly method of high-purity $WO_{2.9}$ NPs and ultrathin g-C_3N_4 NSs. The g-C_3N_4 NSs were produced by polymerization and then ultrasonic exfoliation. An impressive electrochemical activity was reported for as-produced heterojunction-based Vo-S-scheme $WO_{2.9}$/g-C_3N_4 sensor towards detection of 4-nitrophenol (4-NP) as environmental hormone, with lengthy linear range between 4.0×10^{-7} and 1.0×10^{-4} M as well as narrow LOD value of 1.3×10^{-7} M [76]. Figure 3.3 shows a schematic view of construction process electrochemical sensor for 4-NP.

The g-C_3N_4-based voltammetric sensors reported for environmental analysis are listed in table 3.3.

3.2.4 Voltammetric sensors constructed by C_6 composites

Fullerenes, due to their special potential in photochemistry, electrochemistry, and electrophysics, are structural units in materials science and biology. Fullerenes, C_{60} is the most prevalent form, with a 32-faceted body consist of 26-membered rings (6-MR) and 12 five-membered rings (5-MR) containing sp^2-hybridized carbon atoms, whose diameter is about 0.7 nm. After the discovery of C_{60}, its application in catalysis was widely studied. This material has exhibited many advantages, including large specific surface area, great electronic conductivity, inert behavior, good biocompatibility, excellent adsorption capacity relative to organic molecules and stable construct [77]. This material lacks metal impurities and is relatively easy to fulfill and deliver to reproducible electrocatalytic behaviors. Furthermore, C_{60}-based composites could contain polymers, metal NPs, carbon nanomaterials, metallic oxides, and others, with excellent electrocatalytic activity, effective electron transfer, and appreciable stability and impressive sensitivity following synergistic impacts of their chemical traits [78, 79].

Rather *et al* utilized C_{60}-β-cyclodextrin (CD) conjugate (CD-C_{60}), fabricated by 1,3-dipolar cycloaddition to ultra-sensitively detect p-nitrophenol. The CD-C_{60} was then anchored on a GCE surface (CD-C_{60}/GCE), which indicated exceptional

Table 3.3. Different analytical indices of g-C_3N_4-based voltammetric sensors reported for environmental analysis.

Electrochemical sensor	Analyte	Analytical method	LOD	Linear range	References
g-C_3N_4/Nafion/GCE	Cd^{2+}	DPASV	0.5 nM	1.0 nM–100.0 μM	[70]
Ru_0/PANI@g-C_3N_4/GCE	Bisphenol A	DPV	180.0 pM	0.01–1.1 μM	[71]
Au/CeO_2/g-C_3N_4 modified carbon paper electrode	Phenol	CV	2.33 μM	10.0–90.0 μM	[72]
Bi_2Te_3@g-C_3N_4 BNs/GCE	Ractopamine	DPV	1.77 nM	0.015–456.4 μM	[73]
Pt/g-C_3N_4/PAn NCs	Hg^{2+}	DPV	0.014 nM	1.0–500.0 nM	[74]
PdNPs/g-C_3N_4/GC electrode	Hg^{2+}	DPV	0.009 μg L^{-1}	0.01–15.0 μg L^{-1}	[75]
Vo-S-scheme $WO_{2.9}$/g-C_3N_4/ GCE	4-nitrophenol	DPV	1.3×10^{-7} M	4×10^{-7} to 1×10^{-4} M	[76]

performance to detect p-nitrophenol because of C_{60}-mediated synergetic impact. The p-nitrophenol detection was evaluated by SWV over different concentrations (between 2.8×10^{-9} and 4.2×10^{-7} M), and the LOD value was 1.2×10^{-9} M [80].

Mazloum-Ardakani *et al* tried to simultaneously detect hydroxylamine and hydrazine via the modification of GCE with a C_{60}-functionalized CNTs/IL(1-butyl-3-methylimidazolium tetrafluoroborate). According to their observations, the oxidation peaks of hydroxylamine and hydrazine on the modified electrode surface were distinct at a potential difference of 450 mV. These circumstances are enough to detect hydroxylamine and hydrazine present in the specimens using C_{60}–CNTs/IL/GCE. There was an elevation in the linear electrocatalytic currents with the concentrations of hydroxylamine (1.0–300.0 μM) and hydrazine (0.05–700.0 μM). The LOD values were 17 ± 2 Nm for hydrazine and 28 ± 2 nM for hydroxylamine [81].

Teadoum *et al* reported the modification of GCE with a fullerene/MWCNTs/Nafion composite to detect carbendazim, as a toxic fungicide. The CV method was employed to analyze the voltammetric response of carbendazim on the bare and modified GCEs. Then, the analyte was sensitively detected by an electroanalytical procedure on the basis of SWV. A linear relationship in the optimized circumstances was found between the peak current and different carbendazim concentrations (2.0×10^{-8} to 3.5×10^{-7} M), with a thin LOD of 1.7×10^{-8} M and a LOQ of 5.57×10^{-8} M [82].

In a study by Han *et al* graphene, carbon nanofiber CNTs, and fullerene were dispersed in aqueous solution of chitosan (Chit), followed by GCE modification for anodic stripping voltammetric analysis of Cu(II), Hg(II), Cd(II) and Pb(II). The GCE modified with C_{60}-Chit (C_{60}-Chit/GCE) exhibited the highest activity towards the simultaneous detection of study analytes, probably due to robust adsorption affinity, large surface area, impressive electrical conductivity, and the specific crystalline nature of C_{60}. The findings of DPASV portrayed the analytical profiles for Cu(II), Hg(II), Cd(II) and Pb(II), respectively, as follows: (I) peak voltages of −0.11, +0.14, −0.82, and −0.58 V (vs SCE); (II), LOD values of 14.0 nM (0.9 ppb), 3.0 nM (0.6 ppb), 21.0 nM (2.4 ppb) and 1.0 nM (0.2 ppb); and (III) linear ranges of 0.05 to 6.0 μM, 0.01 to 6.0 μM, 0.5 to 9.0 μM, and 0.005 to 6.0 μM [83].

In a study by Zhu *et al* a new electrochemical sensor was fabricated on the basis of platinum NPs (Pt NPs) and C_{60} through Pt/C_{60} modification on a pyrolytic graphite electrode (PGE) surface (Pt/C_{60}/PGE) to selectively detect HQ and CC. The sensor was able to successfully distinguish the HQ and CC oxidations, exhibiting strong catalytic performance and robust oxidation current. The DPV evidence also revealed a separated peak-to-peak potential difference in the HQ and CC oxidation to 101 mV for Pt/C_{60}/PGE from 76.9 mV for PtNPs/PGE. Highly selective Pt/C_{60}/PGE was able to detect HQ and CC simultaneously in the optimized circumstances, with the best behavior in a broad linear range between 50.0 and 1500.0 μM for CC as well as between 50.0 and 1100.0 μM for HQ. The LOD value was 2.97 μM for CC and 2.19 μM for HQ [84].

The C_{60}-based voltammetric sensors reported for environmental analysis are listed in table 3.4.

Table 3.4. Different analytical indices of C_{60}-based voltammetric sensors reported for environmental analysis.

Electrochemical sensor	Analyte	Analytical method	LOD	Linear range	References
CD-C_{60}/GCE	p-nitrophenol	SWV	1.2×10^{-9} M	2.8×10^{-9}–4.2×10^{-7} M	[80]
C60–CNTs/IL/GCE	Hydrazine	DPV	17 ± 2 nM	0.05–700.0 μM	[81]
	Hydroxylamine		28 ± 2 nM	1.0–300.0 μM	
Fullerene/MWCNT/Nafion/GCE	Carbendazim	SWV	1.7×10^{-8} M	2.0×10^{-8}–3.5×10^{-7} M	[82]
C_{60}-Chit/GCE	Hg(II)	DPASV	3.0 nM	0.01–6.0 μM	[83]
	Cu(II)		14.0 nM	0.05–6.0 μM	
	Pb(II)		1.0 nM	0.005–6.0 μM	
	Cd(II)		21.0 nM	0.5–9.0 μM	
Pt/C_{60}/PGE	Catechol	DPV	2.97 μM	50.0–1500.0 μM	[84]
	Hydroquinone		2.19 μM	50.0–1100.0 μM	

3.2.5 Voltammetric sensors constructed by hollow sphere and porous carbon composites

An ultra-sensitive sensor was produced by Zeinu *et al* on the basis of hollow sphere (HS) bismuth oxide (Bi) doped mesoporous carbon (PC) aerogel nanocomposite arising from stable biomass substance to detect pico-molar-level Cd^{2+} and Pb^{2+} simultaneously. They collected, in this nanocomposite material, the merits of a PC matrix arising from large surface area biomass for the pre-enrichment of analyte, exceptional electroanalytical performance of HS-Bi structure to sensitively detect Cd^{2+} and Pb^{2+}. This electrode material in optimized circumstances revealed that the LOD value was 1.72 pM for Pb^{2+} and 1.58 pM for Cd^{2+} in ambient circumstances, the minimum LOD for both Cd^{2+} and Pb^{2+} was related to activated carbon. In addition, there were two broad linear ranges (0.5–10.0 pM and 10.0–100.0 pM) because of adsorption dynamics in dilute electrolytes [85].

A new nanostructure of N-doped hollow carbon spheres decorated with rGO (NHCNS@RG) was fabricated by Yi *et al* who employed silica@polydopamine as a template with the aid of the self-assembly method. Because of electrostatic interaction, the NHCNS@RG nanostructure possessed admirable features of both RG and NHCNS, so that the NHCNS can disrupt the tendency of RG to accumulate and have great electrocatalytic performance, as well as the RG can expand the contact area and provide numerous normalized edge plate structures and active sites. The NHCNS@RG hybrids could electrochemically detect parachlorophenol (PCP) in optimized circumstances, with a LOD value of 0.01 μM and a linear range of 0.03 to 38.00 μM [86].

Qin *et al* utilized a SiO_2-mediated polybenzoxazine (PB) coating method to dope hollow porous carbon spheres with nitrogen (N-HPCSs). Through the *in situ* plating combination of Bi film with the N-HPCSs, sensitive Bi-N-HPCSs were produced to electrochemically detect Cd (II) exploiting DPASV. The Bi-N-HPCSs in optimized circumstances had a linear response to Cd(II) in the range between 0.5 and 50 mg L^{-1}. The LOD value was 0.16 mg L^{-1} for Cd(II) detection, meaning 31 times lower than the safety values for drinking water recommended by the United States Environmental Protection Agency (USEPA) [87].

Long *et al* doped hollow carbon spheres with nitrogen (NCs) utilizing a SiO_2-mediated template method via a carbon precursor of dopamine. Then, an electrochemical sensor was developed by the NCs to detect CC. The electrode conductivity can be significantly elevated via N-doping approach. Hence, the NCs were applied for the modification of SPE, the results of which showed an admirable sensing activity. The NCs/SPE in the optimized circumstances could detect CC with a linear range of 0.5–300 μM and narrow LOD of 0.14 μM [88].

In a study by Yang *et al* a new three-dimensional (3D) multi-doped HCS (N-Co-Fe-HCS) with 30 nm thick layer was produced, which had numerous active sites of heteroatom (N) and transition metals (Co and Fe), through SiO_2-mediated template method, low cost self-adsorption/reduction, *in situ* self-polymerization, and carbonization approaches. Moreover, they prepared multiple HCS composites of the same class (N-Co-HCS, N-HCS, and N-Fe-HCS) via the same protocol and precursor but

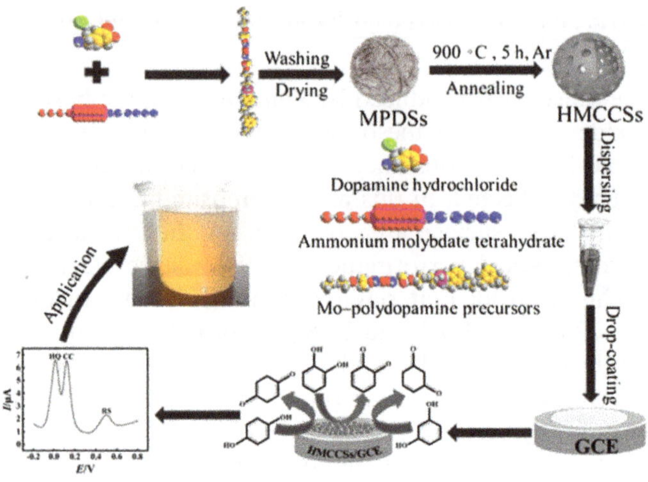

Figure 3.4. Production processes of HMCCSs/GCE. Reprinted from [90]. (Copyright 2019, Springer-Verlag GmbH Austria, part of Springer Nature.) With permission of Springer.

various doping elements. Such differences presented various active sites from HCSs and subsequently electrocatalytic performances for the isomers of dihydroxybenzene. The N-Co-Fe-HCS sensor possessed both exceptional quantitative and qualitative detection potentials for HQ and CC, with the LOD of 75 nM and linear range of 0.5 to 500 μM for CC, the LOD of 80 nM and linear range of 0.5 to 1500 μM for HQ [89].

Hollow molybdenum-dopamine spheres thermally annealed to generate hollow Mo_2C/C spheres (HMCCSs) were produced by Ren et al The spheres were applied to modify GCE to simultaneously determine CC, HQ, and RS. Figure 3.4 presents the production process of HMCCSs/GCE. There were clear oxidation peaks at 0.10, −0.004, and 0.44 V (versus SCE) for CC, HQ, and RS, individually. The linear responses to CC, HQ, and RS were reported at the concentrations of 2.0–2000.0, 0.3–1000.0, and 3.0–600.0 μM, with relative LODs of 0.19, 0.12, and 1.1 μM, successively [90].

The hollow sphere and porous carbon-based voltammetric sensors reported for environmental analysis are listed in table 3.5.

3.3 Conclusion and perspectives

Admirable profiles of carbon nanomaterials, including CNTs, g-C_3N_4, GOs, hollow spheres, C_{60} and porous carbon, have provided unparalleled grounds to electrochemically detect different environmental pollutants. This chapter discusses new advances in the construction of electrochemical sensors using different carbon composites along with voltammetric detection approaches to control environmental contaminants. Electrochemical protocols using carbon composites, when compared with conventional approaches, have been the concern of many researchers because of their green nature and cost-effectiveness. The predominant reason for using carbon composites is their ability to allow for increasing electron transfer rates in

Table 3.5. Different analytical indices of hollow sphere and porous carbon-based voltammetric sensors reported for environmental analysis.

Electrochemical sensor	Analyte	Analytical method	LOD	Linear range	References
Bi/PC/Ch GCE	Pb^{2+}	SWASV	1.72 pM	0.5–10.0 pM	[85]
	Cd^{2+}		1.58 pM	10.0–100.0 pM	
NHCNS@RG/GCE	Parachlorophenol	DPV	0.01 μM	0.03–38.00 μM	[86]
Bi-N-HPCSs/GCE	Cd(II)	DPASV	0.16 mg L^{-1}	0.5–150 mgL^{-1}	[87]
NCs/SPE	Catechol	DPV	0.14 μM	0.5–300 μM	[88]
N-Co-Fe-HCS/GCE	Catechol	DPV	75 nM	0.5–500.0 μM	[89]
	Hydroquinone		80 nM	0.5–1500.0 μM	
HMCCSs/GCE	Catechol	DPV	0.19 μM	2.0–2000.0 μM	[90]
	Hydroquinone		0.12 μM	0.3–1000.0 μM	
	Resorcinol		1.1 μM	3.0–600.0 μM	

most cases, thereby enhancing the sensitivity of the analyte. It should be noted that there is a possibility of further increasing this sensitivity through chemical modification of the surface and even the surface architecture of such carbon materials. The surface modification should allow for further functionalization of surfaces by combining carbon materials with ionic liquids, nanoparticles, polymers and so on, thereby increasing the usability of these sensors. Despite such exceptional progress, there are still many challenges in this area. The first challenge is to build new carbon-based composites with better performance. The second challenge is to maximize the application of electrochemical sensors based on carbon composites by integrating detection techniques for multiplexed target pollutions. The third challenge is the need to build inexpensive, portable, and mobile-based detection instruments for on-site and real-time detection. The introduction of commercial carbon-based equipment to the market is not unexpected.

References

[1] Fang X, Zong B and Mao S 2018 Metal–organic framework-based sensors for environmental contaminant sensing *Nano-micro Lett.* **10** 64

[2] Su S, Chen S and Fan C 2018 Recent advances in two-dimensional nanomaterials-based electrochemical sensors for environmental analysis *Green Energy Environ* **3** 97–106

[3] Rassaei L, Marken F, Sillanpää M, Amiri M, Cirtiu C M and Sillanpää M 2011 Nanoparticles in electrochemical sensors for environmental monitoring *TrAC Trends Anal. Chem* **30** 1704–15

[4] Cernat A, Tertiş M, Săndulescu R, Bedioui F, Cristea A and Cristea C 2015 Electrochemical sensors based on carbon nanomaterials for acetaminophen detection: a review *Anal. Chim. Acta* **886** 16–28

[5] Yu J, Wang X, Kang Q, Li J, Shen D and Chen L 2017 One-pot synthesis of a quantum dot-based molecular imprinting nanosensor for highly selective and sensitive fluorescence detection of 4-nitrophenol in environmental waters *Environ. Sci. Nano* **4** 493–502

[6] Burratti L, Ciotta E, De Matteis F and Prosposito P 2021 Metal nanostructures for environmental pollutant detection based on fluorescence *Nanomaterials* **11** 276

[7] Cai S, Lao K, Lau C and Lu J 2011 "Turn-on" chemiluminescence sensor for the highly selective and ultrasensitive detection of Hg^{2+} ions based on interstrand cooperative coordination and catalytic formation of gold nanoparticles *Anal. Chem.* **83** 9702–8

[8] Tang H, Zhu C, Meng G and Wu N 2018 Surface-enhanced Raman scattering sensors for food safety and environmental monitoring *J. Electrochem. Soc.* **165** B3098

[9] Song D, Yang R, Long F and Zhu A 2019 Applications of magnetic nanoparticles in surface-enhanced Raman scattering (SERS) detection of environmental pollutants *J. Environ. Sci.* **80** 14–34

[10] Yin J, Jiang Z, Chang G and Hu B 2005 Simultaneous on-line preconcentration and determination of trace metals in environmental samples by flow injection combined with inductively coupled plasma mass spectrometry using a nanometer-sized alumina packed micro-column *Anal. Chim. Acta* **540** 333–39

[11] Zhou Q, Xiao J, Wang W, Liu G, Shi Q and Wang J 2006 Determination of atrazine and simazine in environmental water samples using multiwalled carbon nanotubes as the

adsorbents for preconcentration prior to high performance liquid chromatography with diode array detector *Talanta* **68** 1309–15

[12] Zhao Y, Bai X L, Song T, Zhang G L, Yuan Y C, Liu Y M and Liao X 2019 Determination of environmental estrogens and bisphenol A in water samples by ultra-high performance liquid chromatography coupled to Q-Exactive high resolution mass spectrometry after magnetic solid-phase extraction *Microchem. J.* **151** 104212

[13] Wang Y and Hu S 2016 Applications of carbon nanotubes and graphene for electrochemical sensing of environmental pollutants *J. Nanosci. Nanotechnol.* **16** 7852–72

[14] Brett C M 2001 Electrochemical sensors for environmental monitoring. Strategy and examples *Pure Appl. Chem* **73** 1969–77

[15] Švorc Ľ, Rievaj M and Bustin D 2013 Green electrochemical sensor for environmental monitoring of pesticides: determination of atrazine in river waters using a boron-doped diamond electrode *Sens. Actuators B Chem* **181** 294–300

[16] Zhang Y, Zeng G M, Tang L, Li Y P, Chen Z M and Huang G H 2014 Quantitative detection of trace mercury in environmental media using a three-dimensional electrochemical sensor with an anionic intercalator *Rsc Adv* **4** 18485–92

[17] Vinodha G, Cindrella L and Shima P D 2019 Graphene oxide based highly sensitive electrochemical sensor for detection of environmental pollutants and biomolecules *Mater. Res. Expr* **6** 085548

[18] Hanafi R, Mayasari R D, Masmui A, Raharjo J and Nuryadi R 2019 Electrochemical sensor for environmental monitoring system: a review *AIP Conf. Proceed.* **2169** 030007

[19] Adhikari B R, Govindhan M and Chen A 2015 Carbon nanomaterials based electrochemical sensors/biosensors for the sensitive detection of pharmaceutical and biological compounds *Sensors* **15** 22490–508

[20] Silah H, Erkmen C, Demir E and Uslu B 2021 Modified indium tin oxide electrodes: electrochemical applications in pharmaceutical, biological, environmental and food analysis *TrAC Trends Anal. Chem* **141** 116289

[21] Wang H, Yuan X, Zeng G, Wu Y, Liu Y, Jiang Q and Gu S 2015 Three dimensional graphene based materials: synthesis and applications from energy storage and conversion to electrochemical sensor and environmental remediation *Adv. Colloid Interface Sci.* **221** 41–59

[22] Asadian E, Ghalkhani M and Shahrokhian S 2019 Electrochemical sensing based on carbon nanoparticles: a review *Sens. Actuators B Chem* **293** 183–209

[23] Power A C, Gorey B, Chandra S and Chapman J 2018 Carbon nanomaterials and their application to electrochemical sensors: a review *Nanotechnol. Rev.* **7** 19–41

[24] Tiwari J N, Vij V, Kemp K C and Kim K S 2016 Engineered carbon-nanomaterial-based electrochemical sensors for biomolecules *ACS Nano* **10** 46–80

[25] Hanrahan G, Patil D G and Wang J 2004 Electrochemical sensors for environmental monitoring: design, development and applications *J. Environ. Monit.* **6** 657–64

[26] Tyagi D, Wang H, Huang W, Hu L, Tang Y, Guo Z, Ouyang Z and Zhang H 2020 Recent advances in two-dimensional-material-based sensing technology toward health and environmental monitoring applications *Nanoscale* **12** 3535–59

[27] Bansod B, Kumar T, Thakur R, Rana S and Singh I 2017 A review on various electrochemical techniques for heavy metal ions detection with different sensing platforms *Biosens. Bioelectron.* **94** 443–55

[28] Gupta V K, Jain R, Radhapyari K, Jadon N and Agarwal S 2011 Voltammetric techniques for the assay of pharmaceuticals—a review *Anal. Biochem.* **408** 179

[29] Hoyos-Arbeláez J, Vázquez M and Contreras-Calderón J 2017 Electrochemical methods as a tool for determining the antioxidant capacity of food and beverages: a review *Food Chem* **221** 1371–81
[30] Kim Y J, Kim Y, Novoselov K and Hong B H 2015 Engineering electrical properties of graphene: chemical approaches *2D Mater.* **2** 042001
[31] You Y, Zeng W, Yin Y X, Zhang J, Yang C P and Zhu Y 2015 Hierarchically micro/mesoporous activated graphene with a large surface area for high sulfur loading in Li–S batteries *J. Mater. Chem.* A **3** 4799–802
[32] Bonaccorso F, Sun Z, Hasan T and Ferrari A C 2010 Graphene photonics and optoelectronics *Nat. Photonics* **4** 611–22
[33] Fuchs J N and Goerbig M O 2008 Introduction to the physical properties of graphene *Lect. Notes* **10** 11–2
[34] Wu S, He Q, Tan C, Wang Y and Zhang H 2013 Graphene-based electrochemical sensors *Small* **9** 1160–72
[35] Vasseghian Y, Dragoi E N, Moradi M and Khaneghah A M 2021 A review on graphene-based electrochemical sensor for mycotoxins detection *Food Chem. Toxicol.* **148** 111931
[36] Shao Y, Wang J, Wu H, Liu J, Aksay I A and Lin Y 2010 Graphene based electrochemical sensors and biosensors: a review *Electroanalysis* **22** 1027–36
[37] Radhakrishnan S, Krishnamoorthy K, Sekar C, Wilson J and Kim S J 2014 A highly sensitive electrochemical sensor for nitrite detection based on Fe_2O_3 nanoparticles decorated reduced graphene oxide nanosheets *Appl. Catal. B Environ* **148** 22–8
[38] Bai X, Huang X, Zhang X, Hua Z, Wang C and Qin Q 2014 TiO_2–graphene nanoparticle based electrochemical sensor for the bimodal-response detection of 4-chlorophenol *RSC Adv* **4** 13461–68
[39] Yu G, Zhang W, Zhao Q, Wu W, Wei X and Lu Q 2016 Enhancing the sensitivity of hexachlorobenzene electrochemical sensor based on nitrogen-doped graphene *Sens. Actuators B Chem* **235** 439–46
[40] Tang H, Chen J, Zeng Y, Li Z, Huang H and Li L 2016 An electrochemical sensor for 1-naphthylamine based on a novel composite of cyclodextrin-graphene and molecularly imprinted poly (vinylferrocene) *Anal. Methods* **8** 1681–89
[41] Sha R, Puttapati S K, Srikanth V V and Badhulika S 2017 Ultra-sensitive phenol sensor based on overcoming surface fouling of reduced graphene oxide-zinc oxide composite electrode *J. Electroanal. Chem.* **785** 26–32
[42] Manasa G, Mascarenhas R J, Satpati A K, Basavaraja B M and Kumar S 2018 An electrochemical Bisphenol F sensor based on ZnO/G nano composite and CTAB surface modified carbon paste electrode architecture *Colloids Surf B Biointerfaces* **170** 144–51
[43] Wei P, Zhu Z, Song R, Li Z and Chen C 2019 An ion-imprinted sensor based on chitosan-graphene oxide composite polymer modified glassy carbon electrode for environmental sensing application *Electrochim. Acta* **317** 93–101
[44] Baghayeri M, Alinezhad H, Fayazi M, Tarahomi M, Ghanei-Motlagh R and Maleki B 2019 A novel electrochemical sensor based on a glassy carbon electrode modified with dendrimer functionalized magnetic graphene oxide for simultaneous determination of trace Pb (II) and Cd (II) *Electrochim. Acta* **312** 80–8
[45] Migliorini F L, Sanfelice R C, Mercante L A, Facure M H and Correa D S 2019 Electrochemical sensor based on polyamide 6/polypyrrole electrospun nanofibers coated with reduced graphene oxide for malathion pesticide detection *Mater. Res. Express* **7** 015601

[46] Al-Hamry A, Ega T K, Pašti I A, Bajuk-Bogdanović D, Lazarevic-Pašti T, Rodriguez R D *et al* 2019 Electrochemical sensor based on reduced graphene oxide/PDAC for dimethoate pesticide detection *2019 5th Int. Conf. Nanotechnol Instrument. Measur.* pp 1–7

[47] Wang Y, Liang Y, Zhang S, Wang T, Zhuang X, Tian C, Luan F, Ni S-Q and Fu X 2021 Enhanced electrochemical sensor based on gold nanoparticles and MoS_2 nanoflowers decorated ionic liquid-functionalized graphene for sensitive detection of bisphenol A in environmental water *Microchem. J.* **161** 105769

[48] Mnyipika S H, Munonde T S and Nomngongo P N 2021 MnO_2@ reduced graphene oxide nanocomposite-based electrochemical sensor for the simultaneous determination of trace Cd (II), Zn (II) and Cu (II) in water samples *Membranes* **11** 517

[49] Alim S, Vejayan J, Yusoff M M and Kafi A K M 2018 Recent uses of carbon nanotubes & gold nanoparticles in electrochemistry with application in biosensing: a review *Biosens. Bioelectron* **121** 125–36

[50] Sireesha M, Jagadeesh Babu V, Kranthi Kiran A S and Ramakrishna S 2018 A review on carbon nanotubes in biosensor devices and their applications in medicine *Nanocomposites* **4** 36–57

[51] Alam A U and Deen M J 2020 Bisphenol A electrochemical sensor using graphene oxide and β-cyclodextrin-functionalized multi-walled carbon nanotubes *Anal. Chem.* **92** 5532–39

[52] Kour R, Arya S, Young S J, Gupta V, Bandhoria P and Khosla A 2020 Recent advances in carbon nanomaterials as electrochemical biosensors *J. Electrochem. Soc.* **167** 037555

[53] Wan Q, Yang P, Cai H, Song and Yang N 2013 Voltammetry of nanomolar leveled environmental hazards on the polymer/CNT coated electrodes *J. Electroanal. Chem.* **689** 252–56

[54] Zhang J, Niu Y, Li S, Luo R and Wang C 2014 A molecularly imprinted electrochemical sensor based on sol–gel technology and multiwalled carbon nanotubes–Nafion functional layer for determination of 2-nonylphenol in environmental samples *Sens. Actuators B Chem.* **193** 844–50

[55] Hassan R Y, Kamel M S, Hassan H N and Khaled E 2015 Voltammetric determination of mercury in biological samples using crown ether/multiwalled carbon nanotube-based sensor *J. Electroanal. Chem.* **759** 101–6

[56] Messaoud N B, Ghica M E, Dridi C, Ali M B and Brett C M 2017 Electrochemical sensor based on multiwalled carbon nanotube and gold nanoparticle modified electrode for the sensitive detection of bisphenol A *Sens. Actuators B Chem* **253** 513–22

[57] Rajkumar C, Thirumalraj B, Chen S M, Veerakumar P and Lin K C 2018 Voltammetric determination of catechol and hydroquinone using nitrogen-doped multiwalled carbon nanotubes modified with nickel nanoparticles *Microchim. Acta* **185** 395

[58] Sipa K, Brycht M, Leniart A, Urbaniak P, Nosal-Wiercińska A and Pałecz B 2018 β-Cyclodextrins incorporated multi-walled carbon nanotubes modified electrode for the voltammetric determination of the pesticide dichlorophen *Talanta* **176** 625–34

[59] Balram D, Lian K Y and Sebastian N 2018 Synthesis of a functionalized multi-walled carbon nanotube decorated ruskin michelle-like ZnO nanocomposite and its application in the development of a highly sensitive hydroquinone sensor *Inorg. Chem. Front.* **5** 1950–61

[60] Koyun O, Gorduk S, Gencten M and Sahin Y 2019 A novel copper (II) phthalocyanine-modified multiwalled carbon nanotube-based electrode for sensitive electrochemical detection of bisphenol A *New J. Chem.* **43** 85–92

[61] Manasa G, Bhakta A K, Mekhalif Z and Mascarenhas R J 2019 Voltammetric study and rapid quantification of resorcinol in hair dye and biological samples using ultrasensitive maghemite/MWCNT modified carbon paste electrode *Electroanalysis* **31** 1363–72
[62] Wu W, Jia M, Zhang Z, Chen X, Zhang Q and Zhang W 2019 Sensitive, selective and simultaneous electrochemical detection of multiple heavy metals in environment and food using a lowcost Fe_3O_4 nanoparticles/fluorinated multi-walled carbon nanotubes sensor *Ecotoxicol. Environ. Saf.* **175** 243–50
[63] Le Hai T, Hung L C, Phuong T T B, Ha B T T, Nguyen B S and Hai T D 2020 Multiwall carbon nanotube modified by antimony oxide (Sb_2O_3/MWCNTs) paste electrode for the simultaneous electrochemical detection of cadmium and lead ions *Microchem. J.* **153** 104456
[64] Bashir N, Akhtar M, Nawaz H Z R, Warsi M F, Shakir I and Agboola P O 2020 A high performance electrochemical sensor for Pb^{2+} ions based on carbon nanotubes functionalized $CoMn_2O_4$ nanocomposite *Chem. Select* **5** 7909–18
[65] Ghodsi J and Rafati A A 2017 A voltammetric sensor for diazinon pesticide based on electrode modified with TiO_2 nanoparticles covered multi walled carbon nanotube nano-composite *J. Electroanal. Chem.* **807** 1–9
[66] Lyth S M, Nabae Y, Moriya S, Kuroki S, Kakimoto M A and Ozaki J I 2009 Carbon nitride as a nonprecious catalyst for electrochemical oxygen reduction *J. Phys. Chem* C **113** 20148–51
[67] Wang D P, Tang Y and Zhang W D 2013 A carbon nitride electrode for highly selective and sensitive determination of lead (II) *Microchim. Acta* **180** 1303–8
[68] Kumar P, Boukherroub R and Shankar K 2018 Sunlight-driven water-splitting using two-dimensional carbon based semiconductors *J. Mater. Chem* A **6** 12876–931
[69] Yola M L and Atar N 2018 Gold nanoparticles/two-dimensional (2D) hexagonal boron nitride nanosheets including diethylstilbestrol imprinted polymer: electrochemical detection in urine samples and validation *J. Electrochem. Soc.* **165** H897
[70] Gao W, Wang X, Li P, Wu Q, Qi F and Wu S 2016 Highly sensitive and selective detection of cadmium with a graphite carbon nitride nanosheets/Nafion electrode *RSC Adv.* **6** 113570–75
[71] Ponnaiah S K, Prakash P and Muthupandian S 2019 Ultrasonic energy-assisted *in situ* synthesis of Ru_0/PANI/g-C_3N_4 nanocomposite: application for picomolar-level electro-chemical detection of endocrine disruptor (bisphenol-A) in humans and animals *Ultrason. Sonochem* **58** 104629
[72] Yin H Y, Zheng Y F and Wang L 2020 Au/CeO_2/g-C_3N_4 nanocomposite modified electrode as electrochemical sensor for the determination of phenol *J. Nanosci. Nanotechnol.* **20** 5539–45
[73] Rajaji U, Chen T W, Chinnapaiyan S, Chen S M and Govindasamy M 2020 Two-dimensional binary nanosheets (Bi_2Te_3@ g-C_3N_4): application toward the electrochemical detection of food toxic chemical *Anal. Chim. Acta* **1125** 220–30
[74] Mahmoudian M R, Alias Y, Woi P M, Yousefi R and Basirun W J 2020 An electrochemical sensor based on Pt/g-C_3N_4/polyaniline nanocomposite for detection of Hg^{2+} *Adv. Powder Technol.* **31** 3372–80
[75] Han J, Fu R, Jin C, Li Z, Wang M and Yu P 2020 Highly sensitive detection of trace Hg^{2+} via PdNPs/g-C_3N_4 nanosheet-modified electrodes using DPV *Microchem. J.* **152** 104356
[76] He X, Bai S, Jiang J, Ong W J, Peng J and Xiong Z 2021 Oxygen vacancy mediated step-scheme heterojunction of $WO_{2.9}$/g-C_3N_4 for efficient electrochemical sensing of 4-nitro-phenol *Chem. Eng. J. Adv.* **8** 100175
[77] Yáñez-Sedeño P, Campuzano S and Pingarrón J M 2017 Fullerenes in electrochemical catalytic and affinity biosensing: a review *J. Carbon Res.* **3** 21

[78] Pilehvar S and De Wael K 2015 Recent advances in electrochemical biosensors based on fullerene-C60 nano-structured platforms *Biosensors* **5** 712–35

[79] Chen G, Wang X and Wang L 2020 Application of carbon based material for the electrochemical detection of heavy metal ions in water environment *Int. J. Electrochem. Sci.* **15** 4252–63

[80] Rather J A, Debnath P and De Wael K 2013 Fullerene-β-cyclodextrin conjugate based electrochemical sensing device for ultrasensitive detection of p-nitrophenol *Electroanalysis* **25** 2145–50

[81] Mazloum-Ardakani M, Alireza K and Laleh H 2015 Simultaneous determination of hydrazine and hydroxylamine based on fullerene-functionalized carbon nanotubes/ionic liquid nanocomposite *Sens. Actuators B Chem* **214** 132–37

[82] Teadoum D N, Noumbo S K, Arnaud K T, Ranil T T, Mvondo Zé A D and Tonle I K 2016 Square wave voltammetric determination of residues of carbendazim using a fullerene/multiwalled carbon nanotubes/nafion/coated glassy carbon electrode *Int. J. Electrochem* **2016** 7839708

[83] Han X, Meng Z, Zhang H and Zheng J 2018 Fullerene-based anodic stripping voltammetry for simultaneous determination of Hg (II), Cu (II), Pb (II) and Cd (II) in foodstuff *Microchim. Acta* **185** 274

[84] Zhu Y, Huai S, Jiao J, Xu Q, Wu H and Zhang H 2020 Fullerene and platinum composite-based electrochemical sensor for the selective determination of catechol and hydroquinone *J. Electroanal. Chem.* **878** 114726

[85] Zeinu K M, Hou H, Liu B, Yuan X, Huang L and Zhu X 2016 A novel hollow sphere bismuth oxide doped mesoporous carbon nanocomposite material derived from sustainable biomass for picomolar electrochemical detection of lead and cadmium *J. Mater. Chem* A **4** 13967–79

[86] Yi Y, Zhu G, Sun H, Sun J and Wu X 2016 Nitrogen-doped hollow carbon spheres wrapped with graphene nanostructure for highly sensitive electrochemical sensing of parachlorophenol *Biosens. Bioelectron.* **86** 62–7

[87] Qin D, Wang L, Gao S, Wang Y, Mamat X, Li Y, Wagberg T, Cheng H and Hu G 2018 N-doped hollow porous carbon spheres/bismuth hybrid film modified electrodes for sensitive voltammetric determination of trace cadmium *Electroanalysis* **30** 1906–12

[88] Long W, Xie Y, Shi H, Ying J, Yang J, Huang Y, Zhang H and Fu L 2018 Preparation of nitrogen-doped hollow carbon spheres for sensitive catechol electrochemical sensing *Fuller. Nanotub. Carbon Nanostr.* **26** 856–62

[89] Yang H, Li S, Yu H, Zheng F, Lin L, Chen J, Li Y and Lin Y 2019 *In situ* construction of hollow carbon spheres with N, Co, and Fe co-doping as electrochemical sensors for simultaneous determination of dihydroxybenzene isomers *Nanoscale* **11** 8950–58

[90] Ren H, Zhang Y, Liu L, Li Y, Wang D, Zhang R, Zhang W, Li Y and Ye B-C 2019 Synthesis of hollow Mo_2 C/carbon spheres, and their application to simultaneous electrochemical detection of hydroquinone, catechol, and resorcinol *Microchim. Acta* **186** 306

Chapter 4

Graphene-based sensing devices for soil moisture analysis

B Chethan, V Prasad and S Thomas

Today nanomaterials have acquired primary advantages because of their size which is in the range 1–100 nm. Due to the unique characteristics of nanomaterials, namely: high electrical conductivity, thermal stability, and mechanical strength, they find widespread applications in the field of batteries, magneto resistance, electrochemical, and physical sensors. Among these nanomaterials, carbon-based nanomaterials have attained widespread applications since the last century, because of their astonishing physical and chemical properties. Carbon materials are widely available from coal deposits and have a wide range of applications from medicine to rocket technology. Among these carbon nanomaterials, the allotropic form of carbon, graphene, attracted most attention from researchers due to its interesting properties investigated over the last two decades. Graphene was a boon to material science researchers due to its wide variety of applications. Today, research in the field of graphene is looking for new graphene derivatives with better functionalization and modified surfaces. These graphene derivatives like graphene oxide, reduced graphene oxide, graphene quantum dots and modified graphene oxide are extensively used in the field of soil moisture sensors.

Water is the primary source of irrigation. Effective and efficient irrigation helps to preserve water, decreases the use of fertilizers and meliorates crop quality. The successful cultivation of crops mainly depends on monitoring of the soil moisture in the field. Thus, regular monitoring soil moisture is essential for innovative farming. In this regard development of soil moisture monitoring sensors have been accelerated. The selection of the correct probes to use is essential to soil moisture sensors. The graphene-based soil moisture sensors gained the attention of the society due to its real time reading, accuracy, and stability.

In this regard, this chapter provides an insight into the classification of nanoparticles, carbon nanoparticles, graphene, and their synthesis and characterization

doi:10.1088/978-0-7503-5127-0ch4

techniques in general. As typical cases, this chapter provides information about the importance of graphene-based soil moisture sensors, the required qualities of a potential sensor and the mechanism of the soil moisture sensor.

4.1 Introduction

In the present scenario, smart sensors have been promoted due to their widespread applications [1]. These sensors find applications ranging from the preparation of buckets to rockets. To improve the quality of life, smart sensor technology is essential for day-to-day life. More quality research is being undertaken to construct sensors from nanomaterials with good electrical, chemical, mechanical, and thermal properties [2]. The noble metals, ferrites, carbon, and polymer nanoparticles are commonly used to develop sensing materials. Among the nanomaterials, carbon-based nanomaterials are currently extensively used. Carbon-based nanoparticles (molar mass: 12.01 g mol^{-1}; melting and boiling point above 3500 °C) appear as a black powder and are spherical in nature [3]. Due to their advantageous electrical and chemical properties they have widespread applications in a variety of industries. Of the carbon-based nanomaterials, graphene, possessing a 2D honeycomb structure, shows excellent thermal and electrical qualities. The more advanced research work is going on in the field of synthesis graphene since the last decade. Due to its typical nanoporous structure, large active surface area, high mechanical strength, high electrical and thermal conductivity, and maximum stability, it has a variety of applications in batteries, super capacitors, memory devices, bio-medical applications, in engineering, and in sensor fields [4]. There are very many methods namely: spin coating, layer-by-layer assembly, and filtration methods which can be used to modify the surface structure of graphene. Today many research articles are being published about the synthesis of graphene and its utilization for the fabrication of sensing devices.

Agriculture is the primary source of income of rural communities. The quality of crop cultivation mainly depends on the water content of the soil. Plants absorb the moisture from the soil and give better yield [5]. Thus, measurement and monitoring of soil moisture content and humidity is essential and necessary. But, the salt content, fluctions in temperature, reduced stability, variations in the pH of the soil, and corrosion decreases the performance of the sensor. Commercially available soil moisture sensors, galvanic soil moisture sensors, pulse sensors, and time domain reflectometry sensors are also not free from these defects [6]. Hence, highly sensitive, stable, independent of temperature drift, corrosion free, rapid response and recovery, low-cost soil moisture sensing devices are required to meet the demands of todays farmers. The advances in MEMS technology (micro electro mechanical systems), ensuring the improvement of sensor fabrication technology, helps researchers to fabricate potential soil moisture sensors.

Today researchers are working more on polymer-based sensors namely: polyaniline, polypyrrole, PEDOT: PSS and PDMS materials. Nevertheless, these polymers are unstable, degrade in a short duration, and have poor sensitivity [7]. Thus, researchers are looking for a suitable nanomaterial which can fulfill all the requirements of a superior soil moisture sensor. The astonishing features of graphene have pompted the research community to study graphene-based soil moisture sensors. The derivatives of the graphene viz., reduced graphene oxide, modified graphene oxide, graphene oxide (GO), graphene quantum dots (GQD) and graphene ribbons offer high conductivity, maximum thermal stability, and large active surface areas have made them potential materials to use as soil moisture sensors [8]. Among these, GO is most suitable for the water vapor detection, because it is rich in oxygen functional groups. The basal plane of GO contains epoxide and hydroxyl, and the edge plane contains carbonyl and carboxyl groups. These four functional groups make GO hydrophilic and suitable for soil moisture sensing.

This chapter thoroughly showcases the importance of graphene materials and their use in production of soil moisture sensors. This chapter provides an overview of the classification of nanoparticles, the types of carbon-based nanomaterials, synthesis methods of carbon-based nanoparticles, characterization of carbon materials and discusses the use of graphene in the fabrication of soil moisture sensor devices, with the fundamental and industrial application of them in many disciplines. The later sections of this chapter discuss the various types of soil moisture sensors. Thus, this chapter gives an overview of the qualities of graphene-based soil moisture sensors.

4.2 Classification of nanoparticles

Based on the dimension, nanoparticles (NP) are classified into three types [9].

One-dimensional nanoparticles

Thin film and monolayer are one-dimensional nanoparticles and these classes of materials have been used in chemistry, biology, electronics, and engineering fields for many centuries. This type of monolayer nanoparticle is usually used in catalysis and solar cell applications. It has widespread applications in storage devices, chemical sensors, and magneto-optic devices [10].

Two-dimensional nanoparticles

Graphene oxide and carbon nanotubes (CNT) are the two-dimensional nanoparticles are in the size range of 1 nm in diameter and 100 nm in length. CNT nanoparticles are of two types: single walled carbon nanotubes (SWCNT) and multiwalled carbon nanotubes (MWCNT) [11].

Three-dimensional nanoparticles

Dendrimers, quantum dots (QD), fullerenes (carbon 60) are the three-dimensional nanoparticles. These nanoparticles have applications in nanoelectronics, medical fields, anticancer drug industries, and tissue engineering.

4.3 Synthesis techniques of nanoparticles

Nanoparticles can mainly be synthesized by two methods (figure 4.1):

1. bottom-up method
2. top-down method

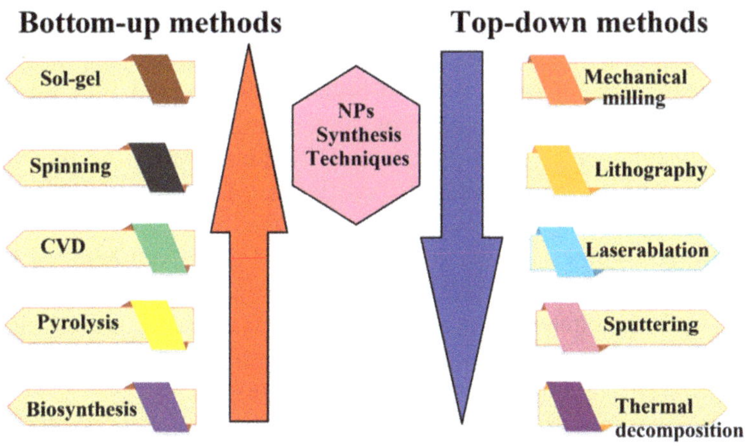

Figure 4.1. Flow chart depicting the different types of nanoparticle synthesis techniques.

1. **Bottom-up methods:** constructive approach and build up approach are the alternate names for this approach; it follows the hierarchy: atoms–clusters–nanoparticles. This approach contains: sol–gel, spinning, chemical vapor deposition (CVD), pyrolysis, and biosynthesis methods.

Sol–gel: Sol–gel is the most important bottom-up method because it is a very simple method and many of the metal NPs can be prepared by this method. There are four stages in the sol–gel process, they are hydrolysis, condensation, growth of particles, and agglomeration of particles. Sol: a colloidal solution consisting of solid suspended in liquid. Gel: solid molecules present in a solvent. It is a wet-chemical process containing a precursor chemical solution. Metal oxide, nitrates, chlorides are commonly used precursor solutions in this process [12]. In this process the chemical solution is dissipated in the host solution by stirring. The NPs are recovered by sedimentation, filtrate, and by centrifugation techniques. The NPs are obtained, after removing the moisture by heating in a hot air oven [13].

Spinning: The NPs are prepared through the spinning disc reactor, this type of synthesis is called spinning. The spinning chamber consists of a disk and the chamber is maintained in the inert condition to avoid the reaction of the material with the oxygen present in it [14]. There is special provision for controlling the temperature, speed, precursor ratio, and liquid flow rate. In the spinning method the atoms/molecules fuse together, and then the solution is filtered and dried to get NPs [15].

Chemical vapor deposition (CVD): Chemical vapor deposition is asophisticated technique used to prepare thin films. It is the process in which a solid is deposited on a heated surface using the gas phase; by combining the gas molecules with the reactant mixture, deposition is done on the glass substrate [16]. A typical system comprises a deposition chamber, gas supply cylinders, and an exhaust system. In this process we can obtain highly pure, uniform, and hard NPs [17].

Pyrolysis: Pyrolysis is the industrial method used for large-scale production. The precursor solution in liquid form is burnt in the furnace at very high pressure to get NPs [18]. Here, by a combustion method NPs are synthesized. The furnace used in this method, has a requirement to control the temperature; some furnaces use lasers and plasma to produce higher temperatures [19]. This method is cost effective, less time consuming, efficient, and easy to synthesize. Hense it finds application in industrial synthesis.

Biosynthesis: Biosynthesis is the eco-friendly and green synthesis of NPs. In this method we can produce non-toxic and biodegradable NPs [20]. Plant extracts, animal extracts, fungi, bacteria, microorganisms, leaf extracts, and cellulose are used with precursor solution to synthesize NPs. These NPs find application in bio-medical, pharmaceutical, and anti-cancers applications.t

2. **Top-down methods:** The destructive approach or reduction approach are the alternative names for this method which follows the hierarchy of bulk to simple scale production of NPs. This method contains: mechanical milling, nano-lithography, laser ablation, sputtering, and thermal decomposition methods.

Mechanical mixing: Mechanical milling is one of the most commonly used methods to produce NPs. Among all the top-down methods this one has acquired maximum attention. The various sample elements are milled in the inert condition to avoid the atmospheric air reaction. The important aspect in this method is deformation of the size of the particles. Rotation milling leads to a decrease in the particle size and cold-welding leads to an increase in the particle size. There are many types of mixing methods. Namely, high-energy milling, attrition ball milling, planetary ball milling, vibrating ball milling, low energy tumbling milling, and high energy ball milling [21].

Nanolithography: In this method, 1D materials which have dimensions of the order 1 to 100 nm are synthesized. In this process, the required shape and structure is printed on the light sensitive material precursor [22]. Lithography is the technique of synthesizing clusters of NPs with a certain design and shape. The major disadvantage of this method is that very sophisticated instruments are required and it a cost is associated. There are different types of nanolithography techniques, e.g., electron beam, scanning probe, multi- photon, and optical lithography [23].

Laser ablation: Laser ablation synthesis is a common method for producing various NPs from different solutions. The metal, which is immersed in solution, is irradiated and condensed and from this NPs are extracted [24].

Sputtering: This is the very simple process of depositing the thin layer of NPs accompanied by annealing [25]. The shape of the nanoparticles can be determined on the basis of the thickness, temperature, and annealing time [26].

Thermal decomposition: This is the endothermic chemical decomposition process. In this method heat is used to break the bonds in the composite. The decomposition temperature is the temperature at which chemicals decompose. The metals undergo decomposition from the NPs at a decomposition temperature. In this way we can obtain the NPs by [27]. Carbon nanoparticles are mainly synthesized from sol–gel,

CVD, and pyrolysis in the bottom-up method and mechanical mixing is mainly used in the top-down method.

4.4 Carbon nanoparticles and their derivatives

The important classes of carbon NPs are graphene, carbon nano tubes (CNT), fullerenes (C60), carbon nanofibers, and carbon black. The physical properties of these NPs are listed in table 4.1.

Graphene-based NPs: Graphene is an allotrope form of carbon and is a hexagonal network of honeycomb structured two-dimensional material [28]. Graphene has evolved as an exotic material in the twentyfirst century. It has gained the attention of the world due to its attractive thermal, optical, mechanical, physical, and charge transport properties [29]. Sheets are around 1 nm thick.

Carbon Nanotubes (CNT): This is a class of materials which is one of the allotropy forms of carbon which is made of a hexagonal two-dimensional lattice of C-atoms, joined together to form a hollow cylinder [30]. These nanotubes have a diameter of 0.7 nm for single-wall carbon nanotubes (SWCNTs) and about 100 nm for multi-wall carbon nanotubes (MWCNT). The length of the nanotubes varies from micrometers to millimeters [31].

Fullerenes: These are the SP^2 hybridized carbon atoms which are held together in a spherical shape. The single layers are 8.3 nm in diameter whereas the multi-layer fullerenes have diameters of about 35 nm. Each spherical structure of fullerene contains 20 to 1500 carbon atoms [32].

Carbon nanofiber: These have similar structures CNTs but are not in the form of regular cylindrical tubes, they are wound around each another to form cone or cup shaped tubes. Graphene nanofoils are used to produce these nanofibers [33].

Carbon black: These are the amorphous materials made up of carbon atoms. These are spherical in shape with diameters of about 10 to 60 nm.

Table 4.1. Chemical and physical properties of carbon-based NPs.

Sl. No.	Material	Chemical and physical properties	References
1.	Graphene NPs	Thermally stable, electrically conductive, light adsorption capacity, high strength.	[34]
2.	Carbon nanotubes NPs	Elastic, flexible, high tensile strength, extremely high thermally and electrically conductive.	[35]
3.	Fullerenes NPs	Transmits light based on intensity, conductor, super conductor.	[36]
4	Carbon nanofiber NPs	High mechanical strength, more electrical and thermal shielding.	[37]
5.	Carbon black NPs	Resistive to UV degradation, more electrical conductivity, high strength.	[38]

4.5 Synthesis of graphene using Hummers' method

There are very many significant methods used to synthesize graphene worldwide. Among them, Hummers' method is an important and industrial synthesized method.

Hummers' method (improvised): In this method, sulfuric acid and phosphoric acid are taken in the ratio of 9:1(360:40 mL) in a conical flask. To this mixture graphite flakes (3.0 g) are added and stirred for 1 h. The oxidizing agent potassium permanganate (18.0 g) was then added by keeping the reaction mixture temperature below 0 °C. After 1 h the temperature of the reaction mixture was raised to 50 °C and stirred for 12 h [39]. Then the reaction mixture was cooled to room temperature and then ice-cold water (400 mL) and hydrogen peroxide (3 mL) was added to it. The reaction mixture was then centrifuged and the residue was filtered through Whatman filter paper in a succession filtration jar. The residue was washed with water (200 mL), 30% hydrochloric acid and ethanol (200 mL) three times. Then the filtrate collected was dried in a hot air oven overnight. The dark brown powder obtained is graphene oxide.

The graphene oxide was mixed with DI water (200 mL) and sonicated for 40 min, then a very small amount of sulfur-laced reducing agent ($NaHSO_3$, $Na_2S{\cdot}9H_2O$, SO_2) was added and stirred for 5 h at room temperature. The mixture was again washed with DI water multiple times and graphene was obtained as a black powder by a succession filtration process.

4.6 Properties and characterization techniques of graphene

Properties: Graphene is a semi-metallic material having hexagonal structure. The atoms are arranged like a honeycomb lattice structure [40]. The three C-atoms in the graphene structure are SP^2 hybridized. The bonding energy of the carbon atoms is about 5.9 eV, the C-atoms are separated by the σ bond with bond length of 1.4 Å [41]. The half-filled P-subshells, form a π-bond with the nearby C-atoms [42, 43]. The graphene sheets are usually one atom thick and piled up on one another to form the 3D structure. The graphene sheets are bonded by Van der Waal's force; hence they are slippery in nature.

Characterization techniques:

1. **Size:** Particle size measurement is one of the most important concepts. It helps to determine the size of the particle and confirm whether it is nano or not. The particle size and its distribution can be confirmed using scanning electron microscopy (SEM) and transmission electron microscopy (TEM) techniques.

   Scanning electron microscopy: The SEM technique is used to determine the shape, size, and surface morphology of the graphene nanoparticles. The SEM images give very little information about the surface morphology, agglomeration of particles, and porosity of the samples. For the study, the conducting gold was sputtered onto the sample and surface morphological pictures were captured.

Transmission electron microscopy : The data obtained using this technique is the same as SEM but it operates with different principles. For TEM characterization the samples must be ultra-thin so it is a time consuming and complex process. The nano size of the particle can be confirmed by TEM analysis.

Atomic force microscopy (AFM): This is called scanning force microscopy. Very high-resolution images of the order of fractions of nanometers are pictured. The working principle of atomic force microscopy is that it physically scans the sample using a probe tip and measures the ultra-high resolution particle size [44, 45]. This technique provides accurate particle size.

2. **Surface area:** The surface to volume ratio of graphene can be thoroughly examined by Brunauer–Emmett–Teller (BET) studies. For the liquid phase nuclear magnetic resonance spectroscopy (NMR) is used for analysis.
3. **Composition:** The actual purity of graphene can be determined by knowing its chemical composition. The secondary elements may hinder the potential and efficiency of the graphene nanoparticles. So, it is essential to know the composition of the graphene NPs [46]. The composition of the graphene NPs can be determined by XPS and EDAX techniques.
4. **Surface morphology:** The NPs possess different surface shapes like spherical, oval, flat, cylindrical, flake, plate, rod, needle, conical, and irregular, some surfaces are crystalline, amorphous, and semi-crystalline in nature. These surface properties play an important role in determining its properties and efficiency. For the graphene NPs SEM and TEM techniques are used to understand the surface morphology [47].
5. **Crystallography:** To understand the crystallographic nature of the atoms and molecules powder XRD or NMR techniques are used [48]. The different characterizations used to study graphene are listed in table 4.2.

Table 4.2. Various methods for characterizing graphene NPs.

Sl. No.	Phase of availability	Characterization techniques Graphene NPs
1	Size	Transmission electron microscope and laser diffraction.
2	Surface area	Brunauer–Emmett–Teller studies
3	Composition	X-ray photoelectron spectroscopy and EDAX
4	Surface morphology	Scanning electron microscope and atomic force microscopy
5	Crystallography	Powder x-ray or neutron diffraction

4.7 Soil moisture sensors types

The sensor is a small device that converts a physical impulse into an electrical signal. The variation in impulse can be detected by a computer interfaced with the sensor. It is easier to measure an electrical impulse than physical one. Once the data has been obtained it is analyzed and information is gathered. Sensors can detect various physical parameters, namely: pressure, temperature, and displacement, and chemical parameters such as ions, molecules, concentration, and interaction between molecules. Scientists are trying to develop a cost effective, small, and very efficient sensor to cope with today's needs (figure 4.2).

Soil moisture sensors are broadly classified into two types [49]:

- **Tensiometric sensors:** measure how tightly the soil particles retain the water. Examples: tensiometers and granular matrix sensors.
- **Volumetric sensors:** measure the actual water volume in the soil. Examples: time domain and frequency domain reflectometry.

1. **Tensiometric sensors**

Tensiometers: These are simple soil moisture sensors (figure 4.3) consisting of a simple ceramic tip connected to a vacuum gauge through a pipe. When the porous ceramic tip is inserted into the soil, the soil water pressure is transmitted to the tensiometer sensor [11]. These sensors do not measure the

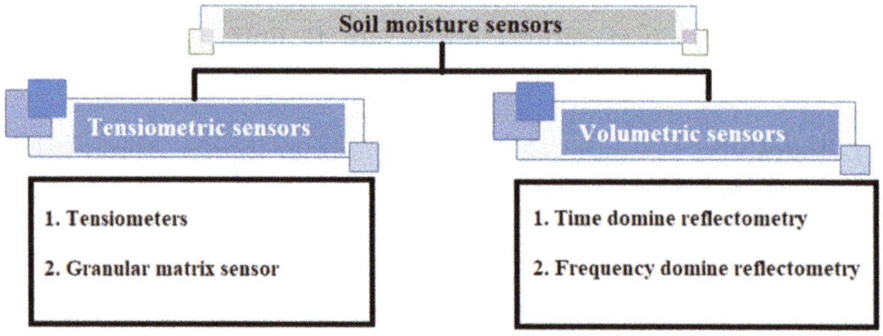

Figure 4.2. Flow chart illustrating the different types of soil moisture sensors.

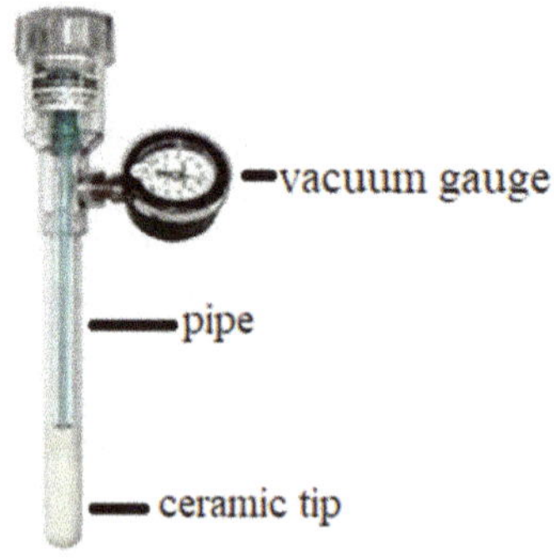

Figure 4.3. A Tensiometer sensor.

moisture content of the soil directly. Instead by using pressure gauge sensors they measure the moisture content. The response time of these sensors is usually very high: about 1 to 2 h.

Advantages:
- Low-cost and user friendly.
- Suitable for sandy and low texture soil.

Disadvantages:
- Regular maintenance is essential to remove air bubbles.
- It may get damaged due to fluctuating temperature.

Granular matrix sensor: These sensors (figure 4.4) are fabricated from a porous ceramic material having two electrodes in it. The two electrodes are made up of a gypsum wafer and granular matrix. When this sensor is inserted into the soil, the granular matrix absorbs water from the soil. As the water content of the granular matrix changes with the moisture content of the soil, the variations are immediately recorded by the electrodes. The difference in the measurement of the electrical resistance between the two electrodes is also recorded. The variation in the electrical resistance is proportional to the moisture content in the soil [50].

Advantages:
- Very low cost and maintenance is not required.
- As the temperature changes, there is very negligible change in the efficiency of sensor.

Disadvantages:
- The efficiency of the sensor changes from soil to soil.
- Less accuracy in sand and low texture soil because of air gaps.

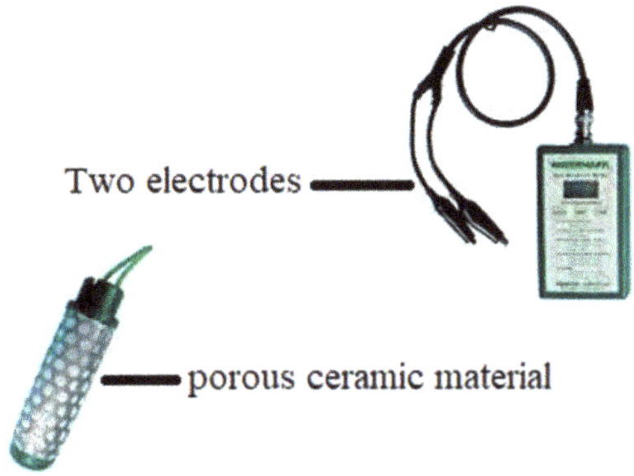

Figure 4.4. A granular matrix sensor.

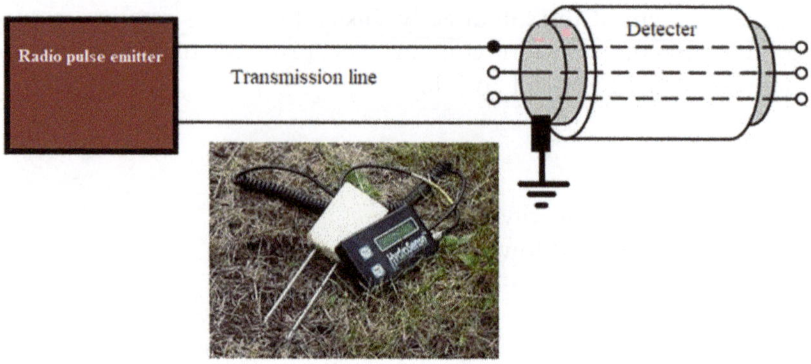

Figure 4.5. A time domain reflectometry sensor.

2. **Volumetric sensors**:

Time domain reflectometry (TDR): In these types of sensors (figure 4.5), a radio pulse is emitted from the source to the transmission line, the velocity of the reflected beam is then measured from the detector. The time period of the reflected pulse is directly proportional to the water content in the soil. These types of sensors are very quick, i.e., about 25 s [51].

Advantages:
- They respond quickly compared to tensiometric sensors.
- There is no variation in the performance of the sensor from soil to soil.

Disadvantages:
- These sensors are costlier.
- It only measures the soil moisture in the locality of the sensor.

Frequency domain reflectometry (FDR): these sensors (figure 4.6) work in a similar way to TDR sensors, the change in the capacitance is directly proportional to the volumetric content of water in the soil [52].

Advantages:
- The measurement accuracy is ±1%.
- It can be used in any type of soil; the performance is same for all types of soil.

Disadvantages:
- It is very expensive compared to a TDR sensor.
- The air gap formation can hinder the efficiency of the sensor.

4.8 Deliberate qualities of soil moisture sensor and review of graphene-based soil moisture sensors

The sensing parameters such as sensing response, LOD, linearity, hysteresis, response and recovery timing, and stability play a very important role in judging

Figure 4.6. A frequency domain reflectometry sensor.

the efficiency, reliability, validity, and potentiality of graphene-based soil moisture sensors.

The noble qualities to judge the potentiality of the sensor are;

Sensing response: The resistance/capacitance values measured over the working gravimetric moisture range are used to calculate the sensing response of the soil moisture sensor.

Limit of detection (LOD): This is calculated by dividing the standard deviation of the regression line by sensitivity.

Linearity (R^2): Linearity is calculated by logarithmic resistance/capacitance v/s gravimetric water content plot in the working relative gravimetric moisture range. High linearity indicates that the sensor can be operated in soils with low as well as high water content and so is a good sensor [53].

Timing behavior: To determine the timing behavior two glass jars, one with low gravimetric water content and the other with high gravimetric water content, were used. The graphene NPs sensing film was quickly transferred from low to high water content and vice-versa (a switching time of about 1 s in both the cases) and the corresponding $T_{response}$ and $T_{recovery}$ were noted [54].

Hysteresis: The difference between the humidification (adsorption) and desiccation (desorption) processes.

Stability: For the stability measurement, the sensing response of the graphene material is usually measured over a period of time with inspection of the sensing response at regular intervals [55].

Graphene-based soil moisture sensors review:

Today many more graphene-based soil moisture sensors are fabricated by the scientific community. Here are some reviews of graphene-based soil moisture sensors (table 4.3).

Hemen Kalita *et al* [56], synthesized graphene quantum dots (GQDs) using the electrochemical method with $LiClO_4$ in propylene carbonate as the electrolyte at room temperature. The size of these GQDs can be varied by applying various potentials,

Table 4.3. The graphene-based soil moisture sensors.

Sl. No.	Sample	Type of soil	Timing behavior	Stability	Reference
1	Graphene quantum dots	Red soil, black cotton soil	180 s	1 Month	[56]
2	Graphene oxide (GO)	Red soil, black cotton soil	100–120 s	4 Months	[57]
3	Graphene quantum dots	White clay, Bentonite	120–180 s	4 Months	[58]

varying oxidation, and reduction time. In their paper, GQD microsensors were fabricated to study the soil moisture sensor. The red soil (silt loam) and black soil (clayey) were tested, the sensor showed good sensing responses of 99% and 97% respectively. The sensor has shown good response times of 180 s for both red and block soil.

Vinay S Palaparthy *et al* [57], fabricated a robust graphene oxide for soil moisture sensor. They characterized the graphene oxide from AFM, RAMAN, and XPS techniques. The sensing response of 340% and 370% was observed for the 1%–55% of soil moisture for red and block soil respectively. The prepared GO sensor shows a good response time of 100–120 s to detect the soil moisture measurement. In every soil moisture sensor, variations in temperature and salt concentration decreases the efficiency of the sensor. Hence, they observed the sensing performance in diurnal temperature and salt concentration. The GO sensor showed a 6% change in the sensing performance when the temperature varied from 25 °C to 65 °C and the sensor showed a 4% variation when salt concentration changed from 0 mol to 0.35 mol. They also discussed the sensing mechanism for soil moisture sensor.

Maryam Shojaei Baghini *et al* [58], fabricated GQDs. By using MEMS fabrication process the inter-digitated electrodes were drawn. The soil moisture sensing was done for two different soils namely, white clay (passive soil) and bentonite clay (an active soil). For the white soil the conductivity changes from 0.06×10^{-6} to $0.68\times10^{-6}\ \Omega^{-1}$ for the gravimetric change in the water content from 4% to 45%. For the bentonite soil the conductivity changes from 0.06×10^{-6} to $0.47\times10^{-6}\ \Omega^{-1}$ over the gravimetric moisture range from 11% to 90%. The response time for the prepared GQDs is about 2–3 min. They also discussed the chain reaction attributed to Grotthuss chain reaction.

Artificial soil moisture setup

To test the potential of the soil moisture sensor, it is essential to record its values for lower water content and higher water content in the soil [59]. To prepare the soil with different water content the following steps are to be followed: the soil collected from the agriculture farm is allowed to dry at 110 °C for 1 day. After the complete removal of water from the soil eight jars have 10 g of the soil is added to each of them and then from 4 mL to 12 mL of water is added to the jars, and left for 48 h. Then through gravimetric analysis, the % water content in the eight jars is measured at room temperature and room humidity. This type of measurement can be done for

Table 4.4. Soil moisture content in different soils.

Soil texture	% Soil moisture content
Sand	7
Loamy soil	12
Loam	20
Clay loam	27
Silty clay loam	28
Silt clay	35

different types of soil. The gravimetric water content in the soil can be measured using equation (4.1).

$$\% = \text{Gravimetric water content} = \frac{\text{weight of the soil with water} - \text{dryweight of the soil}}{\text{dryweight of the soil}} \tag{4.1}$$

The % water content in the soils is listed in table 4.4.

4.9 Soil moisture mechanism for graphene-based soil moisture sensors

The real picture of the mechanisms for soil moisture sensing can be described in three consecutive steps:

1. chemisorptions,
2. first and second steps of physisorption, and
3. condensation.

The complete mechanism for sensing is illustrated in figure 4.7.

Step I: Chemisorption
At the beginning of the process, water molecule diffuse into H^+ (proton) and OH^- (hydroxide ion) due to the self-ionization process [60]. In the initial stage, water vapor is chemically adsorbed onto the active sites of the sensing material at lower water content. The formerly ionized OH^- ions is adsorbed on cations of the active surface which are of higher charge density and firm electric field, thus a primary layer is formed, known as the chemisorbed layer.

$$H_2O \Leftrightarrow H^+ + OH^- \tag{4.2}$$

Step II: Physisorption
The left-out H^+ ions which conduct through the clogging channel leads to a slight increase in conductivity. The further increase in water content tends to form the immobile physisorbed layer, this layer contains two hydroxyl groups forming two-hydrogen bonding with one water molecule [61]. When moisture is very high, many water molecules pile up on one another forming many physiosorbed layers on the

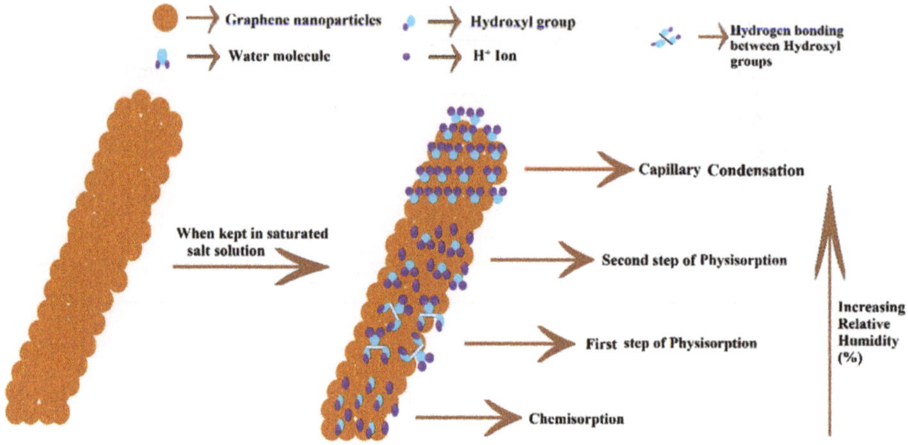

Figure 4.7. A humidity sensing mechanism.

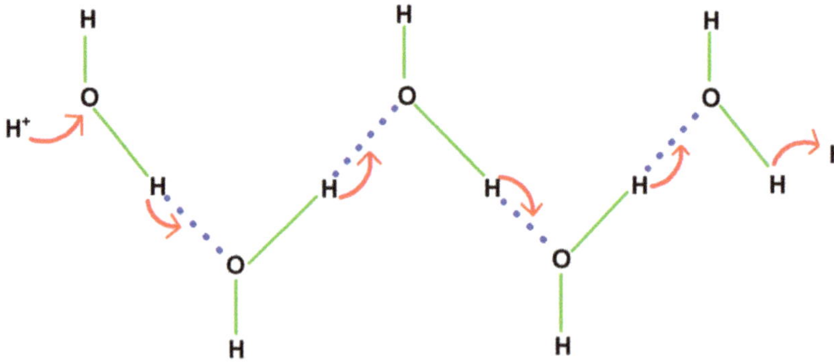

Figure 4.8. Conventional representation of Grotthus mechanism.

first physiosorbed layer. The water molecules have single bonded with one hydroxyl group forming bulk water (H_3O^+). The so formed bulk water is very unstable so it dissociates to form H_2O and H^+ ions. Thus, the formed protons hop from one water molecule to another in accordance with Grotthus chain reaction [62]. This mechanism is illustrated in figure 4.8.

$$2H_2O \rightarrow H_3O^+ + OH^- \tag{4.3}$$

$$H_3O^+ \rightarrow H_2O + H^+ \tag{4.4}$$

Step III: Condensation

The surplus protons disseminate through the hydrogen bonding of the water molecules. These protons conduct through neighboring water molecules by breaking and making bonds. Hence these protons act as a dominant charge carrier in rich

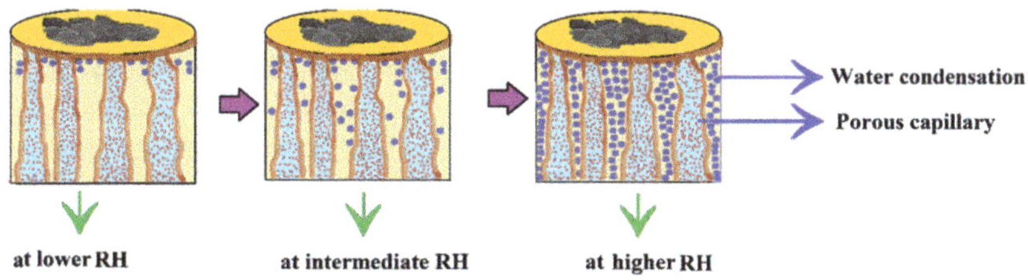

Figure 4.9. The water adsorption process on the sensing surface.

water content environments. In very high moisture conditions, the water molecules condense in a capillary pore and this ensures the decrease in the surface resistance of the sample [63, 64]. The water adsorption process on the sensing surface is depicted in figure 4.9.

4.10 Conclusion

The noble metal of the 20th century is graphene, the transition of the graphene into different forms has enhanced the dynamicity of its properties and found widespread applications. There are many groups in India that are working on graphene-based soil moisture sensors. Achieving the greatest sensitivity, dynamic response and recovery time, best stability with eco-friendly material and low-cost fabrication method, for soil moisture sensing is still a challenge. But the research community is continuously, consistently, and chiefly working on these parameters to fabricate an efficient, reliable, versatile, cheaper graphene-based sensing device for a soil moisture sensor. This chapter throws light on graphene and the application of graphene to soil moisture sensing.

Acknowledgements

The authors acknowledge Department of Physics, Indian Institute of Science, Bengaluru, Karnataka, India for their support and encouragement in carrying out research development activities.

References

[1] Manjunatha J G 2020 A surfactant enhanced graphene paste electrode as an effective electrochemical sensor for the sensitive and simultaneous determination of catechol and resorcinol *Chem. Data Collect.* **25** 100331

[2] Gleiter H 2000 Nanostructured materials: basic concepts and microstructure *Acta Mater.* **48** 1–29

[3] Pushpanjali P A, Manjunatha J G, Nagarajappa H, D'Souza E S, Charithra M M and Prinith N S 2021 Voltammetric analysis of antihistamine drug cetirizine and paracetamol at poly (L-Leucine) layered carbon nanotube paste electrode *Surf. Interfaces* **24** 101154

[4] Hareesha N, Manjunatha J G, Amrutha B M, Sreeharsha N, Basheeruddin A, Md and Khalid A 2021 A fast and selective electrochemical detection of vanillin in food samples on

the surface of poly (glutamic acid) functionalized multiwalled carbon nanotubes and graphite composite paste sensor *Colloids Surf. A: Physicochem. Eng. Asp.* **626** 127042

[5] Manjunatha J G, Deraman M, Nur Hamizah B and Talib I A 2014 Selective detection of dopamine in the presence of uric acid using polymerized phthalo blue film modified carbon paste electrode *Adv. Mater. Res.* **895** 447–51

[6] Hareesha N and Manjunatha J G 2021 Electro-oxidation of formoterol fumarate on the surface of novel poly (thiazole yellow-G) layered multi-walled carbon nanotube paste electrode *Sci. Rep.* **11** 12797

[7] Chethan B, RajPrakash H G, Ravikiran Y T, Vijayakumari S C and Thomas S 2019 Polypyyrole based core–shell structured composite based humidity sensor operable at room temperature *Sens. Actuators B Chem.* **296** 126639

[8] Tigari G and Manjunatha J G 2019 Electrochemical preparation of poly(arginine)-modified carbon nanotube paste electrode and its application for the determination of pyridoxine in the presence of riboflavin: an electroanalytical approach *J. Anal. Test.* **3** 331

[9] Hareesha N, Manjunatha J G, Pushpanjali P A, Prinith Subbaiah N, Charithra M M, Sreeharsha N, Basheeruddin Asdaq S M and Md Khalid A 2021 Electrochemical sensing of antibiotic drug amoxicillin in the presence of dopamine at simple and selective carbon paste electrode activated with cetyltrimethylammonium bromide surfactant *Monatsh. Chem.* **153** 31–8

[10] Pal S L, Jana U, Manna P K, Mohanta G P and Manavalan R 2011 Nanoparticle: An overview of preparation and characterization *J. Appl. Pharm. Sci.* **06** 228–34

[11] Amrutha B M, Manjunatha J G, Aarti S B and Nagarajappa H 2021 Sensitive and selective electrochemical detection of vanillin at graphene-based poly (methyl orange) modified electrode *J. Sci.: Adv. Mater. Devices* **6** 415–24

[12] Ramesh S 2013 Sol–gel synthesis and characterization of nanoparticles *J. Nanosci.* **2013** 929321

[13] Mann S, Burkett S L, Davis S A, Fowler C E, Mendelson N H, Sims S D, Walsh D and Whilton N T 1997 Sol–gel synthesis of organized matter *Chem. Mater.* **9** 2300–10

[14] Tai C Y, Tai C, Chang M and Liu H 2007 Synthesis of magnesium hydroxide and oxide nanoparticles using a spinning disk reactor *Ind. Eng. Chem. Res.* **46** 5536–41

[15] Mohammadi S, Harvey A and Boodhoo K V K 2014 Synthesis of TiO_2 nanoparticles in a spinning disc reactor *Chem. Eng. J.* **258** 171–84

[16] Bhaviripudi S, Mile E, Iii S A S, Zare A T, Dresselhaus M S, Belcher A M and Kong J 2007 CVD synthesis of single-walled carbon nanotubes from gold nanoparticle catalysts *J. Am. Chem. Soc.* **1296** 1516–7

[17] Adachi M, Tsukui S and Okuyama K 2003 Nanoparticle synthesis by ionizing source gas in chemical vapor deposition nanoparticle synthesis by ionizing source gas in chemical vapor deposition *J. Appl. Phys.* **77** 4–7

[18] Kammler B H K, Mädler L and Pratsinis S 2001 Flame synthesis of nanoparticles *Chem. Eng. Technol.* **24** 583–96

[19] Amato R D, Falconieri M, Gagliardi S, Popovici E, Serra E, Terranova G and Borsella E 2001 Synthesis of ceramic nanoparticles by laser pyrolysis: from research to applications *J. Anal. Appl. Pyrolysis* **104** 461–9

[20] Kuppusamy P, Yusoff M M and Govindan N 2014 Biosynthesis of metallic nanoparticles using plant derivatives and their new avenues in pharmacological applications - an updated report *Saudi Pharm. J.* **24** 473–84

[21] Yadav T P, Yadav R M and Singh D P 2012 Mechanical milling: a top down approach for the synthesis of nanomaterials and nanocomposites *Nanosci. Nanotechnol.* **2** 22–48
[22] Pimpin A and Srituravanich W 2011 Review on micro- and nanolithography techniques and their applications *Mod. Eng. Technol.* **16** 37–55
[23] Hulteen J C, Treichel D A, Smith M T, Duval M L, Jensen T R and Duyne R P 1999 Nanosphere lithography: size-tunable silver nanoparticle and surface cluster arrays *J. Phys. Chem.* B **103** 3854–63
[24] Amendol V and Meneghetti M 2009 Laser ablation synthesis in solution and size manipulation of noble metal nanoparticles *Phys. Chem. Chem. Phys.* **11** 3805–21
[25] Shah P and Gavrin A A 2006 Synthesis of nanoparticles using high-pressure sputtering for magnetic domain imaging *J. Magn. Magn. Mater.* **301** 118–23
[26] Lugscheider E, Bärwulf S, Barimani C, Riester M and Hilgers H 1998 Magnetron-sputtered hard material coatings on thermoplastic polymers for clean room applications *Surf. Coatings Technol.* **108** 398–402
[27] Salavati-niasari M, Davar F and Min F N 2008 Synthesis and characterization of metallic copper nanoparticles via thermal decomposition *Polyhedron* **27** 3514–8
[28] Ankamwar B and Surti F 2012 Water soluble graphene synthesis *Chem. Sci. Trans.* **1** 500–07
[29] Zhang D, Wang D, Li P, Zhou X, Zong X and Dong G 2018 Facile fabrication of high-performance QCM humidity sensor based on layer-by-layer self-assembled polyaniline/graphene oxide nanocomposite film *Sens. Actuators, B Chem.* **255** 1869–877
[30] Ibrahim K S 2013 Carbon nanotubes-properties and applications: a review *Carbon Lett.* **14** 131–44
[31] Aqel A, El-Nour K M M A, Ammar R A A and Warthan A A 2013 Carbon nanotubes, science and technology part (I) structure, synthesis and characterisation *Arab. J. Chem.* **5** 1–23
[32] Ealias A M and Saravanakumar M P 2017 A review on the classification, characterisation, synthesis of nanoparticles and their application *IOP Conf. Ser.: Mater. Sci. Eng.* **263** 032019
[33] Prinith N S, Manjunatha J G and Nagarajappa H 2021 Electrochemical validation of L-tyrosine with dopamine using composite surfactant modified carbon nanotube electrode *J. Iran. Chem. Soc.* **18** 3493–503
[34] Huang X, Boey F and Zhang H U A 2010 Graphene–nanoparticle composites *COSMOS* **6** 159–66
[35] De M F L, Volder, Tawfick S H, Baughman R H and Hart A J 2019 Carbon nanotubes: present and future commercial applications *Science* **339** 535–39
[36] Tenne R 2002 Fullerene-like materials and nanotubes from inorganic compounds with a layered (2-D) structure *Colloids Surf.,* A **208** 83–92
[37] Zhang L, Aboagye A, Kelkar A, Lai C and Fong H 2014 A review: carbon nanofibers from electrospun polyacrylonitrile and their applications *J. Mater. Sci.* **49** 463–80
[38] Fawole O G, Cai X and Mackenzie A R 2016 Gas flaring and resultant air pollution: a review focusing on black *Environ. Pollut.* **216** 182–97
[39] Marcano D C, Kosynkin D V, Berlin J M, Sinitskii A, Sun Z and Slesarev A 2010 Improved synthesis of graphene oxide *ACS Nano* **22** 4806–814
[40] Kim Y J, Kim Y, Novoselov K and Hong B H 2015 Engineering electrical properties of graphene: chemical approaches *2D Mater.* **2** 042001
[41] You Y, Zeng W, Yin Y X, Zhang J, Yang C P and Zhu Y 2015 Hierarchically micro/mesoporous activated graphene with a large surface area for high sulfur loading in Li–S batteries *J. Mater. Chem.* A **3** 4799–802

[42] Bonaccorso F, Sun Z, Hasan T and Ferrari A 2010 Graphene photonics and optoelectronics *Nat. Photon.* **4** 611–22

[43] Raril C and Manjunatha J G 2018 Sensitive electrochemical analysis of resorcinol using polymer modified carbon paste electrode: a cyclic voltammetric study *Anal. Bioanal. Electrochem.* **10** 488–98

[44] Muhlen A Z, Muhlen E Z, Niehus H and Mehnert W 1996 Atomic force microscopy studies of solid lipid nanoparticles *Pharm. Res.* **13** 1411–6

[45] Shi H G, Farber L, Michaels J N, Dickey A, Thompson K C, Shelukar S D, Hurter P N, Reynolds S D and Kaufman M J 2003 Characterization of crystalline drug nanoparticles using atomic force microscopy and complementary techniques *Pharm. Res.* **20** 479–84

[46] Sharm V and Rao L J M 2014 An overview on chemical composition, bioactivity and processing of leaves of *Cinnamomum tamala Crit. Rev. Food Sci. Nutr.* **54** 433–48

[47] Hodoroaba V, Rades S and Unger W E S 2014 Inspection of morphology and elemental imaging of single nanoparticles by high- resolution SEM/EDX in transmission mode *Surf. Interface Anal.* **46** 945–48

[48] Yano F, Hiraoka A, Itoga T, Kojima H, Kanehori K and Mitsui Y 1996 Influence of ion implantation on native oxidation of Si in a clean-room atmosphere *Appl. Surf. Sci.* **100-1** 138–42

[49] Francesca V, Osvaldo F, Stefano P and Paolo P R 2010 Soil moisture measurements: Comparison of instrument performances *J. Irrig. Drain. Eng.* **136** 81

[50] Berrada A, Hooten M T, Cardon E G and Broner I 2001 Assessment of irrigation water management and demonstration of irrigation scheduling tools in the full-service area of the Dolores project: 1996–2000 *Technical Report, TR01–7* (Colorado State University) **7** 9

[51] Cepuder P, Evett S, Heng L K, Hignett C, Laurent J P and Ruelle P 2008 Field estimation of soil water content. Training Course Series No. 30 of International Atomic Energy Agency

[52] Susha Lekshmi S U, Singh D N and Shojaei B M 2014 A critical review of soil moisture measurement *Measurement* **54** 92–105

[53] Chethan B, Raj Prakash H G, Ravikiran Y T, Vijayakumari S C, Ramana C H V V, Thomas S and Kim D 2019 Enhancing humidity sensing performance of polyaniline/water soluble graphene oxide composite *Talanta* **196** 337–44

[54] Chethan B, Ravikiran Y T, Vijayakumari S C, Rajprakash H G and Thomas S 2018 Nickel substituted cadmium ferrite as room temperature operable humidity sensor *Sens. Actuators A Phy.* **280** 466–74

[55] Manjunatha S, Machappa T, Ravikiran Y T, Chethan B and Sunilkumar A 2019 Polyaniline based stable humidity sensor operable at room temperature *Phys. B Condens.* **561** 170–78

[56] Kalita H, Palaparthy V S, Baghini M S and Aslam M 2020 Electrochemical synthesis of graphene quantum dots from graphene oxide at room temperature and its soil moisture sensing properties *Carbon* **165** 17

[57] Palaparthy V S, Kalita H, Surya S G, Baghini M S and Aslam M 2018 Graphene oxide-based soil moisture microsensor for *in situ* agriculture applciations *Sens. Actuators: B. Chem.* **273** 1660–669

[58] Kalit H, Palaparthy V S, Baghini M S and Aslam M 2016 Graphene quantum dot soil moisture sensor *Sens. Actuators* B **233** 582–90

[59] Surya S G, Yuvaraja S, Varrla E, Baghini M S, Palaparthy V S and Salama K N 2020 An in-field integrated capacitive sensor for rapid detection and quantification of soil moisture *Sens. Actuators: B. Chem.* **321** 128542

[60] Zhao J, Liu Y, Li X, Lu G, You L, Liang X, Liu F, Zhang T and Du Y 2013 highly sensitive humidity sensor based on high surface area mesoporous $LaFeO_3$ prepared by a nanocasting route *Sens. Actuators, B Chem.* **181** 802–9

[61] Chen Q, Nie M and Guo Y 2018 Controlled synthesis and humidity sensing properties of CdS/polyaniline composite based on CdAl layered double hydroxide *Sens. Actuators, B Chem.* **254** 30–5

[62] Jeseentharani V, Reginamary L, Jeyaraj B, Dayalan A and Nagaraja K S 2012 Nanocrystalline spinel $Ni_xCu_{0.8-x}Zn_{0.2}Fe_2O_4$: a novel material for humidity sensing *J. Mater. Sci.* **47** 3529–34

[63] Chethan B, Raj Prakash H G, Ravikiran Y T, Vijayakumari S C and Thomas S 2020 Polypyrrole based core–shell structured composite-based humidity sensor operable at room temperature *Talanta* **215** 120906

[64] Chethan B, Megha R, Raj Prakash H G, Ravikiran Y T, Vijayakumari S C and Thomas S 2021 Nanostructures for humidity sensing and photocatalytic applications *Nanoscale Processing* (Amsterdam: Elsevier) pp 327–52

Chapter 5

Carbon composite material as a sensor for pharmaceutical sample analysis

Çiğdem Kanbeş Dindar, Fatma Ağın and Bengi Uslu

Pharmaceutical products have become an indispensable part of our daily lives and are in everyday use. In general, qualitative and quantitative analysis of pharmaceutical preparations used to prevent or treat and cure disease is critical to the quality of human health. In this regard, electrochemical sensors are increasingly being used in pharmaceutical drug analysis due to their numerous benefits, including ease and speed of application and low cost. However, to monitor drug efficacy and potential side effects in complex matrix environments, it is necessary to develop selective, sensitive, and reliable analytical techniques in electrochemical pharmaceutical drug analysis. For this reason, carbon-based materials are the most used materials for surface modification in the enhancement of electrochemical sensor development studies in recent years due to their unique electrochemical properties, good biocompatibility, low cost, and easy synthesis. Moreover, carbon composite structures outperform single-component carbon-based sensors in terms of electrochemical performance due to new functionalities gained through synergistic interactions. Thus, this chapter focuses on the importance of carbon composites in electrochemical sensor studies for pharmaceutical drug analysis. In particular, the composition and structure of carbon composites used in drug analysis are defined and discussed. Finally, the applications of carbon composite modified electrodes in drug analysis are presented to show the advantages, challenges, and future prospective of modification in electrochemical sensor studies.

5.1 Introduction

Pharmaceuticals are finished dosage forms containing an active drug molecule or drug molecules, usually formulated with one or more excipients, used for the prevention, diagnosis, treatment, improvement, or modification of a function in humans. Pharmaceuticals are classified according to their therapeutic effects such as

doi:10.1088/978-0-7503-5127-0ch5

analgesics, antibiotics, anti-inflammatories, antidepressants, antihypertensive, etc [1]. When applied to the body, they affect physiological function by interacting and binding with certain proteins and cells, altering the biological structure [2]. Drugs analysis are performed in formulation, stability studies, toxicology, quality control, pharmacological testing and determining the dose in the treatment in animals and humans. These studies require reliable and validated analytical methods to quantify drugs in complex environments such as formulation and biological samples. A variety of analytical methods such as electrochemical [3], spectrophotometry [4], and chromatography [5] have been developed for the regular monitoring of pharmaceuticals.

Electroanalytical methods have advantages over other methods in terms of low cost, high sensitivity, simplicity of the procedure, and test time [6]. In addition, the applications of electroanalytical techniques to pharmaceutical and biomedical analysis have greatly increased in the last decade, as biological reactions and electrochemical reactions at the solution electrode interface follow the same electron transfer pathways [7–9]. Many important enzymatic, physiological, and other biochemical processes operate on the mechanism of the electrochemical oxidation/reduction principle [10, 11].

Sensors are devices that detect changes in a physical stimulus and convert them into an assessable signal that can be recorded or measured [12]. The amount of target compound can be measured by the sensor taking a physical amount and converting it into a signal (e.g., electrical, optical, and mechanical) suitable for processing [13]. The sensor has an active part called a transducer that converts the measurement of physical events into an electrical signal. When it converts a measurable amount into electrical current or voltage, it can be called a sensor. Sensors can be classified according to their output signals [12–14]. Among them, electrochemical sensors are more advantageous than others as the electrodes in this system can detect the materials which are present within the host without damaging the host system [15].

Electrochemical sensors enable the collection of chemical information, such as proton release or electron transfer, and convert it into an analytically useful signal, such as current or potential variation [16]. Electrochemical sensors are considered to be sensitive sensing tools due to their fast response, simple design, negligible sample preparation steps, high sensitivity, and selectivity in complex matrices [17–19]. Research on electrochemical sensors has increased notably because these devices have features such as information obtaining, low environmental impact, and surface modification of the sensor with chemically active species [16–20]. Researchers have been investigating the innovative and novel materials for developing sensors and in this way developing the electrochemical performance of the sensor in terms of detection limits and sensitivity [17].

Electrode modification is a method carried out using various methods (physical, chemical, etc) and materials, for example, nanomaterials, polymers, and biological materials to increase sensitivity, stability, selectivity, and the response of a sensor [21–24]. The electrode modification ensures the detection of an analyte in complex matrices with a very low limit of detection values [21, 25, 26]. Recently, carbon-based materials such as graphene, nanotube, carbon fiber, and fullerene, etc play an important role in electrode modification [2, 27]. The composite materials contain a combination of two or more materials in which the components exhibit their unique characteristics with their unique chemical, physical, and mechanical properties [28]. The carbon-based composite materials are often used as modification agents for electrodes because of their high electrical conductivity, stability, and high surface area. These materials can improve sensitivity and thus decrease the detection limits of drug analysis [21].

In the following section, the studies of sensors prepared with carbon composite materials for the analysis of pharmaceuticals are explained detail.

5.2 Fabrication of carbon composite sensor for electrochemical drug analysis

There are many preparation methods used throughout the design of electrochemical sensors to modify with carbon composite materials in drug analysis, but in this section, electrochemical deposition, drop casting, and dip-coating methods, which are the most used, were explained by giving examples from the literature. Also, some procedures of modified electrodes production have a combination of these methods. Briefly, the most generally utilized methods for fabricating carbon composite-based electrodes for drug analysis are given in this section.

5.2.1 Electrochemical deposition

Electrodeposition is an electrochemical method that has been established to deposit conductive (nano)materials, such as carbon composite materials, metals, and conducting polymers on the working electrode by controlling the given electrochemical environment within the system, such as current or voltage between the electrode system, and the electrolyte being used. The electrochemical deposition process provides fast and easy preparation of a stable and sensitive electrode surface [29]. Also, by using the electrochemical deposition method, a different structure of carbon composite materials can be obtained on the electrode surface.

For illustration, Habibi *et al* propounded an electrochemical sensor for the detection of sertraline hydrochloride as an antidepressant drug by electrochemically modifying a copper-based nitrogen- and sulfur-doped graphene (SNDGr) and metal–organic framework (Cu-MOF) nanocomposite on a pencil graphite electrode (PGE) using a layer-by-layer deposition method. In the first step of electrochemical coating, to begin the preparation of SNDGr/PGE, 0.1 M $(NH_4)_2SO_4$ and 6×10^{-3} M melamine solution was submerged in Pt rod as the auxiliary electrode, and PGE as the working electrode. Following that, PGE was electrochemically treated in an electrolyte solution for 60 s using a constant potential electrolysis treatment of 10 V, and SNDGr was deposited as a thin layer on the electrode surface. In the presence of $(NH_4)_2SO_4$ and melamine mixture solution, the graphite portion of PGE separated between the layers of the PGE structure by the penetration of SO_4^{2-}, NH^{2-} ions, and solvent molecules, leading to the formation of Gr layers. Furthermore, the microstructural stresses began to rise so dramatically as a result of the intervening ions that the fracture Gr plate exhibited a new edge design. Then, in second step of electrochemical deposition, 0.577 g $Cu(NO_3)_2{\cdot}3H_2O$ dissolved in 5 mL DMF as Cu(II) ion source and 0.12 g triethylamine hydrochloride solution and 0.0024 g disodium 5,50-bitetrazole-1-ide solution dissolved in 5 mL DMF were mixed vigorously for 2 h. After sonicating the final solution for 5 min, the working electrode SNDGr/PGE was submerged in it, and Cu-MOF electrodeposition on SNDGr/PGE was completed by providing an adequate cathodic potential (−1.4 V) for 360 s (chronoamperometry). The prepared sensor has been successfully used with appropriate recovery values for the determination of sertraline hydrochloride in real samples [30].

In another study, Kalambate *et al* fabricated coated with nafion (NAF) graphene (Gr)–platinum nanoparticles (PtNP) on glassy carbon electrode (GCE) mainly by electrochemical deposition. First, the GCEs were polished with alumina powder (1.0 μm and 0.3 μm, α-Al_2O_3). The produced graphene oxide (GrO) was ultrasonically exfoliated in water for 2 h to generate homogeneous GrO suspensions with a 10.0 mg L^{-1} GCE was immersed in GrO solution and coated as Gr on the electrode surface with five CV scans potential range between +0.6 V to −1.5 V. PtNPs on Gr-modified GCE by electrochemical reduction, they were immersed in a 10 mM solution of H_2PtCl_6 (in 0.1 M NaCl) and were stirred magnetically by generating an electroaccumulation potential of −0.7 V for 40 s. Finally, NAF modification was applied to the surface of GCE by drop-casting (5.0 μL, 0.1%) and dried at room temperature [31].

5.2.2 Drop-casting preparation

Another method commonly used to prepare modified electrodes by immobilization of composite materials on the electrode surface is the drop-casting method [20]. This technique can also be combined with other electrode preparation procedures. For example, Selvi *et al* applied hydrothermal aided synthesis of Mn_2O_3 combined with reduced GrO (rGrO) to create a simple azathioprine sensor. The screen-printed electrode (SPCE) was first cleaned with distilled water (DI) water to remove dust particles that could interfere with the measurements on the electrode material. The optimized concentration of 3 mg Mn_2O_3/rGrO dispersed in DI water was sonicated for 30 min Afterward, 6 μL of the modification solution was dripped on the SPCE and dried. The sensor had a lifetime of around 10 days with a signal loss of no more than 2% [32]. Furthermore, Shahrokhian *et al* developed a novel clonazepam electrochemical sensor based on the immobilization of a nanocomposite of cellulose nanofibers/carbon nanoparticles (CNFs/CNPs) on a GCE. To begin, 2.0 mg of CNPs in a 0.1% solution of CNFs in 2 mL of dimethylformamide (DMF) was sonicated for 30 min to form a suspension of CNFs/CNPs. The modified GCE was prepared by dropping the optimum value of 1 μL of the modifier droplet onto the GCE surface with a micropipette and water was eliminated in a 50 °C oven. CFPs/CNPs/GCE were found to be superior to bare GCE in terms of lower detection limit and increased sensitivity, according to the analytical findings using the modified electrode created by this easy electrode fabrication approach. Furthermore, the relative standard deviation of the CV peak currents for six duplicate tests was 2.52%, demonstrating that the modified electrode's response was exceptionally repeatable [33].

5.2.3 Dip-coating preparation

The electrode surface is modified after the typical polishing methods, generally by immersion in a homogeneously dispersed solution of nanocomposite molecules [29]. For example, Rouhani *et al* designed electrode modification with Preyssler nanocapsules (PNCs) and rGrO by the dip-coating method for sensitive detection of ceftizoxime. In all phases of designing the PNCs/rGrO/PGE sensor depicted in figure 5.1, 1.75 mg mL^{-1} of rGrO solution was initially prepared in DMF and sonicated for 30 min to achieve a homogenous dispersion. After this process, the treated PGE was dipped vertically into the abovementioned suspension to form the rGrO layer on the PGE surface, stirred with a mechanical stirrer for 45 min, and allowed to dry at room temperature. Thus, a polymer was formed with synthesized rGrOs to improve reachable sites and electron transport on the PGE surface as the supporting material. To generate the PNC thin film, 1.0 mg L^{-1} mL PNCs solution was prepared in 2.0 mL of chloroform, and the solution was ultrasonically treated before coating. The dip-coating method (1 mm min^{-1}) was used to coat the solution produced in the previous phase of the coating process. As a result, a thin PNC coating was created on the rGrO-modified electrode. Finally, the electrode was dried at room temperature overnight to avoid surface cracking. The built sensor (PNCs/rGrO/PGE) was successfully used in application analyses [34].

Moreover, the carbon composite nanosensor procedure developed including both dip-coating and drop-casting electrode preparation methods in another study. Mao *et al* developed a new electrochemical sensor for metronidazole detection based on multiwalled carbon nanotubes (MWCNTs) and chitosan-nickel complex (CTS). First, the GCE surface was modified by drop-drying 5 μL of a suspension of MWCNTs (2 mg mL^{-1}). Second, the MWCNTs/GCE was immersed in the chitosan solution for 30 min, and the carboxyl group functionalized negatively charged MWCNTs self-assembled with the positively charged protonated chitosan by

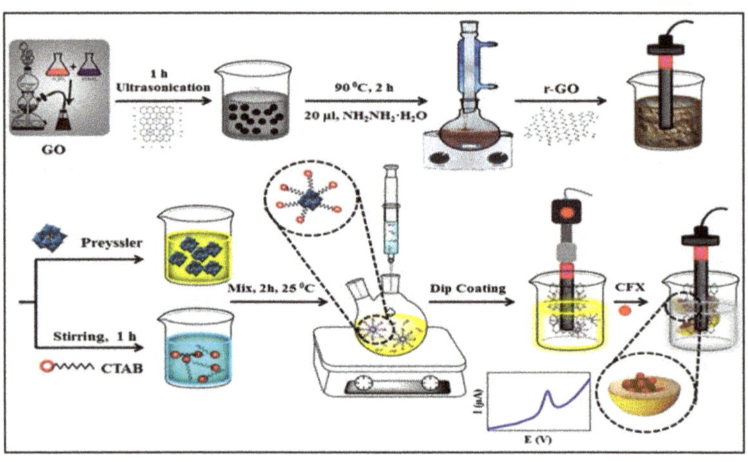

Figure 5.1. Design of Preyssler nanocapsules decorated rGO for detection of Preyssler nanocapsules. Reprinted from [34], copyright 2021, with permission from Elsevier.

electrostatic interactions. Finally, MWCNTs/CTS/GCE were immersed in nickel solution for 30 min and nickel ions formed complexes on chitosan. The prepared sensor has effectively detected metronidazole in real samples [35].

5.3 Application of carbon composite material as a sensor for pharmaceutical sample analysis

In the application of carbon composite material as a sensor for pharmaceutical sample analysis, studies conducted in the last decade using different carbon-based electrodes such as GCE, carbon paste electrode (CPE), SPCE, and PGE have been discussed.

5.3.1 Selected applications of carbon composite with GCE

Glassy carbon, which is the most widely used in electrochemical applications, was defined as a crystallite growth on non-graphitized carbons after examining the chemical and physical properties of graphitizing and non-graphitizing carbons, and investigating the structure of carbons of different origins processed at very high temperatures. Its structure is created by polymerizing a resin using foaming agents and then carbonizing it [3]. Structurally, glassy carbon is carbon in an sp^2 hybrid structure consisting of randomly arranged graphite planes in a complex topology. Its amorphous properties, isotropy, and possible lack of homogeneity are designated by the formation of the structure [7]. Although glassy carbon is used as the electrode material because of its high electrical and mechanical characteristics, extreme chemical inertness, wide potential range, strong resistance to reactive chemicals, impermeability to gases, reasonably reproducible performance, and reusability, analytical performance parameters in pharmaceutical analysis need to be improved [36]. In this context, carbon composite structures come to the fore with many alternatives in order to develop important analytical sensor performance parameters such as selectivity, sensitivity, and stability in the modification of GCEs. Before being modified, the GCEs are polished with a smooth polishing cloth with slurry alumina particles or diamond paste (0.05–1.0 μm) and then washed with ultrapure water. The surface is then electrochemically prepared using CV, typically between +1.0 and 1.0 V, to generate an active and repeatable GCE surface and increase its analytical performance. Selected studies related to carbon composite on glassy carbon electrode are given in table 5.1.

To illustrate, a novel three dimensional manganese nanostructure based carbon nanotube (CNT_S-MnNPs) was prepared (figure 5.2) for the sensitive determination of ascorbic acid (vitamin C) via voltammetric procedure. CNTs as a conductive skeleton were used to anchor the highly electrolytic MnNPs in the composite electrode preparation. Under optimized conditions, the results of CV showed a linear range from 0.06 to 4.0×10^{-3} M and a detection limit of 0.1 μM. The developed composite electrode was successfully used for the determination ascorbic acid in pharmaceuticals [37].

The electrochemical sensor based on MWCNT/cetyltrimethylammonium bromide (CTAB) surfactant composite modified GCE was prepared for the

Table 5.1. Selected applications of carbon composite on GCE in pharmaceuticals.

Drug	Modifier	Method	Medium	Linear range	LOD	Application	References
L-Tryptophan	CMOF/MWCNT	DPV	PBS, pH 6.5	0.4 μM–19 μM	0.11 μM		[45]
D-Tryptophan					0.16 μM		
Nitrofurantoin	f-CNF/CB	Amperometry	PBS, pH 7.0	0.05–104.66 μM	16 nM		[46]
Dopamine	Mn_2O_3/PEDOT/rGO/	Amperometry	PBS, pH 6.60	0.05–135 μM	0.03 μM	Human serum	[47]
Isoniazid	$SmVO_4$/CNF	DPV	PBS, pH 7.0	0.09–800 μM	0.005 μM	Urine, human plasma samples, tablet	[48]
Propranolol	HgS/Gr	DPV	PBS, pH 7.0	0.5–50.0 μM	0.05 μM	Urine, human plasma samples, tablet	[49]
Flutamide	ZnO/Co_3O_4/g-C_3N_4	DPV	PBS, pH 7.0	0.01–98.6 μM	0.00373 μM	Urine, blood serum	[50]
Hydroquinone	P-L-Cys/$Au_{1.5}Pt_1Co_1$ TANPs	DPV	PBS, pH 6.98	0.1–30, 30–200 μM	0.045 μM	Mineral water, tap water, lake water	[51]
Acetaminophen	CNT/NiSex	DPV	PBS, pH 6.0	1–600 μM	0.55 μM	Tablet, urine	[52]
Dopamine				1–500 μM	0.76 μM	Urine	
Theophylline	TiO_2 MPs/GrO	CV	PBS, pH 6.0	0.02 to 209.6 μM	13.26 nM	Tablet, human serum	[53]
Rutin	rGrO/ZIF-8/MIP	DPV	PBS, pH 3.0	0.05–100 and 0.0005–0.05 μM	0.0001 μM	Tablet, orange juice	[44]
Chloramphenicol	g-C_3N_4/$MnWO_4$	DPV	PBS, pH 7.0	4.0–71 nM	1.03 nM	Milk, human blood human serum, sewage samples	[54]

(*Continued*)

Table 5.1. (*Continued*)

Drug	Modifier	Method	Medium	Linear range	LOD	Application	References
Hydroquinone	CNT/PHQ/CE	DPV	PBS, pH 7.0	0.03–100 μM	0.156 nM	Tap water, lake water	[55]
Methotrexate	Pr_2O_3/PS	DPV	PBS, pH 7.0	0.01–236 μM	0.8 nM	Human serum, urine	[56]
Gatifloxacin	GrO/CuS NP/g-C_3N_4	ELC	PBS, pH 7.5	100–0.01 μM	3.5 nM	Mouse plasma	[57]
L-tyrosine	CuS NS/Cs/FMWCNTs	DPV	PBS, pH 7.0	0.08–1.0 μM	4.9 nM.	Pig serum samples	[58]
Paracetamol	PEDOT/PSSLi/MWCNT	DPV	phosphate–citrate buffer, pH 3.0	1.5–500 μM	0.08 μM	Tablet	[59]
Acetaminophen	AgNPs/PHEMA/CMK-1	DPV	PBS, pH 7.0	0.015–7.5 μM	0.005 μM	Tablet, human urine, serum samples	[60]
Domperidone			PBS, pH 8.0	0.03–10.8 μM	0.01 μM	Tablet, human urine, serum samples	
Tizanidine	Gr/SiO_2	SWV	BRB, pH 6.5	5 nM–10 μM	40.05 nM	Tablet	[61]
Pyrazinamide	MWCNT/GrO	DPV	PBS, pH 7.0	37.5–1800 μM	5.54 μM	Human serum samples, tablet	[62]
Clonazepam	CNF/CNPs	LSV	BRB, pH 2.0	0.1–10 μM	0.08 μM	Human serum samples, tablet	[33]
5-fluorouracil	AuNP/MWCNT	DPV	BRB, pH 8.0	0.03–10 μM	20 nM	Injection, artificial urine samples	[39]
4-aminoantipyrine	MWCNT/CTAB	DPV	PBS, pH 3.0	0.005–0.04 μM	16.3 nM	Human urine, human plasma samples	[38]
Paracetamol	WO_3NPs/CNTs	SWV	PBS, pH 7.0	0.001–0.2 μM	0.0554 nM	Tablet	[63]

Paracetamol	NAF/PtNPs/Gr	AdSSWV	PBS, pH 6.0	160–8.2 μM	0.106 nM	Pharmaceutical formulations, human urine, blood serum samples	[31]
Mefenamic acid	OD/FMWCNT/*p*-ATT	Amperometry	PBS, pH 7.2	40–5000 nM	0.09 nM	Tablet samples	[64]
Paracetamol	NAF/Gr	SWV	PBS, pH 2.5	8.3–0.051 nM	1.2 nM	Tablet samples	[65]
Aspirin				0.87–17 μM	0.65 nM		
Caffeine				0.26–10 μM	0.38 nM		
Carbendazim	β-CD/rGrO	DPV	PBS, 7.0	0.1–40 μM	18.6 nM		[42]
Febuxostat	ZnO_2/Gr	SWV	BRB, pH 6.5	10–400 nM	1.84 nM	Pharmaceutical formulations, wastewater	[40]
Tramadol	PdVCPILNF	CV, LSV, CHASWV	PBS, pH 7.0	0.05–200 μM	0.015 μM	Tablet, human blood plasma, urine samples	[43]
Aspirin	P4VP/MWCNT	DPV	PBS, pH 7.4	0.04–350 μM	4.42 nM	Tablet	[66]
Caffeine				2.0–200 μM	1.19 nM		
Dopamine	NCNTs/IL	SWV	PBS, pH 7.0	0.001–40 μM		Injections	[67]
Atropine	CNS/NAF/Ru$(bby)_3^{+2}$	ECL	PBS, pH 8.0	0.5–250 μM	0.17 μM	Injections	[68]
Acetaminophen	P4VP/Gr	DPV	PBS, pH 7.0	0.04–300 μM	3.2 nM	Tablet, urine samples	[69]
Isoniazid	CuO-NPs/GrO	Amperometry	PBS, pH 7.0	0.15–500 μM	0.0234 μM	Human serum, urine samples	[70]
	FePc/*f*-MWCNT	Amperometry	PBS, pH 7.4	5–476 μM	0.56 μM	Saliva, blood plasma samples	[71]

(*Continued*)

Table 5.1. (*Continued*)

Drug	Modifier	Method	Medium	Linear range	LOD	Application	References
Gemcitabine	rGrO/BiNPs	DPVLSV	PBS, pH 3.0	0.1–51.1 μM2.1–61.1 μM	0.05 μM0.7 μM	Pharmaceutical formulations, spiked urine samples	[41]
Ascorbic acide	CNT/Mn_2O_3	CV	50 mM NaH_2PO_4. H_2O	60–400 μM	0.1 μM	Pharmaceuticals	[37]
Brucine	SWNT/NAF	SWV	PBS, pH 7.2	1 nM–8 μM	0.11 nM	Pharmaceuticals, human serum, urine samples	[72]

Abbreviations: AgNPS: silver nanoparticles; AuNP: gold nanoparticles; $Au_{1.5}Pt_1Co_1$ TANPs: gold-platinum-cobalt ternary alloy nanoparticles; BiNPs: bismuth nanoparticles; BRB: Britton Robinson buffer solution; β-CD: β-cyclodextrin; CB: carbon black hybrid composite; CE: benzo-12-crown-4; CMOFs: chiral metal–organic frameworks; CNF: carbon nanofibers; CNS: carbon nanospheres CNT: carbon nanotube; CMK-1: mesoporous carbon CTAB: cetyltrimethylammonium bromide; CuS: copper sulfide nanoparticles; CuS NS: copper sulfide nanosheets; Cs: chitosan; Co_3O_4: cobalt tetraoxide nanomaterials; CV: cyclic voltammetry; DPV: differential pulse voltammetry; f-CNF: functionalized carbon nanofibers; f-MWCNT: functionalized multiwalled carbon nanotube; FMWCNT: carboxylated multiwalled carbon nanotube; FePc: iron phthalocyanine; g-C_3N_4: graphitic carbon nitride; Gr: graphene; HgS: mercury sulfide; IL: ionic liquid; CHA: chronoamperometry; LCV: linear sweep voltammetry; Mn_2O_3: manganese oxide nanoparticle; $MnWO_4$: manganese tungstate; MWCNT: multiwalled carbon nanotube; MIP: molecular imprinted polymer; NAF: nafion; $NiSe_x$; nickel metal–organic framework materials-74, NCNTs: nitrogen doped carbon nanotubes; OD: 1,8 octanediamine; *p*-ATT: nanostructure conducting polymer; PdVCPILNF: Pd nanoparticles loaded on vulcan carbon/conducting polymeric ionic liquid composite nanofibers; PEDOT: poly(3,4-ethylenedioxythiophene); PHEMA: poly (2-hydroxyethyl methacrylate); PSSLi: poly(4-lithium styrene sulfonic acid); P-L-Cys: Poly-L-cysteine; PHQ: poly-hydroquinone; P4VP: poly-4-vinylpyridine; Pr_2O_3: praseodymium oxide; PS: polystyrene; PtNPs: platinium nanoparticles; $Ru(bpy)3^{2+}$; tris(2,2'-bipyridyl) ruthenium(II) rGrO: reduced graphene oxide; SWV: square wave voltammetry; TiO_2 MPs: rutile-type titanium dioxide nanoparticles; $SmVO_4$: samarium vanadate nanospheres: ZnO: zinc oxide nanoparticles; ZIF-8: zeolitic imidazolate framework; WO_3NPs: tungsten oxide nanoparticle.

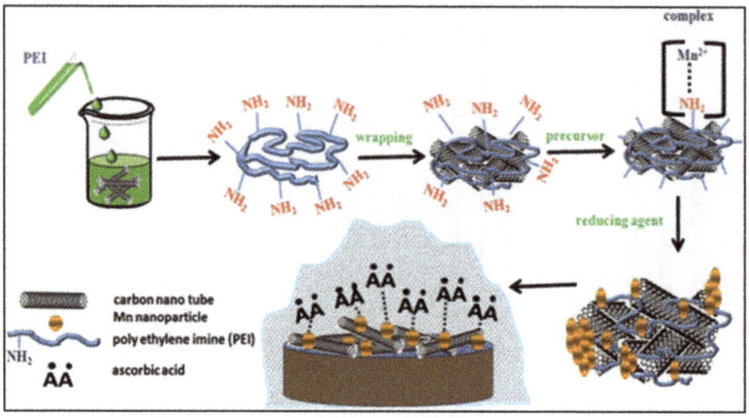

Figure 5.2. Illustration of synthesis procedure of CNTs-MnNPs composite. Reprinted from [37], copyright 2017, with permission from Elsevier.

determination of 4-aminoantipyrine. The CV results showed that the oxidation process of 4-aminoantipyrine was irreversible and diffusion controlled. A differential pulse voltammetric (DPV) method was applied to detect of 4-aminoantipyrine over the concentration of 0.005–0.04 μM with a detection limit of 0.163 nM in phosphate buffer solution (PBS), pH 3.0. The prepared sensor was used to detect the amount of 4-aminoantipyrine in human urine and plasma samples. The composite modified electrode showed good selectivity and good reproducibility [38].

Satyanarayana *et al* described a gold nanoparticle decorated MWCNT (GNP/MWCNT) composite sensor for the analysis of 5-fluorouracil. The GNP/MWCNT composite was fabricated with a simple one-pot synthesis in the presence of chitosan, which acts as both a reducing and stabilizing agent. The electrochemical ability of the GNP/MWCNT composite modified electrode produced for the detection of 5-fluorouracil was investigated by CV, electrochemical impedance spectroscopy (EIS), and DPV. Under optimized conditions, the DPV procedure for the determination of 5-fluorouracil showed a linear relationship between the peak current and concentration in the range of 0.03–10 μM with the detection limit of 20 nM. The developed DPV procedure was used for the determination of 5-fluorouracil from pharmaceuticals and artificial urine [39].

Jain *et al* developed a novel Gr/zinc nanoflowers nanocomposite modified GCE (Gr/ZnOnFs/GCE) for the voltammetric analysis of febuxostat. A homogeneous suspension of Gr and synthesized ZnOnFs in DMF was prepared and then the composite electrode was prepared via dropping the suspension on the GCE surface. A validated simple, sensitive, and rapid square wave voltammetric (SWV) method with a linear range of 10–40 ng mL^{-1} and detection limit of 1.84 ng mL^{-1} was described for the electro-reduction reaction of febuxostat on the Gr/ZnOnFs/GCE. The proposed sensor was applied to determine febuxostat from pharmaceutical formulation and wastewater with acceptable recovery and reproducibility [40].

An electrochemical sensor consisting of electro-reduced GrO-bismuth nanoparticle composite (rGrOBiNPs) was fabricated for the determination of gemcitabine hydrochloride, in this study. The rGrO/BiNPs/GCE was fabricated by dropping a suspension of bismuth nitrate-GrO on a GCE followed by electro-reduction in the potential range of 0.6 V to 1.7 V. Under optimized conditions, the gemcitabine peak current altered linearly between 0.1–51.1 μM and 2.1–61.1 μM in DPV and linear sweep voltammetry (LSV), respectively. The applicability of the rGrO/BiNPs/GCE was successfully utilized for the determination of gemcitabine hydrochloride from pharmaceuticals and spiked urine samples [41].

In another study, NAF/platinum nanoparticle Gr-modified GCE (NAF/PtNP/Gr/GCE) was formed, first by electrochemical reduction of GrO and hexachloroplatinic acid on a GCE surface to form a Gr-PtNP composite, and then coating the nanocomposite structure with NAF (figure 5.3). The electrochemical behavior of paracetamol and domperidone was investigated via CV, EIS, adsorptive stripping square wave voltammetry (AdSSWV) and chronocoulometry. AdSSWV was used for the subnanomolar determination of paracetamol and domperidone simultaneously on NAF/PtNP/Gr/GCE. Under optimized conditions, a linear working range of 1.6 nM–8.20 μM was obtained for paracetamol and domperidone in PBS, pH 6.0. The detection limit values were calculated as 0.106 nM and 0.402 nM for paracetamol and domperidone, respectively. The modified electrode and validated voltammetric method were applied to detect paracetamol and domperidone from pharmaceuticals, blood serum, and human urine samples [31].

A graphene-based electrochemical sensor was prepared to analyze electro-oxidation of carbendazim by Song *et al.* The synthesis of nanocomposites consisting of β-cyclodextrin-functionalized reduced GrO (β-CD/rGrO) was carried out using hydrazine at room temperature. The synthesized nanocomposites were used to modify the surface of the GCE for electrochemical analysis of carbendazim with CV and DPV. The linear range of different carbendazim concentrations on β-CD/rGrO-

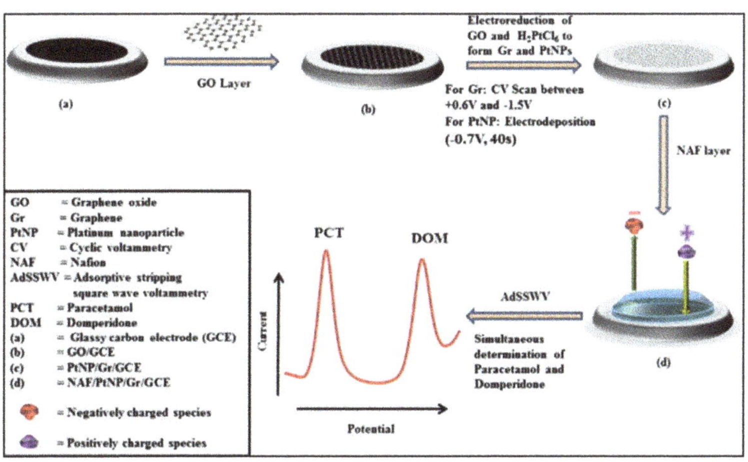

Figure 5.3. Preparation of practical electrochemical sensors for the simultaneous determination of paracetamol and domperidone. Reprinted from [31], copyright 2015, with permission from Elsevier.

modified GCE was determined from 10 nM to 40 μM with a limit of detection (LOD) of 18.6 nM in PBS, pH 7.0 by DPV [42].

In this study a selective and sensitive sensor was fabricated by modifying the GCE with Pd nanoparticles loaded on vulcan carbon/conducting polymeric ionic liquid composite nanofibers (PdVCPILNF) for the electrochemical analysis of tramadol. The CV, LSV, chronoamperometry (CHA) and SWV methods were used to investigate the electro-oxidation behavior of tramodol and the electrochemical studies of the prepared compoiste electrode. The results obtained CV, LSV, CHA, and SWV methods indicated a linear range from 0.05 to 200 μM and LOD of 0.015 μM. The fabricated sensor was used for the determination of tramadol in pharmaceutical and biological samples with good recovery [43].

Moreover, in recent years, one of the interesting research topics in electrochemical sensor development has been molecular imprinted polymer (MIP) technology, which promotes combining it with other materials to tune the sensor's properties for instance large surface area, high catalytic efficiency, thermal, electrical, and mechanical stability. For example, Jaouhari *et al* designed a novel MIP sensor for rutin detection by modifying the surface GCE with rGrO/zeolitic imidazolate framework (ZIF-8) composite. Because of its very stable and electrically conductive nature, rGrO was utilized in electrochemical signal amplification in this work to achieve a highly sensitive MIP sensor. It has also been used in MIP electrode modification due to its large surface area, tunable high porosity ZIF-8 structures, and selective pores specific to the target molecule. In addition to all these, a synergistic effect was created as a result of the combination of these structures. First, they indicate an increase in surface area and electrochemical conductivity from the data obtained after electrochemical characterization. After optimization of sensor preparation parameters, the LOD of rutin was obtained to be 0.0001 μM with a linear range of 0.0005–0.05 μM. The MIP sensor was successfully used in real samples such as tablets and orange juice [44].

5.3.2 Selected applications of carbon composite with CPE

Graphite powder consisting of high-purity and μm-sized particles is mixed with various water-immiscible organic binders such as mineral oils, paraffin oil, silicone oils, and bromonophthalene for the production of carbon paste electrodes. The amount of binder used in the preparation of carbon paste affects the reactivity of the electrode. As the amount of binder increases, the electron transfer rate decreases, and the reactivity increases [3]. Although CPEs have advantages, such as simple and convenient preparation, the paste surface can be easily renewed so that no pretreatment is required, and low cost, they have relatively lower repeatability. The homogeneity of carbon paste affects the reproducibility. Therefore, it is necessaray to make sure that the mixture is homogeneous during the preparation phase. The prepared carbon paste is usually compressed into a glass or teflon pipes with a diameter of 2–4 mm and a conductive wire is placed inside to provide electrical conductivity. The main disadvantages of CPEs are the easy solubility of organic binders in solutions containing organic solvents, the difficulty of obtaining reproducible results, and the long calibration process required for sensitive methods. Modification on electrode surface can increase selectivity, slow surface degradation, better binding of the species to be analyzed to the electrode surface, increase of the electrocatalytic properties of the electrode, and reduced interference in complex samples [7, 73, 74]. Selected studies related with carbon composite on CPE are given in table 5.2.

Kalambate *et al* developed a selective and sensitive composite sensor based on the use of carbon nanofibers and modified silver nanoparticles modified CPEs (CNF-AgNPs-CPE) (figure 5.4) for the electroanalytical analysis of methylergometrine maleate from real samples. The methylergometrine maleate electrocatalytic response on CNF-AgNPs-CPE was analyzed via CV, DPV, and EIS. The oxidation peak current of methylergometrine maleate indicated a rectilinear relationship between 0.04 and 20 μM with a LOD of 7.1 nM with DPV. The proposed composite sensor was utilized to direct analysis of methylergometrine maleate from pharmaceutical dosage forms, serum, and urine samples, with satisfying recovery results [75].

Brahman *et al* prepared a new electrochemical sensor by first dropping fullerene-C_{60} and MWCNTs on CPE and then electrochemically depositing copper nanoparticles (CuNPs) on modified CPE to form the nanocomposite film. The SWV method was utilized for the electrochemical determination of paracetamol with CuNPs/C_{60}/MWCNTs/CPE. The composite film enabled the prepared sensor to exhibit good selectivity and high sensitivity for paracetamol detection. Under optimized conditions, the calibration curve of paracetamol indicated linear response in the paracetamol concentration range of 4.0–400 nM, with the LOD of 0.073 nM. The applicability of the fabricated composite sensor was investigated for the determination of paracetamol from serum, plasma, and urine samples and good recovery results were obtained [76].

CPE was modified with GrO nanosheets, MWCNTs, and pyrogallol (PG), thus fabricating a new highly sensitive composite sensor which was used for the sensitive determination of omeprazole by Mohamed *et al.* The electrochemical oxidation behavior of omeprazole was analyzed with CV and DPV methods on MWCNTs/GrO/PG/CPE. Under optimized conditions, the current response of omeprazole showed two linear dynamic ranges of 2.0×10^{-4}–6.0 μM and 6.0–100.0 μM, with a LOD of 0.0102 nM using DPV. The developed composite sensor was applied successfully to determine of the omeprazole from dosage forms and human serum samples [77].

Ardakni and Khoshroo prepared nanocomposite sensor based on coumarin derivative (7-(1,3-dithiolan-2-yl)-9,10-dihydroxy-6H-benzofuro[3,2-c]chromen-6-on), (DC)-titanium dioxide NPs (TiO_2) and ionic liquid (IL) for the electroanalytical determination of levodopa and carbidopa. This new IL-TiO_2NPs modified CPE because of its increased conductivity and offered very large current response for electroactive species. The electrocatalytic activity of the composite electrode was found to be effective in lowering the anodic overpotential for the oxidation of levodopa and complete resolution of its anodic wave from carbidopa with CV and DPV studies. The oxidation response of levodopa and carbidopa based on DPV showed the dynamic range between 0.10–900 μM and 20.0–900.0 μM respectively.

Table 5.2. Selected applications of carbon composite on CPE in pharmaceuticals.

Drug	Modifier	Method	Medium	Linear Range	LOD	Application	Ref
Ascorbic acid	CNF/AgNPS	DPV	PBS, pH 7.0	20–450 μM	0.029 μM	Human urine and blood	[79]
Dopamine				2–320 μM	0.034 μM		
Paracetamol				1–340 μM	0.019 μM		
Rizatriptan	$SWCNT/ZrO_2$/DES	DPV	PBS, pH 6.0	0.003–100 μM	0.7 nM	Tablet	[80]
Paracetamol				0.08–100 μM	9.0 nM		
Nimesulide	BNC	SWV	PBS, pH 9.2	0.01 μM–0.55 μM	11 nM	Tablet, human serum,	[81]
Linagliptin	gCN/βCD	DPV	PBS, pH 7.0	0.01–50 μM	3 nM	Serum	[82]
Acyclovir	γ-Fe_2O_3/Bent	SWV	PBS, pH 4.2	50 nM to 800 nM	1.55 nM	Tablet, urine	[83]
Norfloxacin	MWCNT/CPE/rGO/p-AHNSA/AuNP	DPV	PBS, pH 5.5	0.03–1.0/1.0–50.0 μM	0.016 μM	Tablet, human plasma	[84]
Atorvastatin	ZnO/NS	DPV SWV	PBS, pH 4.0	7.52–2.5 nM	0.70 nM 0.12 nM	Tablet, urine	[85]
Methyldopa in the presence of diclofenac	CuO-NPs	DPV	PBS, pH 6.0	2.0–600.3 μM	0.51 μM	Tablet, human serum samples	[86]
Methylergometrine maleate	CNF/AgNPs	DPV	BRB, pH 7.0	0.04–20 μM	7.1 nM	Pharmaceutical, urine, blood serum samples	[75]
Paracetamol	CuO-NPs/C_{60}/MWCNTs	SWV	PBS, pH 6.0	4.0–400 nM	0.073 nM	Serum, plasma, urine samples	[76]
Sulfadiazine	rGO/CeO_2NPs	SWV	PBS, pH 7.4	3–10 μM 10–1000 μM	0.17 μM	Tablets	[87]
Tramadol	GrO/MWCNTs	DPV	BRT, pH 9.0	2 nM–110 μM	0.15 nM	Tablets, human blood plasma	[88]
Omeprazole	MWCNTs/GrO/PG	DPV	BRT, pH 7.0	200–6.0 μM 6.0–100 μM	0.0102 nM	Pharmaceuticals, human serum samples	[77]

(*Continued*)

Table 5.2. (*Continued*)

Drug	Modifier	Method	Medium	Linear Range	LOD	Application	Ref
Flutamide	FC/MWCNTs	SWV	ACT, pH 4.5	0.1–110 μM	0.001 μM	Human serum, plasma, urine samples	[74]
Dacarbazine	PolyATD/CNP	DPV	BRB, pH 6.0	0.05–24.0 μM	35 nM	Pharmaceticals, artificial urine	[89]
Ibuprofen	MWCNT	DPV	0.5 M Na_2SO_4/ 0.5 M PBS	2.36–242 μM	9.1 μM	Pharmaceuticals	[90]
Levofloxacin	MCNTs/MIP	DPV	BRB, pH 5.0	0.003–0.440 μM	0.8 nM	Pharmaceuticals, human serum, urine samples	[91]
Methylergometrine maleate	NAF/AgNPs	DPV	PBS, pH 7.0	0.040–20 μM	7.1 nM	Pharmaceuticals, human urine, blood serum samples,	[75]
Paracetamol, Cetirizine and Phenylephrine	MWCNT/PtNPs	AdDPV	PBS, pH 5.5	0.351–56.1 μM 0.19–193 μM 0.29–56.9 μM	0.279 nM 0.586 nM 0.283 nM	Pharmaceuticals,blood serum, urine samples	[92]
Ascorbic acid	BCB/MWCNT	Amperometry	BRB, pH 5.0	0.1–350 μM	0.05 μM	Pharmaceutical vitamin preparation	[93]
Ascorbic acid Acetaminophen Tyrosine	Au-Pd/rGO	DPV	BRB, pH 7.0	0.03–9.50 μM	15.7 μM 11.1 μM 7.6 μM	Pharmaceuticals, human serum, urine samples	[94]
Rivastigmine	CNF/AuNP	AdDPV	PBS, pH 7.0	0.2–600 μM	0.53 nM	Pharmaceuticals formulations, blood serum, urine samples	[95]
Daidzein	$ZrOCl_2$	LSV	PBS, pH 2.5	30 nM–2 μM	0.1 nM	Pharmaceuticals, human uric samples	[96]
Epinephrine	MWCNTs	Amperometry	PBS, pH 7.0	0.8–9.0 μM/ 10.0–100 μM	0.8 μM	Blood serum, adrenaline injection	[97]
Tadalafil	TiO_2/MWCNT	SWV	BRB, pH 3.0	0.27–15.2 μM	0.08 μM	Pharmaceuticals, human serum samples	[98]

Piroxicam	FMWCNTs/β-CD	Potentiometry		1.0 μM –10000 μM	0.7 μM	Pharmaceuticals	[99]
Isoniazid	CoHCF/ MWCNTs	Amperometry	PBS, pH 6.5	5.0–1400 μM	1.3 μM	Pharmaceuticals	[100]
Levodopa	MWCNT/CHNPs	DPV	PBS, pH 7.5	0.1–300 μM	0.075 μM	Pharmaceuticals, human urine samples	[101]
Melatonin				0.01–50 μM	0.004 μM		
Ascorbic acid	MWCNTs/CHIT/ ABPE	LSV	0.1 M HCl	0.18–70.5 μM	0.14 μM	Pharmaceuticals	[102]
Rutin				0.012–6.1 μM	0.006 μM		
Captopril	Gr/FC	DPV	PBS, pH 7.0	1.0–430 μM	0.87 μM	Pharmaceuticals, human urine samples	[103]
Hydrochlorothiazide				0.5–390 μM	0.38 μM		
Levodopa	IL/TiO_2NPs	DPV	PBS, pH 7.0	0.10–900.0 μM	41 nM	Pharmaceuticals, human serum and urine samples, water samples	[73]
Carbidopa				20.0–900.0 μM	0.38 μM		
Dopamine	Poly (beryllon II)/ nanowires- $LaPO_4$	DPV	PBS, pH 7.0	0.10–31.60/ 39.38–403.28 μM	0.03 μM	Pharmaceuticals	[104]
Uric acid				0.82–682.94 μM	0.23 μM		
Cefoperazone	$BMIMPF_6$/Gr/ AuNP/CeO_2 NPs	CFFTAV	PBS, pH 4.0	2000.0–0.120 μM	0.012 nM	Pharmaceuticals	[105]

Abbreviations: AdDPV: adsorptive differentials pulse voltammetry; AgNPS: silver nanoparticles; AuNP: gold nanoparticles; Au-Pd: gold-paladium nanoparticle; BCB: brilliant cresyl blue; BNC: bentonite clay; β-CD: β-cyclodextrin; CFFTAV: coulometric fast fourier transformation admittance voltammetry; CoHCF; cobalt hexacyanoferrate; CHIT/ABPE: chitosan composite film modified acetylene black paste; CHNPs: cobalt hydroxide nanoparticles; CNF: carbon nanofibers; CNPs: carbon nanoparticles; CuO-NPs: copper oxide nanoparticles; C_{60}: fullerence; CV: cyclic voltammetry; DES: deep eutectic solvent; DPV: differential pulse voltammetry;FC: ferrocene; gCN: graphitic carbon nitride; Gr: graphene; IL: ionic liquid; La_2O_3: lanthanum oxide; LCV: linear sweep voltammetry; MCNTs: magnetite decorated carbon nanotubes; MWCNT: multiwalled carbon nanotube; MIP: molecular imprinted polymer; NAF: Nafion; NS: nanosilica; p-AHNSA: poly 4-amino-3-hydroxy-1-naphthalenesulfonic acid; p-ATT: nanostructure conducting polymer; PtNPs: platinium nanoparticles; rGrO: reduced graphene oxide; SWCNT: single walled carbon nanotube; SWV: square wave voltammetry; TiO_2: titanium dioxide; ZnO: zinc oxide nanoparticles; ZrO: zirconium dioxide; γ-Fe_2O_3/bent: iron oxide nanoparticles and bentonite clay.

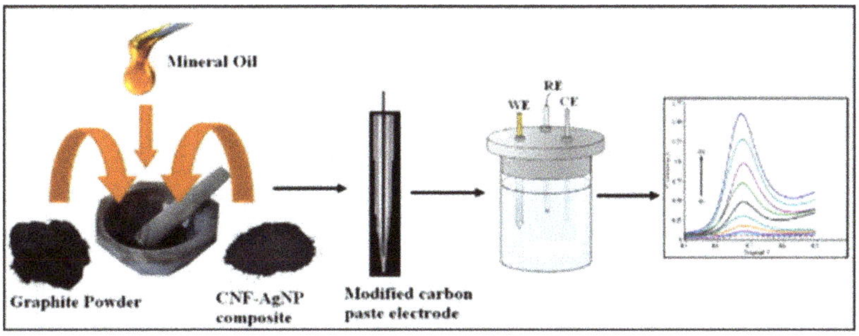

Figure 5.4. Preparation and application of selective carbon composite paste based methylergometrine maleate electrochemical sensor. Reprinted from [75], copyright 2016, with permission from Elsevier.

The LOD values were calculated as 41 nM and 0.38 μM for levodopa and carbidopa, respectively. The proposed composite electrode utilized for the determination of levodopa and carbidopa in pharmaceuticals, human serum and urine samples, and water samples with good recovery results [73].

A selective composite sensor based on molecularly imprinted magnetite decorated carbon nanotubes (MCNTs/MIP) was prepared for the analysis of levofloxacin by Khoshsafar *et al.* The MIPs were prepared using MCNT as support, levofloxacin as template, ethylene glycol dimethacrylate as crosslinker and methacrylic acid as a functional monomer. The prepared MCNTs/MIP were suspended in levofloxacin solution and then gathered on the surface of CPE by a constant magnet, placed within the CPE, and then the voltammetry signals were recorded by the composite sensor. The response currents were linearly related to levofloxacin concentrations in the range 0.003–0.440 μM with a detection limit of 0.8 nM, under optimized conditions. The proposed composite sensor was used for the determination of levofloxacin in pharmaceuticals, human serum and urine samples with satisfactory results [78].

5.3.3 Selected applications of carbon composite with SPCE

Screen-printed electrodes are generally made by screen printing a film electrode onto a plastic or ceramic layer. In the curtain printing technique, a spot of ink is sprayed onto the surface using a curtain pen. Carbon-based inks are used for the working electrodes and silver-based inks are used for the reference electrodes. Normally, a small drop of the solution covers the entire electrode surface and is sufficient for the determination of the substance. The main advantages of screen-printed electrodes are that they are simple to manufacture and easy to modify. Since they are disposable electrodes, they are not susceptible to contamination and offer fewer surface problems [106]. Furthermore, as shown in table 5.3, SPCE has been effectively used in drug analysis by being modified with various composites.

In this study, a sensitive and selective azathioprine sensor was fabricated using a nanocomposite composed of MWCNTs and Mn_3O_4 micro cubes and (Mn_3O_4 MCs). An SPCE was modified with MWCNTs/Mn_3O_4 MCs for the electrochemical analysis of azathioprine. The linear range of 0.045–2530 μM was obtained with a detection limit of 4.3 nM by the amperometric method. An amperometric sensor

Table 5.3. Selected applications of carbon composite on SPCE in pharmaceuticals.

Drug	Modifier	Method	Medium	Linear Range	LOD	Application	Ref
Azathioprine	Mn_2O_3/rGO	DPV	PBS, pH 7.0	0.009–573.5 μM	4 nM^{-1}	Human blood serum and urine samples	[32]
Promethazine	GrO-Zn/ZnO	DPV	PBS, pH 7.0	0.05–177.45 μM	0.3 nM	Tablet, serum, river water	[111]
Rohypnol	AGrO/CuO-NPs	DPV	BRT, pH 7.0	0.4–140 μM	0.13 μM	Fruit juice	[112]
Flutamide	$ZnMn_2O_4$/rGrO	DPV	PBS, pH 7.0	0.05–3.5 μM	8 nM	Human urine	[113]
Nilutamide	$Cu_2V_2O_7$/rGrO	DPV	PBS, pH 7.0	0.001–15 μM	0.00459 nM	Blood serum, urine samples	[114]
Guaifenesin	Ni-SiO_2/MLG	Amperometry	PBS, pH 7.0	0.1–317.5 μM	5.7 nM	Tablet	[109]
Diclofenac	PtNFs/rGrO	DPV	PBS, pH 7.0	0.1–100 μM	40 nM	Human urine samples	[115]
Chloramphenicol	Sr-ZnO/rGrO	LSW	PBS, pH 7.0	0.190–410.84 μM 653.06–2847.14 μM	0.131 μM	Milk and powdered milk samples	[116]
Metronidazole	Dy(VO_4)/*f*-CNF	LSV	PBS, pH 7.0	1.5–1036.9 μM	6 nm	Human blood sample, urine sample, lake water	[117]
Chlorpromazine	MIP/MWCNT	SWV	BRT, pH 3.0	75 nM–25 μM	29.6 nM	Tablet, spiked human urine samples	[118]
Phenylephrine Chlorpheniramine Dextromethorphan	MnHCF/Cs	SWV	PBS, pH 8.0	0.06–255 μM 0.045–242 μM 0.062–240 μM	2.18 μM 3.63 μM 9.10 μM	Pharmaceutical formulations, human blood serum and urine samples	[119]
Cabergoline	La_2O_3/Co_3O_4	DPV	PBS, pH 7.0	0.1–100.0 μM	0.04 μM	Tablet, urine	[120]
Dopamine Epinephrine	aGrO/CB/OMNiDIP	DPASV	PBS, pH 7.2	0.115–5.909 nM 0.079–1.307 nM	0.028 nM 0.017 nM	Blood serum, urine, pharmaceutical samples	[121]
Ractopamine	Fe_3O_4/rGrO	DPV	PBS, pH 7.0	10–100 μM	13 nM	Pork samples	[122]

(Continued)

Table 5.3. (*Continued*)

Drug	Modifier	Method	Medium	Linear Range	LOD	Application	Ref
Solatol	AuNPs/GQD-SH/MIP	DPV	BRT, pH 10.0	0.1–250 μM	0.035 μM	Tablet and human blood serum,2019	[123]
Dopamine 5-hydroxytryptamine	Gr/p-AHNSA	SWV	PBS, pH 7.2	0.05–100 μM 0.05–150 μM	2 nM 3 nM	Pharmaceuticals, human urine, blood samples	[124]
Nilutamide	β-CD/AuNP/GrO	DPV	PBS, pH 7.0	0.01–193 μM	0.4 nM	Tablets, human serum samples	[108]
Epirubicin	AuNPs/Fe_3O_4/ SiO_2/DABCO	LSW	PBS, pH 7.0	0.07–1.0 μM 1.0–21.0 μM	0.04 μM	Human blood serum sample	[125]
Ifosfamide	GQD/MIP	DPV	PBS, pH 7.0	0.25–121.35 nM	0.11 nM	Injection, human urine, blood samples	[126]
Azathioprine	MWCNTs/Mn_3O_4MCs	Amperometry	PBS, pH 7.0	0.045–2530 μM	4.3 nM	Tablets, human serum samples	[107]
Clenbuterol	GrS/NAF/ GMP-BSA-CLB	DPV	PBS, pH 7.5	0.5–200 nM	0.22 nM	Pork samples	[127]
Dopamine	IL-OPPF	DPV	PBS, pH 7.0	1.0 μM–2.5 mM	0.5 μM	Ampoule	[128]

Abbreviations: AGrO: amine-functionalized graphene oxide sheet; AuNP: gold nanoparticles; BSA-CLB: clenbuterol coated bovine albumin; β-CD: β-cyclodextrin; CB-OMNiDIP: carbon black-one monomer molecularly imprinted polymers; $Cu_2V_2O_7$: copper vanadate nanoparticles; CuO-NPs: copper oxide nanoparticles; Cs: chitosan; Co_3O_4: cobalt tetraoxide nanomaterials; DABCO: charged diazoniabicyclo [2.2.2] octane dichloride; DPASV: deferential pulse anodic stripping voltammetry; Dy(VO_4): dysprosium vanadate; DPV: differential pulse voltammetry; IL: ionic liquid; Fe_3O_4: iron oxide; *f*-CNF: oxidized carbon nanofiber; Gr: graphene; GrO: graphene oxide; GrS: graphene sheets; GMP: Fe_3O_4-Au nanoparticles; GQD-SH: thiol graphene quantum dots; GQD: graphene quantum dots; La_2O_3: lanthanum oxide; LCV: linear sweep voltammetry; MLG: multilayer graphene nanosheets; Mn_2O_3: manganese oxide nanoparticle; MnHCF: manganese hexacyanoferrate; MWCNT: multiwalled carbon nanotube; MIP: molecular imprinted polymer; NAF: nafion; Ni-SiO_2 nickel-silicon dioxide nanospheres; OPPF: octylpyridinum hexafluorophosphate; p-AHNSA: poly 4-amino-3-hydroxy-1-naphthalenesulfonic acid; PtNFs: platine nanoflowers; rGO: reduced graphene oxide; Sr-ZnO: strontium doped zinc oxide SWV: square wave voltammetry; $ZnMn_2O_4$: zinc manganese oxide nanoparticles.

was used to detect azathioprine in pharmaceuticals and human serum samples with satisfactory recovery results [107].

Karthik *et al* described an electrochemical method based on a composite sensor prepared from β-cyclodextrin, gold nanoparticles, and GrO (β-CD/AuNP/GrO) for the determination of nilutamide. To reduce gold ions an alkaline glucose solution was used as the reducing agent. The prepared nanocomposite was utilizied to modify a SPCE, and the modified SPCE was applied to detect nilutamide by the DPV method. The linear range was obtained for the peak current of nilutamide versus concentration range between 0.01 and 193 μM. The LOD value was calculated to be 0.4 nM from calibration curve. The practical applicability of the composite sensor was studied for the determination of nilutamide from dosage forms and human serum samples and good recovery results obtained for the determination nilutamide with β-CD/AuNP/GrO/SPCE [108].

Huang *et al* designed Ni-SiO_2 nanosphere-decorated multilayer Gr nanosheets (Ni-SiO_2/MLG) composite modifier for pharmaceutical drug guaifenesin sensing by amperometry techniques. Among carbon-based composite materials, Gr's non-polar and hydrophobic characteristics produce very strong π–π stacking linkages to the benzene ring resulting in exceptional electrocatalytic activity for many medicinal compounds. Because of their extremely heterogeneous electron transport qualities and high sensitivity, Ni and Ni-related materials, which are employed in a variety of applications such as batteries, fuel cells, and electrochromic devices, were chosen to replace the electrode surface for electrochemical investigation in this study. Also, SiO_2 has excellent potential reinforcement capability and mechanical properties; for this reason, it was utilized as a transducer material for detection of guaifenesin. Afterwards, utilizing CV and electrochemical impedance spectroscopy, disposable SPCE were modified with a Ni-SiO_2/MLG modifier to investigate guaifenesin oxidation behavior (figure 5.5). As seen figure 5.5, bare SPCE was found to have low sensing capacity and electrocatalytic activity against guaifenesin because of its poor sensitivity, consistent with the low oxidation peak current (I_{pa} = 4.53 μA) at 1.253 V peak potential, while MLG/SPCE had a higher oxidation peak current

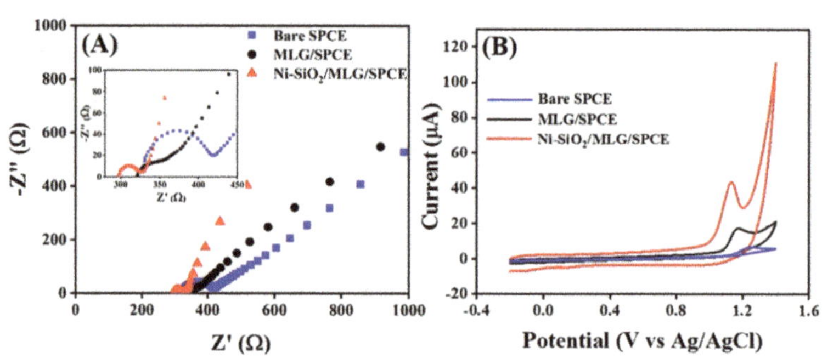

Figure 5.5. (A) Electrochemical impedance properties of different electrodes in 5 mM [Fe (CN)6]$^{4-/3-}$ with 0.1 M KCl electrolyte, the inset shows the enlarged view of the Nyquist plot. (B) CVs of electrochemical oxidation of guaifenesin at different modified electrodes in 200 μM guaifenesin containing PBS (pH 7.0) as electrolyte at 50 mV s^{-1}. Reprinted from [109], copyright 2021, with permission from Elsevier.

(I_{pa} = 12.80 μA; E_{pa} = 1.176 V) and a Ni-SiO_2/MLG sensing electrode modified SPCE to oxidation GFN (I_{pa} = 33.77 μA, E_{pa} = 1.134 V) with high oxidation peak current. Considering the semicircle diameter from the low-frequency region to the high-frequency region, it was found that the Ni-SiO_2/MLG/SPCE electrode had the lowest Rct (charge transfer resistance) values, indicating that the contact resistance between the electrolyte and the electrode interface decreased. Thus, it facilitates the faster transport of electrolyte ions over the conductive network (figure 5.5(A)). Also, the sharp oxidation peak of guaifenesin shows fast electron transfer behavior in Ni-SiO_2/MLG/SPCE. The improved electrochemical performance may result from the combination of metal-doped semiconductors and Gr nanolayers. As results, the linear range of 0.1–317.5 μM, an LOD of 5.7 nM, and high sensitivity of 408.88 μA mM^{-1} cm^{-2} were received. The sensor showed a stability of approximately 30 days with a signal decrease of not more than 5% [109].

One research topic of interest is, for example, the preparation of electrochemical sensors with SPCE for sensitive and reliable measurements when the sample amount is low, to follow-up a course of treatment for cancer. For example, pemetrexed is a cancer drug used in the treatment of lung cancer and mesothelioma, and it is also used with other drugs in some treatments. Ozçelikay *et al* created an electrochemical molecular imaging method for the selective, sensitive, and small quantity of sample determination of pemetrexed, which can cause major side effects and allergy symptoms. SPCE was modified with carbon quantum dots to improve sensitivity, and then electrochemically polymerized with monomer *o*-phenylenediamine (o-PD) to generate a MIP layer in this application. Consequently, an analytically useful concentration range of 5–100 nM was obtained, and the LOD was 1.61 nM. Also, the selectivity of the MIP sensor was examined in the presence of molecules that have similar structures or are part of pemetrexed such as guanine, sulpride, and ascorbic acid [110].

5.3.4 Selected applications of carbon composite with graphite and PGE

Graphite, a polymorph of the element carbon, is one of the stable forms of carbon at surface temperatures and pressures. Synthetic graphite is achieved by heating different hydrocarbon sources to temperatures of 2800 °C or higher. Although graphite electrodes are good conductors of electricity and their electrochemical behavior is similar to mercury electrodes, they are not as widely used as other carbon-based electrodes [3, 7]. The reason for this is that early researchers utilized a wax-coated graphite electrode for voltammetric investigations, and resurfacing by breaking the lower end of the rod necessary due to film development resulted in poor reproducibility of the current measurement [129]. However, disposable PGE is one type of electrode that solves this problem. In the late 1990s, graphite pencils, which are often used for writing, were classified as inexpensive, easily accessible, and renewable electrodes. Natural graphite (75%–80%) is mixed with organic binder (13%) and spindle oil (8%) to make PGEs. This resulting mixture is then compressed into the shape of a pencil stick, dried at 100 °C–300 °C, and sintered at 1000 °C. PGEs have been successfully used as a voltammetric sensor for electrochemical measurements in the analysis of various drugs because inexpensive and easily available PGEs are preferred due to their superior properties such as high electrochemical reactivity, mechanical strength, easy renewability, and easy modification. Selected studies related to carbon composite on PGE are given in table 5.4.

For example, Dehnavi *et al* presented an effective electrochemical sensor using nanostructured TiO_2-MWCNT as a nanocomposite modifier on a pencil graphite electrode for the detection of propranolol drugs in real samples such as tablets, blood serum, and urine. Characterization measurements of the nanocomposite indicated the presence of TiO_2 nanoparticles in the clustered structure formed on the carbon nanotube surfaces without deforming the nanotubes. The formation of this composite structure is thought to depend on the affinity of Ti atoms to form C–O–Ti or O=C–O–Ti bonds with –OH and –COOH groups on functionalized MWCNTs. Furthermore, the PGE was modified with TiO_2/MWCNT nanocomposite material solution by the dip-coating method. The linear range of concentrations for propranolol determination was 85–650 nM, the detection limit was 31 nM [130].

Pashazadeh *et al* designed a new, sensitive electrosensor platform based on electrotreated PGE (ETPGE) modified with Cu-MOFs/carbon nitride quantum dots (CNQDs) nanocomposite for determination of isoniazid in a different medium. Pure MOFs have weak electrical conductivity, low mechanical stability, decreased mass permeability, and poor electrocatalytic activity, while having a high surface area when modified to the electrode. Because of these limitations, MOF-based modified electrodes do not facilitate effective charge transfer in electrochemical processes. CNQDs with qualities including large particle sizes, non-toxicity, high stability, low cost, suitable band structure, and ease of fabrication were employed in this work to overcome these challenges. The electrode modified with Cu-MOFs/CNQDs/ETPGE exhibited the highest current value of the anodic peak of isoniazid oxidation. Isoniazid determination was carried out using DPV in the concentration range of 0.1–65 μM with an LOD of 0.0675 μM [131].

In another work, Nagarajan *et al* developed tin adjoined chitosan polymer matrix coated pencil graphite electrode (Sn/Cs/PGE) by using the first simple immersion method to electrochemical sensing of riboflavin. Tin can be easily incorporated into CS polymeric structure because it has a positive charge in the tin chloride solution. Thereby increasing the net electrons, charge carriers, and electronic conductivity of the polymer. The linear range of concentrations for riboflavin detection was 10 to 1200 nm, the detection limit was 5.56 nM. When analytical parameters of this sensor were compared with recently published other sensors, results indicated the lowest LOD value was obtained in this study. Also, the long-term stability of the electrode sensor, the change in drug oxidation was evaluated by the SWV method for 15 days, and no significant difference was found in the oxidation current of the drug [132].

Oghli *et al* presented a very interesting carbon composite-based electrochemical sensor designed for the simultaneous determination of two important drugs, including sumatriptan and paroxetine. In this study, the combination of rGrO and polyoxometallates such as phosphotungstic acid with electrocatalytic properties develops electroactive composites that increase the electrochemical sensitivity of the electrode as a modifier while sensor selectivity was increased by using MIP. Also, the

Table 5.4. Selected applications of carbon composite on PGE in pharmaceuticals.

Drug	Modifier	Method	Medium	Linear range	LOD	Application	References
Propranolol	TiO_2/MWCNT	AdDPV	PBS, pH 1.4	0.085–6.5 μM	0.021 μM	Human urine and blood serum	[130]
Sumatriptan	PWA/rGO/MIP	AdDPV	BRT, pH 7.0	0.02–3.0 μM	4.0 nM	Tablet, urine, serum samples	[6]
Paroxetine				0.005–2.2 μM	0.7 nM		
Riboflavin	Sn/Cs	SWV	PBS, pH 7.0	10–1200 nM	5.56 nM	Tablet, milk powder	[132]
Ascorbic acid	AuNP-p(l-met)	DPV	BRT, pH 5.0	10–2000 μM	3.03 μM	Tylol hot C, human serum	[133]
Acetaminophen				3.18–3000 μM	0.95 μM		
Chlorpheniramine maleate				9.74–718 μM	2.92 μM		
Caffeine				8.3–2000 μM	2.50 μM		
Ceftizoxime	PNC_S/rGO	DPV	BRT, pH 3.0	0.01–30 nM	1.8 pM	Tablet, blood serum	[34]
Tramadol	CuO-NPs/pPy	SWV	BRT, pH 8.5	5.0 nM–380 μM	1.0 nM	Tablet, injected samples	[134]
Fexofenadine	NDG/MIP	DPV	PBS, pH 8.0	0.5 nM–7.8 μM	0.15 nM	Tablet, urine, blood sample	[135]
Sertraline	Cu-MOF/SNDGr	DPV	PBS, pH 7.0	0.05–2.67 μM	0.038 μM	Tablet, human blood serum	[30]
Methyldopa	PMo/GrO	DPV	PBS, pH 2.8	0.49 nM–0.1 μM	0.12 nM	Human blood serum, urine, milk samples	[136]
Isoniazid	Cu-MOFs/CNQDs	DPV	PBS, pH 7.0	0.1–65.0 μM	0.0675 μM	Tablet, human blood serum	[131]
Levofloxacin	Azure-B	DPV	PBS, pH 7.0	2–125 μM	1.2 μM	Tablet, human serum and urine	[137]
Tramadol	fMWCNTs/AuNPs	DPV	PBS, pH 7.0	0.012–0.1 μM 0.1–3 μM	0.005 μM	Tablet, urine	[138]
Levodopa	AgNPs/DHPA/GrO	DPV	PBS, pH 6.0	0.01 nM–10 μM. 3.0 nM–0.1 μM	76 nM	Tablet, blood serum, urine	[139]
Hydrochlorothiazide	TyB	DPV	PBS, pH 7.0	0.5–7 μM	0.1327 μM	Tablet	[140]

Flutamide Cyproterone	Nano-AB	SWV	PBS, pH 7.0	20.3 nM–0.33 μM 34.4 nM–1.07 μM	1.39 nM 4.74 nM	Tablet, human serum and urine	[141]
Lorazepam	Polypyrrole/AuNP/MIP	SWV	PBS, pH 7.0	0.2–2.0 nM 2.0–20.0 nM	0.09 nM	Tablet, plasma, urine	[142]
L-ascorbic acid D-ascorbic acid	C-dots/FSA/MIP	DPV	PBS, pH 7.0	6.0–165.0 nM 6.0–155.0 nM	0.001 nm 0.002 nm	Pharmaceuticals, blood plasma	[143]
Ledipasvir	MnO_2	SWV	PBS, pH 4.0	0.025–3.60 μM	4.5 nM 9.2 nM	Tablet, rat plasma	[144]
Aceclofenac	MWCNTs	DPV	PBS, pH 7.0	1–60 μM	2.6 nM	Tablet, human urine	[145]
Ranitidine	AuNP/MIP/f-MWCNT	CV	PBS, pH 6.8	0.05–2.0 μM	0.02 μM	Human urine	[146]
Lomustine	MF	SWV	BRT, pH 2.0	380 nM–23.7 μM	0.81 nM	Human blood and urine	[147]
Phenothiazine	MIP	DPV	BRT, pH 2.0	1–300 μM 0.5–10 mM	30 μM	Beef, sheep, chicken, fish, plasma, serum	[148]
Lamivudine Zidovudine	HCs/MIP	DPV	PBS, pH 7.4	8.92–75.92 nM 6.82–115.99 nM	2.23 nM 1.26 nM	Pharmaceuticals, human blood serum	[149]

Abbreviations: AdDPV: adsorptive differential pulse voltammetry; AgNPS: silver nanoparticles; AuNP: gold nanoparticles; AuNP-p(l-met): gold nanoparticles-decorated poly(L-methionine); CNQDs: carbon nitride quantum dots; Cu-MOF: copper-based metal–organic framework; CuO-NPs: copper oxide nanoparticles; Cs: chitosan; DHPA: Dawson hetero-polyacid; DPV: differential pulse voltammetry; f-MWCNT: multiwalled carbon nanotube; f-MWCNT: functionalized multiwalled carbon nanotube; FSA: ferrocene sulfonic acid; HCs: hollow core shells nanoparticles; MF: mercury film; MnO_2: manganese dioxide nanoparticles; MWCNT: multiwalled carbon nanotube; MIP: molecular imprinted polymer; NAF: nafion; nano-AB: banoacetylene black; PMo: phosphomolibdate; PNCs: preyssler nanocapsules; pPy: polypyrrole; PWA: phosphotungstic acid; SNDGr: sulfur and nitrogen co-doped graphene; Sn: Tin; Ty-B: trypan blu; TiO_2: titanium dioxide.

sol–gel process was used to apply a MIP to the electrode surface to stabilize the MIP and improve the permeability of the target molecules. Methacrylic acid was used as the monomer and ethylene glycol dimethacrylate was used as the crosslinker to create MIP for each analyte. Consequently, analytically useful concentration ranges of 20–300 nM and 5–220 nM were obtained, and the LOD values were 4.0 nM and 0.7 nM for sumatriptan and paroxetine, respectively. Moreover, the developed sensor was successfully used for the simultaneous detection of these drugs in real samples. The interference study results showed that, due to the high level of electroactive modifiers (PWA/rGrO) on the electrode surface, the use of the MIP/sol–gel layer on the electrode considerably decreased interfering species access to the electrode surface, confirming the sensor's high selectivity for the desired analyte [6].

5.3.5 Selected applications carbon composite-based on other electrodes

Although not frequently employed in the creation of carbon composite-based electrochemical nanosensors in drug analysis, electrodes with inert metals such as platinum, gold, and ruthenium are used. In recent years, methods based on the deposition of metals that increase the catalytic activity on noble metal electrodes at more positive potentials than the Nernst equilibrium potential are frequently used. The most important advantage of metal electrodes is their high conductivity [7]. In addition, other advantages of metal electrodes are that they have low ground current and a rich surface area, are low in cost, allow different detection and detection applications, and give repeatable results. Due to the oxide layer formed on the surface of noble metal electrodes, its use in stripping analysis is limited. The physical and chemical behavior of electrode surfaces with noble metal structures changes when they are modified. For example, Kutluay and Aslanoglu proposed a square wave adsorptive stripping voltammetric method for the determination of bromhexine using MWCNTs functionalized by nickel nanoparticles (NiNPs) on Pt electrode. According to CV measurements, a bare Pt electrode exhibits a poor bromhexine oxidation peak at 1025 mV, NiNPs/MWCNT/Pt modified electrode results exhibits a sharp and well-defined anodic bromhexine peak at 997 mV. The authors observed that the nanosensors can be used in a wide concentration range 5.0×10^{-6} to 2.3×10^{-4} M with LOD 5.0×10^{-6} M [150].

Microelectrodes, which are promising electrode types in electrochemistry, have miniaturized working electrodes and at least one of the dimensions is less than 25 micrometers. The small size of these electrodes has important advantages such as examining microscopic fields, measuring local concentration problems, determining microflow systems, and working with very small amounts (μL) of sample [151–153]. For example, Chang *et al* designed a GrO-modified carbon fiber microelectrode (CFME) by electrodeposition with CV for dopamine detection in brain tissue. When compared to Gr and other carbon materials, GrO is less conductive; nonetheless, the oxygen functional groups in the edge plane are more favorable for enhancing dopamine sensitivity than other carbon nanomaterials. Under optimized conditions, the sensitivity of the modified microelectrode was a linear range from 25 nM to 1 mM, and a LOD of 11 nM. Furthermore, the improved GrO/CFME was effectively employed to assess electrically generated dopamine release in mouse brain slices [151].

Also, in recent years, researchers have prepared a paste electrode by mixing carbon-based nanomaterials with a binder such as nujol (mineral oil), paraffin oil, silicone oil, and bromonaphthalene [7]. For example, Pushpanjali *et al* created an electrochemical sensor for the detection of cetirizine in the presence of paracetamol using a carbon nanotube paste electrode modified with a thin layer of electrochemically polymerized L-Leucine. The anodic peak current obtained by the CV of cetirizine increased linearly with concentration ranges of 5 μM to 50 μM and with the LOD of 0.17 μM. When compared to some values in the literature and modified electrode results, the electrode developed in this study performed better, whereas other electrodes/techniques demonstrated lower LOD and higher sensitivity than the electrode used in this study, but these electrodes are more expensive and the techniques require lengthy procedures [154]. Another study included electrochemical Voltaren® detection in the presence of tryptophan by utilizing poly(l-methionine) coated carbon nanotube paste electrodes. Because the surface of the bare carbon nanotube paste electrode is brittle, the surface of the carbon nanotube electrode was stabilized by electropolymerization of L-methionine, a conductive substance, in this work. Voltaren® concentration was measured by DPV in 0.2 M PBS pH 7.0. The authors discovered a linear relation of peak current with Voltaren® concentration in the 2–50 μM, with LOD of 0.10 μM. Also, the developed carbon composite electrode was applied with Voltaren® analysis in pharmaceutical formulations with appropriate recovery values [155].

5.4 Conclusion

For many pharmaceuticals, electrochemical qualitative and quantitative analyses using bare electrodes fail to produce the desired sensitivity, selectivity, and repeatability of analytical findings. However, when employed in the drug's research-development, manufacture, and treatment processes, sensitive, rapid, and repeatable methodologies are required. As a consequence, in recent years, researchers have conducted several studies on the manufacture of carbon composite-based electrochemical sensors, since the composite material carries the characteristics of the molecules that comprise it, and these molecules may be exploited to make desired modifications. Furthermore, carbon composites can be formed in a variety of combinations, including metal–carbon, carbon–carbon, carbon–polymer, and metal–organic frameworks, as well as in structures such as nanorods, nanoflowers, nanospheres, and nanotubes, and not only the superior properties of all these molecules, but also the synergistic effects that occur when they come together, can be utilized.

This chapter book has focussed on the importance of carbon composites in electrochemical sensor studies for pharmaceutical drug analysis. In particular, the composition and structure of the electrode-based carbon composite used in drug analysis have been discussed, and the most commonly used composite electrode preparation methods in drug analysis have been mentioned. How the electrochemical sensors using carbon composite modified electrodes in drug analysis in the literature improve with application examples was then explained.

Although intensive studies have been carried out by researchers in recent years, there is no electrochemical sensor drug analysis that has been commercialized or accepted by international guidelines such as The International Council for Harmonisation of Technical Requirements for Pharmaceuticals for Human Use and pharmacopeias. In this perspective, various conclusions may be derived based on the current literature review about the development of carbon composite-based electrochemical sensors. First of all, carbon composite-based electrochemical sensors are commonly employed in GCE due to their durability, conductivity, low porosity, and exceptional stability under stress conditions, as well as their large application potential range. However, the sensor life is approximately 30 days when the carbon composite is prepared with these electrodes. Thus, it is not possible to sell a product with such a short shelf life or it is impractical to modify the electrode before use. Secondly, one of the most important parameters in the validation of drug analyses in pharmaceutical formulations is stability studies. These investigations look into whether the drug can be evaluated reliably in the presence of impurities. However, no research has been conducted to evaluate if it can be detected in the situation of probable impurities in electrochemical sensor experiments. Furthermore, investigations that examine the presence of both the drug and its metabolites, which are critical for drug monitoring throughout therapy, are not carried out. To overcome these challenges, improvement of multicomponent electrochemical systems based on carbon composite materials may be a feasible alternative. Moreover, extensive research has been conducted in recent years on metal–organic frameworks, bimetallic nanomaterials, and Gr-based nanomaterials to improve electrode sensitivity and stability and to develop cost-effective carbon-based composite material replacements. However, the explanations for the mechanism of the improvement in the drug response made by these components in the carbon composite material are unclear. As a consequence, the mechanism of action of the carbon composites to be employed in the modification should be understood more thoroughly. Finally, some drug studies need sensitive, reliable, and accurate measurement with extremely small sample amounts in biological fluids such as blood, urine, and cells. For this reason, carbon composite-based electrochemical sensor studies using microelectrodes, SPCE, and lab-on-chip technology are promising.

In conclusion, even considering the many advances in the fabrication and application of sensors composed of carbon composite materials in the field of drug analysis in the last ten years, there is still much to be explored with carbon composite materials.

Acknowledgments

Çiğdem Kanbeş Dindar thanks the Scientific and Technological Research Council of Turkey (TUBITAK) under the BIDEB/2218-National Postdoctoral Research Fellowship Program for Turkish Citizens.

References

[1] Sousa C P, Ribeiro F W P, Oliveira T M B F, Salazar-Banda G R, de Lima-Neto P, Morais S and Correia A N 2019 Electroanalysis of pharmaceuticals on boron-doped diamond electrodes: a review *ChemElectroChem* **6** 2350–78

[2] Pushpanjali P A, Manjunatha J G and Hareesha N 2021 An overview of recent developments of carbon-based sensors for the analysis of drug molecules *J. Electrochem. Sci. Eng.* **11** 161–77

[3] Uslu B and Ozkan S A 2007 Electroanalytical application of carbon based electrodes to the pharmaceuticals *Anal. Lett.* **40** 817–53

[4] Parameswara Rao K and Rao M C 2016 Spectrophotometric methods in the analysis of drugs in pure and dosage forms *Int. J. Chem. Sci.* **14** 2389–96

[5] Nebot C, Gibb S W and Boyd K G 2007 Quantification of human pharmaceuticals in water samples by high performance liquid chromatography—tandem mass spectrometry *Anal. Chim. Acta* **598** 87–94

[6] Hassan Oghli A and Soleymanpour A 2021 Ultrasensitive electrochemical sensor for simultaneous determination of sumatriptan and paroxetine using molecular imprinted polymer/sol–gel/polyoxometalate/rGO modified pencil graphite electrode *Sens. Actuators B Chem.* **344** 130215

[7] Cavalheiro É T G, Brett C M A, Oliveira-Brett A M and Fatibello-Filho O 2012 Bioelectroanalysis of pharmaceutical compounds *Bioanal. Rev.* **4** 31–53

[8] Lima H R S, da Silva J S, de Oliveira Farias E A, Teixeira P R S, Eiras C and Nunes L c.c. 2018 Electrochemical sensors and biosensors for the analysis of antineoplastic drugs *Biosens. Bioelectron.* **108** 27–37

[9] Ren S, Zeng J, Zheng Z and Shi H 2021 Perspective and application of modified electrode material technology in electrochemical voltammetric sensors for analysis and detection of illicit drugs *Sens. Actuators, A Phys.* **329** 112821

[10] Srivastava A K, Upadhyay S S, Rawool C R, Punde N S and Rajpurohit A S 2018 Voltammetric techniques for the analysis of drugs using nanomaterials based chemically modified electrodes *Curr. Anal. Chem.* **15** 249–76

[11] Kim E, Leverage W T, Liu Y, White I M, Bentley E and Payne G F 2014 Redox-capacitor to connect electrochemistry to redox-biology *Analyst* **139** 32–43

[12] Bogue R 2004 Nanosensors : a review of recent progress *Sens. Rev.* **28** 3–8

[13] Kurbanoglu S and Ozkan S A 2018 Electrochemical carbon based nanosensors: a promising tool in pharmaceutical and biomedical analysis *J. Pharm. Biomed. Anal.* **147** 439–57

[14] Bogue R 2009 Nanosensors : a review of recent research *Sens. Rev.* **29** 310–5

[15] Dhahi T H S, Bin Hashim U D A, Ahmed N M and Mat Taib A 2010 A review on the electrochemical sensors and biosensors composed of nanogaps as sensing material *J. Optoelectron. Adv. Mater.* **12** 1857–62

[16] Porto L S, Silva D N, de Oliveira A E F, Pereira A C and Borges K B 2020 Carbon nanomaterials: synthesis and applications to development of electrochemical sensors in determination of drugs and compounds of clinical interest *Rev. Anal. Chem.* **38** 1–16

[17] Kalambate P K, Rao Z, Dhanjai, Wu J, Shen Y, Boddula R and Huang Y 2020 Electrochemical (bio) sensors go green *Biosens. Bioelectron.* **163** 112270

[18] Maduraiveeran G, Sasidharan M and Jin W 2019 Progress in materials science earth-abundant transition metal and metal oxide nanomaterials: synthesis and electrochemical applications *Prog. Mater Sci.* **106** 100574

[19] Wongkaew N, Simsek M, Griesche C and Baeumner A J 2019 Functional nanomaterials and nanostructures enhancing electrochemical biosensors and lab-on-a-chip performances: recent progress, applications, and future perspective *Chem. Rev.* **119** 120–94
[20] Baig N, Sajid M and Saleh T A 2019 Recent trends in nanomaterial-modified electrodes for electroanalytical applications *TrAC—Trends Anal. Chem.* **111** 47–61
[21] Kaya S I, Karabulut T C, Kurbanoglu S and Ozkan S A 2019 Chemically modified electrodes in electrochemical drug analysis *Curr. Pharm. Anal.* **16** 641–60
[22] Mousa M H, Dong Y and Davies I J 2016 Recent advances in bionanocomposites: preparation, properties, and applications *Int. J. Polym. Mater. Polym. Biomater.* **65** 225–54
[23] Cheng W, Tang X, Zhang Y, Wu D and Yang W 2021 Applications of metal-organic framework (MOF)-based sensors for food safety: enhancing mechanisms and recent advances *Trends Food Sci. Technol.* **112** 268–82
[24] Kaya S I, Yıldırım S, Cetinkaya A, Erkmen C, Uslu B and Ozkan S A 2021 Nanomaterial-based electroanalytical sensors for the selected prohibited anabolic agents, hormones and metabolic modulators and their sensitive assays *TrAC, Trends Anal. Chem.* **145** 116457
[25] Tanner E E L and Compton R G 2018 How can electrode surface modification benefit electroanalysis *Electroanalysis* **30** 1336–41
[26] Goyal R N and Bishnoi S 2012 Surface modification in electroanalysis: past, present and future *Indian J. Chem.* **51** 205–25
[27] Walcarius A 2012 Electrocatalysis, sensors and biosensors in analytical chemistry based on ordered mesoporous and macroporous carbon-modified electrodes *TrAC—Trends Anal. Chem.* **38** 79–97
[28] Manea F, Radovan C, Pop A, Corb I, Burtica G, Malchev P, Picken S and Schoonman J 2009 Carbon composite electrodes applied for electrochemical sensors *Sensors for Environment, Health and Security* pp 179–89
[29] Silah H, Erkmen C, Demir E and Uslu B 2021 Modified indium tin oxide electrodes: electrochemical applications in pharmaceutical, biological, environmental and food analysis *TrAC—Trends Anal. Chem.* **141** 116289
[30] Habibi B, Pashazadeh S, Saghatforoush L A and Pashazadeh A 2021 Direct electrochemical synthesis of the copper based metal-organic framework on/in the heteroatoms doped graphene/pencil graphite electrode: highly sensitive and selective electrochemical sensor for sertraline hydrochloride *J. Electroanal. Chem.* **888** 115210
[31] Kalambate P K, Sanghavi B J, Karna S P and Srivastava A K 2015 Simultaneous voltammetric determination of paracetamol and domperidone based on a graphene/platinum nanoparticles/nafion composite modified glassy carbon electrode *Sens. Actuators, B Chem.* **213** 285–94
[32] Selvi S V, Nataraj N, Chen S M and Prasannan A 2021 An electrochemical platform for the selective detection of azathioprine utilizing a screen-printed carbon electrode modified with manganese oxide/reduced graphene oxide *New J. Chem.* **45** 3640–51
[33] Shahrokhian S, Balotf H and Ghalkhani M 2015 Nano composite coating based on cellulose nanofibers/carbon nanoparticles: application to voltammetric determination of clonazepam *J. Solid State Electrochem.* **19** 251–60
[34] Rouhani M and Soleymanpour A 2021 Ultrasensitive electrochemical determination of trace ceftizoxime using a thin film of Preyssler nanocapsules on pencil graphite electrode surface modified with reduced graphene oxide *Microchem. J.* **165** 106160
[35] Mao A, Li H, Yu L and Hu X 2017 Electrochemical sensor based on multi-walled carbon nanotubes and chitosan-nickel complex for sensitive determination of metronidazole *J. Electroanal. Chem.* **799** 257–62

[36] de Souza Vieira L 2022 A review on the use of glassy carbon in advanced technological applications *Carbon N. Y.* **186** 282–302

[37] Hameed S, Munawar A, Khan W S, Mujahid A, Ihsan A, Rehman A, Ahmed I and Bajwa S Z 2017 Assessing manganese nanostructures based carbon nanotubes composite for the highly sensitive determination of vitamin C in pharmaceutical formulation *Biosens. Bioelectron.* **89** 822–8

[38] Gowda J I, Buddanavar A T and Nandibewoor S T 2015 Fabrication of multiwalled carbon nanotube-surfactant modified sensor for the direct determination of toxic drug 4-aminoantipyrine *J. Pharm. Anal.* **5** 231–8

[39] Satyanarayana M, Goud K Y, Reddy K K and Gobi K V 2015 Biopolymer stabilized nanogold particles on carbon nanotube support as sensing platform for electrochemical detection of 5-Fluorouracil in-vitro *Electrochim. Acta* **178** 608–16

[40] Jain R, Sinha A, Karolia P and Lateef Khan A 2016 Zinc oxide nanoflowers based graphene nanocomposite platform for catalytic studies of febuxostat *Int. J. Electrochem. Sci.* **11** 10223–37

[41] Tandel R, Teradal N, Satpati A and Jaldappagari S 2017 Fabrication of the electrochemically reduced graphene oxide-bismuth nanoparticles composite and its analytical application for an anticancer drug gemcitabine *Chin. Chem. Lett.* **28** 1429–37

[42] Pham T S H, Fu L, Mahon P, Lai G and Yu A 2016 Fabrication of β-cyclodextrin-functionalized reduced graphene oxide and its application for electrocatalytic detection of carbendazim *Electrocatalysis* **7** 411–9

[43] Fathirad F, Mostafavi A and Afzali D 2016 Electrospun Pd nanoparticles loaded on Vulcan carbon/conductive polymeric ionic liquid nanofibers for selective and sensitive determination of tramadol *Anal. Chim. Acta* **940** 65–72

[44] El Jaouhari A, Yan L, Zhu J, Zhao D, Zaved Hossain Khan M and Liu X 2020 Enhanced molecular imprinted electrochemical sensor based on zeolitic imidazolate framework/reduced graphene oxide for highly recognition of rutin *Anal. Chim. Acta* **1106** 103–14

[45] Wei X, Li L, Lian H, Cao X and Liu B 2021 Grain-like chiral metal-organic framework/multi-walled carbon nanotube composited electrosensing interface for enantiorecognition of Tryptophan *J. Electroanal. Chem.* **886** 115108

[46] Haidyrah A S, Sundaresan P, Venkatesh K, Ramaraj S K and Thirumalraj B 2021 Fabrication of functionalized carbon nanofibers/carbon black composite for electrochemical investigation of antibacterial drug nitrofurantoin *Colloids Surf. A Physicochem. Eng. Asp.* **627** 127112

[47] Promsuwan K, Soleh A, Saisahas K, Saichanapan J, Kanatharana P, Thavarungkul P, Guo C, Li C M and Limbut W 2021 Discrimination of dopamine by an electrode modified with negatively charged manganese dioxide nanoparticles decorated on a poly(3,4-ethylenedioxythiophene)/reduced graphene oxide composite *J. Colloid Interface Sci.* **597** 314–24

[48] Nataraj N and Chen S M 2021 Samarium vanadate nanospheres integrated carbon nanofiber composite as an efficient electrocatalyst for antituberculosis drug detection in real samples *Colloids Surf. A Physicochem. Eng. Asp.* **617** 126385

[49] Ahmadi-Kashani M and Dehghani H 2021 A new electrochemical sensing platform based on HgS/graphene composite deposited on the glassy carbon electrode for selective and sensitive determination of propranolol *J. Pharm. Biomed. Anal.* **194** 113653

[50] Umesh N M, Antolin Jesila J, Wang S F, Govindasamy M, Alshgari R A, Ouladsmane M and Asharani I V 2021 Fabrication of highly sensitive anticancer drug sensor based on

heterostructured ZnO-Co_3O_4 capped on carbon nitride nanomaterials *Microchem. J.* **167** 106244

[51] Zhang B, She N, Du J, Zhang M, Fang G and Wang S 2021 Nanocomposites based on quasi-networked Au1.5Pt1Co1 ternary alloy nanoparticles and decorated with poly-L-cysteine film for the electrocatalytic application of hydroquinone sensing *Ecotoxicol. Environ. Saf.* **207** 111251

[52] Yang W, Guo H, Xue R, Zhao X, Guan Q, Fan T, Zhang L, Yang F and Yang W 2021 0.2CNT/NiSex composite derived from CNT/MOF-74 as electrode material for electrochemical capacitor and electrochemical sensor *Microchem. J.* **168** 106519

[53] Chen T W, Chinnapaiyan S, Chen S M, Hossam Mahmoud A, Elshikh M S, Ebaid H and Taha Yassin M 2020 Facile sonochemical synthesis of rutile-type titanium dioxide microspheres decorated graphene oxide composite for efficient electrochemical sensor *Ultrason. Sonochem.* **62** 104872

[54] Vilian A T E, Oh S Y, Rethinasabapathy M, Umapathi R, Hwang S K, Oh C W, Park B, Huh Y S and Han Y K 2020 Improved conductivity of flower-like $MnWO_4$ on defect engineered graphitic carbon nitride as an efficient electrocatalyst for ultrasensitive sensing of chloramphenicol *J. Hazard. Mater.* **399** 122868

[55] Atta N F, Galal A and El-Gohary A R M 2020 New insight for simultaneous determination of hazardous di-hydroxybenzene isomers at crown ether modified polymer/carbon nanotubes composite sensor *J. Hazard. Mater.* **388** 122038

[56] Mutharani B, Ranganathan P, Chen S M and Sireesha P 2019 Ultrasound-induced radicals initiated the formation of inorganic–organic Pr2O3/polystyrene hybrid composite for electro-oxidative determination of chemotherapeutic drug methotrexate *Ultrason. Sonochem.* **56** 410–21

[57] Jiang W, Huang L, Zhang D, Wang Y and Pan G 2021 Graphene-based composites for electrochemical sensor fabrication and their application for drug detection *Int. J. Electrochem. Sci.* **16** 21037

[58] Zhu Q, Liu C, Zhou L, Wu L, Bian K, Zeng J, Wang J, Feng Z, Yin Y and Cao Z 2019 Highly sensitive determination of L-tyrosine in pig serum based on ultrathin CuS nanosheets composite electrode *Biosens. Bioelectron.* **140** 111356

[59] Sipa K, Socha E, Skrzypek S and Krzyczmonik P 2017 Electrodes modified with composite layers based on poly(3,4-ethylenedioxythiophene) as sensors for paracetamol *Anal. Sci.* **33** 287–92

[60] Khoshsafar H, Rofouei M K, Bagheri H and Kalbasi R J 2018 Ordered mesoporous carbon/poly (2-hydroxyethyl methacrylate)/Ag nanoparticle composite modified glassy carbon electrode; an amplified sensor for simultaneous determination of acetaminophen and domperidone *Electroanalysis* **30** 2454–61

[61] Sinha A, Dhanjai and Jain R 2015 Electrocatalytic determination of α 2-adrenergic agonist tizanidine at graphene-silicon dioxide nanocomposite sensor *Mater. Res. Bull.* **65** 307–14

[62] Mani S, Cheemalapati S, Chen S M and Devadas B 2015 Anti-tuberculosis drug pyrazinamide determination at multiwalled carbon nanotubes/graphene oxide hybrid composite fabricated electrode *Int. J. Electrochem. Sci.* **10** 7049–62

[63] Baytak A K, Duzmen S, Teker T and Aslanoglu M 2015 A novel composite electrode based on tungsten oxide nanoparticles and carbon nanotubes for the electrochemical determination of paracetamol *Mater. Sci. Eng.* C **57** 164–70

[64] Rajalakshmi K and John S A 2015 Selective determination of mefenamic acid in the presence of 1000-fold excess paracetamol and caffeine using a multiwalled carbon nanotube-polymer composite electrode *Anal. Methods* **7** 3506–11

[65] Yiğit A, Yardim Y, Çelebi M, Levent A and Şentürk Z 2016 Graphene/Nafion composite film modified glassy carbon electrode for simultaneous determination of paracetamol, aspirin and caffeine in pharmaceutical formulations *Talanta* **158** 21–9

[66] Ghadimi H, Tehrani R M A, Basirun W J, Ab Aziz N J, Mohamed N and Ab Ghani S 2016 Electrochemical determination of aspirin and caffeine at MWCNTs-poly-4-vinylpyridine composite modified electrode *J. Taiwan Inst. Chem. Eng.* **65** 101–9

[67] Joshi A, Chavan S N, Mandal D and Nagaiah T C 2016 Ionic liquid and nitrogen doped carbon nanotubes composite material for sensitive and selective detection of dopamine *Electroanalysis* **28** 2373–81

[68] Zhang A, Miao C, Shi H, Xiang H, Huang C and Jia N 2016 A novel solid-state electrochemiluminescence sensor for atropine determination based on $Ru(bpy)_3^{2+}$/carbon nanospheres/Nafion composite film *Sens. Actuators, B Chem.* **222** 433–9

[69] Ghadimi H and Tehrani R M A 2016 Sensitive voltammetric determination of acetaminophen at poly(4-vinyl pyridine)/graphene composite modified electrode *Anal. Bioanal. Res.* **3** 111–21

[70] Balamurugan T S T, Manibalan K, Chen S-M, Balasubramnian P and Huang S-T 2017 High sensitive electrochemical quantification of isoniazid in biofluids using copper particles decorated graphene oxide nano composite *Int. J. Electrochem. Sci.* **12** 9150–60

[71] Spindola R F, Zanin H, Macena C S, Contin A, de Cássia Silva Luz R and Damos F S 2017 Evaluation of a novel composite based on functionalized multi-walled carbon nanotube and iron phthalocyanine for electroanalytical determination of isoniazid *J. Solid State Electrochem.* **21** 1089–99

[72] Savalia R and Chatterjee S 2017 Sensitive detection of brucine an anti-metastatic drug for hepatocellular carcinoma at carbon nanotubes—nafion composite based biosensor *Biosens. Bioelectron.* **98** 371–7

[73] Mazloum-Ardakani M and Khoshroo A 2013 Nano composite system based on coumarin derivative-titanium dioxide nanoparticles and ionic liquid: determination of levodopa and carbidopa in human serum and pharmaceutical formulations *Anal. Chim. Acta* **798** 25–32

[74] Brahman P K, Suresh L, Reddy K R and Bondili J S 2017 An electrochemical sensing platform for trace recognition and detection of an anti-prostate cancer drug flutamide in biological samples *RSC Adv.* **7** 37898–907

[75] Kalambate P K, Rawool C R, Karna S P and Srivastava A K 2016 Highly sensitive and selective determination of methylergometrine maleate using carbon nanofibers/silver nanoparticles composite modified carbon paste electrode *Mater. Sci. Eng.* C **69** 453–61

[76] Brahman P K, Suresh L, Lokesh V and Nizamuddin S 2016 Fabrication of highly sensitive and selective nanocomposite film based on CuNPs/fullerene-C60/MWCNTs: an electrochemical nanosensor for trace recognition of paracetamol *Anal. Chim. Acta* **917** 107–16

[77] Mohamed M A, Yehia A M, Banks C E and Allam N K 2017 Novel MWCNTs/graphene oxide/pyrogallol composite with enhanced sensitivity for biosensing applications *Biosens. Bioelectron.* **89** 1034–41

[78] Khoshsafar H, Bagheri H, Rezaei M, Shirzadmehr A, Hajian A and Sepehri Z 2016 Magnetic carbon paste electrode modified with a high performance composite based on

molecularly imprinted carbon nanotubes for sensitive determination of levofloxacin *J. Electrochem. Soc.* **163** B422–7
[79] Santhosh A S, Sandeep S, James Bound D, Nandini S, Nalini S, Suresh G, Swamy N K, Jothiramalingam R and Arokiyaraj S 2021 A multianalyte electrochemical sensor based on cellulose fibers with silver nanoparticles composite as an innovative nano-framework for the simultaneous determination of ascorbic acid, dopamine and paracetamol *Surf. Interfaces* **26** 101377
[80] Bavandpour R, Rajabi M, Karimi-Maleh H and Asghari A 2021 Application of deep eutectic solvent and SWCNT-ZrO2 nanocomposite as conductive mediators for the fabrication of simple and rapid electrochemical sensor for determination of trace anti-migration drugs *Microchem. J.* **165** 106141
[81] Prabhu K, Malode S J, Veerapur R S and Shetti N P 2021 Clay-based carbon sensor for electro-oxidation of nimesulide *Mater. Chem. Phys.* **272** 124992
[82] Gandomi F, Marzi Khosrowshahi E, Sohouli E, Aghaei M, Saleh Mohammadnia M, Naghian E and Rahimi-Nasrabadi M 2020 Linagliptin electrochemical sensor based on carbon nitride-β-cyclodextrin nanocomposite as a modifier *J. Electroanal. Chem.* **876** 114697
[83] Shetti N P, Malode S J, Nayak D S, Naik R R, Kuchinad G T, Reddy K R, Shukla S S and Aminabhavi T M 2020 Hetero nanostructured iron oxide and bentonite clay composite assembly for the determination of an antiviral drug acyclovir *Microchem. J.* **155** 104727
[84] Liu Z, Jin M, Cao J, Wang J, Wang X, Zhou G, van den Berg A and Shui L 2018 High-sensitive electrochemical sensor for determination of Norfloxacin and its metabolism using MWCNT-CPE/pRGO-ANSA/Au *Sens. Actuators, B Chem.* **257** 1065–75
[85] Bukkitgar S D, Shetti N P and Kulkarni R M 2018 Construction of nanoparticles composite sensor for atorvastatin and its determination in pharmaceutical and urine samples *Sens. Actuators, B Chem.* **255** 1462–70
[86] Ghoreishi S M, Saeidinejad F, Behpour M and Masoum S 2015 Application of multivariate optimization to electrochemical determination of methyldopa drug in the presence of diclofenac at a nanostructured electrochemical sensor *Sens. Actuators, B Chem.* **221** 576–85
[87] Ebrahimi M, Nikoofard H, Faridbod F, Dezfuli A S, Beigizadeh H and Norouzi P 2017 A ceria NPs decorated graphene nano-composite sensor for sulfadiazine determination in pharmaceutical formulation *J. Mater. Sci., Mater. Electron.* **28** 16704–12
[88] Mohamed M A, Atty S A, Salama N N and Banks C E 2017 highly selective sensing platform utilizing graphene oxide and multiwalled carbon nanotubes for the sensitive determination of tramadol in the presence of co-formulated drugs *Electroanalysis* **29** 1038–48
[89] Satyanarayana M, Yugender Goud K, Koteshwara Reddy K and Vengatajalabathy Gobi K 2017 Conducting polymer-layered carbon nanotube as sensor interface for electrochemical detection of dacarbazine in-vitro *Electrocatalysis* **8** 214–23
[90] Rivera-Hernández S I, Álvarez-Romero G A, Corona-Avendaño S, Páez-Hernández M E, Galán-Vidal C A and Romero-Romo M 2016 Voltammetric determination of ibuprofen using a carbon paste—multiwalled carbon nanotube composite electrode *Instrum Sci. Technol.* **44** 483–94
[91] Khoshsafar H, Soc J E, Khoshsafar H, Bagheri H, Rezaei M, Shirzadmehr A and Hajian A 2016 Magnetic carbon paste electrode modified with a high performance composite based on molecularly imprinted carbon nanotubes for sensitive determination of levofloxacin

magnetic carbon paste electrode modified with a high performance composite based on mole *Electrochem. Soc.* **163** 422–7

[92] Kalambate P K and Srivastava A K 2016 Simultaneous voltammetric determination of paracetamol, cetirizine and phenylephrine using a multiwalled carbon nanotube-platinum nanoparticles nanocomposite modified carbon paste electrode *Sens. Actuators, B Chem.* **233** 237–48

[93] Žabčíková S, Vu D L, Červenka L, Tambor V and Vašatová M 2016 Determination of ascorbic acid in pharmaceutical preparation and fruit juice using modified carbon paste electrode *Potravinarstvo* **10** 255–9

[94] Tadayon F, Vahed S and Bagheri H 2016 Au–Pd/reduced graphene oxide composite as a new sensing layer for electrochemical determination of ascorbic acid, acetaminophen and tyrosine *Mater. Sci. Eng.* C **68** 805–13

[95] Kalambate P K, Biradar M R, Karna S P and Srivastava A K 2015 Adsorptive stripping differential pulse voltammetry determination of rivastigmine at graphene nanosheet-gold nanoparticle/carbon paste electrode *J. Electroanal. Chem.* **757** 150–8

[96] Li Y, Li Y, Gao J, Wang L, Zou L and Ye B 2015 A novel strategy of electrochemically treated $ZrOCl_2$ doped carbon paste electrode for sensitive determination of daidzein *Electroanalysis* **27** 1719–25

[97] Pradhan P, Mascarenhas R J, Thomas T, Namboothiri I N N, D'Souza O J and Mekhalif Z 2014 Electropolymerization of bromothymol blue on carbon paste electrode bulk modified with oxidized multiwall carbon nanotubes and its application in amperometric sensing of epinephrine in pharmaceutical and biological samples *J. Electroanal. Chem.* **732** 30–7

[98] Demir E, Inam R, Ozkan S A and Uslu B 2014 Electrochemical behavior of tadalafil on TiO_2 nanoparticles–MWCNT composite paste electrode and its determination in pharmaceutical dosage forms and human serum samples using adsorptive stripping square wave voltammetry *J. Solid State Electrochem.* **18** 2709–20

[99] Khaled E, Kamel M S, Hassan H N A, Haroun A A, Youssef A M and Aboul-Enein H Y 2012 Novel multi walled carbon nanotubes/β-cyclodextrin based carbon paste electrode for flow injection potentiometric determination of piroxicam *Talanta* **97** 96–102

[100] Li X, Chen Z, Zhong Y, Yang F, Pan J and Liang Y 2012 Cobalt hexacyanoferrate modified multi-walled carbon nanotubes/graphite composite electrode as electrochemical sensor on microfluidic chip *Anal. Chim. Acta* **710** 118–24

[101] Babaei A, Taheri A R and Khani Farahani I 2013 Nanomolar simultaneous determination of levodopa and melatonin at a new cobalt hydroxide nanoparticles and multi-walled carbon nanotubes composite modified carbon ionic liquid electrode *Sens. Actuators, B Chem.* **183** 265–72

[102] Deng P, Xu Z and Li J 2013 Simultaneous determination of ascorbic acid and rutin in pharmaceutical preparations with electrochemical method based on multi-walled carbon nanotubes-chitosan composite film modified electrode *J. Pharm. Biomed. Anal.* **76** 234–42

[103] Gholivand M B and Khodadadian M 2013 Simultaneous voltammetric determination of captopril and hydrochlorothiazide on a graphene/ferrocene composite carbon paste electrode *Electroanalysis* **25** 1263–70

[104] Zhou Y, Zhang H, Zhang J, Liu T and Tang W 2013 Chemical electrochemically sensitive determination of dopamine and uric acid based on poly (beryllon II)/nanowires-$LaPO_4$ modified carbon paste electrode *Sens. Actuators B. Chem.* **182** 610–7

[105] Norouzi P, Larijani B, Faridbod F and Ganjali M R 2013 Determination of cefoperazone based on nano-composite electrode using coulometric FFT admittance voltammetry *Int. J. Electrochem. Sci.* **8** 6118–30
[106] Mohamed H M 2016 Screen-printed disposable electrodes: pharmaceutical applications and recent developments *TrAC, Trends Anal. Chem.* **82** 1–11
[107] Muthumariappan A, Govindasamy M, Chen S M, Sakthivel K, Mani V, Chen T W and Selvaraj S 2017 Determination of non-steroidal anti-inflammatory drug (NSAID) azathioprine in human blood serum and tablet samples using multi-walled carbon nanotubes (MWCNTs) decorated manganese oxide microcubes composite film modified electrode *Int. J. Electrochem. Sci.* **12** 7446–56
[108] Karthik R, Karikalan N, Chen S M, Gnanaprakasam P and Karuppiah C 2017 Voltammetric determination of the anti-cancer drug nilutamide using a screen-printed carbon electrode modified with a composite prepared from β-cyclodextrin, gold nanoparticles and graphene oxide *Microchim. Acta* **184** 507–14
[109] Huang S J, Kannaiyan S, Venkatesh K, Cheemalapati S, Haidyrah A S, Ramaraj S K, Yang C C and Karuppiah C 2021 Synthesis and fabrication of Ni-SiO_2 nanosphere-decorated multilayer graphene nanosheets composite electrode for highly sensitive amperometric determination of guaifenesin drug *Microchem. J.* **167** 106325
[110] Ozcelikay G, Karadas-Bakirhan N, Taskin-Tok T and Ozkan S A 2020 A selective and molecular imaging approach for anticancer drug: pemetrexed by nanoparticle accelerated molecularly imprinting polymer *Electrochim. Acta* **354** 136665
[111] Sebastian N, Yu W C, Hu Y C, Balram D and Yu Y H 2022 Morphological evolution of nanosheets-stacked spherical ZnO for preparation of GO-Zn/ZnO ternary nanocomposite: a novel electrochemical platform for nanomolar detection of antihistamine promethazine hydrochloride *J. Alloys Compd.* **890** 161768
[112] Mohammadnia M S, Naghian E, Ghalkhani M, Nosratzehi F, Adib K, Zahedi M M, Nasrabadi M R and Ahmadi F 2021 Fabrication of a new electrochemical sensor based on screen-printed carbon electrode/amine-functionalized graphene oxide–Cu nanoparticles for Rohypnol direct determination in drink sample *J. Electroanal. Chem.* **880** 114764
[113] Venkatesh K, Muthukutty B, Chen S M, Karuppiah C, Amanulla B, Yang C C and Ramaraj S K 2021 Nanomolar level detection of non-steroidal antiandrogen drug flutamide based on ZnMn2O4 nanoparticles decorated porous reduced graphene oxide nanocomposite electrode *J. Hazard. Mater.* **405** 124096
[114] Sharma T S K and Hwa K Y 2021 Rational design and preparation of copper vanadate anchored on sulfur doped reduced graphene oxide nanocomposite for electrochemical sensing of antiandrogen drug nilutamide using flexible electrodes *J. Hazard. Mater.* **410** 124659
[115] Kimuam K, Rodthongkum N, Ngamrojanavanich N, Chailapakul O and Ruecha N 2020 Single step preparation of platinum nanoflowers/reduced graphene oxide electrode as a novel platform for diclofenac sensor *Microchem. J.* **155** 104744
[116] Selvi S V, Nataraj N and Chen S M 2020 The electro-catalytic activity of nanosphere strontium doped zinc oxide with rGO layers screen-printed carbon electrode for the sensing of chloramphenicol *Microchem. J.* **159** 105580
[117] Muthukutty B, Arumugam B, Chen S M and Ramaraj S K 2021 Low potential detection of antiprotozoal drug metronidazole with aid of novel dysprosium vanadate incorporated

oxidized carbon nanofiber modified disposable screen-printed electrode *J. Hazard. Mater.* **407** 124745
[118] Motaharian A, Hosseini M R M and Naseri K 2019 Determination of psychotropic drug chlorpromazine using screen printed carbon electrodes modified with novel MIP-MWCNTs nano-composite prepared by suspension polymerization method *Sens. Actuators, B Chem.* **288** 356–62
[119] Rajpurohit A S and Srivastava A K 2019 Simultaneous electrochemical sensing of three prevalent anti-allergic drugs utilizing nanostructured manganese hexacyanoferrate/chitosan modified screen printed electrode *Sens. Actuators, B Chem.* **294** 231–44
[120] Mohammadi S Z, Beitollahi H, Allahabadi H and Rohani T 2019 Disposable electrochemical sensor based on modified screen printed electrode for sensitive cabergoline quantification *J. Electroanal. Chem.* **847** 113223
[121] Fatma S, Prasad B B, Jaiswal S, Singh R and Singh K 2019 Electrochemical simultaneous analysis of dopamine and epinephrine using double imprinted One MoNomer acryloylated graphene oxide-carbon black composite polymer *Biosens. Bioelectron.* **135** 36–44
[122] Poo-arporn Y, Pakapongpan S, Chanlek N and Poo-arporn R P 2019 The development of disposable electrochemical sensor based on Fe_3O_4-doped reduced graphene oxide modified magnetic screen-printed electrode for ractopamine determination in pork sample *Sens. Actuators B Chem.* **284** 164–71
[123] Roushani M, Jalilian Z and Nezhadali A 2019 Screen printed carbon electrode sensor with thiol graphene quantum dots and gold nanoparticles for voltammetric determination of solatol *Heliyon* **5** e01984
[124] Raj M, Gupta P, Goyal R N and Shim Y B 2017 Graphene/conducting polymer nanocomposite loaded screen printed carbon sensor for simultaneous determination of dopamine and 5-hydroxytryptamine *Sens. Actuators, B Chem.* **239** 993–1002
[125] Hashkavayi A B and Raoof J B 2017 Design an aptasensor based on structure-switching aptamer on dendritic gold nanostructures/Fe_3O_4@SiO_2/DABCO modified screen printed electrode for highly selective detection of epirubicin *Biosens. Bioelectron.* **91** 650–7
[126] Bali Prasad B, Kumar A and Singh R 2017 Synthesis of novel monomeric graphene quantum dots and corresponding nanocomposite with molecularly imprinted polymer for electrochemical detection of an anticancerous ifosfamide drug *Biosens. Bioelectron.* **94** 1–9
[127] Yang X, Wu F, Chen D Z and Lin H W 2014 An electrochemical immunosensor for rapid determination of clenbuterol by using magnetic nanocomposites to modify screen printed carbon electrode based on competitive immunoassay mode *Sens. Actuators, B Chem.* **192** 529–35
[128] Ping J, Wu J and Ying Y 2010 Development of an ionic liquid modified screen-printed graphite electrode and its sensing in determination of dopamine *Electrochem. Commun.* **12** 1738–41
[129] Annu S S, Jain R and Raja A N 2020 Review—pencil graphite electrode: an emerging sensing material *J. Electrochem. Soc.* **167** 037501
[130] Dehnavi A and Soleymanpour A 2021 Titanium dioxide/multi-walled carbon nanotubes composite modified pencil graphite sensor for sensitive voltammetric determination of propranolol in real samples *Electroanalysis* **33** 355–64
[131] Pashazadeh S and Habibi B 2020 Determination of isoniazid by a copper-based metal-organic frameworks/carbon nitride quantum dots modified pencil graphite electrode as a highly sensitive and selective sensor *J. Electroanal. Chem.* **876** 114493

[132] Nagarajan S and Vairamuthu R 2021 Electrochemical detection of riboflavin using tin-chitosan modified pencil graphite electrode *J. Electroanal. Chem.* **891** 115235

[133] Eskiköy Bayraktepe D, İnal E K and Yazan Z 2021 Preparation and characterization of a pencil graphite electrode modified with gold nanoparticles decorated poly (l-methionine) and its use in the simultaneous sensitive electrochemical analysis of ascorbic acid, acetaminophen, chlorpheniramine maleate, and caffeine *Microchem. J.* **171** 106812

[134] Arabali V, Malekmohammadi S and Karimi F 2020 Surface amplification of pencil graphite electrode using CuO nanoparticle/polypyrrole nanocomposite; a powerful electrochemical strategy for determination of tramadol *Microchem. J.* **158** 105179

[135] Oghli A H and Soleymanpour A 2021 Pencil graphite electrode modified with nitrogen-doped graphene and molecular imprinted polyacrylamide/sol–gel as an ultrasensitive electrochemical sensor for the determination of fexofenadine in biological media *Biochem. Eng. J.* **167** 107920

[136] Dehnavi A and Soleymanpour A 2020 Highly sensitive voltammetric electrode for the trace measurement of methyldopa based on a pencil graphite modified with phosphomolibdate/graphene oxide *Microchem. J.* **157** 104969

[137] Vinay M M, Nayaka Y A, Yatisha R O, Basavarajappa K V, Manjunatha P and Purushothama H T 2020 Development of Azure-B modified pencil graphite electrode as an electrochemical sensor for the investigation of Levofloxacin in pharmaceutical and biological samples *Chem. Data Collect.* **28** 100441

[138] Kolahi-Ahari S, Deiminiat B and Rounaghi G H 2020 Modification of a pencil graphite electrode with multiwalled carbon nanotubes capped gold nanoparticles for electrochemical determination of tramadol *J. Electroanal. Chem.* **862** 113996

[139] Rouhani M and Soleymanpour A 2020 Preparation of Dawson heteropolyacid-embedded silver nanoparticles/graphene oxide nanocomposite thin film used to modify pencil graphite electrode as a sensor for trace electrochemical sensing of levodopa *Mater. Sci. Eng.* C **117** 111287

[140] Purushothama H T and Arthoba Nayaka Y 2019 Pencil graphite electrode based electrochemical system for the investigation of antihypertensive drug hydrochlorothiazide: an electrochemical study *Chem. Phys. Lett.* **734** 136718

[141] Ibrahim H and Temerk Y 2020 A novel disposable electrochemical sensor based on modifying graphite pencil lead electrode surface with nanoacetylene black for simultaneous determination of antiandrogens flutamide and cyproterone acetate *J. Electroanal. Chem.* **859** 113836

[142] Rezaei B, Boroujeni M K and Ensafi A A 2014 A novel electrochemical nanocomposite imprinted sensor for the determination of lorazepam based on modified polypyrrole@sol–gel@gold nanoparticles/pencil graphite electrode *Electrochim. Acta* **123** 332–9

[143] Pandey I and Jha S S 2015 Molecularly imprinted polyaniline-ferrocene-sulfonic acid-Carbon dots modified pencil graphite electrodes for chiral selective sensing of D-Ascorbic acid and L-Ascorbic acid: a clinical biomarker for preeclampsia *Electrochim. Acta* **182** 917–28

[144] Abdel-aal F A M, Rageh A H, Said M I and Saleh G A 2018 ε-MnO2-modified graphite electrode as a novel electrochemical sensor for the ultrasensitive detection of the newly FDA approved Hepatitis C antiviral drug ledipasvir *Anal. Chim. Acta* **1038** 29–40

[145] Manjunatha P, Arthoba Nayaka Y, Chethana B K, Vidyasagar c.c. and Yathisha R O 2018 Development of multi-walled carbon nanotubes modified pencil graphite electrode for the

electrochemical investigation of aceclofenac present in pharmaceutical and biological samples *Sens. Bio-Sens. Res.* **17** 7–17

[146] Rezaei B, Lotfi-Forushani H and Ensafi A A 2014 Modified Au nanoparticles-imprinted sol–gel, multiwall carbon nanotubes pencil graphite electrode used as a sensor for ranitidine determination *Mater. Sci. Eng.* C **37** 113–9

[147] Temerk Y, Ibrahim M, Ibrahim H and Kotb M 2016 Adsorptive stripping voltammetric determination of anticancer drug lomustine in biological fluids using *in situ* mercury film coated graphite pencil electrode *J. Electroanal. Chem.* **760** 135–42

[148] Nezhadali A, Rouki Z and Nezhadali M 2015 Electrochemical preparation of a molecularly imprinted polypyrrole modified pencil graphite electrode for the determination of phenothiazine in model and real biological samples *Talanta* **144** 456–65

[149] Prasad B B and Singh K 2017 Molecularly imprinted polymer-based core-shells (solid vs hollow) @ pencil graphite electrode for electrochemical sensing of certain anti-HIV drugs *Sens. Actuators, B Chem.* **244** 167–74

[150] Kutluay A and Aslanoglu M 2014 Nickel nanoparticles functionalized multi-walled carbon nanotubes at platinum electrodes for the detection of bromhexine *Sens. Actuators, B Chem.* **192** 720–4

[151] Chang Y and Venton B J 2020 Optimization of graphene oxide-modified carbon-fiber microelectrode for dopamine detection *Anal. Methods* **12** 2893–902

[152] Vitale F *et al* 2018 Fluidic microactuation of flexible electrodes for neural recording *Nano Lett.* **18** 326–35

[153] Wang K, Frewin C L, Esrafilzadeh D, Yu C, Wang C, Pancrazio J J, Romero-Ortega M, Jalili R and Wallace G 2019 High-performance graphene-fiber-based neural recording microelectrodes *Adv. Mater.* **31** 1805867

[154] Pushpanjali P A, Manjunatha J G, Hareesha N, Souza E S D, Charithra M M and Prinith N S 2021 Voltammetric analysis of antihistamine drug cetirizine and paracetamol at poly(L-Leucine) layered carbon nanotube paste electrode *Surf. Interfaces* **24** 101154

[155] Pushpanjali P A, Manjunatha J G and Srinivas M T 2020 Highly sensitive platform utilizing poly(l-methionine) layered carbon nanotube paste sensor for the determination of voltaren *FlatChem* **24** 100207

Chapter 6

Recent innovations in voltammetric techniques

Milan Z Momčilović

6.1 Introduction

Voltammetric techniques present a sophisticated analytical tool which is valuable in the reliable determination of various sorts of analytes since they have been shown to be highly sensitive, simple to operate, and efficient. The main advantage of voltammetry is based in its practicality since this technique requires low-cost equipment and can be successfuly applied both in laboratory and field conditions. Miniaturization of instruments used in voltammetric assay is the core of such properties.

It is known that organic compounds are routinely subjected to liquid and gas chromatographic separation followed by their photometric, thermal, or mass detection while inorganic species are usually determined by spectroscopic methods such as, for instance, inductively coupled plasma atomic emission spectroscopy and atomic absorbance or emission spectroscopy. As a contrast to these robust and expensive techniques which often demand time for sample preparation, voltammetry in the form of portable analysis has drawn attention as a most suitable substitute that can be used in almost any case. In addition, among the emerging alternatives that in simplified form can be applied for determination of wide range of analytes, voltammetry stands out especially in terms of its sensitivity and selectivity.

Generally, voltammetric techniques are regularly practiced by applying defined potential sweep and measuring the resulting currents. Pulses or oscillations are often imposed on the sweep resulting in a desired increase in sensitivity and additional data sets which can be related to versatile electrode processes [1].

The typical voltammetric set up is composed of three electrodes immersed in an electrochemical cell containing a solution of the analyte and background electrolyte, and a potentiostat which measures and displays electrochemical parameters of the system. A potentiostat is a device with an electronic circuit which consists of four components: the electrometer, the current/potential converter, the control amplifier, and the signal. Three electrodes in the system have different roles: the working

doi:10.1088/978-0-7503-5127-0ch6

electrode is an electrode where reactions of interest take place, the counter electrode (usually platinum coil or wire) which closes the circuit and allows electricity to flow through the cell, and the reference electrode (usually Ag/AgCl or saturated calomel electrode) which measures and controls the potential at the working electrode without passing any current [2].

So, the current is measured and analyzed as the potential at the working electrode is varied. Depending on the strategy of applying a potential difference several voltammetric techniques are established. Polarography was the first technique that was invented but due to its impracticality and the use of mercury many other options emerged over time such as linear sweep, square wave, staircase, cyclic voltammetry, and normal and differential pulse voltammetry. All these techniques are distinguished by the unique way of applying potential gradient in time. Stripping voltammetry is another variation which involves preconcentration of an analyte on the surface of the working electrode followed by subsequent redox reactions while current with analytical function is recorded. Three electrode voltammetric system also covers techniques with chromogenic and luminescent chemosensors [3] along with immunosensing and enzymatic biosensing [4].

As working electrodes in voltammetry, glassy carbon electrodes (GCE) and carbon paste electrodes (CPE) are the most frequently employed. These electrodes are very well known and recognized in everyday laboratory work and at the same time offer the possibility of simple comparison of the obtained results.

GCE as a commercially available working electrode exibits good electrical conductivity, good mechanical and dimensional stability, temperature resistance, low porosity, and wide potential range [5]. The performance of the electrode is often enhanced by deposition of materials with sensing atributes at the electrode surface.

In contrast to GCE, a carbon paste electrode is usually a custom made electrode which is known for its high sensitivity useful in the determination in many organic analytes. CPE is designed as a piston-driven holder with a paste packed inside of a short Teflon rod, polyethylene syringe or a glass tube electrically contacted via a conducting wire. Carbon paste is usually a mixture of graphite powder and liquid binder (e.g., silicone or paraffin oil, tricresyl phosphate). However, other suitable powdered materials can be introduced during the mixing of the paste in order to enhance the electrode response. CPEs are distinguished by their easy preparation, wide range of applicable potentials, economics, low current in the background, excellent surface features, mainly low toxicity, etc [6].

Many scientific papers convey novel ideas for functionalization of commercial working electrodes and their application in determination of analytes of interests. Such modification often implies involvment of nanomaterials with specific features which in some degree alter the pristine electrode surface and its functionality. Generally, the usable potential range of the electrodes depends on several factors, such as: solvent decomposition, decomposition of the supporting electrolyte, and electrode dissolution or formation of a layer of an insulating/semiconducting substance on its surface. In addition, solid electrodes can be adversely affected by poisoning through contact with solutions which contain harmful interferences [7].

This comprehensive study is dedicated to the most innovative ideas in the field of linear sweep, square wave, cyclic, and differential pulse voltammetry along with their stripping derived forms. Typically, application of voltammetry in purely analytical context will be outlined especially in scenarios where novel nanomaterials have been used for electrode modification and coupled to common voltammetric principles. In addition, new approaches in electrode fabrication in order to make them more convenient for specific use, more sensitive and more efficient are herein emphasized, as well.

6.2 Linear sweep voltammetry

In linear sweep voltammetry (LSV) a linear potential wave is applied while the scan rate is kept constant throughout the sweep. Voltammograms obtained using this technique give information about redox system, namely, analyte concentrations, diffusion constants, and rate constants. The linear potential wave is applied in the form of a function given by equation (6.1):

$$E_{(\tau)} = \pm \nu\tau + E_{(\tau=0)} \tag{6.1}$$

where $E(\tau)$ is the applied potential with respect to time, ± sign shows whether the reaction is oxidative or reductive, and ν is the scan rate.

Although well established decades ago this technique is still evolving and is being used to determine various sorts of compounds in aqueous media. One of the key specifics about application of this technique is the novel concepts of working electrodes.

For instance, a simple carbon nanotube-modified stencil printed carbon electrode as a one-shot electrode can be used for the determination of antidepressant drug vortioxetine (VOR) in pharmaceutical formulations and biological fluids by the LSV [8]. In this work, a regular potentiostat was used while apart from the mentioned working electrode, Pt wire was used as auxiliary electrode and standard Ag/AgCl was used as a reference electrode. This method is also distinguished by good accuracy and precision.

Great advantage herein is that polishing and smoothing steps, commonly unavoidable for other conventional solid electrodes, are not necessary for this procedure. Determination of vortioxetine in spiked human serum at low concentrations is possible due to higher sensitivity of the electrode. The explanation lays in the fact that bulk-modified disposable electrodes can replace the time-consuming drop casting and costly electrodeposition steps which are usually used for the preparation of multiple electrodes. It was revealed that under defined experimental conditions, this method shows linearity over the concentration range from 3×10^{-8} to 5×10^{-6} M. The detection limit is 1×10^{-8} M in aqueous solution and biological fluids. Figure 6.1 shows an evident increase in peak height in terms of obtained current influenced by the modified electrode. Regularity in peak heights correlated to vortioxetine concentrations is observed at figure 6.2 where corresponding calibrating curves are displayed.

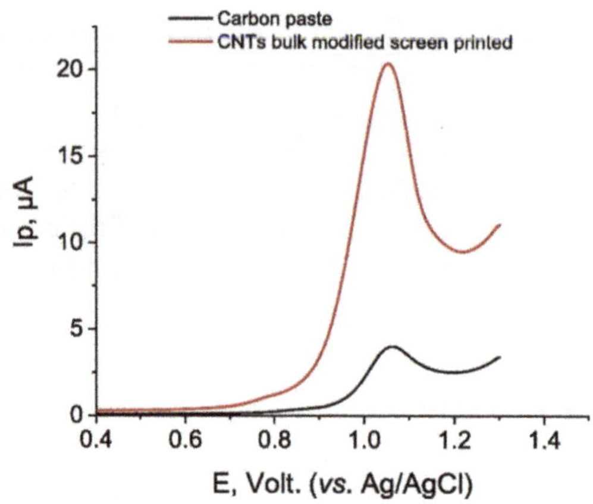

Figure 6.1. Linear sweep voltamograms for the electrochemical oxidation of 1×10^{-5} M VOR at the CP and the CNTs screen-printed electrodes. The measurements were carried out in BR buffer at pH 5 and at scan rate of 100 mV s^{-1}. (Reprinted from [8], copyright 2021, with permission from Elsevier.)

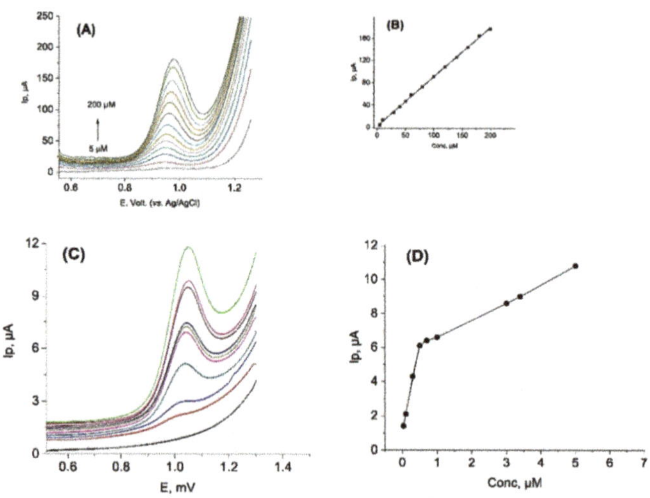

Figure 6.2. LSV of VOR and the corresponding calibration curve for the CP electrode (A and B) and the CNTs screen-printed electrode (C and D). The measurements were carried out in BR buffer at pH 5 and scan rate 100 mV s^{-1}. (Reprintedf rom [8], copyright 2021, with permission from Elsevier.)

Extraction, separation, and speciation analysis of chromium ions can be realized by the simple and fast solid phase micro extraction method using magnetic dextran (Sephadex G-150) as a sorbent coupled to method involving LSV [9]. When magnetic composites are employed as an adsorbent for so called, magnetic solid phase extraction (M-SPE), they are often surface modified by different kinds of materials. Consequently, special surface modification of magnetic nanoparticles is of huge significance. The retained Cr (VI) ions on the magnetic dextran sorbents can be eluted and analyzed by LSV at the gold nanoparticles modified screen-printed carbon electrode (M-SPCE). Figure 6.3 shows regular ratios in a current versus

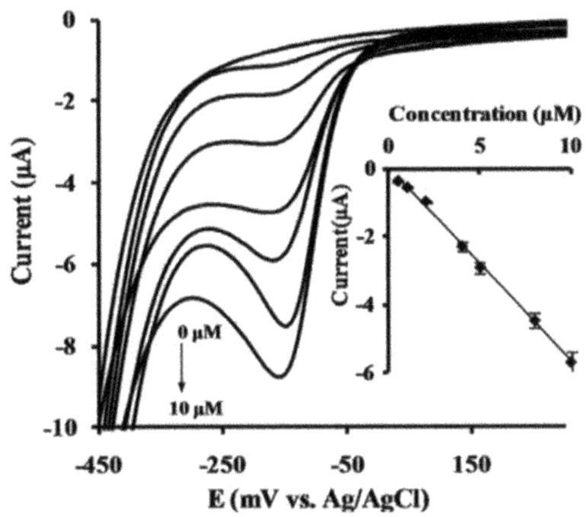

Figure 6.3. Linear sweep voltammograms (after extraction process) at AuNPs@ SPCE for different Cr (VI) concentrations in acetate buffer (pH 4.75): 0.0, 0.5, 1.0, 2.0, 4.0, 5.0, 8.0, and 10.0 μM. The inset shows the corresponding analytical curve obtained using the reduction peak currents of these voltammograms. Vertical bars indicate the relative standard deviation (n = 3). Magnetic solid phase extraction conditions: sample pH: 3.0, sorbent amount (mg): 10, adsorption time (min): 5.0, desorption time (min): 1.0, elution volume (mL): 0.25 mL of 0.1 M NaOH. (Reprinted from [9], copyright 2019, with permission from Elsevier.)

potential diagram for linear sweep voltammetry. These electrodes allow the selective analysis of chromium (VI) ions in the presence of relatively high Cr (III) concentrations which is a base for the method designed for speciation of these analytes. This sensor showed a wide linear range (0.5–10 μM), lower detection limit (0.01 μM), acceptable reproducibility, selectivity, and stability in analyses of tea, coffee, and mineral water samples. Precision and accuracy of the M-SPCE@LSV technique are satisfactory. It is concluded that the described method could be applied to the determination of trace levels of chromium ions in biological, pharmaceutical, and environmental samples.

Microelectrodes have become a part of the standard electroanalytical set-ups due to their advanced mass transport characteristics, their ease of handling, and their reusability. Recently, a novel approach for carrying out linear sweep voltammetry at a microdisc electrode has attracted some attention. In this method a semi-circular potential wave creates a transiently infinitely fast scan rate at the mid-point of the potential window and slower scan rates everywhere else. The infinitely fast scan rate results in a significant increase in the current response while the slow scan rates in other areas reduce the current response [10]. Conclusively, this method can be applied for selective detection of analytes in the presence of a species in a mixture.

This approach can be used for determination of the standard electrochemical rate constant via its application at a microdisc electrode. Uchida and coworkers investigated a one-electron reduction reaction of one reactant to its electrochemical product with varying dimensionless kinetic parameters (K_0) and evidenced that the semi-circular potential wave approach can measure K_0 values one order greater than the values that can be measured via the conventional linear potential wave technique [11].

The presented data allows experimentalists to determine the rate constant of the redox systems under study with ease by comparing their values with that in the figures.

Experimentally, an analyst measures multiple linear sweep voltammograms with a semi-circular potential wave using various E_{centre} values and the peak current of each resulting voltammogram is determined. E_{centre} is the average of the potential window. The absolute maximum dimensional peak current is then converted to dimensionless coordinates using the known transformation equations. Then, the E_{centre} at which it is observed is recorded.

From figure 6.4, one can also estimate the θ_{shift} at which the experimentally obtained dimensionless absolute maximum peak current is observed. θ_{shift} is a dimensionless formal potential of the electrochemical couple. Using E_{centre} and θ_{shift}, the formal potential can be calculated and further use to determine the rate constant.

There is an interesting novel voltammetric method based on a semi-circular potential sweep (SCV) applied to the simultaneous electroanalytical determination of two components with similar oxidation potentials [12]. Here their resolution is precluded in conventional voltammetric methods including linear sweep, square wave, and pulse voltammetries. In a semi-circular potential sweep the instantaneous scan rate varies from zero to near infinite (within the limitations of the controlling electronics) at the mid-point of the scan (figure 6.5). This approach is used in two component mixtures such as ascorbic acid/acetaminophen, glucose/ethanol and hydroquinone/catechol which are important in biological assays. Semi-circular sweep voltammetry is capable of giving nice convenience over the other frequently used voltammetric techniques in terms of signal resolution and sensitivity. Figure 6.6 shows the application of this technique for ascorbic acid (AA) determination in the presence of acetaminophen (AC). Conditions of the experiments are provided in the description of the figure.

It is concluded that this novel approach is characterized by the fact that SCV resolves signals that are strongly overlapped and are not distinguished using other

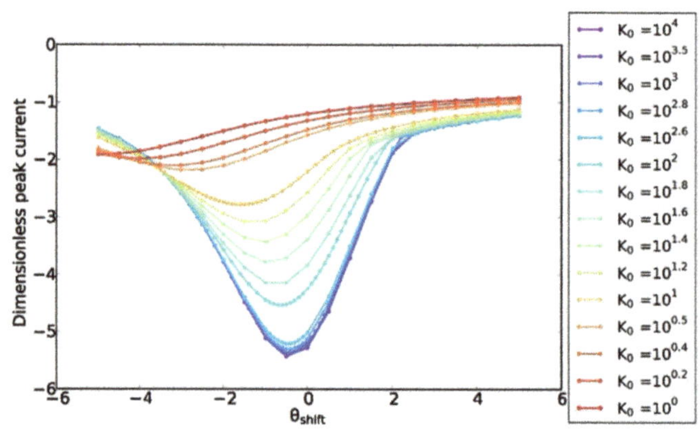

Figure 6.4. Dimensionless peak currents plotted against corresponding θ_{shift} values for varying K_0 using $\sigma = 3.89$. ($\alpha = 0.5$). (Reprinted from [11], copyright 2019, with permission from Elsevier.)

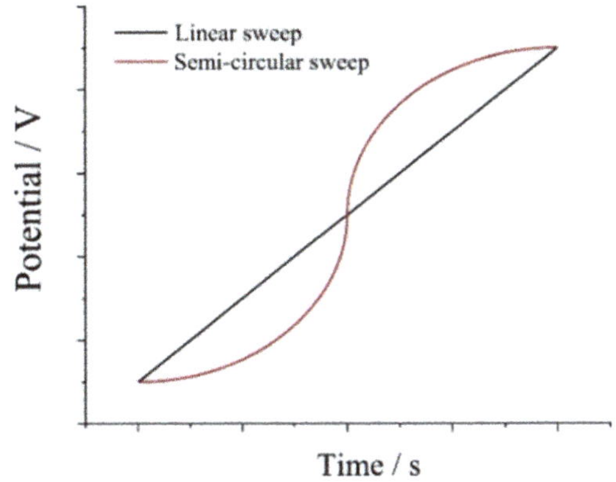

Figure 6.5. Waveforms of cyclic voltammetry (black line) and semi-circular sweep voltammetry (red line) at the same average scan rate. (Reprinted from [12], copyright 2021, with permission from Elsevier.)

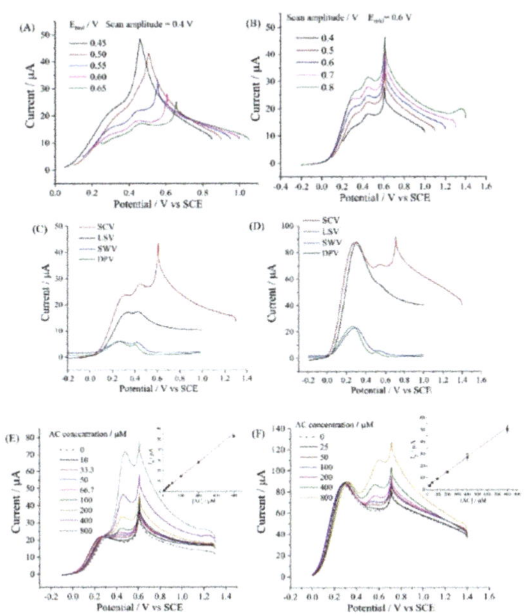

Figure 6.6. Semi-circular sweep voltammetric responses of 100 μM AC in the presence of 1 mM AA at the GC electrode recorded at potential windows with (A) varying E_{mid} (B) varying scan amplitudes. Semi-circular sweep, linear sweep, square wave, and differential pulse voltammetric responses of 100 μM AC in the presence of (C) 1 mM and (D) 5 mM AA. Semi-circular voltammetric responses of (E) 10–800 μM AC in the presence of 1 mM AA and (F) 25–800 μM AC in the presence of 5 mM AA. The E_{mid} of the potential windows used in SCV at the potential was 0.6 V vs. SCE in (C) and (E) and 0.7 V vs. SCE in (D) and (F). The scan amplitude and the average of scan rate used in (C)–(E) were 0.7 V and 50 mVs^{-1}, respectively. Insets of (E) and (F): plots of subtracted peak currents obtained from semi-circular voltammograms versus AC concentration. (Reprinted from [12], copyright 2021, with permission from Elsevier.)

voltammetric techniques which is crucial in resolution of the peaks. Also, SCV amplifies its peak current and improves the sensitivity of detection.

6.3 Cyclic voltammetry

Cyclic voltammetry (CV) is an electrochemical technique performed by applying a linearly cycled potential to the working electrode which is immersed in an electrolyte solution containing one or more analytes. Such a triangular potential wave leads to the defined current response which can be used for determination of the analytes' concentrations. This technique is classically a three electrode set up widely used for determination of many compounds including heavy metals, pesticides, hormones, medicaments, and other classes of chemicals in almost every type of sample. Although this technique is almost 100 years old, it is in its modern interpretations still one of the most popular options among the contemporary analytical methods. Moreover, knowledge from this field is constantly being enriched by new intelligence related to the detection of new compounds.

In one of many examples, CV was employed to deposit copper nanostructures on a film of multiwall carbon nanotubes (MWCNTs) modified pencil graphite electrode (MWCNTs/PGE). This composite sensor (CuNS–MWCNT/PGE) exerted synergistic electrocatalytic activity for the oxidation of hydrazine in phosphate buffer [13]. Such a possibility is exploited for hydrazine determination. The attributes of this sensor are simple preparation, low cost, outstanding sensitivity, selectivity, and reproducibility. The catalytic currents of square wave voltammetry has a linear correlation with the hydrazine concentration in the range 0.1 to 800 μM with a low detection limit of 70 nM which is used as a base for the determination. Amperometric oxidation current shows a linear correlation with hydrazine concentration in the range 50 to 800 μM with the detection limit of 4.3 μM. Work with real samples characterized this sensor with good reproducibility, long-term stability, and its selective response to hydrazine with no interferences.

The tungsten pure metal electrode can be used for selective phosphate-ion determination [14]. One should first determine the modification potential of the reaction to form phosphotungstic acid ($H_3O_{40}PW_{12} \times H_2O$) by cyclic voltammetry. Then, the modified electrochemical method should be applied in a three electrode system. The tungsten electrode should be electrodeposited under a constant potential certified by CV. Concentration of phosphate ions are to be determined by such a modified tungsten electrode by potentiometry in a two-electrode system. So, it is here demonstrated that CV can be used as a pretreatment step in order to prepare potentiometric sensor for phosphate ions. The detection limit of this electrode for phosphate ions are reported as follows: 10^{-6} M from pH 7 to pH 8 and 10^{-5} M from pH 9 to pH 10. Comparison of this method to conventional methods for phosphate determination showed results close to real values.

Recently, a new application of a glassy carbon electrode was elaborated in s case when it was modified with gadolinium sulfide (Gd_2S_3) and reduced graphene oxide (RGO) composite and used in the detection of pesticide carbofuran (CF) [15]. This novel composite was prepared by the hydrothermal procedure and used as GCE modified sensor towards environmental and food samples. Electrochemical impedance spectroscopy and cyclic voltammetry demonstrated that this Gd_2S_3/RGO sensor was highly sensitive and selective to cf. It exhibited a low detection limit of 0.0128 μM and linearity in the range 0.001 to 1381 μM.

It is also shown that cyclic voltammetry and square wave voltammetry can be used for the selective and rapid determination of bisphenol A in standard samples and in thermal paper samples by means of a silver nanoparticle-exfoliated graphite nanocomposite (AgNPs-EG) sensor [16]. It is reported that silver nanoparticles (AgNPs) were synthesized separately, loaded onto the expanded layers of exfoliated graphite (EG), compressed into a 0.6 cm pellet, and used as a sensor for the voltammetric assay. By applying defined experimental conditions, the oxidation peak current is a linear function of bisphenol A concentrations. This is used for the concentration range 5.0 to 100 μM with excellent coefficient of determination and low limit of detection such as 0.23 μM. "An interesting fact is that this electrode is resistant to fouling with good reproducibility (RSD = 2.62%, $n = 5$) by mechanical polishing of the electrode on emery paper. Also, this method shows good accuracy with 93.1 to 113% recovery.

Recent progress in electrochemical sensing has led to the evolving of miniature sensors for the determination of the manganese ion (Mn^{2+}). Boselli and coauthors developed a sensor which relies on a cathodic stripping voltammetry (CSV) electroanalytical technique performed on a miniaturized platinum working electrode [17]. Construction of the sensor is given in the figure 6.7 and explained in its

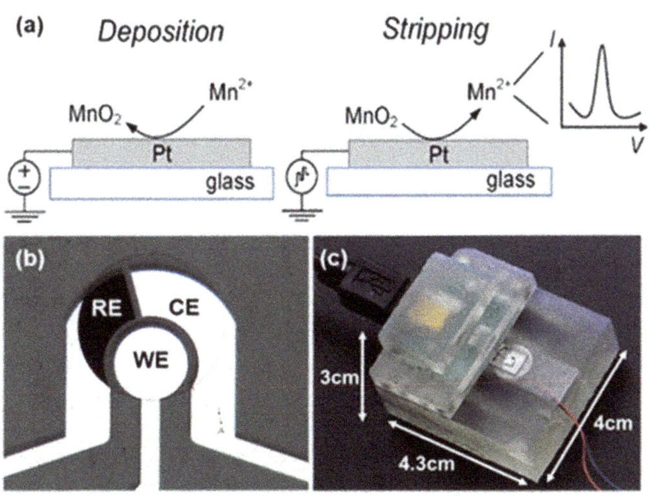

Figure 6.7. Sensing electrode with thin platinum film for the determination of manganese in drinking water. (a) Scheme of the principle employed in CSV (deposition phase with oxidation of Mn^{2+}) which is followed by formation of a layer of MnO_2 on the surface of the Pt working electrode, and the stripping phase to reduce Mn^{4+} back to Mn^{2+} off the electrode. (b) Photograph of the three electrode micro fabricated sensor with a 1 mm diameter circular Pt WE, a Pt CE, and a Ag/AgCl RE (WE = working electrode, RE = reference electrode, CE = counter electrode). (c) 3D printed interface housing for the sensor. (Reprinted with permission from [17]. Copyright 2021, American Chemical Society.)

description below. The cathodic stripping voltammetry parameters were as follows: deposition at 0.7 V for 900 s, with square wave parameters of 70 ms period, 25 mV amplitude, and 4 mV step potential. This sensor coupled to a miniature potentiostat was used for determination of Mn^{2+} concentrations in drinking water. Inductively coupled plasma mass spectrometry was employed as a standard technique for validation. Analysis was focused in the range 0.03 ppb to 5.3 ppm. In addition, a compact small-sized interface to integrate agitation function and improve the limit of quantification to 6 ppb was devised. Agreement of 100% and acceptable accuracy (~70%) and precision (91%) when benchmarked with ICP-MS were achieved.

Special attention has been focused on the phenomenon that voltammetric systems for simultaneous detection of biomolecules can be embodied as a smartphone supported device. A smartphone integrated electrochemical system contains a disposable sensor, a coin-size detector, and a smartphone with an Android application installed on it. In a paper by Ji and coauthors, such a system was used for reliable, effective, and rapid determination of ascorbic acid, dopamine, and uric acid in artificial urine [18]. Reduced graphene oxide and gold nanoparticles were modified on screen-printed electrodes and employed as sensors. The detector of the system measures the resulting current which rises in combined cyclic and differential pulse voltammetry. The smartphone has the role of communicating with the detector, calculating data, and plot voltammograms in real time. Three peak currents with analytical importance at 120, 10, and 220 mV which correspond to ascorbic acid, dopamine, and uric acid, respectively, were observed (figure 6.8). The authors have proved that this smartphone integrated voltammetric technique could be easily applied to test the activity of defined biomolecules with linearity in a wide concentration range, high sensitivity, and selectivity. The range of linearity and detection limits are 20–375 μmol L^{-1} and 1.04 μmol L^{-1} for ascorbic acid, 1–160 μmol L^{-1} and 0.29 μmol L^{-1} for dopamine, and 25–200 μmol L^{-1} and 5.4 μmol L^{-1} for uric acid, respectively. The resulting voltammogram of the smartphone based system and the commercial electrochemical workstation in redox

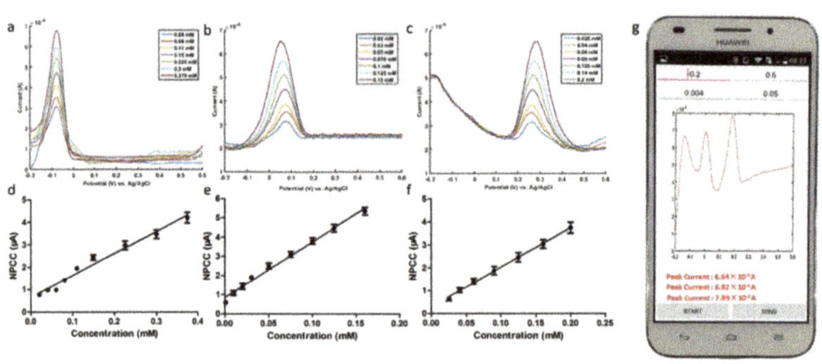

Figure 6.8. The diagram of differential pulse voltammetry measurement using the system for (a) ascorbic acid, (b) dopamine, and (c) uric acid at different concentrations. The dose-dependent curve of (d) ascorbic acid, (e) dopamine, and (f) uric acid. (g) Real-time differential pulse voltammetry detection on the smartphone screen. (Reprinted from [18], copyright 2018, with permission from Elsevier.)

couple solution for both cyclic and differential pulse voltammetry are almost identical which is an indicator of the reliability of this set up.

Novel voltammetric sensors are also being developed for the determination of polycyclic aromatic hydrocarbon. Determination of anthrone by cyclic voltammetry and differential pulse voltammetry is possible at the electrochemical sensor constructed using carbon paste electrode modified with octyl phenol ethoxylate surfactant (OPEMCPE) and carbon paste electrode modified with sodium lauryl sulfate (SLSMCPE) [19]. Both electrodes offer greater stability, rapid electron transfer rate, good reproducibility and repeatability, and sensitivity in comparison to the bare unmodified form of the electrode. The electrode reaction is reversible with one electron and one proton participating in the transfer process. This kinetics of this process is defined as diffusion controlled.

Surfactant modified graphite paste electrodes are a promising tool for determination of hormones such as estriol with dopamine and uric acid [20, 21], medications such as resorcinol [22], dyes such as indigotine [23] and Alizarin Red-S [24], food supplements such as curcumin [25], vitamins such as riboflavin [26], and other chemicals with specific purpose such as phloroglucinol [27]. In all of the cases, cyclic and differential pulse voltammetry has been shown to be suitable techniques for quantitative determination of these chemically diverse compounds.

Graphite paste electrodes modified by electro polymerization is another convenient option for effective determination of amino acids such as tyrosine [28], neurotransmitter such as dopamine [29], vitamins such as riboflavin [30], antihistamine drugs such as cetirizine [31], and compounds found in foodstuff such as vanillin and caffeine [32]. Electro polymerization is commonly used technique for the formation of polymeric film at the surface of bare electrode where the analyst can control the thickness of the polymer film by optimizing the number of scanning segments and the potential window. Generally, the bare electrode is cycled in the solution of a compound that is planned for polymerization such as amino acids DL-valine, adenine, threonine or vitamin niacin as seen in the above mentioned cases. During the continuous potential scanning, the current response goes on increases, signifying the formation of polymeric film and the enhancement in the conductivity of the electrode surface.

CV can also be successfully used for testing the reversibility of a reaction, the formal reduction potential of a species, electron transfer kinetics, energy levels of semiconducting polymer, etc.

6.4 Differential pulse voltammetry

In differential pulse voltammetry (DPV), a voltage profile is applied with small, increased voltage pulse superimposed over the voltage ramp while the current is sampled just before the pulse is applied and just before the pulse ends. The difference between the two recorded currents is plotted versus the base potential. In this way, enhancement of sensitivity is achieved in contrast to normal pulse voltammetry.

One of the valuable features offered by DPV is the possibility to study biochemical reactions. In a work by Fini and Kerman, DPV was used as an analytical tool to analyze nitrite reductase activity (NRA) of hemoglobin in a potential range of +0.3 to −1.3 V (vs. Ag/AgCl) [33]. This method provided fast measurements of NRA by using low quantity of reagents at the electrode surface. Two different regimes for reduction of nitrite by hemoglobin in its Fe (II) and Fe (I) states were identified. The reduction in Fe (II) state was selected as a means for comparing the NRA of hemoglobin (Hb) and hemoglobin-S (Hb-S) under native-like conditions in a didodecyldimethyl ammonium bromide liquid crystal film. Highly sensitive DPV measurements was successfully employed for understanding the direct mechanism of hemoglobin redox reactions utilizing a simplified mediator-free platform.

Triclosan is antimicrobial agent whose levels in commercial products can be determined by DPV. As shown by Su *et al* the core–shell structure of bacteriochlorin-supported gold nanoparticles can be synthesized for further chemical modification of electrodes which are applied as a highly sensitive sensor for determination of triclosan at sub-micromolar level [34]. Such a modified electrode is characterized by its high sensitivity, including large surface-confined dyes, good molecular recognition and efficient interaction, good conductivity, good adsorption characteristics, and high accumulation efficiency. This electrode has superb electrocatalytic effects towards the triclosan oxidation by lowering of the oxidation overpotential and an enhancement of the oxidation current responses in comparison with bare electrodes. DPV in this case expresses major analytical features, including a wide linear range, low detection limit and high sensitivity. The proposed method can be applied by using the standard addition method for the determination of triclosan in commercial toothpaste and hand wash as real personal care products with good recovery values.

Simultaneous quantification of several heavy metal ions is possible using DPV. In a paper by Sun *et al* a novel electrochemical sensor based on indium tin oxide (ITO) glass electrode modified by hydroxyapatite (HAp) film was fabricated by the electrodeposition method and applied for simultaneous detection of Pb^{2+}, Cu^{2+} and Hg^{2+} in tap water [35]. Linearity of the measurements can be achieved in the ranges 1×10^{-7} to 1×10^{-6} M and 3×10^{-6} to 1×10^{-5} M for Pb^{2+}, 1×10^{-7} to 1×10^{-6} M and 1×10^{-6} to 1×10^{-5} M for Cu^{2+}, and 1×10^{-7} to 1×10^{-5} M for Hg^{2+}. Low limits of detection (S/N=3) such as: 1×10^{-12}, 1×10^{-12}, 8×10^{-8} M for Pd^{2+}, Cu^{2+} and Hg^{2+}, respectively, confirm that this idea stands a chance for a regular practice of heavy metal determination along with its outstanding anti-interference capabilities and selectivity.

Similarly, DPV can be used for simultaneous determination of Cu^{2+}, Pb^{2+} and Cr^{3+} in real water and food samples as shown by Ghorbani *et al* [36]. Thid group of authors developed magnetic dispersive solid phase micro extraction as a clean-up step which reduces the effect of interfering species in the test with real samples. As working electrodes in DPV, four variants of glassy carbon electrode modified with derivatives of β-cyclodextrin amino acids (methionine, valine, phenylalanine, and tryptophan) were used. This technique was chosen in contrast to cyclic voltammetry for sharper and better analyte peaks along with resolutions improved with decreasing background current. It was noticed that DPV showed suitable linearity range with high repeatability and good stability in experiments. The limit of detection and limit of quantification were lower than 0.15 and 0.5 ng mL^{-1} for all analytes, respectively. Upon analysis of several real samples, it can be concluded that this method is a suitable, sensitive, and accurate procedure for the determination of Cu^{2+}, Pb^{2+} and Cr^{3} in aqueous media with relative recoveries of higher than 86.4.

There is an interesting example of application of DPV to carry out in-depth research of the electro-Fenton process. In a work by Arellano *et al* degradation of the ionic liquid p-xylene-bis(n-pyridinium bromide) was analyzed and herein DPV allowed monitoring of the evolution of Fe (III) reduction as redox probe [37]. This enabled examination of the degradation process of p-xylene-bis(n-pyridinium bromide), the effect of the potential and main intermediates which were formed. DPV can be employed to evaluate the application of sequential current intensities along the electro-Fenton process as an alternative to reduce the energy consumption keeping the same efficiency. It was established that currents of 50 and 600 mA were optimal for testing a bi-electrical current system. This was based on the profile of the voltammetric curves and the determination of hydroxyl radicals. Hence, a current of 600 mA can be applied for 30 min followed by 50 mA for 60 min, and vice versa. Both trials achieved a total organic carbon abatement around 60%, the former system being less energy consuming (0.24 kWh g^{-1} TOC) than the latter system (5.90 kWh g^{-1} TOC). This work emphasizes the potential of the DPV technique to profoundly study the electro-Fenton process and represent an example that proves wide application area of DPV.

Another interesting application of DPV is found in screening the samples of seized drugs. 3,4-methylenedioxymethamphetamine (MDMA), known as Ecstasy, can be determined in a drug sample by DPV and applying boron-doped diamond electrode (BDDE) as the working sensing electrode [38]. The proposed method is based on the oxidation of the MDMA at a BDDE in Britton Robinson buffer solution pH 10 as the supporting electrolyte. Determination of MDMA in this way is characterized with good linearity in the range 1.0 to 500 μmol L^{-1} and a low limit of detection (0.3 μmol L^{-1}) (figure 6.9). In addition, this method shows perfect

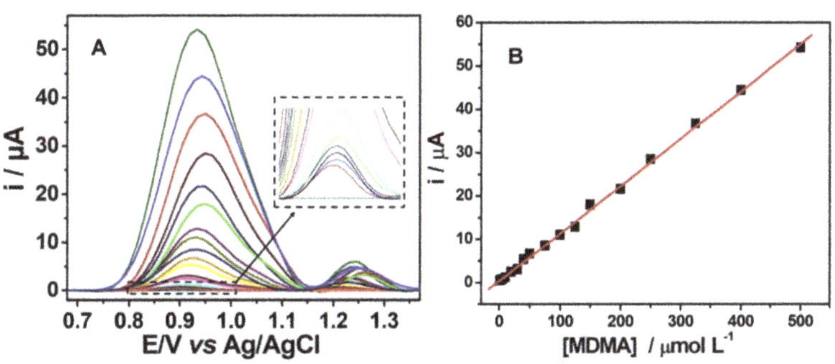

Figure 6.9. (A) Voltammograms obtained by DPV at the BDDE for MDMA standard solutions (1.1 to 500 μmol L^{-1}). The inset is a zoom-in of lower measurable signals. Scan rate of 10 mV s^{-1} and amplitude of 100 mV. (B) Calibration curve obtained with data from A. (Reprinted from [38], copyright 2020, with permission from Elsevier.)

selectivity with defined experimental conditions when nine other synthetic drugs, including amphetamine, methamphetamine, and N-methyldiethanolamine are present among the seized samples.

Recently, DPV has found a significant role in determination of pharmaceuticals in water samples. So called β-adrenenergic receptor blockers are one of the most frequently prescribed pharmaceutics all over the world and their presence in natural waters is of huge interest. Pindolol (Pin), acebutolol (Ace) and metoprolol (Met) are representatives of this group of medicaments which can be determined in aqueous media by applying modified carbon paste electrode (CPE) as working electrode in DPV. In a work of Silva and coauthors, hexagonal mesoporous silica (HMS) and amino-functionalized HMS (HMS-NH_2) were prepared and incorporated into carbon paste electrodes for a modification purpose [39]. Initially, the dependence of the resulting current with the type and pH of the electrolyte support along with the accumulation time was investigated in order to optimize the experimental conditions for further analyses. Three well-defined peaks in the voltammograms that appear at about +0.85 V (Pin), +1.11 V (Ace) and +1.45 V (Met) are of analytical importance. It is established that prepared sensor exhibits good reproducibility, wide linear ranges and very low detection limits (0.1, 0.046, and 0.23 μmol L^{-1} for Pin, Ace, and Met, respectively) when real samples such as tap, mineral, river, and sewage effluent waters were assayed. Also, this method has facile sample treatment and has recoveries in the range 97% to 112%.

To conclude, DPV is a powerful technique capable of determining heavy metals, antimicrobial agents, biochemically active molecules, and for studying the mechanisms of electrode processes.

6.5 Square wave voltammetry

Square wave voltammetry starts at a defined potential, and is pulsed negatively then positively by a potential (ΔE_p>), before repeating a positive and negative pulse around a new potential which adds a step (E_{step}). So, potentials changes in this manner:

$$E_{start} \rightarrow E_{start} + \Delta E_p \rightarrow E_{start} - \Delta E_p \rightarrow E_{start} + \Delta E_p + E_{step} \rightarrow E_{start} - \Delta E_p + E_{step}$$
$$\rightarrow E_{start} + \Delta E_p + 2E_{step} \rightarrow E_{start} + \Delta E_p - 2\,E_{step}\ldots$$

The current is usually recorded towards the end of each pulse, and the three different values plotted. These values are the forward current (where ΔE_p > is added), the reverse current (where ΔE_p > is subtracted), and their difference is plotted.

As pointed out by Mirčeski *et al* theoretical and experimental results in SW voltammetric behavior is controlled by the two time parameters, namely the frequency and the overall potential scan rate [40].

Square wave voltammetry can be applied for quantitative detecting of phenol in industrial flowing wastewater samples. The mechanism of detection is based on the oxidation of phenol into hydroquinone which shows an oxidation peak at 0.1 V in

0.1 M phosphate buffer at pH 7. This peak is stable and does not shift with changing phenol concentration. With a preconcentration step, the limit of detection for phenol can reach 0.1 ppb. Furthermore, the sensor shows good sensitivity for phenol between 30 and 240 ppb in a milifluidic device [41].

In a paper by Altunkaynak *et al* SWV has been identified as a practical alternative to spectroscopic and chromatographic methods in clinical quantification of vildagliptin (VLDG), one of the drugs used in the treatment of diabetes [42]. The method can be performed without any pretreatment and separations in commercial tablet and urine samples. For the first time, on the pencil graphite electrode (PGE) as a working electrode for sensing VLDG, a good linear correlation was obtained between the anodic current signal and concentration in the range 2.94 to 49.98 μM in phosphate buffer solution at pH 9.0. The limit of detection and relative standard deviation were calculated as 82 nM (24.88 ng L^{-1}) and 2.95%, respectively. So, SWV is here labeled as a less time-consuming, much cheaper, simple, sensitivem and selective electroanalytical method for determination of this medicament. Figure 6.10 shows voltammograms for determination of VLDG in tablet of medicament Vitaptin which contains it in a urine sample.

Parabens are biocides used as preservatives in food, cosmetics, and pharmaceuticals [43]. It was recently shown that SWV can be used for simultaneous determination of methyl, ethyl, propyl, and butyl parabens in sweetener samples without any previous pretreatment [44]. A simple bare glassy carbon electrode was used for the study. As concluded from electrocatalytic oxidation probe, the highest sensitivity of the working electrode was achieved in phosphate buffer at pH 4.5. Since the response potentials are very near each other and overlap to some degree, selectivity of the electrode for these species is limited. In order to overcome this issue unfolded

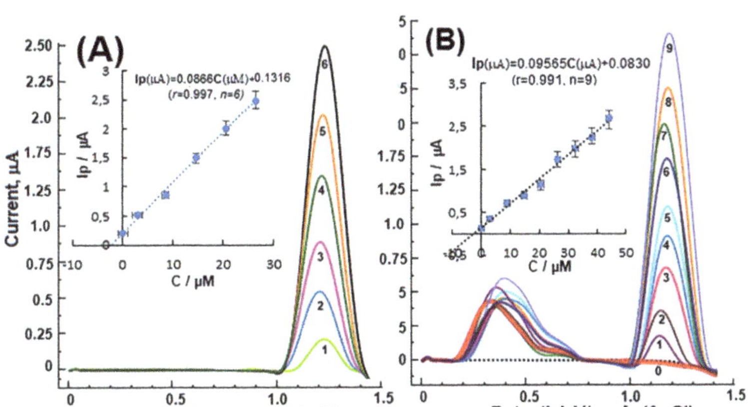

Figure 6.10. SWV curves of real samples; (A) (1) the diluted samples of Vitaptin tablet formulation and after standard additions of (2): 2.94, (3): 8.82, (4):14.71, (5): 20.58 and (6): 26.46 μM VLDG in PBS (9.0) and (B) the urine sample; (0): blank urine, (1): spiked with VLDG to reach the final concentrations of 1.47 μM, after standard additions of (2):2.94, (3): 8.82, (4): 14.71, (5): 20.58, (6): 26.46, (7): 32.37, (8) 38.22 and (9) 44.11. Inset depicts the result of analysis by standard addition method. Dashed line, supporting electrolyte. (Reprinted from [42], copyright 2021, with permission from Elsevier.)

partial least squares with residual bilinearization calibration model was constructed and evaluated using a validation set obtained using a Taguchi design. The second-order advantage succeeded and led to excellent results. This analysis is characterized with good linear response in the range 0.78 to 4.48 μmol L^{-1}, and low LOD values such as 0.29 for MeP, 0.38 for EtP, 0.36 for PrP and 0.39 μmol L^{-1} for BuP. Also, it has high selectivity, sensibility, and accuracy for these parabens.

Apart from classical sensing tasks, square wave voltammetry found application in biomedicinal research. In a work by Otarola *et al* SWV was applied for testing delivery of nonsteroidal anti-inflammatory drug piroxicam *in vitro* [45]. Nanostructured lipid carriers (NLC) based gel was developed as a potential topical system for piroxicam delivery. SWV was selected as a good alternative technique to study the release of piroxicam from the carrier system and allowed to determine the entrapment efficiency and the release profile of the NLCs. The study was carried out by using a Franz diffusion cell and an artificial cellulose nitrate membrane. From the SWV study, it was also possible to understand the release kinetic mechanism of the proposed carrier system. There are also recent reports of where SWV was used to investigate organized systems, such as vesicles, micelles, and liposomes [46, 47].

6.6 Stripping voltammetry

Stripping voltammetry is voltammetric technique which involves preconcentration of an analyte on the working electrode followed by subsequent redox reactions while current with analytical function is recorded. It has several variants depending of the type of an electrode and the pattern of applying potential.

Stripping voltammetry can be used for determination of copper in aviation biokerosene as reported by Primo *et al* [48]. The authors developed a simple, sensitive, and low-cost electroanalytical method on an electrode modified with a composite formed from multiwalled carbon nanotubes and pectin with their ratio as the crucial point. Square wave anodic stripping voltammetry (SWASV) with such a working electrode has a linear working range for copper from 1.0×10^{-8} to 1.0×10^{-6} mol L^{-1} while limits of detection and quantification were 9.0×10^{-9} mol L^{-1} and 3.0×10^{-8} mol L^{-1}, respectively. Also, this electrode did not evidence significant interference from the solution containing the following ions: Fe^{+3}, Pb^{+2}, Ni^{+2}, Zn^{+2}, Cd^{+2}, Ca^{+2}, K^{+}, SO_4^{-2}, Cl^{-}, NO^{3-}, Na^{+}, Mg^{+2}, and Li^{+}. It can be reused up to 40 consecutive times without losing its analytical signal. The preconcentration step is the most important parameter in stripping techniques. In this case, it is found that potential and time of electrodeposition significantly affect the electrode response. The anodic peak current for copper stripping increases toward less negative potentials and reaches its maximum at −0.70 V while at more negative potentials, Cu^{+2} reduction is possibly affected due to solvent discharge which impairs copper electrodeposition on the electrode surface. After the potential of −0.70 V, the electrode response decreases considerably up to −0.50 V. In conclusion, the optimal potential for conducting the preconcentration procedure is −0.70 V.

There is a major challenge to determine copper concentration when it is in the same solution with arsenic (III). The issue in stripping voltammetry for these two analytes is based in the overlapping of their oxidation peaks. Nodehi *et al* designed special nanostructured sensors in order to uplift their electrochemical performance which with adjusting of the measurement conditions, especially the pH of the solution, led to good results for testing concentrations in pool and drinking water samples [49]. A glassy carbon electrode modified with a new nanocomposite based on poly methyldopa along with gold nanoparticles immobilized on the surface of magnetic graphene oxide (GCE/GO/Fe_3O_4@PMDA/AuNPs) can significantly increase the sensitivity of the Cu^{2+} and As^{3+} determination. Optimization of the measurement conditions by square wave stripping voltammetry (SWSV) resulted in great peak distinguishing, while the peak currents increased around ten times in comparison to the bare electrode. The proposed electrode is characterized with: low detection limits (S/N $\geqslant$ 3) of 0.15 ppb for As^{3+} and 0.11 ppb for Cu^{2+}, good linearity over a wide concentration range of 5–500 ppb for As^{3+} and 0.5–750 ppb for Cu^{2+} and good recovery values in any case.

One of the most important issues in the field of monitoring pesticide concentrations in water and food is establishing reliable electrochemical method which would serve the purpose. Differential pulse stripping voltammetry found its role in this area. Recently, it was shown that this technique can be efficiently used for the determination of pesticide clomazone in aqueous media by novel electrochemical sensor based on Pt supported multiwalled carbon nanotubes [50]. Clomazone is determined in 0.1 M phosphate buffer solution at pH 7.0 in the concentration range from 0.61 to 20.56 ppb, with limit of quantification of 0.61 ppb and limit of detection of 0.38 ppb. The same authors employed glassy carbon electrode functionalized by MoO_2 nanoparticle–decorated multiwalled carbon nanotubes and examined as a working electrode for pesticide oxyfluorfen detection by the same technique [51]. DPSV showed good reproducibility (RSD <2.4%) in the concentration range from 2.5 to 34.5 ppb, with $r = 0.99$ and a limit of detection of 1.5 ppb. In both cases interference study proved usability of this technique in dealing with real samples.

Salunke and coworkers designed an electrode with gold nanoparticles (AuNPs) functionalized single polypyrrole nanowire (PpyNW) to detect arsenic (III) in water samples [52]. The chronocoulometry and cyclic voltammetry methods were used to synthesize PpyNW and AuNPs, respectively. After fixing PpyNW on the electrode, the surface was functionalized with AuNPs. The sensing performance of the AuNPs/PpyNW sensor was monitored using an anodic stripping voltammetry method. As shown in figure 6.11, useful dependence is established between concentration and the analytical response. From the obtained data, the sensor showed high sensitivity of 63.13 mA mM^{-1} (i.e. 0.8417 mA/ppb) and an ultra-low detection limit of 0.005 mM (i.e. 0.37 ppb) during arsenic (III) detection in solution. The limit of detection for the sensor was calculated to be 0.0032 mM (0.24 ppb) and 0.0050 mM (0.37 ppb) for laboratory-grade water and adulterated tap water, respectively.

In an interesting study by Wang and coauthors on-site determination of cadmium in crayfish by dilute acid rapid extraction and anodic stripping voltammetry was presented [53]. The standard curve was achieved with crayfish matrix reference material. For the enrichment potential in stripping voltammetry, a value of −1300 mV was chosen while 110 s was chosen as the enrichment time. There was a good linear relationship between

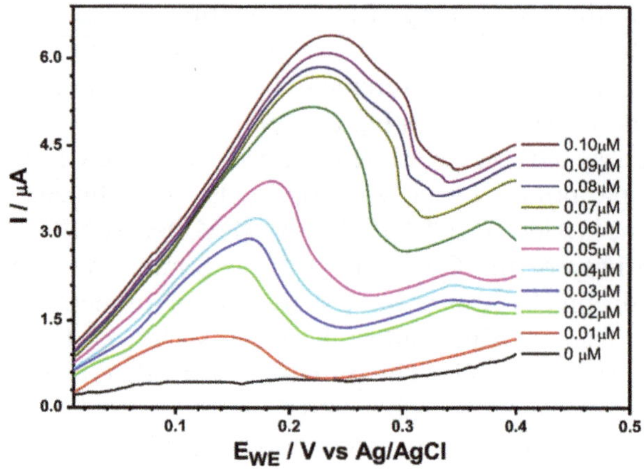

Figure 6.11. Anodic stripping current obtained at AuNPs/PpyNW sensor in an electrolyte solution containing 1.0 M HCl with increasing concentration of arsenic (III). (Reprinted from [52], copyright 2021, with permission from Elsevier.)

the cadmium content and peak current intensity in the range from 0.005 to 1.0 mg kg^{-1}. The instrument calibration of crayfish matrix reference materials are of great significance in elimination of the key influence of matrix effect and then obtain satisfactory detection results.

6.7 Concluding remarks

Ideas and realizations presented in this chapter clearly demonstrate that voltammetric techniques are mighty tools to analyze compounds of various structure, physical, and chemical characteristics and diverse purpose and place of origin. The main advantage of voltammetric techniques over chromatographic and spectrophotometric techniques is that they are faster, require less expensive equipment, use less solvent, and may be applied in flow systems containing complex matrix.

It is acknowledged that harmful or poisonous chemicals can be determined at the sub-micromolar range by applying already established principles of voltammetry and specially functionalized working electrodes. Extensively developed nanomaterials with unique properties are crucial for obtaining the desired selectivity and sensitivity of the measurement. This way, a wide range of metal ions, pesticides, medicaments, dyes, and other chemical species can be determined in many types of aqueous samples with great accuracy and low limits of detection and quantification. An analyte preconcentration step which precedes the voltammetric determination can improve the sensitivity of the electrode. Hence, stripping derivative versions are here to upgrade existing voltammetric practice.

However, practical usage of the described voltammetric options in modern analytical chemistry is still to be tested in an extensive experimental work which certainly raises many question. Voltammetric systems with successful performances reflected in terms of stability of the electrode surface, repeatability of the measurements after many times, and intrinsic affinity towards selected analytes stand a chance of being commercialized. There is a general impression that both the industrial market and individual customers have a need for small portable devices which will determine levels of analytes of interest in a fast and easy manner. And this is the path which voltammetry will follow in the future.

A special place in this context takes the development of medical diagnostics for, so called, point of care testing. There is a huge potential for voltammetric techniques to be involved in sophisticated pocket devices which will easily be used to determine, e.g., hormone, neurotransmitter, or disease indicator levels in human blood, urine, saliva, or sweat. Also, wide application is to be expected in many industries, hygienic laboratories, environment, and almost any on spot quality control procedures.

Acknowledgements

This chapter is part of the work supported by the Ministry of Education, Science and Technological Development of the Republic of Serbia.

References

[1] Brett C M A and Brett A M O 1993 *Electrochemical Experiments: Electrochemistry. Principles, Methods and Applications* (Oxford: Oxford University Press) ch 7 p 129

[2] Momčilović M Z, Milićević J S and Ranđelović M S 2021 Recent advances in electrochemical determination of pesticides *J. Nanosci. Nanotechnol.* **21** 5795–811

[3] Hu S, Ouyang W, Guo L, Lin Z, Jiang X, Qiu B and Chen G 2017 Facile synthesis of Fe_3O_4/g-C_3N_4/HKUST-1 composites as a novel biosensor platform for ochratoxin A *Biosens. Bioelectron.* **92** 718–23

[4] Wang W, Wang X, Cheng N, Luo Y, Lin Y, Xu W and Du D 2020 Recent advances in nanomaterials-based electrochemical (bio)sensors for pesticides detection *TrAC, Trends Anal. Chem.* **132** 116041

[5] Smarzewska S, Jasinska A, Ciesielski W and Guziejewski D 2017 First electroanalytical studies of profluralin with square wave voltammetry using glassy carbon electrode *Electroanalysis* **29** 244–8

[6] Kalijadis A, Dordević J, Papp Z, Jokić B, Spasojević V, Babić B and Trtic-Petrović T 2017 A novel carbon paste electrode based on nitrogen-doped hydrothermal carbon for electrochemical determination of carbendazim *J. Serb. Chem. Soc.* **82** 1259–72

[7] Yoshitake H, Kikkawa T, Muto G and Ota K 1995 Poisoning of surface hydrogen processes on a Pd electrode during electrochemical reduction of carbon dioxide *J. Electroanal. Chem.* **396** 491–8

[8] El Henawee M, Saleh H, Attia A K, Hussien E M and Derar A R 2021 Carbon nanotubes bulk modified printed electrochemical sensor for green determination of vortioxetine hydrobromide by linear sweep voltammetry *Measurement* **177** 109239

[9] Filik H and Avan A A 2019 Dextran modified magnetic nanoparticles based solid phase extraction coupled with linear sweep voltammetry for the speciation of Cr(VI) and Cr(III) in tea, coffee, and mineral water samples *Food Chem.* **292** 151–9

[10] Uchida Y, Kätelhön E and Compton R G 2018 Linear sweep voltammetry with non-triangular waveforms at a microdisc Electrode *J. Electroanal. Chem.* **823** 465–73

[11] Uchida Y, Kätelhön E and Compton R G 2019 Sweep voltammetry with a semi-circular potential waveform: electrode kinetics *J. Electroanal. Chem.* **835** 60–6

[12] Wang Y and Compton R G 2021 Semi-circular sweep voltammetry. Bio-analytical applications *Biosens. Bioelectron.* **179** 113083
[13] Heydari H, Gholivand M B and Abdolmaleki A 2016 Cyclic voltammetry deposition of copper nanostructure on MWCNTs modified pencil graphite electrode: an ultra-sensitive hydrazine sensor *Mater. Sci. Eng.* C **66** 16–24
[14] Xu K, Li Y and Li M 2021 Potentiometric phosphate ion sensor based on electrochemical modified tungsten electrode *ACS Omega* **6** 13795–801
[15] Mariyappan V, Keerthi M and Chen S M 2021 Highly selective electrochemical sensor based on gadolinium sulfide rod-embedded rGO for the sensing of carbofuran *J. Agric. Food Chem.* **69** 2679–88
[16] Tsekeli T R, Sebokolodi T I, Karimi-Maleh H and Arotiba O A 2021 A silver-loaded exfoliated graphite nanocomposite anti-fouling electrochemical sensor for bisphenol A in thermal paper samples *ACS Omega* **6** 9401–9
[17] Boselli E, Wu Z, Friedman A, Henn B C and Papautsky I 2021 Validation of electrochemical sensor for determination of manganese in drinking water *Environ. Sci. Technol.* **55** 7501–9
[18] Ji D, Liu Z, Liu L, Low S S, Lu Y, Yu X, Zhu L, Li C and Liu Q 2018 Smartphone-based integrated voltammetry system for simultaneous detection of ascorbic acid, dopamine, and uric acid with graphene and gold nanoparticles modified screen-printed electrodes *Biosens. Bioelectron.* **119** 55–62
[19] Prinith N S and Manjunatha J G 2019 Surfactant modified electrochemical sensor for determination of anthrone—a cyclic voltammetry *Mater. Sci. Energy Technol.* **2** 408–16
[20] Charithra M M, Manjunatha J G G and Raril C 2020 Surfactant modified graphite paste electrode as an electrochemical sensor for the enhanced voltammetric detection of estriol with dopamine and uric acid *Adv. Pharm. Bull.* **10** 247–53
[21] Raril C and Manjunatha J G 2018 Carbon nanotube paste electrode for the determination of some neurotransmitters: a cyclic voltammetric study *Mod. Chem. Appl.* **6** 1000263
[22] Manjunatha J G 2017 A new electrochemical sensor based on modified carbon nanotube-graphite mixture paste electrode for voltammetric determination of resorcinol *Asian. J. Pharm. Clin. Res.* **10** 295–300
[23] Raril C, Manjunatha J G, Nanjundaswamy L, Siddaraju G, Ravishankar D K, Fattepur S and Niranjan E 2018 Surfactant immobilized electrochemical sensor for the detection of indigotine *Anal. Bioanal. Electrochem.* **10** 1479–90
[24] Amrutha B M, Manjunatha J G, Bhatt A S, Rarila C and Pushpanjali P A 2019 Electrochemical sensor for the determination of Alizarin Red-S at non-ionic surfactant modified carbon nanotube paste electrode *Phys. Chem. Res.* **7** 523–33
[25] Raril C, Manjunatha J G and Tigari G 2020 Low-cost voltammetric sensor based on an anionic surfactant modified carbon nanocomposite material for the rapid determination of curcumin in natural food supplement *Instrum Sci. Technol.* **48** 561–82
[26] Tigari G and Manjunatha J G 2020 A surfactant enhanced novel pencil graphite and carbon nanotube composite paste material as an effective electrochemical sensor for determination of riboflavin *J. Sci.: Adv. Mater. Devices* **5** 56–64
[27] Tigari G and Manjunatha J G 2020 Optimized voltammetric experiment for the determination of phloroglucinol at surfactant modified carbon nanotube paste electrode *Instrum. Exp. Tech.* **63** 750–7
[28] Raril C, Manjunatha J G, Ravishankar D K, Fattepur S, Siddaraju G and Nanjundaswamy L 2020 Validated electrochemical method for simultaneous resolution of tyrosine, uric acid,

and ascorbic acid at polymer modified nano-composite paste electrode *Surf. Eng. Appl. Electrochem.* **56** 415–26

[29] Prinith N S, Manjunatha J G and Raril C 2019 Electrocatalytic analysis of dopamine, uric acid and ascorbic acid at poly(adenine) modified carbon nanotube paste electrode: a cyclic voltammetric study *Anal. Bioanal. Electrochem.* **11** 742–56

[30] Manjunatha J G, Raril C, Hareesha N, Charithra M M, Pushpanjali P A, Tigari G, Ravishankar D K, Mallappaji S C and Gowda J 2020 Electrochemical fabrication of poly (niacin) modified graphite paste electrode and its application for the detection of riboflavin *Open Chem. Engin. J.* **14** 90–8

[31] Girish T, Manjunatha J G, Pushpanjali P A, Prinith N S, Ravishankar D K and Siddaraju G 2021 Poly (DL-valine) electro-polymerized carbon nanotube paste sensor for determination of antihistamine drug cetirizine *J. Electrochem. Sci. Eng.* **11** 27–38

[32] Pushpanjali P A, Manjunatha J G, Tigari G and Fattepur S 2020 Poly(niacin) based carbon nanotube sensor for the sensitive and selective voltammetric detection of vanillin with caffeine *Anal. Bioanal. Electrochem.* **12** 553–68

[33] Fini H and Kerman K 2020 Revisiting the nitrite reductase activity of hemoglobin with differential pulse voltammetry *Anal. Chim. Acta* **1104** 38–46

[34] Su Z L, You H C, Cheng S H and Lin C Y 2020 Fabrication of bacteriochlorin shell/gold core nanoparticles for the sensitive determination of trichlosan using differential pulse voltammetry *Anal. Chim. Acta* **1123** 44–55

[35] Sun M, Li Z, Wu S, Gu Y and Li Y 2018 Simultaneous detection of Pb^{2+}, Cu^{2+} and Hg^{2+} by differential pulse voltammetry at an indium tin oxide glass electrode modified by hydroxyapatite *Electrochim. Acta* **283** 1223–30

[36] Ghorbani M, Pedramrad T, Aghamohammadhasan M, Seyedin O, Akhlaghi H and Lahoori N A 2019 Simultaneous clean-up and determination of Cu(II), Pb(II) and Cr(III) in real water and food samples using a magnetic dispersive solid phase microextraction and differential pulse voltammetry with a green and novel modified glassy carbon electrode *Microchem. J.* **147** 545–54

[37] Arellano M, Oturan N, Oturan M A, Pazos M, Sanromán M A and González-Romero E 2020 Differential pulse voltammetry as a powerful tool to monitor the electro-Fenton process *Electrochim. Acta* **354** 136740

[38] Teófilo K R, Arantes L C, Marinho P A, Macedo A A, Pimentel D M, Rocha D P, de Oliveira A C, Richter E M, Munoz R A A and dos Santos W T P 2020 Electrochemical detection of 3,4-methylenedioxymethamphetamine (ecstasy) using a boron-doped diamond electrode with differential pulse voltammetry: simple and fast screening method for application in forensic analysis *Microchem. J.* **157** 105088

[39] Silva M, Morante-Zarcero S, Pérez-Quintanilla D and Sierra I 2019 Simultaneous determination of pindolol, acebutolol and metoprolol in waters by differential-pulse voltammetry using an efficient sensor based on carbon paste electrode modified with amino-functionalized mesostructured silica *Sens. Actuators* B **283** 434–42

[40] Mirceski V, Stojanov L and Ogorevc B 2019 Step potential as a diagnostic tool in square-wave voltammetry of quasi-reversible electrochemical processes *Electrochim. Acta* **327** 134997

[41] Yi Z and Kun-Lin Y 2021 Quantitative detection of phenol in wastewater using square wave voltammetry with pre-concentration *Anal. Chim. Acta* **1178** 338788

[42] Altunkaynak Y, Yavuz O and Levent A 2021 Firstly electrochemical examination of vildagliptin at disposable graphite sensor: sensitive determination in drugs and human urine by square-wave voltammetry *Microchem. J.* **170** 106653

[43] Bolujoko N B, Unuabonah E I, Alfred M O, Ogunlaja A, Ogunlaja O O, Omorogie M O and Olukanni O D 2021 Toxicity and removal of parabens from water: a critical review *Sci. Total Environ.* **792** 148092

[44] Dantas M S R, Lourenço A S, Silva A C, Bichinho K M and Araujo M C U 2021 Simultaneous determination of methyl, ethyl, propyl, and butyl parabens in sweetener samples without any previous pretreatment using square wave voltammetry and multiway calibration *Food Chem.* **365** 130472

[45] Otarola J J, Cobo Solis A K C, Farias M E, Garrido M, Correa N M and Molina P G 2020 Piroxicam-loaded nanostructured lipid carriers gel: design and characterization by square wave voltammetry *Colloids Surf.* A **606** 125396

[46] Cobo Solis A K, Correa N M and Molina P G 2017 Determination of benzyl-hexadecyldimethylammonium 1,4 bis(2ethylhexyl)sulfosuccinate vesicle permeability by using square wave voltammetry and an enzymatic reaction *Langmuir* **33** 12080–6

[47] Cobo Solis A K, Correa N M and Molina P G 2016 Electrochemical and photophysical behavior of 1-naphthol in benzyl-n-hexadecyldimethylammonium 1,4-bis (2-ethylhexyl) sulfosuccinate large unilamellar vesicles *Phys. Chem.* **18** 15645–53

[48] Primo C M, Buffon E and Stradiotto N R 2021 A carbon nanotubes-pectin composite for electrochemical determination of copper in aviation biokerosene by anodic stripping voltammetry *Fuel* **302** 121180

[49] Nodehi M, Baghayeri M and Veisi H 2021 Preparation of GO/Fe_3O_4@PMDA/AuNPs nanocomposite for simultaneous determination of As^{3+} and Cu^{2+} by stripping voltammetry *Talanta* **230** 122288

[50] Ranđelović M S, Momčilović M Z, Milićević J S, Durović-Pejcev R D, Mofarah S S and Sorrel C C 2019 Voltammetric sensor based on Pt nanoparticles suported MWCNT for determination of pesticide clomazone in water samples *J. Taiwan Inst. Chem. Eng.* **105** 115–23

[51] Milićević J S, Ranđelović M S, Momčilović M Z, Zarubica A R, Mofarah S S, Matović B and Sorrel C C 2020 Multiwalled carbon nanotubes modified with MoO_2 nanoparticles for voltammetric determination of the pesticide oxyfluorfen *Microchim. Acta* **187** 429

[52] Salunke R S, Nakate Y T, Umar A, Nakate U T, Ahmad R and Shirale D J 2021 Anodic stripping voltammetry analysis of gold nanoparticles functionalized one-dimensional single polypyrrole nanowire for arsenic sensing *Surf. Interfaces* **23** 100895

[53] Wang Y K, Ma K, Yin B L, Fu C, Chen M and Cai Y Q 2021 Rapid on-site determination of cadmium in crayfish by dilute acid extraction and anodic stripping voltammetry *Meas.: Sens.* **18** 100126

Chapter 7

Carbon-based electrodes for forensic sample analysis

H C Charan Kumar, R Shilpa, S Ananda and Srikantamurthy Ningaiah

In this chapter, we discuss the recent advances in the electrochemical application of carbon-based electrodes with special emphasis on forensic analysis. The type of carbon electrode materials like graphite, glassy carbon, diamond, fullerene, carbon fiber, etc, used in biosensors play a major role in electroanalytical chemistry, because of their chemical inertness, physicochemical properties, low cost, and rich surface chemistry.

7.1 Introduction

Among the most promising methods, electrochemistry has aroused interest and has been shown to meet most requirements [1–3]. In particular, the development of modified electrodes starting from bare carbon materials (glass carbon, boron-doped diamond, graphene, and screen-printed carbon electrodes) has produced important results and interesting applications [4–7]. Surface modification is the key to achieving superior performance in terms of sensitivity and specificity, although the significant additional cost in terms of increased labor and final equipment price is a limiting factor. In addition, the short-term stability and reproducibility of modified electrodes have recently been emphasized as an important issue, which hinders further applications, especially in industry [8–10]. Bare electrodes without functionalization represent an interesting alternative, especially when high sensitivity is not required. This method uses a simpler system, which reduces production and use costs, and proves long-term stability [11]. Among the different types of bare electrodes, the glassy carbon electrode (GCE) has been widely studied due to its relatively low cost, chemical inertness, and wide anode potential window [12, 13]. Recently, some successful examples of electrochemical sensors based on bare GCE have been reported in the literature, especially in the field of pharmaceutical preparation analysis [14–17]. Electrochemical techniques are attractive as strategies

doi:10.1088/978-0-7503-5127-0ch7

for the investigation of several biologically active compounds. Their attraction stems from their cheap systems and clean principles, which have made them an advantageous approach for biomolecules evaluation and amongst carbon electrodes, Graphene paste electrodes (GPEs) have great significance. The simple, easy fabrication, getting a brand new surface, and deficient history, contemporary feasibility to incorporate one-of-a-kind substances during the past practice, and porous surface are some advantages. In this way, over recent years, GPEs containing exclusive modifiers have been organized and implemented with the purpose of numerous analytes [18]. Electrochemical techniques consisting of cyclic voltammetry (CV) and differential pulse voltammetry (DPV) effectively rely on working electrodes. From the earlier literature, a number of works have detected electroactive riboflavin (RL) with the aid of different sensors, which includes modified glassy carbon and nanoparticle-carbon composite electrodes with high llimit of detection (LOD), low sensitivity, and complex instruction. The carbon-based electrodes monitor a good sensing action, in that the carbon paste electrode (CPE) is quickly available, selective, sensitive, and stable, and it demonstrates good results for the ultra-sensitive investigation of RL, ascorbic acid (AA), and tyrosine (TR) through floor amendment [19]. Different methods were suggested for the individual analysis of norepinephrine (NE) and tyramine (TYM), which include optical, electrochemical, and chromatographic strategies. Of these methods, electroanalytical techniques find more relevance inside the development of sensors due to their attractive specificity together with cost effectiveness, ease of managment, higher sensitivity, simultaneous resolution of different biomolecules and opportunities for miniaturization and many other advantages. Raoof *et al* [20] described the use of functionalized multiwalled carbon nanotube changed glassy carbon electrodes for the quantification of TYM, where the sensing was accomplished using diffusion managment. Later, Kuzuk *et al* [21] used poly (three -methyl thiophene) deposited platinum electrodes for the detection of TYM. Electro-oxidation of NE at single wall carbon nanotubes, changed glassy carbon electrode and graphene-modified palladium electrode were also mentioned. It has been observed that the simultaneous determination of NE and TYM changed into executed most effective through HPLC approach [22]. For this type of electrode, polymer deposited electrodes (running electrodes) could be a powerful device in electrochemical strategies for the determination of electroactive species, due to film thickness, permeation, and rate switch transport that can be controlled by adjusting the electrochemical parameters. In the future, the modified electrodes which act as sensors can be used in the disciplines of medicine and biotechnology [23]. Potential applications of glassy carbon electrodes are in quality control in the food industry. For this purpose, caffeine has been identified as a model analyte. Caffeine (CAF) is mainly found in coffee, and its concentration varies with the type of coffee beans, the degree of roasting, the type of brewing and the temperature of use [24, 25]. However, in recent years, the number of caffeine-containing beverages on the market has increased significantly, not only consumed by adults, but also by increasing numbers of young people. CAF is a physiological stimulant. As an adenosine A receptor antagonist, it has been shown to have positive and negative health effects [26]. When the intake of

CAF is moderate (less than 400 mg per day), it will have positive effects, such as increasing concentration and reducing fatigue [27], and, in those doses, its use has also been shown to be helpful for the treatment of breathing problems and system diseases [28]. However, high doses of CAF may cause cardiovascular and calcium balance problems [29], and there is still controversy about what is a safe intake of CAF, especially for children and young people [30]. In other commercially available beverages, the content of caffeine may vary greatly, from 0.30 gL^{-1} in energy drinks, or 0.1 gL^{-1} in cola drinks, or as tablets (usually 200 mg per tablet) or supplements with the formulation ratio 65 mg/tablet. Because of the widespread existence, use, and biological effects of CAF, routine analysis of its content in all commercial beverages is the main task of its manufacturers. At some point in the voltammetric assessment, carbon paste sensors have been extensively selected for the sensitive and selective identity of bioactive molecules due to their high reproducibility, the opportunity of change through numerous modifiers, execution of regular and well-resolved voltammograms and inexpensive and smooth training procedures. In addition, electrochemical techniques show excessive sensitivity and selectivity for bioactive molecules at the level of primarily conductive polymer based sensors. Electrochemically conductive polymer layers have a broad appeal and have been utilized as chemical modifiers for the development of change sensors because of their extreme stability, repeatability, reproducibility, sensitivity, long-lasting affinity toward the electrode exterior and excessive dynamic surface websites. Therefore, the poly (DLPA) is modified on the outside of carbon paste for voltammetric sensing of RF, in the previous literature various analytical strategies were mentioned for the evaluation of RF, including chemiluminescence, spectrophotometry, mass spectrometry, liquid chromatography, ultraviolet-visible spectroscopy, fluorescence, high overall performance liquid chromatography, capillary electrophoresis, and surface plasmon resonance. All these techniques require expensive equipment, are time-consuming, difficult to operate, have a high intake of reagents and solvents and higher LOD with lower sensitivity. Therefore, these strategies are less suitable for the detection of RF. The electrochemical methods display some essential application over those analytical tools, along with rapid reaction, clean operation, high sensitivity, selectivity, precision, and balance with low LOD. The working sensors learning the use of different materials is vital for the detection of the redox nature of RF. A number of the running sensor substances for RF detection are graphite, silver stable amalgam, glassy carbon, pencil graphite, carbon nanotubes (CNTs), mercury, nickel, and so on. However, a composite form of CNTs and functionalized carbon nanofiber (FCNF) fabric has been used as a running material for RF detection. Today, CNTs and FCNF materials are used as top sensing materials for biomolecule detection with extra interest. CNT fabric exhibits a few crucial functions together with highly energetic surface, high chemical, and thermal balance, better electrical conductivity, better mechanical energy, and remarkable digital behaviors. Additionally, FCNF fabric has an exceptional shape and porous carbon nanofibers which offer a first-rate electrochemical performance along with precise cycle stability, excessive reversible ability, and chemical bonding flexibility. Hence, CNT and FCNF substances are used as the main sensing fabric in composite form for the evaluation of the redox nature of RF [31]. A carbon paste electrode changed with electropolymerization of glycine exhibited an attractive performance for the detection and determination of an anticancer drug, Mitoxantrone (MTX). Cyclic voltammetry (CV) and differential pulse voltammetry (DPV) have been utilized in a mixed manner to pick out the electrochemical characteristics and to optimize the situations for the detection of MTX. The electrochemical behavior of MTX was investigated and a well-defined oxidation height with excessive sensitivity was determined on the film electrode. Poly (glycine) modified carbon paste electrode (PGMCPE) significantly more desirable the oxidation peak contemporary of MTX as a result of the extraordinary properties of glycine film [32]. Dyes are important materials, utilized as synthetic coloring in food, beauty products, clothing, paper, medicines, furnishings, images, paints, leather products, digital materials, and also in the staining of samples in forensic investigations. Several estimates have suggested that more than 10 000 dyes are used in many industries, and more than 70 micro-tons of synthetic artificial colorants used within the worldwide market. Unfortunately, 10% to 50% of these dyes are wasted at some stage in the dyeing and completing processes, and huge quantities of wasted dyes are directly released into the environment. To a certain extent these dyes are noxious and mutagenic, with serous impacts on the environment which include oxygen deficiency, impeded sunlight penetration (resulting in reducing of photosynthetic activity), and version in organic oxygen call for, pH, the salinity of the soil, and chemical oxygen call for, and affecting fundamental makes use of consisting of irrigation and ingesting water. Those defects cause many deadly diseases amongst living organisms. Even less than 1.0 mg of dye in a single liter of water can have hazardous effects on the aquatic environment [33, 34]. Carbon nanotubes are frequently used in the electroanalysis of bioactive molecules because of their unique properties including history, lengthy capacity domains, excessive electric powered conductivity, compatibility, balance, and smooth surface amendment, and so forth. [15–22]. Electrochemical strategies are a powerful and flexible analytical approach that gives excessive sensitivity, accuracy, and comparatively low-value instrumentation and are used by most scientists, chemists, and chemical engineers to study the dynamics and response. The electrochemical approach CV is an extensively used method in many areas of electroanalytical chemistry for acquiring qualitative facts about electrochemical reactions. The strength of CV is its potential to provide information about the thermodynamics redox approaches, absorption approaches, and kinetics of heterogeneous electron transfer reactions with low value, easy, and rapid way of reading the biologically energetic molecule. CV is the frequently done test in an electrochemical study which measures the ensuing current (anodic and cathodic height modern-day) when a capability is applied at the operating electrode in both ahead and opposite course [35]. Graphene is studied internationally because of its distinct properties of high thermal conductivity, more advantageous surface region, 0 bandgaps, permeability to gases, and good electron mobility at room temperature. At room temperature, graphene carriers show core mobility, preserve their structure over some micrometers, and are massless. Presently, graphene is considered because of its particular thermal, digital, optical, and mechanical properties. Electropolymerization of natural dyes to achieve a surface like graphene is a dependable device for recognizing electroactive species. Electropolymerising natural dyes help to acheive surfaces shape a homogeneous polymer film at the electrode floor that could control charge transport. Subsequently, those electrodes show improved electrocatalysis, have an antifouling nature, and prevent undesirable reactions with the electrode [36].

7.2 Electrode material

Electrode materials play a vital role in electrode technology. The selection of the electrode is mainly based on the potential range where the electrolyte-solvent does not undergo electrolysis. However, when selecting a specific electrode material, the catalytic characteristics of the electrode and its design must also be considered [37, 38]. The greatest motivation for electrocatalytic research comes from the search for cheaper catalytic electrode materials [39, 40]. An electrode process includes at least three steps

A. sort of adsorption of reactants (chemical step)
B. charge transfer (electrochemical step) and
C. product desorption (chemical step)

Since the changes around the electrode during the electrolysis process are affected by the electrode potential and the electrode double-layer structure, the characteristics of the electrode become more important, and the double-layer structure itself is also affected by the electrode potential [41, 42]. The utilization of carbon, by and large, manages carbonium particles while that of platinum gives revolutionary intermediates. It has seen that the capacity of the carbon anode to advance the age of carbonium particles is due to the presence of paramagnetic focus at its surface which would hinder the desorption of the at first shaped revolutionaries. As a general rule, carbon cathodes are fit to two-electron oxidation, giving cationic species. A few considerations ought to be applied to the determination of cathode materials for use in natural amalgamation. Following are the boundaries to be considered when choosing a terminal:

1. It should be a decent conductor
2. the surface should be a powerful driver if possible
3. it should not experience the ill effects of synthetic or electrochemical assault and
4. it should be of inflexible development

Platinum, carbon, and lead dioxide are the most generally utilized terminals for the anodic cycle. Nonetheless, titanium dioxide, ruthenium oxide, gold, and changed terminals have acquired significance in the present electro manufactured and insightful works [43].

7.2.1 Carbon and graphite

Both carbon and graphite have been utilized broadly as anodes as a result of their overall protection from oxidation. Carbon glue anodes have been portrayed exhaustively by Adams [44]. Graphite anodes are somewhat permeable, which might be a benefit if a high surface region is required, however porosity might become an issue due to blocking with insoluble residues. There are different kinds of graphite produced which contrast with each other in the strategy for creation and temperature treatment. These are sensibly safe and good conductors. Likewise, a dark slop is framed because of mechanical corruption. About high anode chances, the precious stone design is artificially assaulted by electrolysis in the pores, prompting mechanical breakdown. Techniques for further developing the utilization and interference of graphite have been accounted for. A type of graphite known as pyrolytic graphite, even though at present costly, has the advantageous properties of being impenetrable to fluids and gases. It is an exceptionally requested translucent type of graphite and as a result, is anisotropic. Pyrolytic graphite has been utilized effectively in various natural electrochemical examinations [45]. Other uncommon graphite materials, particularly lustrous carbon [46] and a similar material called glassy carbon [47], have been utilized as anodes. They are marginally permeable and subsequently may endure debasement like that of the more normal graphite.

7.2.2 Graphite structure

Graphite is a carbon allotrope completely comprised of carbon molecules. Graphite is semimetal having a planar layered construction. In each layer, the carbon particles are organized in a honeycomb hexagonal grid with particle detachment of 0.142 nm, and adistance between the two planes of 0.335 nm [48, 49]. In 1789, Abraham Gottlob Werner first used the name ‘graphite’ which means composing stone. Graphite is also called plumbago or dark lead. Graphite is created during transformation by the decrease of sedimentary mixtures containing carbon. It is found in metamorphic rocks, molten rocks, and meteors. Graphite is additionally connected with minerals like micas, quartz, tourmalines, and calcite. There are two chief types of graphite, the alpha and beta structures. Alpha has hexagonal structure and beta has rhombohedral structure, however, they have similar physical and compound properties. Graphite is an electric conduit and can direct power because of enormous electron delocalization inside the planes of carbon layers. The electric conduction is expected to move free valence electrons. Notwithstanding, graphite has uses in light circular segment anode and carbon amplifiers as a tension sensor [50]. Regular graphite is, for the most, part used in refractories, extended graphite, batteries, brake linings, steelmaking, foundry confronting molds, pencils, and greases [51]. Graphene is probably the most grounded substance, which happens normally in graphite and has fantastic properties and practical applications.

Anodes are made in two grades for the steel business. The normal grade is for the most part used in somewhat low-power applications, and the superior grade is used for cutting-edge super high-power heaters. This further developed grade offers specific benefits as far as strength and current-conveying limit. Graphite altered anodes address a cutting-edge way to deal with the electrochemical framework. The two significant methodologies for the combination of GMEs are to comprehend the system of electrons moving at the cathode surface and to concentrate on the synthetic particulars of the adjusted anode which is inaccessible at the unmodified graphite terminals [52, 53]. GMEs are by and large characterized as an electron made of a leading or semiconducting material that is either fortified or covered on the cathode surface, to import the conduct of that substance to the terminal. This conductor might be synthetic, electrochemical, optical, or electrical. The graphite-adjusted cathodes have become moderately recognized because of their effortlessness, upgraded selectivity, and affectability towards analyte detection. GMEs have tracked down various significant applications like solar energy transformation and capacity, atomic gadgets, particular electro-natural combination and electro investigation [54, 55].

To overcome these constraints, the terminals should be altered.

Hence the cathodes are adjusted to

1. To acquire principal data about the system of electron movement at an anode surface.
2. To mediate the electron movement between the analyte and the cathode
3. To check the surface properties
4. To reduce a dynamically negative response at an unmodified surface

7.3 Sorption based artificially changed terminals

The technique for cathode alteration is a significant stage in the advancement of artificially adjusted terminals.

7.3.1 Physisorption technique

CMEs dependent on this technique are prepared using the physisorption of unadulterated natural or organometallic buildings on permeable carbon bases like glassy carbon, graphite terminal, standard pyrolytic graphite and basal plane pyrolytic graphite by basic covering with non-fluid arrangement followed by bead evaporation [56]. The anode dependent on carbon is generally utilized due to its low background flows, synthetic idleness, and reasonableness for an assortment of detecting applications.

7.3.2 Chemisorption technique

Chemisorption is adsorptive cooperation between an atom and a surface in which the electron thickness is shared by the adsorbed particles and the surface. The compound film is unequivocally and preferably, irreversibly adsorbed (chemisorbed) onto the terminal surface [57–62]. This methodology for the most part yields monolayer inclusion recollected for this kind of change is simply the substrate coupled with gathered monolayers in which uncorrelated atoms precipitously chemisorbs at explicit locales on the outer layer of the cathode to shape a super cross-section.

7.4 Conclusion

Despite the design, development, versatility, and applications of carbon-based electrodes in other electrochemical biosensors, it still requires proper investigation into the potential analysis of forensic samples with full promise of accuracy and precision.

References

[1] Khashaba P Y, Ali H R H and El-Wekil M M 2017 Development of a novel and cost-effective redox sensor for voltammetric determination of pantoprazole sodium during pharmacokinetic studies *R. Soc. Open Sci.* **4** 170324

[2] Švorc Ľ 2013 Determination of caffeine: a comprehensive review on electrochemical methods *Int. J. Electrochem. Sci.* **8** 5755–773

[3] Gnana kumar G, Amala G and Gowtham S M 2017 Recent advancements, key challenges and solutions in non-enzymatic electrochemical glucose sensors based on graphene platforms *RSC Adv.* **7** 36949–76

[4] Xie T *et al* 2017 A facile molecularly imprinted electrochemical sensor based on graphene: application to the selective determination of thiamethoxam in grain *RSC Adv.* **7** 38884–94

[5] Tyszczuk-Rotko K and Bęczkowska I 2015 Nafion covered lead film electrode for the voltammetric determination of caffeine in beverage samples and pharmaceutical formulations *Food Chem.* **172** 24–9

[6] Velmurugan M, Karikalan N, Chen S and Karuppiah C 2016 core–shell like Cu2O nanocubes enfolded with Co(OH)2 on reduced graphene oxide for the amperometric detection of caffeine *Microchim. Acta* **183** 2713–721

[7] Yiğit A, Yardım Y, Çelebi M, Levent A and Şentürk Z 2016 Graphene/Nafion composite film modified glassy carbon electrode for simultaneous determination of paracetamol, aspirin and caffeine in pharmaceutical formulations *Talanta* **158** 21–9

[8] Sadok I, Tyszczuk-rotko K and Nosal-Wiercińska A 2016 Bismuth particles Nafion covered boron-doped diamond electrode for simultaneous and individual voltammetric assays of paracetamol and caffeine *Sens. Actuators B Chem.* **235** 263–72

[9] Lim J M, Ryu M Y, Kim J H, Cho C H, Park T J and Park J P 2017 An electrochemical biosensor for detection of the sepsis-related biomarker procalcitonin *RSC Adv.* **7** 36562–5

[10] Yu L, Zhang P, Dai H, Chen L, Ma H and Lin M 2017 An electrochemical sensor based on Co3O4 nanosheets for lead ions determination *RSC Adv.* **7** 39611–16

[11] Amidi S, Ardakani Y H, Amiri-Aref M, Ranjbari E, Sepehri Z and Bagheri H 2017 Sensitive electrochemical determination of rifampicin using gold nanoparticles/poly-melamine nanocomposite *RSC Adv.* **7** 40111–8

[12] Švorc Ľ, Tomčik P, Svítková J, Rievaj M and Bustin D 2012 Voltammetric determination of caffeine in beverage samples on bare boron-doped diamond electrode *Food Chem.* **135** 1198–204

[13] Dekanski A, Stevanovic J, Stevanovic R, Branislav Z N and Jovanovic V M 2001 Glassy carbon electrodes: I. Characterization and electrochemical activation *Carbon* **39** 1195–205

[14] Uslu B and Ozkan S A 2007 Electroanalytical application of carbon based electrodes to the pharmaceuticals *Anal. Lett.* **40** 817–53

[15] Alghamdi A F 2014 Electrochemical oxidation behavior of hydrochlorothiazide on a glassy carbon electrode and its voltammetric determination in pharmaceutical formulations and biological fluids *J. Food Drug Anal.* **22** 363–69

[16] Levent A, Altun A, Yardım Y and Zühre S 2014 Sensitive voltammetric determination of testosterone in pharmaceuticals and human urine using a glassy carbon electrode in the presence of cationic surfactant *Electrochim. Acta* **128** 54–60

[17] Sýs M, Švecová B, Švancara I and Metelka R 2017 Determination of vitamin E in margarines and edible oils using square wave anodic stripping voltammetry with a glassy carbon paste electrode *Food Chem.* **229** 621–27

[18] Manjunatha J G 2020 A surfactant enhanced graphene paste electrode as an effective electrochemical sensor for the sensitive and simultaneous determination of catechol and resorcinol *Chem. Data Coll.* **25** 100331
[19] Hareesha N and Manjunatha J G 2020 Elevated and rapid voltammetric sensing of riboflavin at poly(helianthin dye) blended carbon paste electrode with heterogeneous rate constant elucidation *J. Iran. Chem. Soc.* **17** 1507–519
[20] Raoof J B *et al* 2012 Application of a glassy carbon electrode modified with functionalized multi-walled carbon nanotubes as a sensor device for simultaneous determination of acetaminophen and tyramine *Anal. Methods* **4** 1579–87
[21] Kucuk A and Torul O 2018 Voltammetric sensor based on poly (3- methylthiophene) synthesized in dichloromethane for tyramine determination in moldy cheese *Synth. Met.* **237** 2328
[22] Anand S K *et al* December 2020 Individual and simultaneous voltammetric sensing of norepinephrine and tyramine based on poly(L-arginine)/reduced graphene oxide composite film modified glassy carbon electrode *J. Electroanal. Chem.* **878** 114531
[23] Raril C and Manjunatha J G 2018 Sensitive electrochemical analysis of resorcinol using polymer modified carbon paste electrode: a cyclic voltammetric study *Anal. Bioanal. Electrochem.* **10** 488–98
[24] Yu J, Jin H, Gui R and Wang Z 2017 A facile strategy for ratiometric electrochemical sensing of quercetin in electrolyte solution directly using bare glassy carbon electrode *J. Electroanal. Chem.* **795** 97–102
[25] Bell L N, Wetzel C R and Grand A N 1996 Caffeine content in coffee as influenced by grinding and brewing techniques *Food Res. Int.* **29** 785–89
[26] Caporaso N, Genovese A, Canela M D, Civitella A and Sacchi R 2014 Neapolitan coffee brew chemical analysis in comparison to espresso, moka and American brews *Food Res. Int.* **61** 152–60
[27] Harpaz E, Tamir S, Weinstein A and Weinstein Y 2017 The effect of caffeine on energy balance *J. Basic Clin. Physiol. Pharmacol.* **28** 1–10
[28] Clark I and Landolt P H 2017 Coffee, caffeine, and sleep: a systematic review of epidemiological studies and randomized controlled trials *Sleep Med. Rev.* **31** 70–8
[29] Atik A, Harding R, De Matteo R, Kondos-Devcic D, Cheong J, Doyle L W and Tolcos M 2017 Caffeine for apnea of prematurity: effects on the developing brain *Neurotoxicology* **58** 94–102
[30] Zulli A *et al* 2016 Caffeine and cardiovascular diseases: critical review of current research *Eur. J. Nutr.* **55** 1331–343
[31] Varun D N *et al* 2021 Simple and sensitive electrochemical analysis of ribofavin at functionalized carbon nanofber modifed carbon nanotube sensor *Monatsh. Chem* **152** 1183–191
[32] Manjunatha J G June 2018 Highly sensitive polymer based sensor for determination of the drug mitoxantrone' *J. Surf. Sci. Technol.* **34** 74–80
[33] Hareesha N, Manjunatha J G, Amrutha B M, Pushpanjali P A, Charithra M M and Prinith Subbaiah N 2021 Electrochemical analysis of indigo carmine in food and water samples using a poly(glutamic acid) layered multi-walled carbon nanotube paste electrode *J. Electron. Mater.* **50** 1230–238
[34] Pushpanjalia P A *et al* Development of carbon nanotube-based polymer-modified electrochemical sensor for the voltammetric study of Curcumin *Mater. Res. Innov.* **25** 412–20
[35] Raril C and Manjunatha J G 2018 Sensitive and selective analysis of nigrosine dye at polymer modified electrochemical sensor *Anal. Bioanal. Electrochem.* **10** 372–82

[36] Monnappa A B *et al* September 2021 Sensitive and selective electrochemical detection of vanillin at graphene based poly (methyl orange) modified electrode *J. Sci.: Adv. Mater. Devices* **6** 415–24

[37] Bockris J O ' M and Srinivasan S 1959 Electrode kinetic aspects of electrochemical energy conversion *J. Electroanal. Chem.* **11** 350–89

[38] Liebhafsky H A and Cairns E J 1968 *Fuel Cells and Fuel Batteries. Guide to Their Research and Development* (New York: Wiley)

[39] Bockris J O ' M, Damjanovic A and Mannan R 1968 Catalysis of the electrodic hydrogen evolution and dissolution reactions on rationally chosen substrates *J. Electroanal. Chem. Interfacial Electrochem.* **18** 349–61

[40] Bockris J O M, Mannan R and Damjanovic A 1968 Dependence of the rate of electrodic redox reactions on the substrate *J. Chem. Phys.* **48** 1898–904

[41] Pragst F, Janda M and Stibor I 1980 Electrochemistry of pyrylium compounds—III. Synthesis of 4h-pyranes by cathodic alkylation of the 2, 4, 6triphenylpyrylium cation *'Electrochim. Acta* **25** 779–83

[42] Koehl W 1964 Anodic oxidation of aliphatic acids at carbon anodes *J. Am. Chem. Soc.* **86** 4686–90

[43] Parker V and Adams R 1969 Side-chain anodic oxidation. Mechanism of benzyl cation formation *Tetrahedron Lett.* **10** 1721–24

Trasatti S and Buzzanca G 1971 Ruthenium dioxide: a new interesting electrode material. Solid-state structure and electrochemical behavior *J. Electroanal. Chem. Interfacial Electrochem.* **29** A1A5

[44] Ross S D, Finkelstein M and Rudd E J 2013 *Anodic Oxidation: Organic Chemistry: A Series of Monographs* Vol 32 (Amsterdam: Elsevier)

[45] Carr P W and Jordan J 1973 Kinetic titrimetry. Theory of automated linear titrations and application to thermometric endpoints *Anal. Chem.* **45** 634–40

[46] Adams R N 1969 *Electrochemistry at Solid Electrodes* (New York: Marcel Dekker)

[47] Baizer M M 1984 Recent developments in organic synthesis by electrolysis *Tetrahedron* **40** 935–69

[48] Pletcher D, Whyte I, Walsh F and Millington J 1991 Reticulated vitreous carbon cathodes for metal ion removal from process streams part I: mass transport studies *J. Appl. Electrochem.* **21** 659–66

[49] Cipris D 1981 Nonpetroleum feedstock in electro-organic synthesis *J. Electrochem. Soc.* C122–122

[50] Therese G H A and Kamath P V 2000 *Chem. Mater.* **12** 1195

[51] Nyffenegger R M, Craft B, Shaaban M, Gorer S, Erley G and Penner R M 1998 *Chem. Mater.* **10** 1120

[52] Chung D D L 2002 Review graphite *J. Mater. Sci.* **37** 1475–89

[53] Dweiri R and Sahari J 2007 Computer simulation of electrical conductivity of graphite-based polypropylene composites based on digital image analysis *J. Mater. Sci.* **42** 10098–0102

[54] Ndlovu T, Arotiba O A, Sampath S, Krause R W and Mamba B B 2012 Reactivities of Modified and unmodified exfoliated graphite electrodes in selected redox systems *Int. J. Electrochem. Sci.* **7** 9441–453

[55] Khana Z U, Kausara A and Ullahb H 2016 A review on composite papers of graphene oxide, carbon nanotube, polymer/GO, and polymer/CNT: processing strategies, properties, and relevance *Polym. Plast. Technol. Eng.* **55** 559–81

[56] Dell R M and Rand D A J 2001 *Understanding Batteries* (Cambridge: The Royal Society of Chemistry)

[57] Novak Petr S M E and Dietrich G 2010 *Carbons for Electrochemical Energy Storage and Conversion Systems* (CRC Press)
[58] Haik O, Ganin S, Gershinsky G, Zinigard E, Markovsky B, Aurbach D and Halalay I 2011 On the thermal behavior of lithium intercalated graphite *J. Electrochem. Soc.* **158** 913–A923
[59] K-Back C and Prakash J 2011 Consideration of carbon structure effect on the thermal stability of carbon anode for Li-ion rechargeable batteries *Thermodynamic Act* **520** 93–8
[60] Zheng J and Dahn T 1999 *Carbon Materials for Advanced Technologies* (Oxford: Pergamon)
[61] Flandrosis S and Simon B 1999 Carbon materials for lithium-ion rechargeable batteries *Carbon* **37** 165–80
[62] Bernardo P, Dentzer J, Gadion R, Markle W, Goers D, Novak P, Spahr M and Vix-Guterl C 2011 Influence of graphite surface properties on the first electrochemical lithium intercalation *Carbon* **49** 4867–76

Chapter 8

Carbon composite voltammetric sensors for food quality assessment

R Shashanka

Food safety is one of the fundamental requirements of mankind, but the world has witnessed an increasing threat around food adulteration, contamination and it is affects on human health. Therefore, researchers are giving more importance to developing suitable and more accurate methods of assessing food quality. In the past few years, electrochemical sensing interfaces have become substantial, easy, and bonafide techniques to evaluate food quality. Generally, the quality of these sensors mainly depends upon the concentration of the target analyte, the food source, the type of reaction, the type of electrodes used, etc. For food quality assessment one should have knowledge of common contaminants of foods such as pesticides, toxins, additives, animal drug residues, pathogens, organic and inorganic residues and measure their assays. This chapter focuses on the measurement of such contaminants using various carbon composite materials like carbon nanotubes (CNTs), graphenes, graphites, and their derivatives. The excellent sensitivity, reduced surface fouling, maximum surface area, stability, specificity, and biocompatibility has made carbon composite electrodes a very popular choice to determine the food contaminants by voltammetric methods like cyclic voltammetry (CV), linear sweep voltammetry (LSV), differential pulse voltammetry (DPV), and square wave voltammetry (SWV). Carbon composite materials can also be used as modifiers to improve the assessment of food safety further.

8.1 Introduction

Food safety is one of the crucial requirements for all of humananity. In recent years, we have witnessed enhanced cases of food-borne diseases due to the presence of various chemicals in foods. To fulfill the huge food demands for the increased population we are adulterating foods by adding chemicals to increase the yield and

doi:10.1088/978-0-7503-5127-0ch8

to reduce the length of time required to grow crops. This results in an accumulation of a huge amount of chemicals inside our bodies which affects our health significantly. Therefore, food safety is an important concern and food needs to undergo a series of quality assessments starting from its production through its preservation, transportation, and consumption. Hence, many researchers are developing simple, accurate, and reliable methods and techniques to determine the various chemicals present in foods [1, 2]. Food contaminants such as pesticides, toxins, additives, animal drug residues, pathogens, organic, and inorganic residues can cause serious food-borne diseases like botulism, listeriosis, or hemolytic uremia syndrome when we consume them. Therefore, food quality assessment is an important and prioritized concern for consumers to assure food safety and nutritional quality [3]. Therefore, we need quick and accurate measurement of food quality during different stages of food production until consumption [4].

Many conventional analytical methods can determine the chemicals present in food, but their detection limit is not significant and may not detect the series of chemicals present in the food simultaneously in a single step; sometimes a low concentration of chemicals remains undetected and the techniques may be destructive and affect the seal of the packaged food [5]. In recent years researchers have been using various electrochemical methods to overcome the above problems caused by conventional analytical methods. Voltammetric methods like CV [6–10], LSV [11–13], DPV, and SWV, have been widely used in food analysis; these methods can detect the multiple analytes at low concentration in a single step. The efficiency, specificity, reliability, accuracy, and stability of the tests mainly depend upon the type of electrode materials used. Therefore, the type of electrode materials plays an important role in determining the quality of food using the above voltammetric techniques.

The most common and important materials used as working electrodes are carbon materials like CNT, graphenes, carbon dots, boron-doped diamond (BDD), ordered mesoporous carbon (OMC), etc, either as modifiers in carbon paste electrodes or as a whole. The unique electrochemical characteristics have made these electrodes reliable electrochemical sensing electrodes with excellent performances.

Many researchers have reported the preparation, surface modification, and application of carbon-based nanomaterials exhibiting needle, rod, and barrel shapes [14]. The significant chemical electronic properties of carbon materials are due to their heterocyclic state of the C-C bonds. Some of the important characteristic features of the carbon composites are: their refined structure, interface, large surface area, dielectric confinement, macroscopic quantum tunneling effects, etc [15]. Due to these above characteristics, carbon materials are extensively used to fabricate super-sensing devices for a food safety.

This chapter discusses the various characteristics of carbon materials and their applications as sensors in determining the various organic, inorganic chemicals, toxins, pesticides, illegal food additives, animal drug residues, and pathogens present in foods. The recent investigations on the fabrication and formulation of highly-efficient carbon composite materials for food safety assessment are also included.

The safety of food should be prioritized and more research must be undertaken to improve the detection levels, stability, accuracy, repeatability, and efficiency of the carbon-based electrodes.

8.2 Types of carbon nanomaterials

The CNT, graphenes, carbon dots, ordered mesoporous carbon, and BDD are the types of carbon composite materials generally used to assess food safety. All these carbon materials exhibit unique characteristics and they differ from one another in morphologies, surface area, conductivity, surface modifications, etc. Therefore, each carbon material acts differently against the same target analytes. These exceptional differences and unique properties have made carbon composite materials a popular choice for food safety assessment (figure 8.1).

8.2.1 Carbon nanotubes

The Japanese scientist S. Ijima discovered carbon nanotubes (CNTs) in 1991 [16]. The sp^2 hybridization in CNTs allows each carbon atom with three electrons to establish trigonally coordinated bonds with three other carbon atoms. In fact, CNT is a single sheet of graphene rolled into a hollow carbon nanotube and is characterized by rolled graphene sheets piled in cylindrical or tubular formations with a diameter of several nanometers [16, 17]. These carbon nanotubes have proved to be one of the best and most widely used carbon materials employed as a modifier in carbon paste electrodes or as a whole to assess food safety. The efficiency of these types of electrodes mainly depends upon the length, diameter, surface modification,

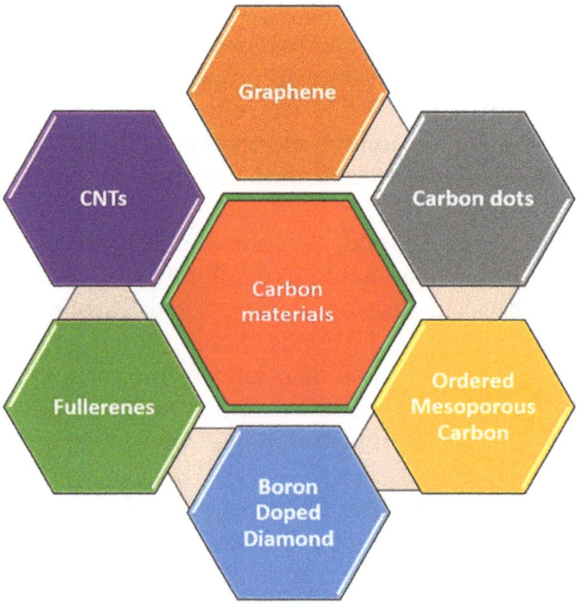

Figure 8.1. Types of carbon nanomaterials.

and the number of layers of CNTs. The CNTs are mainly classified as single-walled carbon nanotubes (SWCNTs) and multi-walled carbon nanotubes (MWCNTs).

SWCNTs have a diameter of 1–3 nm, whereas MWCNTs are 5–25 nm in diameter and are about 10 μm in length [18]. Over the last few years, researchers have reported the unique electronic properties of CNTs (some of them are electrical conductors while others are semiconductors), and also reported their high surface area, easy surface modification, excellent porosity, etc. These properties have made CNTs an excellent candidate for food safety assessment. CNTs can be either used as modifiers in carbon paste electrodes or can be used as a whole to determine various chemicals, illegal additives, drugs, pathogens, and toxins in foods. All of these characteristics of CNTs have made it one of the most researched carbon materials. On the other hand, CNTs also exhibit excellent electrochemical reactivity for significant biomolecules and increase the electron transfer processes [19–21]. The CNT-based electrochemical sensors are very sensitive and show extraordinary surface conductivity. Scientists are therefore attracted to CNTs because they may be used to make a wide range of electrochemical sensors [22–24].

Chen *et al* used MWCNTs to fabricate acetylcholinesterase an (AChE)-based electrochemical sensor to detect pesticide present in food samples [25]. The authors used MWCNTs to increase the surface area of the electrode significantly, thereby, easing the electrochemical polymerization of Prussian blue and also maintaining the excellent enzymatic activity of AChE by decreasing the Michaelis–Menten constant (K_m). They reported that the fabricated sensor depicted significant stability, reproducibility, and a quick response on a series of pesticides in real samples.

Generally, CNTs materials show tremendous catalytic properties and conductivity, this greatly reduces the overpotential and efficiently accelerates the electron transfer in electrochemical reactions compared to ordinary materials. Therefore, Yari *et al* [26] and Goud *et al* [27] reported that the use of CNTs as a modifier increases the sensitivity, wide linear detection range, and fast response of the electrodes. The MWCNT surface modified with molybdenum disulfide nanosheet (MoS_2) coating acts as an excellent hybrid sensor for the determination of chloramphenicol (CAP) [28]. The CAP drug causes suspected carcinogenicity and aplastic anemia in humans. Govindasamy *et al* reported that the MoS_2/MWCNT composites exhibit significant electrochemical properties and display remarkable catalytic ability on the detection of CAP even at lower concentrations [28].

Bhardwaj *et al* [29] used an economic antibody-SWCNT bio-conjugate modified electrode to detect the food-borne pathogen *Staphylococcus aureus* in real food samples. They covalently attached the *anti-S. aureus* antibodies on the surface of SWCNTs and immobilized the surface of the working electrode to detect the target analyte, thereby altering the peak current followed by antigen–antibody complex formation. The DPV method was used to study the selectivity assay using *Escherichia coli B, Bacillus subtilis,* and *Staphylococcus epidermidis* but the fabricated electrode showed its sensory activity towards *S. aureus* only. It also depicted excellent linearity ($R^2 = 0.976$) between the increased peak current and *S. aureus* concentrations within a short detection time of 30 min and a limit of detection of 137 colony forming units (CFU) mL^{-1} in spiked milk samples.

Kara *et al* [30] used SWCNT to fabricate a multijunction sensor specially designed to multiplex detection of food-borne pathogens. Yang and Zhao [31] used synthesized 3-hexadecyl-1-vinyl imidazolium chloride (C16VimCl) to improve the dispersion of MWCNTs, and to obtain MWCNTs modified molecularly imprinted polymer to detect CAP (banned drug). The prepared electrochemical sensor showed a significant response to CAP and exhibited the detection limit of 1.0×10^{-10} mol L^{-1}. Finally, they concluded that the obtained results are satisfactory and the fabricated sensor can be used to determine CAP in real samples as well. Tyramine is a decarboxylation product of tyrosine, found in many food products like wine, beer, beans, fish, banana peel, chocolate, etc [32]. Excessive intake of tyramine-containing food can affect health by producing symptoms like strong migraine headaches [33], vomiting, tachycardia, rashes, hypertonia, palpitations, flushing, etc [34]. Kochana *et al* [35] reported the fabrication of a MWCNT modified graphite electrode to determine Tyramine. The fabricated sensor electrode showed greater sensitivity, repeatability, limit of detection, long-term stability, and reproducibility. The electrocatalytic activity for tyramine oxidation showed a linear range from 6 to 130 μM, and a maximum sensitivity of 486 μA $mM^{-1}cm^{-2}$, and a limit of detection of 1.5 μM were recorded. Kochana *et al* [35] concluded that the fabricated electrode demonstrated a maximum biological affinity towards tyramine present in food product samples.

Yin *et al* [36] prepared polydopamine by monomeric self-polymerization in water and surfaces modified to MWCNTs to detect sunset yellow dye which is a petroleum-derived orange azo dye generally used as a coloring agent in food. The acceptable daily intake of this dye is 0–4 mg kg^{-1}. but excessive intake may cause food intolerance and ADHD-like behavior in children (no scientific evidence).

8.2.2 Graphene and related compounds

Graphene is a 2D allotropic carbon made up of a single layer. There are 0.142 nanometers of space between the carbon hexagons in a hexagonal crystal lattice, linked by s and p bonds in sp^2 hybridization. This material was discovered by Canadian theoretical physicist P R Wallace in 1947 and further studies were carried by A Geim and K Novoselov [37–39]. Graphene was theoretically studied far in the past, but graphene materials have only been practically studied in the past few decades.

Most researchers are working on graphenes and reporting their exceptional properties and wide range of applications. Although graphenes have dynamic mechanical rigidity and thermal stability, their unique electrical properties have attracted the attention of electrochemists. The electrical properties of graphenes significantly differ from the electrical properties depicted by CNTs, fullerenes, graphites, etc [40–44]. Graphene depicts a high degree of light transparency (97.7%) and excellent chemical, physical, thermal, and mechanical properties. At normal temperature, the charge carrier mobility of graphene is around 15 000 to 20 000 $cm^2 Vs^{-1}$ and these characteristics have made graphene an ideal candidate for electrode material in assessing food safety. It is estimated that the mobility of electrons in graphene layers is 100 times greater than that of silicon [45]. As a result, in the future it is expected to replace silicon in the electronic manufacturing sector. Nowadays graphene is a widely used sensor material in food safety assessment due to its high specific surface area and high charge carrier mobility [46].

The two important forms of graphenes are graphene oxides (GO) and reduced graphene oxides (rGO), both differ slightly in their properties. Synthesis of GO is less expensive and takes a relatively short time. The most widely accepted method of preparing GO is the Hummers method [47], this method is simple, takes less time, and does not dissolve hazardous compounds. Potassium permanganate and concentrated sulfuric acid are used as oxidation agents and for peeling off graphite to produce graphene oxide (GO) [47]. Because of its hydrophilic nature, GO is acidic. The GO sheets are easily distributed in water and can be used. GO may be chemically converted back to graphene by employing reduction techniques and thermal reductions [48]. rGO can be prepared by reducing the GO by using various techniques such as electrochemical reduction, chemical, annealing, and laser radiation techniques. The characteristic properties of rGO deteriorate due to the destruction of graphene's basal plane by strong chemicals used for oxidation. Because of this, graphene from graphite is peeled off using appropriate solvents and surfactants to remove the graphite [49]. In some solvents, graphene tends to clump into graphite. Preparing single-layer graphene in solvents that is pure and evenly distributed is a difficult task. For the mechanical peeling of the 2D graphite, adhesive tapes [50] with a lower flaw density are used.

The carbon atoms of the GO can be linked to the hydroxy or epoxy groups and their surface can be modified with carboxyl and carbonyl groups [51]. This surface modification not only increases the electrical conductivity but also the surface area of the GO. Therefore, one of the important applications of GO is the fabrication of chemically modified electrodes in detecting various organic, inorganic chemicals, toxins, animal drugs, pathogens, toxic dyes, additives, etc in foods [51]. Qiu *et al* [52] reported the electrochemical sensor behavior of tartrazine and sunset yellow dyes using the combined form of CNTs and GO electrode materials. The combinational effect of GO and CNT enhanced the anodic peak current of tartrazine and sunset yellow. They reported the LOD of their simultaneous determination of both the dyes at 1×10^{-8} mol L^{-1} and 2.5×10^{-8} mol L^{-1}, respectively [52]. Wang *et al* modified the GO-CNT electrodes with butyl-3-methylimidazolium hexafluorophosphate ([BMIM][PF_6]) to improve the electrical conductivity in detecting amaranth dye in soft drinks. Gorduk [53] determined vanillin (used as a flavoring material) in commercial food products using electrochemically fabricated GO modified electrode by the cyclic voltammetry method. The fabricated electrode depicted tremendous electrocatalytic activity towards vanillin and showed a wide linear range (0.1–1000 μM), a low detection limit (0.05 μM) compared to the unmodified graphite electrode. The author has used the fabricated sensor to detect vanillin in ice-cream, coffee, and cookie samples and obtained high recovery values [53].

Wang *et al* [54] fabricated a graphene nanocomposite modified glassy carbon electrode to detect Clenbuterol in pork. It is used as a food additive for livestock to increase muscle growth and decrease fat deposition. However, the high accumulation of Clenbuterol in animal tissues may cause severe health issues to humans like cardiac palpitation, vomiting, nausea, and chills [54]. The linear sweep voltammetry method is used for the determination of Clenbuterol and it has shown a good linear relationship of the oxidation peak current with respect to the concentrations of the target analyte with a superior detection limit of 2.2×10^{-8} mol L^{-1}. They reported that the fabricated graphene nanocomposite sensor depicted a tremendous electrocatalytic activity towards the oxidation of Clenbuterol present in pork [54].

Ma and Chen [55] reported the electrochemical sensor application of graphene doped gold nanoparticle modified electrode for the detection of Diethylstilboestrol (DES). DES is a synthetic estrogen, generally as a growth stimulant in animals for food and as a treatment for estrogen-deficiency disorders in veterinary medicine [56]. The residue of DES in meat is proven to be highly carcinogenic for humans [55]. The fabricated graphene doped gold nanoparticle modified electrode depicted an excellent electrocatalytic performance for the electrochemical oxidation of DES. The oxidation peak currents of DES showed a linear relationship with their concentrations in the ranges of 1.20×10^{-8} to 1.20×10^{-5}mol L^{-1} with the high detection limit of 9.80×10^{-9} mol L^{-1}. Finally, the authors concluded that the fabricated electrode showed good selectivity and stability in real food samples. Graphene can easily form stable complexes with most of polymers, therefore graphene–polymer composites are mainly used to assess food safety [57]. Qi *et al* [58] reported the use of a thiol graphene modified glassy carbon electrode decorated with gold nanoparticles to determine the tebuconazole in vegetable and fruit samples. Tebuconazole is a triazole fungicide, mainly used in agricultural fields to control the pathogenic fungi on vegetables, fruits, and grains. But, the excessive use of this fungicide causes potential harm to human health [59]. The fabricated thiol graphene modified glassy carbon electrode decorated with gold nanoparticles sensor exhibited a wide linear range from 5.0×10^{-8} mol L^{-1} to 4.0×10^{-4} mol L^{-1} with a detection limit of 1.25×10^{-8} mol L^{-1}. The authors have concluded that the present sensor can detect Tebuconazole by electro-oxidizing the analyte and can also be used in real samples with satisfactory recoveries [59].

8.2.3 Carbon dots

Carbon dots (CDs) are a new class of carbon materials and one of the emerging research areas of recent years. They are composed of tiny carbon atoms with a size of less than 10 nm (0D). Low toxicity, environmental friendliness, low cost, and simple synthetic routes have made carbon dots popular among researchers. One of the interesting properties of CDs is easy surface passivation and functionalization which allows their physicochemical properties to be controlled [60]; as a result of which CDs can be used in electrocatalysis, chemical, and biosensing areas.

Costas-Mora *et al* [61] constructed a nanoprobe using CDs for the detection of methylmercury. Methylmercury is considered to be one of the dangerous neurotoxins which can be accumulated in the human body through the food chain (methylmercury is linked to fish consumption) and affects the brain and nervous system; particularly dangerous to pregnant women and infants [62]. Costas-Mora *et al* [61] reported that the fabricated sensor was proved to be very quick (takes less than 1 min) highly selective for CH_3Hg^+ detection at low concentrations of the analyte with the detection limit of 5.9 nM.

Li *et al* [63] fabricated CD-based electrodes to detect one of the harmful neurotoxin organophosphate pesticides. These pesticides are the most common agrochemicals, used to prevent, control, and eradicate insect pests, thereby boosting the production yield of crops and vegetables. Even the low concentrated residue in plants and vegetables can inactivate acetylcholinesterase (AChE), a key enzyme for neurotransmission in humans [64]. The fabricated electrodes show good sensitivity for rapid detection of the analyte with a detection limit of 0.4 ng mL^{-1}. Dong *et al* [65] used CDs to detect highly sensitive amantadine in foodstuffs. Amantadine is prohibited in many countries and was widely used in the poultry industry due to its efficient antiviral activity and its excessive use can cause nausea, dizziness, etc.

Xu *et al* [66] used CD-embedded molecularly imprinted polymer for the detection of sterigmatocystin (ST) in grains. ST is a secondary fungal metabolite produced by various Aspergillus species. ST is widely distributed in a variety of food and feed (grains of wheat, corn, and millet), and can affect both humans and livestock that consume contaminated foods. The fabricated sensor exhibited excellent electrocatalytic property towards ST and it has proved to be highly selective and sensitive in detecting the target analyte with a detection limit of 0.019 mg L^{-1}.

Li *et al* [67] constructed molecularly imprinted polymer embedded CD sensors for fast and robust detection of tetracycline present in honey. Tetracycline is an antibiotic chemical, excessive accumulation leads to loss of appetite, nausea, rectal discomfort, diarrhea, sore throat, vomiting, headache, mouth sores, black hairy tongue, etc. The fabricated sensors depicted tremendous linearity within 10–200 μg L^{-1} with a low detection limit of 3.1 μg L^{-1}. Similarly, Yang *et al* [68] and Hou *et al* [69] used carbon dots to detect tetracycline in fish and milk respectively.

8.2.4 Ordered mesoporous carbon

The porous carbons are the new class of carbon materials exhibiting huge internal porosity with maximum surface area [51]. This new grade of carbon material is attracting more researchers due to its excellent energy surface heterogeneity, which allows the use of porous carbon materials as electron transfer mediators along with fullerenes and carbon nanotubes [70]. Due to these properties, these carbon materials are mainly used as adsorbent carriers, catalyst supports, hydrogen storage materials, and electrode materials [57]. Mesoporous carbon materials are classified into two main groups: ordered mesoporous carbon (OMC) and disordered mesoporous carbon based upon their orientation or regularity of pores. Generally, OMC exhibits more highly ordered pores and macroporous carbon nanorods than disordered mesoporous carbon materials; as a result of which OMCs are more electrochemically stable and therefore used in electrode materials [57].

Jian and Li-Ping [71] prepared an OMC modified glassy carbon electrode for the detection of amaranth dye. This dye belongs to the azo dye category and is generally added as a coloring agent in wines and soft drinks to make them more attractive. Amaranth is one of the widely used low-cost azo dyes and it has been strictly controlled in many countries because of its suspected carcinogenic effect [72]. Jian and Li-Ping [71] used differential pulse voltammetry to detect amaranth and peak current was found to be in line with the analyte's concentrations (1.0×10^{-7} to 3.0×10^{-6} M) with a low detection of limit (3.2×10^{-8} M). They concluded that the fabricated electrodes are more selective and sensitive than CNT and other carbon-modified electrodes in detecting amaranth in wine samples.

Yang *et al* [73] used an OMC modified glassy carbon electrode to determine the toxic ractopamine present in pork samples using cyclic voltammetry. Ractopamine is generally used as a heart tonic, bronchodilator, and tocolytic in veterinary medicine. The chances of drug residues accumulating in animal tissues are high and may affect the consumer health with frequent vomiting, muscular tremors, nervousness, and cardiac palpitations [73]. Yang *et al* reported that the OMC modified electrode can significantly increase the electrocatalytic oxidation of ractopamine by enhancing the peak current [73]. The peak current of the target analyte showed linearity ranging from 0.085 to 8.0 μM with a detection limit of 0.06 μM respectively. Finally, they concluded that prepared sensors depict high sensitivity and selectivity towards the detection of ractopamine with satisfying recoveries in the range 96.6%–104.5%.

Yang *et al* [74] fabricated an OMC modified three-dimensional molecularly imprinted electrochemical sensor for the detection of para-aminobenzoic acid (p-ABA) in the presence of template metolcarb. Metolcarb is a type of chemical pesticide widely used in agricultural fields to control rice leafhoppers, planthoppers, and fruit flies. The intake of metolcarb can affect the nervous system by inhibiting the growth of acetylcholinesterase [74]. Yang *et al* [74] reported that the fabricated 3D electrode demonstrated the enhanced peak current response during the detection of metolcarb in the linearity of 5.0×10^{-10} to 1.0×10^{-4} mol L^{-1} with a limit of detection of 9.3×10^{-11} mol L^{-1}.

8.2.5 Boron-Doped Diamond

Most carbon materials are highly conductive in nature [75–78]. Another important carbon material which exhibits very interesting and unique properties is boron-doped diamond (BDD). It is one of the special types of carbon material gaining special attention from electrochemical researchers all over the world due to its broad potential window, high stability, high conductivity, very low noise (ten times lower than glassy carbon), chemically inert, biocompatible, and resistant to passivation. Due to these properties, BDD has become one of the best potential electrode materials for detecting various organic, inorganic chemicals, pathogens, veterinary drugs, toxic dyes, pesticides, etc present in foods.

Micheletti *et al* [79] constructed a cathodically pretreated BDD electrode for the voltammetric detection of toxic azo dye carmoisine E-122 in food. The carmoisine E-122 dye is a synthetic red dye that generally gives a red to maroon color and is mostly used in yogurt, candies, chewing gum, sauces, many foods, and beverages [79].

The excessive consumption of this dye can cause hyperactivity in children, alter biochemical markers from renal and hepatic functions, induce oxidative stress, and possible allergies and intolerances [80]. The authors have reported that the differential pulse voltammetric peak for the analyte is linear over the concentration range of 0.059–1.31 μmol L^{-1} and 0.010–0.079 μmol L^{-1} with limits of detection of 7.0 and 3.0 nmol L^{-1}, for the anodic and cathodic processes respectively. Finally, they claimed that the fabricated sensor is highly sensitive and selective in determining the carmoisine E-122 dye in surface water and foodstuffs with accurate results [79].

Chuanuwatanakul *et al* [81] electrochemically investigated the chloramphenicol (an antibiotic) using the BDD electrode. Chloramphenicol has been widely used as a veterinary drug to prevent and treat many bacterial infections because of its efficiency, easy availability, and low cost. The excessive intake of chloramphenicol can cause anemia, leucopenia, agranulocytosis, and aplastic anemia in humans. This drug is banned worldwide and was used for aquaculture farming [82]. The fabricated BDD electrode has shown a linear current response against the concentration of chloramphenicol over the range 0.1–50 μM with a limit of detection of 0.03 μM. The authors have used this sensor for the successful determination of chloramphenicol in sterile eye drops and milk samples by the standard addition method.

Medeiros *et al* [83] used a cathodically pretreated BDD electrode for the simultaneous voltammetric determination of various synthetic dyes used in food. These synthetic dyes are very harmful and are added to food products like sugar candies, sugar-coated pills, jelly beans, powdered drinks, sweets, nutrient-enhanced sports beverages, ice creams, and gelatins to make them visually attractive [84]. These dyes can possibly cause allergic and asthmatic reactions, sometimes followed by the development of cancer and other diseases [83]. Medeiros *et al* [83] simultaneously detected two pairs of synthetic drugs (atrazine and sunset yellow or brilliant blue and sunset yellow) in mixed food products using differential pulse voltammetry method by cathodically pretreated BDD electrode. This resulted in the separation of the reduction peak potentials of atrazine and sunset yellow or brilliant blue and sunset yellow about 150 mV with the detection limit 62.7 nmol L^{-1} and 13.1 nmol L^{-1} or 143 nmol L^{-1} and 25.6 nmol L^{-1}, respectively.

8.3 Conclusions

Due to the variety of carbon materials and unique electronic properties allows them to ne used in the development of new electrochemical or biosensors for the detection of a wide range of organic, inorganic chemicals, illegal food additives, dyes, pathogens, pesticides, veterinary drugs present in the food. Carbon materials like CNTs, graphene, BDD, carbon dots, OMCs are generally used as a modifier to enhance the detection capacity of the various electrodes to assess food safety. The ability to incorporate carbon nanostructures into other materials such as organic polymers or hydrogels, or to couple them to metal or oxide nanoparticles, might boost the sensors' detecting capabilities or improve their performance. Furthermore, carbon-based nanostructures may be chemically built into 2D and 3D structures, giving additional characteristics that increase sensing capabilities. A huge number of theoretical and practical investigations have been carried out by various researchers to explain the detailed studies of carbon composite materials in the food testing-related field. In the near future the world will witness the further development of carbon composite materials in nanotechnology and sensor technology to assess food safety.

Conflicts of interest

The authors declare no conflict of interest.

References

[1] Hareesha N and Manjunatha J G 2020 *J. Iran. Chem. Soc.* **17** 1507–19
[2] Raril C and Manjunatha J G 2018 *Anal. Bioanal. Electrochem.* **10** 372–82
[3] Karlsson R 2004 *J. Mol. Recognit.* **17** 151–61
[4] Sekhon B S 2014 *Nanotechnol. Sci. Appl.* **7** 31–53
[5] Duncan T V 2011 *J. Colloid Interface Sci.* **363** 1–24
[6] Shashanka R, Chaira D and Swamy B E K 2015 *Int. J. Electrochem. Sci.* **10** 5586–98
[7] Shashanka R, Chaira D and Swamy B E K 2015 *Int. J. Sci. Eng. Res.* **6** 1863–71
[8] Reddy S, Swamy B E K, Aruna S, Kumar M, Shashanka R and Jayadevappa H 2012 *Chem. Sens.* **2** 7
[9] Rajendrachari S, Swamy B E K, Reddy S and Chaira D 2013 *Anal. Bioanal. Electrochem.* **5** 455–66
[10] Shashanka R, Chaira D and Swamy B E K 2016 *Int. J. Sci. Eng. Res.* **7** 1275–85
[11] Shashanka R 2018 *Anal. Bioanal. Electrochem.* **10** 349–61
[12] Shashanka R, Chaira D and Swamy B E K 2018 *Arch. Metall. Mater.* **63** 749–63
[13] Shashanka R 2018 *Voltammetry* (London: Intechopen)
[14] Cha C Y, Shin S R, Annabi N, Dokmeci M R and Khademhosseini A 2013 *ACS Nano* **7** 2891–7
[15] Agustin G C, Escarpa A and Carlos D G 2018 *Carbon-based Nanomaterials in Analytical Chemistry* (Cambridge: Royal Society of Chemistry) pp 1–45
[16] Iijima S 1991 *Nature* **354** 56–8
[17] Kumar S, Rani R, Dilbaghi N, Tankeshwar K and Kim K H 2017 *Chem. Soc. Rev.* **46** 158–96
[18] Zhang R, Zhang Y, Zhang Q, Xie H, Qian W and Wei F 2013 *ACS Nano* **7** 6156–161
[19] Manjunatha J G, Deraman M, Basri N H, Mohd Nor N S, Talib I A and Ataollahi N 2014 *C.R. Chim.* **17** 465–76
[20] Manjunatha J G, Deraman M, Basri N H and Talib I A 2018 *Arab. J. Chem.* **11** 149–58
[21] Manjunatha J G 2017 *J. Electrochem. Sci. Eng.* **7** 39–49
[22] Tigari G, Manjunatha J G, Raril C and Hareesha N 2019 *ChemistrySelect* **4** 2168–73
[23] Raril C and Manjunatha J G 2018 *Mod Chem Appl* **6** 1000263
[24] Charithra M M and Manjunatha J G 2020 *J. Electrochem. Sci. Eng.* **10** 29–40
[25] Chen H D, Zuo X L, Su S, Tang Z Z, Wu A B, Song S P, Zhang D B and Fan C H 2008 *Analyst* **133** 1182–6
[26] Yari A and Shams A 2018 *Anal. Chim. Acta* **1039** 51–8
[27] Goud K Y, Kalisa S K, Kumar V, Tsang Y F, Lee S E, Gobi K V and Kim K H 2018 *Biosens. Bioelectron.* **121** 205–22

[28] Govindasamy M, Chen S M, Mani V, Devasenathipathy R, Umamaheswari R, Santhanaraj K J and Sathiyan A 2017 *J. Colloid Interf. Sci.* **485** 129–36

[29] Bhardwaj J, Devarakonda S, Kumar S and Jang J 2017 *Sens. Actuators B-Chem.* **253** 115–23

[30] Yamada K, Choi W, Lee I, Cho B K and Jun S 2016 *Biosens. Bioelectron.* **77** 137–43

[31] Yang G M and Zhao F Q 2015 *Biosens. Bioelectron.* **64** 416–22

[32] Batra B, Lata S, Devi R, Yadav S and Pundir C S 2012 *J. Solid State Electrochem.* **16** 3869–876

[33] Hanington E 1967 *Br. Med. J.* **2** 550–1

[34] Telsnig D, Kassarnig V, Zapf C, Leitinger G, Kalcher K and Ortner A 2012 *Int. J. Electrochem. Sci.* **7** 10476–86

[35] Kochana J, Wapiennik K, Knihnicki P, Pollap A, Janus P, Oszajca M and Kustrowski P 2016 *Anal. Bioanal. Chem.* **408** 5199–210

[36] Yin Z Z, Cheng S W, Xu L B, Liu H Y, Huang K, Li L, Zhai Y Y, Zeng Y B, Liu H Q and Shao Y 2018 *Biosens. Bioelectron.* **100** 565–70

[37] Novoselov K S 2004 *Science* **306** 666–69

[38] Novoselov K S 2005 *Nature* **438** 197–200

[39] Geim A K 2009 *Science* **324** 1530–4

[40] Manjunatha J G and Deraman M 2017 *Anal. Bioanal. Electrochem.* **9** 198–213

[41] Manjunatha J G 2020 *Chem. Data Collect.* **25** 100331

[42] Raril C and Manjunatha J G 2020 *Microchem. J.* **154** 04575

[43] Raril C and Manjunatha J G 2020 *J. Anal. Sci. Technol.* **11** 3

[44] Amrutha B M, Manjunatha J G, Bhatt A S and Nagarajappa H 2021 *J. Sci.: Adv. Mater. Devices* **6** 415–24

[45] Bolotin K I 2008 *Solid State Commun.* **146** 351–5

[46] Lu G, Ocola L E and Chen J 2009 *Nanotechnology* **20** 445502

[47] Hummers W S and Offeman R E 1958 *J. Am. Chem. Soc.* **80** 1339

[48] Stankovich S 2007 *Carbon N. Y.* **45** 1558–65

[49] Lotya M 2009 *J. Am. Chem. Soc.* **131** 3611–20

[50] Lim E K, Kim T, Paik S, Haam S, Huh Y M and Lee K 2015 *Chem. Rev.* **115** 327–94

[51] Lipskikh O I, Korotkova E I, Khristunova Y P, Barek J and Kratochvil B 2018 *Electrochim. Acta* **260** 974–85

[52] Qiu X, Lu L, Leng J, Yu Y, Wang W, Jiang M and Bai L 2016 *Food Chem.* **190** 889–95

[53] Gorduk O 2021 *AKU J. Sci. Eng.* **21** 84–96

[54] Wang L, Yang R, Chen J, Li J, Qu L and Harrington P B 2014 *Food Chem.* **164** 113–8

[55] Ma X and Chen M 2015 *Sens. Actuators* B **215** 445–50

[56] Wang J, Ye H Z, Jiang Z, Chen N S and Huang J L 2004 *Anal. Chim. Acta* **508** 171–6

[57] Pan M, Yin Z, Liu K, Du X, Liu H and Wang S 2019 *Nanomaterials* **9** 1330

[58] Qi P *et al* 2018 *Electrochim. Acta* **274** 406–14

[59] Munoz-Leoz B, Ruiz-Romera E, Antigüedad I and Garbisu C 2011 *Soil Biol. Biochem.* **43** 2176–83

[60] Zheng X T, Ananthanarayanan A, Luo K Q and Chen P 2015 *Small* **11** 1620–636

[61] Costas-Mora I, Romero V, Lavilla I and Bendicho C 2014 *Anal. Chem.* **86** 4536–43

[62] Castoldi A F, Coccini T, Ceccatelli S and Brain M L 2001 *Res. Bull.* **55** 193–203

[63] Li H, Yan X, Lu G and Su X 2018 *Sens. Actuators* B **260** 563–70

[64] Shashanka R, Taslimi P, Karaoglanli A C, Uzun O, Alp E and Jayaprakash G K 2021 *Arab. J. Chem.* **14** 103180

[65] Dong B, Li H, Maria G M, Yu X, Yu W, Wen K, Ke Y, Shen J and Wang Z 2019 *Food Chem.* **294** 347–54
[66] Xu L, Fang G, Pan M, Wang X and Wang S 2016 *Biosens. Bioelectron.* **77** 950–56
[67] Li H, Zhao L, Xu Y, Zhou T, Liu H, Huang N, Ding J, Li Y and Ding L 2018 *Talanta* **185** 542–9
[68] Yang J, Lin Z Z, Nur A Z, Lu Y, Wu M H, Zeng J, Chen X M and Huang Z Y 2018 *Spectrochim. Acta* A **190** 450–6
[69] Hou J, Li H Y, Wang L, Zhang P, Zhou T Y, Ding H and Ding L 2016 *Talanta* **146** 34–40
[70] Ndamanisha J C and Guo L 2012 *Anal. Chim. Acta* **747** 19–28
[71] Jian J and Li-Ping G 2013 *Chin. J. Anal. Chem.* **41** 681–6
[72] Tsuda S J, Matsusaka N, Madarame H, Ueno S, Suas N, Ishida K, Kawamura N, Sekihashi K and Sasaki Y F 2000 *Mutat. Res.* **465** 11–26
[73] Yang X, Feng B, Yang P, Ding Y, Chen Y and Fei J 2014 *Food Chem.* **145** 619–24
[74] Yang Y, Cao Y, Wang X, Fang G and Wang S 2015 *Biosens. Bioelectron.* **64** 247–54
[75] Shashanka R and Swamy B E K 2020 *Phys. Chem. Res.* **8** 1–18
[76] Shashanka R and Swamy B E K 2020 *SN Appl. Sci.* **2** 956
[77] Jayaprakash G K, Swamy B E K, Rajendrachari S, Sharma S C and Flores-Moreno R 2021 *J. Mol. Liq.* **334** 116348
[78] Shashanka R, Jayaprakash G K, Prakashaiah B G, Kumar M and Swamy B E K 2022 *Mater. Res. Innov.* **26** 229–39
[79] Micheletti L, Coldibeli B, Salamanca-Neto C A R, Almeida L C and Sartori E R 2020 *Talanta* **220** 121417
[80] Amin K A, Hameid H A and Abd Elsttar A H 2010 *Food Chem. Toxicol.* **48** 2994–9
[81] Chuanuwatanakul S, Chailapakul O and Motomizu S 2008 *Anal. Sci.* **24** 493–8
[82] Shakila R J, Vyla S A, Kumar R S, Jeyasekaran G and Jasmine G I 2006 *Food Microbiol.* **23** 47
[83] Medeiros R A, Lourencao B C, Rocha-Filho R C and Fatibello-Filho O 2012 *Talanta* **97** 291–7
[84] Nevado J J B, Flores J R and Llerena M J V 1997 *Fresen, J. Anal. Chem.* **357** 989–94

Chapter 9

Recent advances in electrochemical monitoring of epinephrine using carbon-based (bio)sensor devices for clinical applications

Gabriel J Mattos, Renan S Mariano and Elen R Sartori

Epinephrine is an important catecholamine neurotransmitter in the mammalian central nervous system, acting as a cellular chemical messenger. Extreme variations of epinephrine levels are related to mental disorders and metabolic dysfunction, which makes it a suitable biomarker for clinical evaluation. Several analytical methods employing high-performance liquid chromatography coupled to spectrometric systems of detection were extensively described in the literature over the years for the detection and determination of this neurochemical. Electroanalytical methods have been gaining more space recently with new concepts of electrochemical (bio)sensors and technological evolution related to nanomaterials applied as transducers. The current multidisciplinary research in electroanalysis has resulted in remarkable improvement of analytical performance, including high sensitivity and sensibility, extremely low limits of detection, robustness, portability, and specificity. This chapter focusses on the recent progress in the field of electrochemical (bio)sensing of epinephrine, including cutting-edge research for *in vivo* neurochemical monitoring. Given the large number of publications regarding epinephrine sensing using electrochemical protocols, the authors decided to limit the references included in this chapter to those reporting carbon-based electrodes and voltammetric techniques.

9.1 Introduction

Epinephrine belongs to a catecholamine group present in the human body, being the main hormone and neurotransmitter responsible for preparing the body for 'fight and flight' situations. It is secreted in the medullary region of the adrenal glands and released after stimuli from nerve endings [1, 2]. The release of epinephrine increases blood pressure and heart rate, so it is used in emergency medicine, for example, in

doi:10.1088/978-0-7503-5127-0ch9

cardiac arrest to stimulate heart function [3, 4]. Also, epinephrine is administered in anaphylactic shock and as standard medication in cardiopulmonary resuscitation because it increases heart rate. Epinephrine is considered a biomarker and its determination in the blood helps in the diagnosis of diseases such as pheochromocytoma and Parkinson's [1, 5–7]. In this way, due to their clinical properties and importance, several analytical methodologies have been developed over the years for the determination of epinephrine in pharmaceuticals and biological fluids.

The analytical procedures developed for the determination of epinephrine aim to promote a standard for monitoring the biological fluids of patients, whether for toxicological studies, for clinical purposes, or quality control of drugs. In this sense, analytical devices based on electrochemical detection, including sensors and enzymatic biosensors, can be used for epinephrine determination. Hence, the scope of the current review is limited to the determination of epinephrine employing voltammetric techniques and carbon electrode materials developed from 2020 to 2021, with an emphasis on the chemical properties of nanostructured carbon and metallic particles that determine its electrochemical performance. Furthermore, this chapter will discuss the analytical applications reported in the literature and provide a brief history about the discovery of epinephrine and its importance in the clinical field.

9.2 Epinephrine: a brief history

Epinephrine (figure 9.1; pK_a 8.88 at 25 °C) [8] is a classical 'fight or flight' hormone and a neurotransmitter in the central nervous system in mammals [2]. In the adrenal medulla and a few brain regions, the amino acid tyrosine is transformed through a series of reactions to norepinephrine, and then the enzyme phenylethanolamine *N*-methyltransferase catalyzes the methylation of norepinephrine to epinephrine [9]. This catecholamine is stored in adrenal glands and its release is triggered by nervous impulses coming from the brain, functioning primarily to increase cardiac output and raise glucose levels in the blood. It also expands the field of vision by dilating the pupils of the eyes and preparing the body for stressful situations; it helps the individual cope with the stress, manifesting itself in moments of excitement, or situations of emotional stress [2].

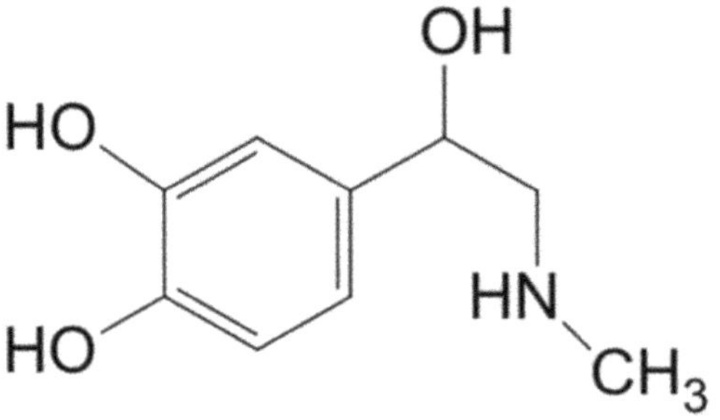

Figure 9.1. Chemical structure of epinephrine.

Epinephrine was the first hormone discovered and it possesses several benefits. Historically and very briefly, epinephrine was isolated as a stable crystalline compound from the adrenal glands of sheep and oxen in 1900 by the biochemist Jokichi Takamine, with the sponsorship of Parke-Davis & Co [10–14]. Takamine observed extremely potent vasopressor properties of epinephrine. He named it adrenaline, where the prefix 'ad' refers to the adrenal glands and the suffix 'ine' is the same as used in the nomenclature of compounds from the amine group. The patent filing was challenged because a natural hormone existing in nature cannot be patented; thus, for purposes of securing the patent, Takamine dropped the 'e', starting to write it as adrenalin. However, researchers found that the crystalline preparation was not pure yet, since epinephrine was shown to be contaminated with norepinephrine, and they pointed out that epinephrine was the most appropriate name for the hormone produced by the adrenal gland. The correct empirical formula was determined by Aldrich in 1901 and gradually the term 'adrenaline' fell out of the scientific literature and was replaced with 'epinephrine', but several researchers continue to use the name adrenaline for historical reasons [10–12, 14].

Epinephrine was the first hormone to be synthesized in the laboratory for clinical uses [15]. Friedrich Stolz, in 1904, produced epinephrine in a synthetic form becoming an economical, reliable, and simple method, leading to large-scale production and avoiding possible animal injuries, which was the source of epinephrine before [12]. It was prepared from a reaction between pyrocatechol and chloroacetyl chloride, followed by a reaction with methylamine to form the ketone that is reduced. The racemic mixture could be separated using tartaric acid to yield commercially viable quantities of purified active epinephrine. Nowadays, this substance is available exclusively in synthetic form [11].

With the synthetic production of epinephrine, local anesthesia took a remarkable step forward. Scientists observed that it increased the analgesic power of cocaine, which at the time was almost exclusively used, and greatly restricted blood loss [16, 17]. In addition, epinephrine helped in the discovery, in 1907, of the neurotransmitter norepinephrine [11], a biosynthetic precursor of epinephrine [10–12]. Several other applications for medical and clinical purposes were emerging for this historic molecule.

9.3 Epinephrine as a biomarker and its clinical uses

Epinephrine stands out in clinical applications being the most important direct-acting hormone, neurotransmitter, and medication. In the human body, epinephrine regulates blood pressure, immune system, heart rate, lipolysis, and glycogen metabolism. Generally, epinephrine is present in the body in small and fluctuating amounts that only increase appreciably during and shortly after a stressful situation. Normal levels of epinephrine in human urine are $<0.11\ \mu mol\ L^{-1}$ (per 24 h) and human plasma are 18–125 pg mL^{-1} [18]. These values may increase during exercise and times of stress or thyroid hormone deficiency. Besides, in the blood and urine of patients with pheochromocytoma (usually a noncancerous tumor that develops in an adrenal gland), there is a high level of epinephrine and norepinephrine [1, 5, 7]. Some disorders, e.g., depression, schizophrenia, and essential hypertension can also produce large amounts of epinephrine, dopamine, and norepinephrine, resulting in increased concentrations in the excreted fluids [19]. The epinephrine concentration in the tear fluid of some patients with Parkinson's disease was reduced [6]. So, because of the high death rates, early recognition of cancer and other diseases is desirable. It is also important to determine epinephrine in blood, plasma, or serum, to aid in medical diagnosis, as a biomarker to identify the presence of some diseases or intoxication and monitor therapeutic administration.

As a medical drug, epinephrine is an indispensable medicine for hemodynamic management in anesthesiology practice and the primary drug administered to revive a patient from a cardiac arrest [3, 4]. Moreover, this drug is used in emergencies to quickly improve breathing, stimulate the heart, raise the blood pressure, reverse hives, and reduce swelling of the face, lips, and throat caused by foods (e.g., nuts and shellfish), insect venom, drugs, or unknown causes [20]. After completing their action, epinephrine metabolizes into metanephrine and vanillylmandelic acid, being eliminated from the body in the urine along with their metabolites [7, 18].

Epinephrine is the only medication that can reverse the symptoms of a severe life-threatening allergic reaction, known as anaphylactic shock [12] because of its mix of actions, cardiovascular and bronchial. It is not practical to administer epinephrine orally because it is rapidly conjugated and oxidized in the wall of the gastrointestinal tract and the liver. Synthetic epinephrine solution in auto-injector form for the emergency treatment of anaphylaxis was first approved for marketing by the US Food and Drug Administration (FDA) in 1987 [21]. Other dosage routes have been explored for its administration, including intravenous or subcutaneous injection and inhalation or nebulization. The sublingual tablet has been studied [22, 23] and has been successful in animal studies; it would be a route of administration of epinephrine with affordability and availability worldwide, besides the absorption from the mucosa in the sublingual area pass directly into the systemic circulation.

While it can be a lifesaving drug, epinephrine can also cause death, becoming a toxic compound [24]. For example, depending on the applied dosage it can cause heart failure. Thus, epinephrine should be administered with caution in patients with some pre-existing diseases; the patient must know their health condition, avoid overdoses of epinephrine and the dosage administered must be the one recommended for the patient. Among several side effects, mental confusion, difficulty in speaking suddenly, and weakness can be cited as symptoms of an overdose [25]. In this sense, it is important that the drug to be released for clinical use undergoes rigorous quality control.

9.4 Analytical methods employed in the quantification of epinephrine

Accurate and selective measurements of epinephrine in pharmaceuticals and biological samples are important for both clinical diagnoses and pathological studies of some diseases. In this way, the development of inexpensive, precise, and accurate analytical methods that allow rapid screening of a large number of samples using simple methods and measurements in complex samples has attracted the attention of many researchers.

The assay of epinephrine in pharmaceuticals is officially established by the European Pharmacopoeia v. 10.6, which recommends titration with perchloric acid in glacial acetic acid and using crystal violet as the indicator [26]. This classic analytical method is simple to apply in the quality control of pharmaceuticals; however, it uses a large volume of solution and it is impossible to determine low concentrations of epinephrine in biological fluid samples. Another simple analytical method to be applied for quality control of epinephrine in pharmaceutical preparations is spectrophotometry [27]. It is not selective for the assays of this analyte in complex samples with the presence of structurally related compounds with similar chemical properties, like norepinephrine and dopamine, resulting in spectrum overlap.

A separation technique widely used to separate mixtures of chemical substances is chromatography with its various types of detectors [28, 29]. It presents several advantages over other existing techniques, mainly in simultaneous determinations or in samples where it is desired to determine the analyte of interest in the presence of several other compounds. The United States Pharmacopoeia describes the high-performance liquid chromatography (HPLC) with spectrophotometric detection as the official method to assay the drug in injections [30]. However, chromatographic methods often require an unreasonably large amount of organic solvents, expensive equipment, internal standards are also needed sometimes, and a sample clean-up prior step, which may preclude the analysis of pharmaceuticals or biological fluids in laboratories or industries of small or medium-size.

In this sense, the use of HPLC with electrochemical detection can be an alternative, ensuring easy maintenance, use of a smaller amount of sample and organic solvent, and sensitivity to the determination (capable of detecting femtomolar concentrations) that implies no preconcentration of the sample [31, 32]. The choice of the oxidation/reduction potential for epinephrine allows more chemical-specific detecting conditions because fewer compounds are electroactive at the same potential, which minimizes interferences and eliminates sample preparation steps [18]. Moreover, electrochemical detection offers the perfect balance of performance and price. Even so, depending on the sample matrix, electrochemical sensors and biosensors can be used directly in the determination of epinephrine [33, 34], without the necessity of coupling the separation procedures of HPLC.

9.5 Voltammetric techniques

Voltammetric techniques have become more widely used over the past decades due to the introduction of new sensing concepts combined with numerous technological innovations related to instrumentation. Furthermore, the measurements are highly reproducible, simple dilution/dissolution of sample, no use of organic solvent, field-portable and cost-effective procedures, speed, inexpensive instrumentation, and also less harmful to the analyst and the environment [35].

Analytical devices as electrochemical sensors and enzymatic biosensors in conjunction with voltammetry have been used for the determination of epinephrine in complex samples [36]. Due to the applied potential and/or choice of electrode material, these electrochemical methods have the ability to selectively determine the target analyte in solution [37]. For example, using the anodically pretreated boron-doped diamond electrode in ammonium buffer solution (pH 9.0) it was possible to determine a quaternary mixture of the antihypertensives amlodipine besylate, amiloride hydrochloride, hydrochlorothiazide, and atenolol with very well-resolved and reproducible oxidation peaks at 0.68, 0.89, 1.04 and 1.21 V (vs. Ag/AgCl (3.0 mol L^{-1} KCl)), respectively [33]. This determination was possible due to differences observed in the oxidation potentials of each drug in this type of electrode, which was not observed in the bare glassy carbon electrode (GCE).

In principle, electrochemical measurements involve the application of electric potential in a suitable working electrode that is inserted into a supporting electrolyte, followed by the measurement of the current resulting from the chemical reaction (oxidation or reduction), which occurs during the measurement [38]. If the analyte is oxidized in the working electrode, the resulting electrons pass through the potentiostat to the auxiliary electrode. On the other hand, if the analyte is reduced the current flows from the auxiliary electrode to the working electrode. In general, the anodic current resulting from the oxidation of analyte at the working electrode has a positive sign and the cathodic current has a negative signal. When there are no electroactive species present in the supporting electrolyte, the current is negligible over most of the potential range. At the most positive or negative potential, the observed current is due to oxidation or reduction, respectively, of the electrolyte in the aqueous solution or the working electrode.

Voltammetry typically uses an electrochemical cell containing three electrodes where the working electrode plays a decisive role in the analysis [39] (figure 9.2(a)). It should ideally be situated as close as possible to the working electrode, for the reaction to occur more effectively. The most widely utilized reference electrodes are the saturated Ag/AgCl (in a saturated KCl solution or 3.0 mol L^{-1} KCl) electrode and the saturated calomel electrode (SCE). The potential of the reference electrode should remain constant during all measurements. The auxiliary electrode should be constructed of an inert material, such as platinum and glassy carbon.

In cyclic voltammetry (CV) the current of redox processes is measured during a cycled potential sweep between two or more set values. The switch of the potential scan can happen multiple times during a single experiment. It is the most widely used technique for acquiring qualitative information about intrinsic properties of substances, such as thermodynamics, the kinetics of electronic-transfer reactions, and it also provides a mathematical analysis of an electron transfer process at the electrode [40–44]. The recording of current as a function of potential is called a voltammogram (figure 9.2(b)).

Voltammetric techniques, such as differential pulse voltammetry (DPV) and square-wave voltammetry (SWV) are preferable to quantify a target analyte in solution due to their greater sensitivity [39, 42, 43, 45]. In these techniques, the potential applied between the working electrode and the auxiliary electrode is varied

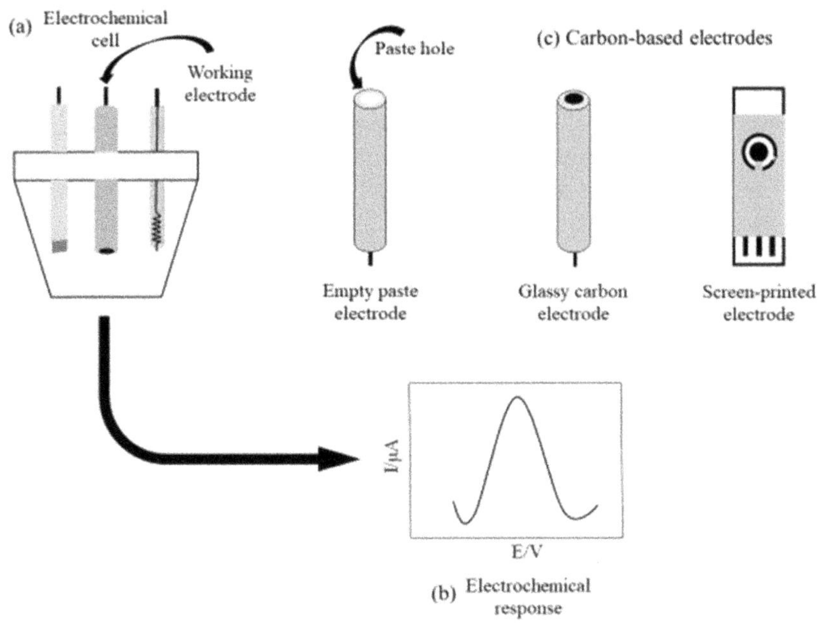

Figure 9.2. General scheme of voltammetry setup and carbon-based devices: (a) electrochemical cell, (b) carbon-based electrodes and (c) electrochemical response.

throughout the measurement and the current from the redox reaction that occurs on the surface of the working electrode is monitored.

Stripping analysis based on SWV or DPV techniques involves the electrochemical preconcentration of the analyte onto a working electrode surface, followed by the stripping step when the species are oxidized or reduced on the electrode and the current is measured [41]. Due to the application of a preconcentration step before the stripping step the technique is highly sensitive, widely used for the determination of analytes in low concentration, mainly in very dilute samples [42, 43, 46]. In addition, it improves the selectivity and stability of the method by isolating the analyte from the sample matrix during the preconcentration step. These methods have been used for measurements of epinephrine in complex samples [47]. In anodic stripping voltammetry, the species are oxidized on the electrode during the stripping step. Otherwise, in cathodic stripping voltammetry, the species are reduced on the electrode surface.

9.5.1 Working electrodes based on carbon materials

Nowadays, modified or non-modified carbon-based electrodes have been used more frequently for the determination of analytes in clinical samples, including epinephrine, since it is an electroactive molecule [48–50]. In most of the works the authors explore the oxidation process of the epinephrine to its quantification, which depends on the pH of the supporting electrolyte and the type of electrode or material used for modifying its surface. An illustration of some carbon-based electrodes can be seen in figure 9.2(c).

Carbon materials in the form of glassy carbon and graphite are inexpensive, usually present long-term stability and a wider potential range than solid metal electrodes. The type and quality of carbon material used to prepare the electrodes, and the composition of the supporting electrolyte influence their properties. GCE presents potential range from −0.8 to 1.2 V and carbon paste electrode (CPE), prepared with graphite powder, from −1.6 to 1.1 V (pH 4.5) [48]. This kind of electrode is prepared by a mixture of graphite powder and a water-immiscible organic liquid (e.g., mineral oil and paraffin oil). The paste is packed into a small Teflon holder with an electrical contact at the back.

An attractive alternative to these conventional electrode substrates is the use of a disposable screen-printed electrode (SPE) that includes working auxiliary and reference in the same substrate surface. It requires a reduced sample amount which avoids the cleaning and polishing of electrode surface, provides high flexibility, has low cost, and high reproducibility [51, 52].

As mentioned, in the human organisms, epinephrine is extensively metabolized through various pathways (oxidation, conjugation, and deamination) and as a result, it is found only in a small percentage of unchanged epinephrine excreted in human urine. In addition, it is worth mentioning that the biomolecules such as ascorbic acid, dopamine, norepinephrine, and uric acid coexist with epinephrine in physiological samples and have the same oxidizing potentials on the traditional working electrodes [53–55]. In this context, modification of carbon-based electrodes should be used to overcome these difficulties. The modification of electrode surface or graphite paste with materials at the nanoscale level leads to enhanced sensor performance with an increase of the electrocatalytic activity, sensitivity, and lower limits of detection [34, 56, 57].

The surface modification of carbon-based electrodes can be done with carbon nanomaterials, metallic nanoparticles, or enzymes. When preparing a modified carbon paste electrode, the modification agent can be dissolved in the pasting liquid or physically mixed with the paste. For the surface modification of GCE or SPE with nanomaterials, Nafion®, dihexadecylphosphate [58], carboxymethylated polysaccharide [59], chitosan [60], poly(allylamine hydrochloride) [61] or organic solvents [62] are commonly used.

Graphene and multiwalled carbon nanotubes (MWCNTs) have been applied to the modification of carbon-based electrodes [63, 64]. Graphene is a two-dimensional single sheet of sp^2 hybridized carbon with high electrical conductivity [65]. The modern improvements for graphene preparations are based on mechanical exfoliation of graphite or Hummer's oxidation [66–71]. Mechanical exfoliation of graphite in a mixture of water-solvent can be used as an alternative technique to separate individual graphene sheets. Then, they can be stabilized and suspended in a solution with the addition of cellulose derivative biopolymer carboxymethyl-cellulose [67]. It shows a very high surface area and conductivity, which makes them very attractive materials for the construction of sensors and biosensors. However, mechanical exfoliation is not effective for large-scale synthesis, so one of the most popular approaches to produce graphene is the use of strong oxidizing agents to obtain graphene oxide, a practical approach for the bulk scale synthesis of graphene. Graphene oxide can be readily synthesized by the oxidation of the natural flake graphite powder. Hummer's method employs $KMnO_4$ e $NaNO_3$ in concentrated H_2SO_4 [66]. This method has been improved/modified to replace $NaNO_3$ with other compounds such as H_2SO_4, H_3PO_4, and double the amount of $KMnO_4$ to avoid the emission of toxic gases [72]. A detailed review of known methods of preparation of graphene oxide by oxidation of graphite is given in Singh *et al* [73]. The structure and chemical composition of graphene oxide are unstable and some methods are proposed for its restoration, based on removing functional groups of the GO particles that are responsible for its high surface reactivity. For this, thermal methods can be used, which employ heating of graphene oxide directly or by irradiation (microwave, infrared visible, or ultraviolet) in a vacuum, inert, or reducing atmosphere. Another method used is chemical reduction, by employing hydrazine hydrate, hydroquinone, potassium alkaline solutions, etc. As described by Dideikin and Vul' [74], the advantage of chemical methods is the opportunity for simultaneous chemical modification while synthesizing the material.

Carbon nanotube (CNT) is a well-known one-dimensional nanostructured material with a hollow structure and presents strong active functional groups obtained by its oxidation in acid mixtures [75]. MWCNTs are comprised of several layers of concentrically ringed graphene with an outer diameter ranging from 2.5 to 30 nm. They have been widely used to modify electrode surfaces due to their properties of increasing the sensibility of electrochemical sensors and the ease of immobilization of enzymes on their surface [36, 76, 77]. The dispersion of CNTs is limited for almost all organic and inorganic solvents due to the smooth surface of carbon nanomaterials, lacking any overhanging bonds. In order to address this problem, chemical treatments with an acid mixture of sulfuric acid and nitric acid at the ratio of 3:1 (V/V) are commonly used [75, 78]. This treatment promotes the formation of defects sites and also introduces oxide groups on the surface (e.g., hydroxyl, carboxylic and/or quinones), increasing their dispersibility [61, 79] and decreasing the cytotoxicity [75].

9.5.2 Electrochemical biosensors based in polyphenol oxidases

The development of electrochemical biosensors has been attracting the attention of several researchers over the last years to detect and quantify phenolic compounds, including epinephrine, in pharmaceuticals and clinical samples [49]. In principle, an electrochemical enzymatic biosensor consists of an enzyme immobilized on the sensor surface, which has a transduction phase and is connected to a device for processing the electronic signal. The enzyme selectively interacts with the target analyte and the biological response is converted into an electrical signal.

Laccase and tyrosinase enzymes are the most used biocatalysts to fabricate biosensors for the determination of epinephrine. Both enzymes are called polyphenol oxidases and catalyze the oxidation of a variety of organic and inorganic substances. Laccase enzyme (EC 1.10.3.2. diphenol:oxygen oxidoreductase) is one of the most used enzymes for the determination of phenolic compounds. It contains four copper as cofactors and present high catalytic versatility to oxidation of phenolic compounds (including mono and polyphenols, amines, methoxyphenols, and aromatic amines), and low catalytic requirements (only O_2 of the air is required for regeneration). In the oxidation catalyzed by laccase, molecular oxygen is the final electron acceptor. The oxidation product is then electrochemically reduced on the biosensor surface and the resulting current is used for its quantification. Thus, the current of reduction is used for the quantification of epinephrine in the samples [33].

In the last two years, only the tyrosinase enzyme (EC 1.14.18.1, monophenol, diphenol:oxygen oxidoreductase) was used to developed biosensors for epinephrine determination. This enzyme contains two copper atoms in their active site that catalyzes two types of reaction that occur sequentially: the first one is the hydroxylation of monophenols to *o*-diphenols and the second is the oxidation of diphenols to quinones with the concomitant reduction of molecular oxygen to water in both steps. Gopal *et al* developed an electrochemical biosensor based on tyrosinase for epinephrine determination [9]. The enzyme used in the present study was extracted and isolated from *Flavadon* sp. fungi, which was isolated from soil contaminated with industrial waste. According to the authors, the enzyme catalyzes the oxidation of the catechol functional group to *o*-quinone functional group. In the next step, there is the reduction of *o*-quinone and further formation of 1,4-Micheal addition product. Thus, this product was reduced and subsequently reversibly oxidized. Additionally, this device was prepared by the drop casting method of enzyme on GCE modified with MWCNTs without the use of a covalent crosslinking reagent, avoiding steps on the preparation of biosensor and the use of many aggressive and toxic chemicals like glutaraldehyde. Apetrei and Apetrei described the immobilization of tyrosinase on single walled carbon nanotubes by a drop-and-dry method followed by crosslinking [80]. The interaction between tyrosinase and the SWCNTs involved physical adsorption of the enzyme by hydrophobic and electrostatic interaction. Some authors have used saturated glutaraldehyde vapor or glutaraldehyde solution prepared in phosphate buffer solution for tyrosinase immobilization by covalent crosslinking [81].

There are a variety of immobilization methods; however, there is no specific one applicable for all known enzymes. It is always important to know the physical and chemical characteristics of the support, as well as the possibility of its regeneration and also immobilization techniques. Moraes *et al* developed a laccase biosensor for epinephrine determination by the addition of laccase enzyme in the carbon black paste [33]. This biosensor presented a stable response over a period of three months when stored at 4 °C when not in use. In addition, the biosensor was able to perform 50 measurements without surface renewing. The addition of the enzyme to the paste makes the biosensor simple to prepare, however, it may be reported in the literature that the enzyme can leach into the solution during measurements.

The immobilization of the enzyme on the surface of a carbon electrode can be through physical interaction (adsorption, encapsulation, etc) or chemical bonding (covalent and covalent crosslinking). Physical adsorption is based on the interaction of the enzyme with the support through hydrogen bonds, Van der Waals forces, hydrophobic interactions and ionic bonds [80]. This method is simple, easy and of low cost, and there is still the possibility of reusing the support after several analyses. In summary, the enzyme remains confined in a membrane located on the electrode surface, which has a controlled porosity for the free diffusion of the analyte and their reaction products in the semipermeable polymeric membranes (nylon, cellulose acetate, etc).

Enzyme immobilization by the covalent bonding method is the most popular chemical immobilization method. Enzymes link to specific reactive groups present on the support (hydroxyls, carbonyls, amines, thiols, etc) through their functional groups that are not essential for their catalytic activity with [82]. These reactive groups on the support can be modified to produce reactive intermediates using crosslinking reagents (glutaraldehyde, cystamine, 1-ethyl-3-(3-dimethylamino-propyl) carbodiimide, N-hydroxysuccinimide), followed by enzymatic coupling to the activated support. These promote stronger bonds between the enzyme and the electrode, guaranteeing greater stability and resistance of the enzyme to variations of pH, temperature, organic solvents, and the enzyme is not leached to the solution.

9.6 Recent voltammetric platforms developed for epinephrine determination

Due to the great importance of epinephrine in pharmaceuticals and clinical areas, several sensors and biosensors have been developed for the determination of epinephrine. From 2020 through 2021, 11 papers have been published on the development of sensors and biosensors for this purpose.

Table 9.1 presents the reported papers which employed carbon-based sensors as working electrodes in the last two years for the determination of epinephrine. Furthermore, this table summarizes the analytical features of these papers, such as voltammetric technique, linear range, and samples analyzed.

Charithra and Manjunatha used poly oxalic acid (POX) to modify the surface of an MWCNTs paste electrode (MWCNTsPE) [83]. The electrode base was prepared by blending 60% carbon nanotube powder and 40% silicon oil, loaded on the crater of the Teflon tube. Next, the poly oxalic acid was electropolymerized on its surface after 10 cycles of CV in phosphate buffer solution (pH 7.0). From impedance studies, it was observed that the surface area of the modified electrode (0.02 cm^2) was greater than the unmodified electrode (0.01 cm^2), implying a superior ability to transfer electrons. In addition, it possesses lower resistance, excellent electrical conductivity, and superior electron transfer rate. With the modification, the sensor showed a decrease of 0.097 V in the potential of epinephrine oxidation, greater peak current intensity over the unmodified electrode, and linearity in the range of epinephrine concentration from 10 to 110 μmol L^{-1} by DPV. Repeatability and reproducibility of the proposed sensor present relative standard deviation (RSD) of 4.2 and 3.5%, respectively. The authors reported that after 40 cycles the sensor retained 97% of its activity. Employing this sensor, epinephrine can be determined in the presence of folic acid and paracetamol without overlap between the oxidation peaks and satisfactory separation of peak potential. But the authors did not analyze

Table 9.1. Analytical features of reported papers for determination of epinephrine.

Sensor	Technique	Electrochemical process	Linear range (μmol L^{-1})	Application	References
MWCNTs paste electrode modified with poly oxalic acid	DPV	Oxidation	10.0–110	Pharmaceutical formulation and human serum	[83]
GCE modified with ZnONPs and MWCNTs	DPV	Oxidation	0.40–2.4	Pharmaceutical formulation and human serum	[84]
GCE modified with MWCNTs and tyrosinase enzyme	DPV	Reduction	3.0–200	Pharmaceutical formulation and human serum	[85]
CPE modified with TiO_2/rGO composite	DPV	Oxidation	5.0–100	Ternary mixture of epinephrine	[86]
GCE modified with graphene oxide nanosheets	DPV	Reduction	10.0–300;300–1300	Human serum	[87]
GCE modified with 2D-Porous NiO-rGO	CV	Oxidation	50.0–1000	Human serum	[88]
GCE modified with graphene nanoribbons	DPV	Reduction	6.4–100	Pharmaceutical formulation	[81]
GCE modified with sulfacetamide film	Anodic stripping DPV (ASDPV)	Oxidation	0.10–1.9	Pharmaceutical, human serum and human urine	[89]
GCE modified with phosphorus-doped carbon microspheres mesoporous	SWV	Oxidation	0.010–6.0	Human serum	[90]
GCE modified with Pd/$BiVO_4$ composite	CV and DPV	Oxidation	0.90–3.4; 6.9–27.5		[91]
SPE modified with chitosan-coated magnetite nanoparticles	DPV	Oxidation	0.10–0.800. 10–0.60	Synthetic cerebrospinal fluid and synthetic urine	[92]

epinephrine in the presence of uric acid and other catecholamines (dopamine and norepenephrine), which are very common substances present in biological samples and with oxidation potential close to that of epinephrine oxidation on carbon-based electrodes. The authors showed the applicability of the sensor in the determination of epinephrine in an injection sample and blood serum samples.

Shaikshavali decored MWCNTs with ZnONPs using an ultra-sonication procedure [84]. The nanocomposite ZnO/MWCNTs was dispersed on the surface of GCE and the sensor showed good electrocatalytic activity to epinephrine oxidation resulting in higher peak currents than those obtained when employing bare GCE. The electrochemical reaction mechanism of epinephrine was controlled by an adsorption process. Determination of this catecholamine endowed linear dynamic range from 0.40 to 240 μmol L^{-1} in phosphate buffer solution (pH 7.0) by DPV. This sensor was able to determine epinephrine in the presence of ascorbic acid and uric acid with good selectivity, wide peak-to-peak separation, and a good limit of detection. In addition, 50 fold higher concentrations of dopamine, paracetamol, serotonin, and inorganic elements such as Fe^{2+}, Ba^{2+}, Zn^{2+}, Na^{+}, and Cu^{2+} did not result in interference on epinephrine response. Regarding the stability of the sensor, the authors comment that there was no apparent disturbance in the electrochemical response and stable behavior of epinephrine current response was observed. This sensor presented a repeatability with an RSD of 4.6% and was applied to epinephrine determination in pharmaceutical formulations and human serum samples with recoveries varying from 98.3 to 104%.

A GCE was modified with MWCNTs and tyrosinase enzyme for the selective determination of epinephrine [85]. MWCNTs were dispersed in ethanol and dropped onto the freshly prepared bare GCE surface; the enzyme was prepared in phosphate buffer solution (pH 6.5) and immobilized onto the MWCNTs/GCE surface by physical adsorption. The nanomaterial increases the rate of electron transfer and thus the active surface area of MWCNT/GCE (0.01208 cm^2) was higher than the bare GCE (0.01021 cm^2). However, the active surface area of the biosensor (0.01021 cm^2) was decreased when compared to that obtained by MWCNT/GCE which was due to the blocking of active sites on the surface of MWCNTs by the enzyme. The authors tested the applicability of this biosensor in pharmaceutical formulations and human serum samples with recovery values of 97.3 to 106.6%. Despite the application in biological samples, they did not present a selectivity study. The determination of epinephrine presented a linear range from 3.0 to 200 μmol L^{-1} in phosphate buffer solution (pH 7.0) by DPV. An RSD of 5.61% was obtained in the reproducibility study and 2.41% for stability study realized over 30 days.

In the CPE modified with TiO_2/rGO composite, developed by Joseph and Thomas, the epinephrine molecule presented a different oxidation potential compared to ascorbic acid and uric acid, showing that the electrode modification with TiO_2/rGO composite improves the selectivity of the CPE as well as electrical conductivity [86]. The use of TiO_2/rGO composite indicates that epinephrine can be determined individually in the presence of these substances or simultaneously since the observed peak-to-peak separation was 0.214 V between ascorbic acid and epinephrine and 0.186 V for epinephrine and uric acid. The linear range of epinephrine concentration varied from 5.0 to 100 μmol L^{-1} by DPV in potassium phosphate buffer solution (pH 7.0). The authors mention that this sensor realized more than a hundred cycles without a significant decrease in current; however, they do not present DPR values.

Other authors have also used graphene nanosheets in the modification of several GCE surfaces for epinephrine determination. Immanuel and Sivasubranian synthesized reduced graphene oxide by the modified Hummer's method, dispersed it in ethanol, and used it to modify the GCE surface, improving its features, with an active surface area of 0.238 cm^2 [87]. To avoid interference from electroactive compounds such as dopamine, uric acid, and ascorbic acid, improving the selectivity of the sensor, the determination of epinephrine was carried out using DPV through the reduction process of epinephrine chrome to leucoepinephrine chrome. Dopamine reduction in this sensor by DPV technique occurs at 0.15 V, different from that observed for epinephrine (at −0.22 V). The two different linear ranges can be employed to quantify epinephrine, 10 to 300 μmol L^{-1} and 300 to 1300 μmol L^{-1} in phosphate buffer solution (pH 7.4) by DPV. This sensor was stored at room temperature and it was able to retain around 80% of the initial current after the end of the week. Determination of epinephrine was carried out in serum samples with recovery values varied from 112 to 70.9%.

Ramu *et al* used 2D-porous NiO nanoparticles firmly attached over reduced graphene oxide to modify the GCE surface presenting high conductivity and charge transfer over bare GCE [88]. 2D-porous NiO-rGO was synthesized by hydrothermal treatment of nickel salt in an aqueous solution with GO suspension and drop casted on the top of GCE. This procedure promotes the reduction of GO to rGO. The formation of NiO and complete reduction GO into rGO by synthesized by a hydrothermal treatment was confirmed employing FTIR and XPS techniques. This sensor exhibited the irreversible oxidation peak of epinephrine at 0.28 V vs. Ag/AgCl, characterized by a diffusion process, which corresponds to the oxidation of epinephrine to epinephrine quinone. The linear range of epinephrine concentration varied from 50 to 1000 μmol L^{-1} by CV in potassium phosphate buffer solution (pH 7.0). Inorganic compounds, such as K^+, Cl^-, Na^+, citrate, and NO_3^- shows anti-interfering ability; however, ascorbic acid and hydroquinone interfere on epinephrine sensing using this sensor. The modified electrode was kept in the refrigerator when it is not used. They did not report the amount of measurement that the sensor performs without modifying the surface. Epinephrine was determined in human serum sample with a recovery percentage of 99.63%.

Graphene nanoribbons were synthesized by the chemical route and used to modify the GCE surface [81]. To minimize interference, the authors used the analytical signal of the reduction peak corresponding to the conversion of epinephrine chrome to leucoepinephrine chrome (−0.25 V) instead of the oxidation peak usually employed to the quantification. A linear concentration range from 6.4 to 100 μmol L^{-1} was used to quantify epinephrine in phosphate buffer solution (pH 8.0) by DPV. Repeatability and reproducibility of this sensor present RSD values of 5.2 and 6.9%, respectively. The sensor response 20 days after preparation displayed 90% of the original response, being kept at 4 °C after preparation.

GCE was modified with sulfacetamide film achieved by applying a potential between +0.4 and + 1.4 V at the scan rate of 100 mVs^{-1} for 20 cycles [89]. For the formation of the film, the amino group of the sulfacetamide oxidizes forming a radical cation and then forming carbon–nitrogen linkages at the surface of GCE; this reaction is based on steric effects as well as diffusion rates. Based on electrochemical studies of sensor behavior in the presence of potassium ferrocyanide, the authors report that the film formed on the surface of GCE leads to an increase in the electron transfer kinetics, exhibiting improvement of redox peak current and electrocatalytic activity. In this sensor, epinephrine oxidizes at 0.230 V vs. Ag/AgCl and presented a linear response in the range from 0.10 to 1.9 μmol L^{-1} in phosphate buffer solution (pH 5.5) by anodic stripping differential pulse voltammetry (ASDPV; $E_{accumulation}$ = +0.05 V and $t_{accumulation}$ = 60 s). The authors justify that the use of this technique was due to the oxidation process of epinephrine in this type of electrode being controlled by adsorption. The sensor was tested in the presence of uric acid, ascorbic acid, glucose, and dopamine, which two well-defined oxidation peaks for uric acid and epinephrine mixture were obtained with peak-to-peak separation of 120 mV. For ascorbic acid and glucose, under the experimental conditions used (in the ratio of 1:10 epinephrine and interferent), no peaks were observed. Dopamine showed serious interference on epinephrine response in the mixture containing the same concentration. The sensor presented good repeatability and reproducibility, with RSD values of 4.6 and 3.4%, respectively, and was stable for 15 days, showing a decrease in current values of 95.4% during this period. The applicability of this sensor was verified in the determination of epinephrine in pharmaceutical sample preparation, human serum, and urine samples.

Another modification of the GCE was performed with carbon microspheres mesoporous doped by P atoms [90]. The sensor features a highly active catalytic surface with fast charge mobility with several active sites, probably due to high surface area (0.02 cm^2 compared with 0.007 cm^2 of GCE), the surface morphology of the rugged surface, holes/grooves on the surface, well-dispersed and interlinked spheres, and curvature ends. In this sensor, the oxidation of epinephrine was controlled by an adsorption process, with a higher current value when compared with bare GCE. The active surface of the modified electrode induces binding with epinephrine as a result of the positive charge of the analyte in the solution and the negatively charged surface of the sensor due to the presence of P atoms. A linear relationship in the concentration range from 0.010 to 6.0 μmol L^{-1} was obtained employing the oxidation peak in phosphate buffer solution (pH 7.0) and SWV technique. The sensor presented repeatability and reproducibility with RSD values of 3.90 and 4.25%, respectively; however, the number of measurements that the sensor performs was not described in the article. Epinephrine has shown a stable response in the presence of compounds such as ascorbic acid, uric acid, glucose, H_2O_2 at the same concentration. Dopamine and norepinephrine resulted in interfering signals (overlap of peak potential). A recovery range of 0.98%–99.5% of epinephrine was obtained in human serum samples, showing the selectivity and suitability of this sensor.

GCE was also modified Pd/BiVO_4 composite [91]. This material was prepared in 5% chitosan and acetone. This slurry was dropped onto the freshly polished GCE and dried in an oven at 80 °C, to make the acetone evaporate completely. The modified sensor exhibited better electrochemical performance with a significantly higher redox peak current than those of bare GCE. By CV, epinephrine presented an oxidation potential at 0.40 V vs. saturated Ag/AgCl electrode. A linear relationship in the concentration range from 0.90 to 3.4 μmol L^{-1} and 6.9 to 27.5 μmol L^{-1} was obtained. The authors report that the current signal of epinephrine was almost ten times greater than that observed for ascorbic acid, L-cysteine, and dopamine, at the same concentration. The sensor has not been applied to real samples. The authors reported that the sensor repeatability was tested in 15 days and that the current value was maintained around the same value.

Mattioli *et al* developed a screen-printed disposable electrode using graphite-polyurethane composite modified with magnetite (Fe_3O_4NPs-SPE) or chitosan-coated magnetite nanoparticles (CHFe_3O_4NPs-SPE) [92]. For both sensors linear response to epinephrine concentration range was achieved from 0.10 to 0.80 μmol L^{-1} at 0.03 V (vs. pseudo-Ag/AgCl), and from 0.10 to 0.60 μmol L^{-1} at 0.0 V (vs. pseudo-Ag/AgCl), respectively to Fe_3O_4NPs-SPE and CHFe_3O_4NPs-SPE, in phosphate buffer solution (pH 7.0) by DPV. The devices were used for the quantification of epinephrine in synthetic urine and cerebrospinal synthetic fluids with recoveries from spiked samples are in the 95.6 to 102.2% range for the CHFe_3O_4NPs-SPE and in the 98.3 to 109% range for MNPs-SPE. The epinephrine peak currents were found to decrease by only 4% for the CHFe_3O_4NPs-SPE, while the Fe_3O_4NP-SPE lost 23% of its epinephrine peak current for five months. CHFe_3O_4NPs-SPE was the most stable and sensitive sensor for epinephrine determination due to the chitosan coating which protects the Fe_3O_4NPs from oxidation under air and at a different applied potential. Furthermore, the authors comment that the presence of some functional groups in chitosan around the Fe_3O_4NPs may improve the electrochemical response of this sensor. The sensor presented an RSD value of 5.0% for repeatability employing CHFe_3O_4NPs-SPE without adsorption of the analyte onto its surface. Uric acid, creatinine, and tryptophan presented negative interference to the epinephrine determination, which was attributed to competition by electroactive sites of the electrodes and even adsorption of oxidation products. Dopamine, presented a positive interference due to its structural similarity with epinephrine, leading to extreme competition with high sensitivity at such devices. The authors comment that the separation method for dopamine and epinephrine is imperative for quantitative purposes, which is disadvantageous for faster analyzes development and the application of these devices in real samples.

9.7 Electrochemical mechanism of epinephrine oxidation

The electrochemical mechanism of epinephrine is a well-known process described in the literature. Hawley *et al* employed electrochemical techniques to study the oxidation pathways of catecholamines *in vitro*, employing a carbon paste electrode [8]. These species present redox peaks that depend on the pH of the reaction medium. Briefly, the first peak corresponds to the oxidation of epinephrine to form epinephrine quinone by a two-electron process. During the reverse potential scan, the peak of reduction of epinephrine quinone is observed and also another cathodic peak as well as two other anodic peaks appear in the subsequent scans. The product from the reduction of epinephrine quinone is a precursor for the intramolecular 1,4 (Micheal) addition reaction to form epinephrine chrome (cyclization product). The third peak is attributed to the reduction of the epinephrine chrome to leucoepinephrine chrome, while the fourth peak corresponds to the oxidation of leucoepinephrine chrome to epinephrine chrome. Next, leucoepinephrine chrome undergoes a dehydration reaction, forming 5,6-dihydroxy-N-methylindole, which is then oxidized to epinephrine (corresponding to the fifth peak). Thus, the epinephrine is regenerated and hence can be oxidized at the electrode surface. At low pH (1.00 mol L^{-1} H_2SO_4) the epinephrine is oxidized to epinephrine quinone and during the reversal scanning potential, its reduction to epinephrine is observed. In this case, due to the acidic conditions, the cyclization reaction is precluded. At pH 3.0, however, sufficient unprotonated quinone is available to allow the cyclization reaction to take place.

Subsequently, Kim *et al* studied the electrochemical behavior of epinephrine at a gold electrode by using CV and spectroelectrochemical techniques [93]. In very low pH solution, epinephrine quinone is unstable enough to be converted to epinephrine. Thus, epinephrine is oxidized to epinephrine quinone and following reduced, the same behavior previously observed. On the other hand, based on the analysis of a series of spectra obtained by *in situ* spectroscopic measurements they observed that epinephrine quinone is formed in the first step of oxidation of epinephrine followed by cyclization of the quinone product to produce leucoepinephrine chrome. The leucoepinephrine chrome is finally converted into epinephrine chrome chemically or electrochemically. In acidic conditions, the formation of a dehydrated form of leucoepinephrine chrome is not likely to occur, as previously proposed by Hawley *et al* [8].

In the most recent articles, the authors present and comment only on the process of oxidation of epinephrine into epinephrine quinone and subsequent reduction of this into epinephrine. All measurements were performed in a phosphate buffer solution with a pH near the physiological. Considering the same potential range studied, only on some sensors the epinephrine has more redox processes. For example, Emran *et al* observed four peaks at low potentials of 0.36, 0.262, 0.05, and −0.06 V vs. Ag/AgCl using a GCE modified with carbon microspheres mesoporous doped by P atoms [90]. These peaks are related to the oxidation of epinephrine to form a correspondent quinone, and the reduction of this quinone; meanwhile, the next steps involve the reduction of a cyclized product from epinephrine chrome to leucoepinephrine chrome and finally, the oxidation of leucoepinephrine chrome to epinephrine chrome (figure 9.3). Whereas, Ahmed *et al* employing a GCE modified with sulfacetamide presented two redox peaks, which the first corresponding to the oxidation of epinephrine to epinephrine quinone around 0.25 V and the second around 0.21 V vs. Ag/AgCl, corresponding to the reduction of epinephrine quinone back to epinephrine [89].

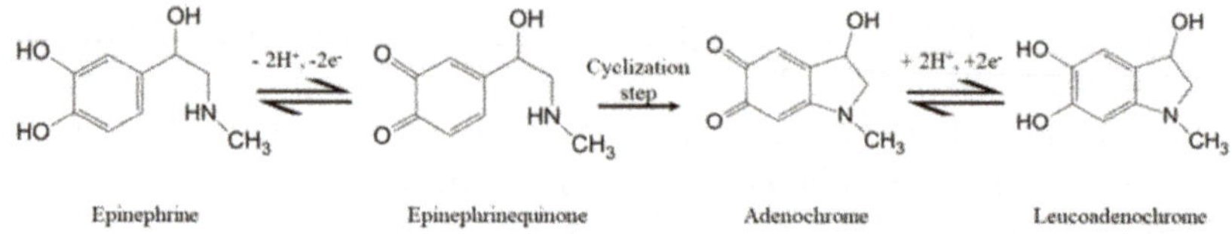

Figure 9.3. Schematic representation of the epinephrine oxidation mechanism [8, 9].

9.8 Recent progress in microelectrodes for *in vivo* electrochemical sensing of epinephrine

Neurotransmitters are involved in chemical messaging through the brain cells during a very dynamic process in living beings, called neurotransmission [94]. As such, *in vivo* real-time monitoring of neurochemical signaling is of particular interest in the diagnosis and treatment of neurological disorders. Being different from the traditional methodologies for building (bio)sensing platforms, microengineering is the cutting-edge method to produce analytical devices that overcome the challenges of measuring species *in vivo* [95]. Overall, current technologies for designing microelectrodes applied in neural recording include metal wires, silicon, and carbon-based materials [96].

Carbon-fiber microelectrodes (CFMEs) have been used in neural interface devices as an important tool in neurochemical research and clinical applications (see scheme in figure 9.4). CFMEs have many technical advantages to solve most of the challenges associated with other electrode designs. Regarding implantable devices applications, the carbon-based fibers are biologically compatible and non-toxic to cells and tissues, which avoid an inflammatory response in the host body [96]. These microfiber-based electrodes with low micron dimensions and soft surfaces are very amenable for implantation procedures and cause less tissue damage than conventional electrodes [97].

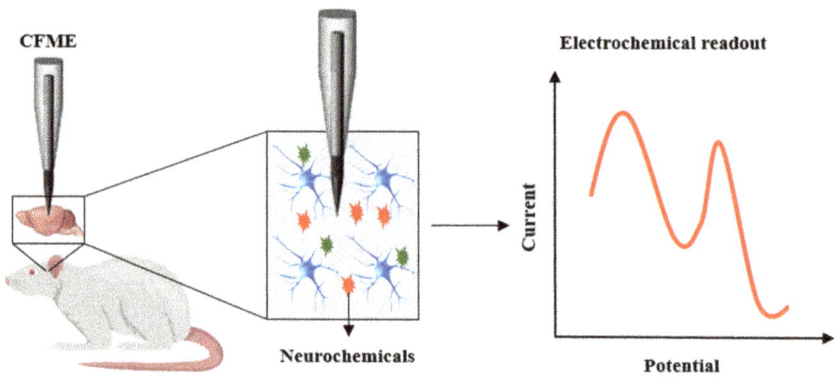

Figure 9.4. An overview of carbon-fiber microelectrodes applied during *in vivo* electrochemical measurements.

The electrochemistry of CFMEs has been well established over the years, especially their ability to perform rapid measurements using fast electrochemical techniques, including fast-scan cyclic voltammetry (FSCV), amperometry, and high-speed chronoamperometry [98]. Microelectrodes allied to FSCV provide excellent temporal resolution to follow the dynamics of neurochemical processes. Differently from conventional CV, FSCV operates at high scan rates of about 400 V s^{-1}, which is 1000 times faster and allows subsecond temporal resolution [99]. Figure 9.5 shows the FSCV waveform and redox reactions for epinephrine. The potential applied changes from 0.1 V to 1.45 V at 400 V s^{-1}, and the frequency is 10 Hz.

Roberts and collaborators [100] have successfully applied CFMEs in intracellular electrochemical cytometry for monitoring exocytosis events in mammalian cells using amperometry and FSCV. The carbon-based nanoelectrodes were designed to allow a selective electrochemical sampling of catecholamines in the cell cytosol. The nanobiosensors were also applied using FSCV to distinguish secretory vesicles containing epinephrine from other catecholamine-containing vesicles in the cell. Despite the structural similarity of epinephrine and norepinephrine, fast-scan cyclic voltammograms have shown that the oxidation of norepinephrine results in a single peak (+0.65 V vs. Ag/AgCl), while two peaks are detected during the oxidation of epinephrine. The first peak (+0.65 V vs. Ag/AgCl) corresponds to the oxidation process of the reduced form to *o*-quinone and the second one (+1.45 V vs. Ag/AgCl) arises from irreversible oxidation of the amine to imine. According to the authors, epinephrine content is highly variable from cell to cell and ~37% of the vesicle impact events resulted in detection of this catecholamine, which is consistent with previous HPLC analysis.

Conventional carbon-based microelectrodes are fundamental bioanalytical tools for real-time neurochemical detection in live brain tissue. However, chemical modifications in their composition can contribute to enhancing sensitivity [96]. Carbon nanotube technology applied to CFMEs has grown over the years. CNTs

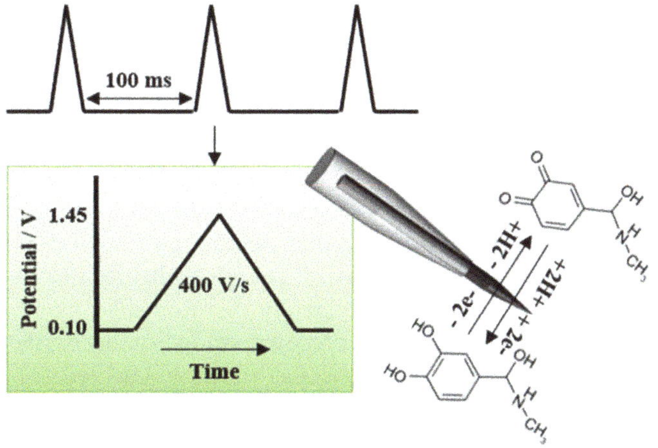

Figure 9.5. Fast-scan cyclic voltammetry waveform during the electrochemical cycle of epinephrine on the surface of a carbon-fiber microelectrode.

offer interesting physical properties for electrochemical analysis in biological systems due to their ability to momentarily trap the analyte into the cavities among the structure, which increases sensitivity [101]. Shao *et al* [102] described the use of carbon nanotubes yarn microelectrodes (CNTYMEs) and cavity carbon nanopipette electrodes (CNPEs) as useful tools in discriminating catecholamines in biological tissue, particularly epinephrine and norepinephrine at micromolar concentrations. According to the authors, two peaks were observed for epinephrine oxidation; the first one at +0.15 V (vs. Ag/AgCl), and the second one at +1.30V (vs. Ag/AgCl), while for dopamine and norepinephrine only one peak was observed. Overall, CNTYMEs and CNPEs were shown as excellent electrochemical tools for discriminating different catecholamines with similar structures in the biological environment. Both electrodes have trapping effects caused by thin-layer structure allied to fast electron transfer kinetics and high temporal resolution, which make them suitable for *in vivo* applications.

Gupta *et al* [103] reported the first use of highly densified carbon nanotubes fibers (HD-CNT*f*) for *in vivo* neurochemicals detection. Microarrays of HD-CNT*f* were applied to detect and quantify dopamine, serotonin, epinephrine, and norepinephrine using CV and SWV. The microelectrodes operated in a wide linear range from 1 nmol L^{-1} to 100 μmol L^{-1}, with a LOD of 64 pmol L^{-1} for epinephrine. The authors evaluated the kinetic advantages of using densified CNT fibers over the nondensified microarrays. Based on the cyclic voltammograms, the redox peaks for all the neurotransmitters were well-identified using both electrodes, but higher peak currents and lower peak-to-peak separation were observed at the densified CNT-based microelectrode. According to the authors, the ideal reversibility and the faster electron transfer kinetics at the HD-CNT*f* are attributed to the increased conductivity by increasing the contact of individual CNTs. The ultrasensitive HD-CNT*f* microelectrode was successfully applied in neurochemical real-time monitoring of living pheochromocytoma cells.

Miniaturization of electroanalytical systems has progressed effectively not only in the application of conventional carbon-based materials. Nanodiamonds (ND) were explored for the first time in neurochemical electrochemistry by Puthongkham and Venton [104]. This carbon nanomaterial has a core diamond crystalline structure and crystal defects on its surface. NDs may prove to be a better option to the traditional nanomaterials currently employed in drug delivery, implantable materials, biosensors, and biomedical robots due to their unique properties such as inertness and low cytotoxicity [105]. According to the authors, the nanodiamond modified carbon-fiber microelectrode (ND/CFME) was able to detect neurochemicals such as epinephrine, norepinephrine, and ascorbic acid at nanomolar levels. Epinephrine was successfully distinguished from another species because of the additional anodic peak corresponding to the oxidative cyclization resulting in leucoaminochrome. The same peak is not detected for dopamine and norepinephrine due to the slower kinetics of the cyclization process for those molecules [106]. Additionally, ND coating also reduces electrochemical fouling and biofouling events, still maintaining enhanced electrochemical readout.

9.9 Conclusion and future perspectives

In this chapter, the authors have described and commented on key highlights of the recent progress in electrochemical monitoring of epinephrine. Overall, the voltammetric sensing of neurochemicals employing carbon-based materials has been experiencing solid growth over the years. As a result of the increasing improvement in micro- and nanoengineering, complex benchtop electrochemical instruments, including conventional potentiostats, can be easily miniaturized to create portable systems for neurosignaling. Materials science advancements allow the research in the direction of carbon nanomaterials as transducers remains very active, especially for graphene, carbon nanotubes, and fullerene. Several research groups have been dedicated to developing new protocols of electrode modification with composites of nanomaterials based on metal and metal oxide nanoparticles to improve analytical properties. The progress in the domain of microelectrodes and disposable paper-based voltammetric devices is also very impressive and brings these devices closer to point-of-care, in-field, and *in vivo* applications. As a consequence of all those advancements, epinephrine has been successfully detected and quantified in complex real samples, including pharmaceutical formulations, human serum, human urine, cerebrospinal fluid, and brain tissue. Despite the recent progress in the field of epinephrine electrochemical sensing, there are still challenges and questions that need to be addressed when moving towards electrochemical devices applied as diagnostic tools. Future perspectives in this field of research may include efforts to make probe sizes at the nanometer scale to allow synaptic measurements; advancements in miniaturization and cytotoxicity to overcome issues of physical and biochemical perturbation in the brain tissue during implantation procedures; and (bio)chemical modification on the electrodes surface to improve selectivity in complex environments.

Acknowledgments

We gratefully acknowledge the financial support from CNPq (grants no. 408591/2018–8; 305320/2019–0; E. R. Sartori), Fundação Araucária do Paraná, and CAPES.

References

[1] Collste P, Brismar B, Alveryd A, Björkhem I, Hårdstedt C, Svensson L and Ostman J 1986 The catecholamine concentration in central veins of hypertensive patients-an aid not without problems in locating phaeochromocytoma *Acta. Chir. Scand. Suppl.* **530** 67–71

[2] Goldstein D S 2012 Noradrenergic neurotransmission *Primer on the Autonomic Nervous System* 2nd edn (New York: Academic) pp 37–43

[3] Tschopp C, Tramèr M R, Schneider A, Zaarour M and Elia N 2018 Benefit and harm of adding epinephrine to a local anesthetic for neuraxial and locoregional anesthesia: a meta-analysis of randomized controlled trials with trial sequential analyses *Anesth. Analg.* **127** 228–39

[4] Callaway C W 2013 Epinephrine for cardiac arrest *Curr. Opin. Cardiol.* **28** 36–42

[5] Page L B, Raker J W and Berberich F R 1969 Pheochromocytoma with predominant epinephrine secretion *Am. J. Med.* **47** 648–52
[6] Kim A R, Nodel M R, Pavlenko T A, Chesnokova N B, Yakhno N N and Ugrumov M V 2019 Tear fluid catecholamines as biomarkers of the parkinson's disease: a clinical and experimental study *Acta Nat.* **11** 99–103
[7] Young W F 2012 Adrenal medulla, catecholamines, and pheochromocytoma 24th edn, vol 2 *Goldman's Cecil Medicine* (Amsterdam: Elsevier) pp 1470–475
[8] Hawley M D, Tatawawadi S V, Piekarski S and Adams R N 1967 Electrochemical studies of the oxidation pathways of catecholamines *J. Am. Chem. Soc.* **89** 447–50
[9] Siegel G J, Agranoff B W, Albers R W and Molinoff P B 1999 *Basic Neurochemistry: Molecular, Cellular and Medical Aspects* 6th edn (New York: Haven Press)
[10] Bennett M 1999 One hundred years of adrenaline: the discovery of autoreceptors *Clin. Auton. Res.* **9** 145–59
[11] Sneader W 2001 The discovery and synthesis of epinephrine. *Drug News Perspect.* **14** 4914
[12] Ted T and Song P L 2015 *Current Opinion in Allergy and Clinical Immunology* pp 323–28
[13] Rao Y 2019 The first hormone: adrenaline *Trends Endocrinol. Metab.* **30** 331–4
[14] Saito S K J 2011 Adrenaline: isolated 111 years ago by an energetic Japanese scientist *Masui* **60** 1331–41
[15] Dalal R and Grujic D 2021 *Epinephrine* (Treasure Island: StatPearls Publishing)
[16] Kennedy D W G, Shaikh Z, Fardy M J, Evans R J and Crean S J V 2004 Topical adrenaline and cocaine gel for anaesthetising children's lacerations. An audit of acceptability and safety *Emerg. Med. J.* **21** 194–6
[17] Grant S A D and Hoffman R S 1992 Use of tetracaine, epinephrine, and cocaine as a topical anesthetic in the emergency department *Ann. Emerg. Med.* **21** 987–97
[18] Yi Z and Brown P R 1991 Chromatographic methods for the analysis of basic neurotransmitters and their acidic metabolites *Biomed. Chromatogr.* **5** 101–7
[19] Nagatsu T 2006 The catecholamine system in health and disease—relation to tyrosine 3-monooxygenase and other catecholamine-synthesizing enzymes *Proc. Jpn. Acad. Ser. B. Phys. Biol. Sci.* **82** 388–415
[20] Wood J P, Traub S J and Lipinski C 2013 Safety of epinephrine for anaphylaxis in the emergency setting *World J. Emerg. Med.* **4** 245
[21] FDA approves first generic version of EpiPen [Internet] 2016 https://fda.gov/news-events/press-announcements/fda-approves-first-generic-version-epipen
[22] Rawas-Qalaji M M, Simons F E R and Simons K J 2006 Sublingual epinephrine tablets versus intramuscular injection of epinephrine: dose equivalence for potential treatment of anaphylaxis *J. Allergy Clin. Immunol.* **117** 398–403
[23] Rachid O, Rawas-Qalaji M and Simons K J 2018 Epinephrine in anaphylaxis: preclinical study of pharmacokinetics after sublingual administration of taste-masked tablets for potential pediatric use *Pharmaceutics.* **10** 24–34
[24] Reingardienė D 2006 The consequence of epinephrine (adrenaline) overdose *Medicina (Kaunas)* **42** 606–9
[25] Benowitz N L 2012 Epinephrine *Poisoning & Drug Overdone* 6th edn, ed K R Olson (New York: McGraw Hill) ch 185
[26] Nahler G 2009 European pharmacopoeia (Eur Ph) *Dictionary of Pharmaceutical Medicine* (Vienna: Springer Vienna) p 69

[27] Gorbunova M V, Gutorova S V, Berseneva D A, Apyari V V, Zaitsev V D, Dmitrienko S G and Zolotov Y A 2019 Spectroscopic methods for determination of catecholamines: a mini-review *Appl. Spectrosc. Rev.* **54** 631–52

[28] Wang Y, Fice D S and Yeung P K F 1999 A simple high-performance liquid chromatography assay for simultaneous determination of plasma norepinephrine, epinephrine, dopamine and 3,4-dihydroxyphenyl acetic acid *J. Pharm. Biomed. Anal.* **21** 519–25

[29] Gámiz-Gracia L, García-Campaña A M, Huertas-Pérez J F and Lara F J 2009 Chemiluminescence detection in liquid chromatography: applications to clinical, pharmaceutical, environmental and food analysis-a review *Anal. Chim. Acta* **640** 7–28

[30] United States Pharmacopeial Convention 1985 United States Pharmacopeia National Formulary XXXIII

[31] Musso N R, Vergassola C, Pende A and Lotti G 1990 HPLC with electrochemical detection of catecholamines in human plasma. A mini-review *J. Liq. Chromatogr.* **13** 1075–90

[32] Bicker J, Fortuna A, Alves G and Falcão A 2013 Liquid chromatographic methods for the quantification of catecholamines and their metabolites in several biological samples—a review *Anal. Chim. Acta* **768** 12–34

[33] Moraes J T, Salamanca-Neto C A R, Švorc Ĺ, Schirmann J G, Barbosa-Dekker A M, Dekker R F H and Sartori E R 2019 Laccase from: *Botryosphaeria rhodina* MAMB-05 as a biological component in electrochemical biosensing devices *Anal. Methods* **11** 717–20

[34] Karimi-Maleh H, Karimi F, Rezapour M, Bijad M, Farsi M, Beheshti A and Shahidi S-A 2019 Carbon paste modified electrode as powerful sensor approach determination of food contaminants, drug ingredients, and environmental pollutants: a review *Curr. Anal. Chem.* **15** 13

[35] Gupta V K, Jain R, Radhapyari K, Jadon N and Agarwal S 2011 Jan Voltammetric techniques for the assay of pharmaceuticals-a review *Anal. Biochem.* **408** 179–96

[36] Beitollahi H, Safaei M and Tajik S 2019 Voltammetric and amperometric sensors for determination of epinephrine: a short review (2013–2017) *J. Electrochem. Sci. Eng.* **9** 27–43

[37] Hussain C M and Keçili R 2020 Electrochemical techniques for environmental analysis *Modern Environmental Analysis Techniques for Pollutants* ed C M Hussain and R Kecili (New Jersey: Elsevier) pp 199–222

[38] Warner J T 2019 Electrochemistry basics *Lithium-Ion Battery Chemistries* 1st edn (Grand Blanc, MI: Elsevier) 17–41

[39] Islam M N and Channon R B 2019 Electrochemical sensors *Bioengineering Innovative Solutions for Cancer* (Amsterdam: Elsevier) pp 47–71

[40] Pletcher D, Greff R, Peat R, Peter L M and Robinson J 2010 Potential sweep techniques and cyclic voltammetry *Instrumental Methods in Electrochemistry* (Southampton: Woodhead Publishing) pp 178–228

[41] Bard A J and Faulkner L R 2001 *Electrochemical Methods: Fundamentals and Applications.* 2nd edn (New York: Wiley)

[42] Gosser D K 1993 *Cyclic Voltammetry: Simulation and Analysis of Reaction Mechanisms.* (New York: Wiley) p 154

[43] Scholz F 2010 *Electroanalytical Methods.* 2nd edn (Berlin: Springer)

[44] Raril C and Manjunatha J G 2018 Sensitive electrochemical analysis of resorcinol using polymer modified carbon paste electrode: a cyclic voltammetric study *Anal. Bioanal. Electrochem.* **10** 488–98

[45] Scholz F 2016 *Pulse Voltammetry in Physical Electrochemistry and Electroanalysis. Theory and Applications* 1st edn pp 463–580

[46] Cheraghi S, Taher M A, Bijad M and Sadeghifar H 2017 A review: stripping voltammetric methods as a high sensitive strategy for trace analysis of ions, pharmaceutical and food samples *Curr. Anal. Chem.* **13** 5–12

[47] Ly S Y 2012 Determination of epinephrine in carp brain cells by square-wave anodic stripping voltammetry with carbon nanotubes *Anal. Lett.* **45** 1197–203

[48] Uslu B and Ozkan S A 2007 Electroanalytical application of carbon based electrodes to the pharmaceuticals *Anal. Lett.* **40** 817–53

[49] Ribeiro J A, Fernandes P M V, Pereira C M and Silva F 2016 Electrochemical sensors and biosensors for determination of catecholamine neurotransmitters: a review *Talanta* **160** 653–79

[50] Dhanjai, Sinha A, Lu X, Wu L, Tan D, Li Y, Chen J and Jain R 2018 Voltammetric sensing of biomolecules at carbon based electrode interfaces: a review *TrAC—Trends Anal. Chem.* **98** 174–89

[51] Couto R A S, Lima J L F C and Quinaz M B 2016 Recent developments, characteristics and potential applications of screen-printed electrodes in pharmaceutical and biological analysis *Talanta* **146** 801–14

[52] García-Miranda Ferrari A, Rowley-Neale S J and Banks C E 2021 Screen-printed electrodes: transitioning the laboratory in-to-the field *Talanta Open.* **3** 100032

[53] Manjunatha J G, Swamy B E K and Deraman M 2013 Simultaneous determination of ascorbic acid, dopamine and uric acid at poly (aniline blue) modified carbon paste electrode: a cyclic voltammetric study *Int. J. Pharm. Pharm. Sci.* **5** 355–61

[54] Manjunatha J G, Swamy K and Mamatha G P 2011 Poly(maleic acid) modified carbon paste electrode for simultaneous detection of dopamine in the presence of uric acid: a cyclic voltammetric study *Anal. Bioanal. Electrochem.* **3** 146–59

[55] Manjunatha J G and Swamy K 2013 Electrochemical studies of dopamine, ascorbic acid and their simultaneous determination at a poly (rosaniline) modified carbon paste electrode: a cyclic voltammetric study *Anal. Bioanal. Electrochem.* **5** 426–8

[56] Adarakatti P S and Kempahanumakkagari S K 2019 Modified electrodes for sensing *Electrochemistry* ed C Banks and S McIntosh (London: The Royal Society of Chemistry) ch2

[57] Bard A J 1983 Chemical modification of electrodes 4th edn *J. Chem. Educ.* **13** 302304

[58] Scremin J, Barbosa E C M, Salamanca-Neto C A R, Camargo P H C and Sartori E R 2018 Amperometric determination of ascorbic acid with a glassy carbon electrode modified with TiO_2–gold nanoparticles integrated into carbon nanotubes *Microchim. Acta* **185** 251

[59] Eisele A P P, Valezi C F, Mazziero T, Dekker R F H, Barbosa-Dekker A M and Sartori E R 2019 Layering of a film of carboxymethyl-botryosphaeran onto carbon black as a novel sensitive electrochemical platform on glassy carbon electrodes for the improvement in the simultaneous determination of phenolic compounds *Sens. Actuators, B Chem.* **287** 18–26

[60] Jiang L, Wang R, Li X, Jiang L and Lu G 2005 Electrochemical oxidation behavior of nitrite on a chitosan-carboxylated multiwall carbon nanotube modified electrode *Electrochem. Commun.* **7** 597–601

[61] Sartori E R and Fatibello-Filho O 2012 Simultaneous voltammetric determination of ascorbic acid and sulfite in beverages employing a glassy carbon electrode modified with carbon nanotubes within a poly(allylamine hydrochloride) film *Electroanalysis.* **24** 627–34

[62] Deroco P B, Lourencao B C and Fatibello-Filho O 2017 The use of modified electrode with carbon black as sensor to the electrochemical studies and voltammetric determination of pesticide mesotrione *Microchem. J.* **133** 188–94
[63] Hareesha N, Manjunatha J G, Raril C and Tigari G 2019 Sensitive and selective electrochemical resolution of tyrosine with ascorbic acid through the development of electropolymerized alizarin sodium sulfonate modified carbon nanotube paste electrodes *ChemistrySelect.* **4** 4559–67
[64] Tigari G and Manjunatha J G 2020 A surfactant enhanced novel pencil graphite and carbon nanotube composite paste material as an effective electrochemical sensor for determination of riboflavin *J. Sci. Adv. Mater. Devices* **5** 56–64
[65] Maiti D, Tong X, Mou X and Yang K 2019 Carbon-based nanomaterials for biomedical applications: a recent study *Front. Pharmacol.* **9** 1–16
[66] Hummers W S and Offeman R E 1958 Preparation of graphitic oxide *J. Am. Chem. Soc.* **80** 1339
[67] Sinar D, Mattos G J and Knopf G K 2020 Disposable hydrogel activated circular interdigitated sensor for monitoring biological fluid pH *IEEE Sens. J.* **20** 14624–31
[68] Bhuyan M S A, Uddin M N, Islam M M, Bipasha F A and Hossain S S 2016 Synthesis of graphene *Int. Nano Lett.* **6** 65–83
[69] Choi W, Lahiri I, Seelaboyina R and Kang Y S 2010 Synthesis of graphene and its applications: a review *Crit. Rev. Solid State Mater. Sci.* **35** 52–71
[70] Catania F, Marras E, Giorcelli M, Jagdale P, Lavagna L, Tagliaferro A, Bartoli M 2021 A review on recent advancements of graphene and graphene-related materials in biological applications *Appl. Sci.* **11** 614
[71] Dhinakaran V, Lavanya M, Vigneswari K, Ravichandran M and Vijayakumar M D 2020 Review on exploration of graphene in diverse applications and its future horizon *Mater. Today Proc.* **27** 824–8
[72] Marcano D C, Kosynkin D V, Berlin J M, Sinitskii A, Sun Z, Slesarev A, Alemany L B, Lu W and Tour JM 2010 Improved synthesis of graphene oxide *ACS Nano* **4** 4806–14
[73] Singh R K, Kumar R and Singh D P 2016 Graphene oxide: strategies for synthesis, reduction and frontier applications *RSC Adv.* **6** 64993–5011
[74] Dideikin A T and Vul' AY 2019 Graphene oxide and derivatives: the place in graphene family *Front. Phys.* **6** https://doi.org/10.3389/fphy.2018.00149
[75] Vardharajula S, Ali S Z, Tiwari P M, Eroğlu E, Vig K, Dennis V A and Singh S A 2012 Functionalized carbon nanotubes: biomedical applications *Int. J. Nanomed.* **7** 5361–74
[76] Saifuddin A R and Juniazah N 2013 Carbon nanotube: a review on structure and their interaction *J. Chem.* **2013** 676815
[77] Anzar N, Hasan R, Tyagi M, Yadav N and Narang J 2020 Carbon nanotube—a review on synthesis, properties and plethora of applications in the field of biomedical science *Sensors Int.* **1** 100003
[78] Rahmam S, Mohamed N M and Sufian S 2014 Effect of acid treatment on the multiwalled carbon nanotubes *Mater. Res. Innov.* **18** S6-196–S6-199
[79] Salamanca-Neto C A R, Olean-Oliveira A, Scremin J, Ceravolo G S, Dekker R F H, Barbosa-Dekker A M, Teixeira M F S and Sartori E R 2020 Carboxymethyl-botryosphaeran stabilized carbon nanotubes aqueous dispersion: a new platform design for electrochemical sensing of desloratadine *Talanta* **210** 120642

[80] Apetrei I M and Apetrei C 2013 Biosensor based on tyrosinase immobilized on a single-walled carbon nanotube-modified glassy carbon electrode for detection of epinephrine *Int. J. Nanomed.*. **8** 4391–8

[81] Sainz R *et al* 2020 Chemically synthesized chevron-like graphene nanoribbons for electrochemical sensors development: determination of epinephrine *Sci. Rep.* **10** 14614

[82] Akyilmaz E, Turemis M and Yasa I 2011 Voltammetric determination of epinephrine by White rot fungi (Phanerochaete chrysosporium ME446) cells based microbial biosensor *Biosens. Bioelectron.* **26** 2590–4

[83] Charithra M M and Manjunatha J G 2021 Electrochemical sensing of adrenaline using surface modified carbon nanotube paste electrode *Mater. Chem. Phys.* **262** 124293

[84] Shaikshavali P, Madhusudana Reddy T, Venu Gopal T, Venkataprasad G, Kotakadi V S, Palakollu V N and Karpoormath R 2020 A simple sonochemical assisted synthesis of nanocomposite (ZnO/MWCNTs) for electrochemical sensing of Epinephrine in human serum and pharmaceutical formulation *Colloids Surf. A Physicochem. Eng. Asp.* **584** 124038

[85] Gopal P, Narasimha G and Reddy T M 2020 Development, validation and enzyme kinetic evaluation of multi walled carbon nano tubes mediated tyrosinase based electrochemical biosensing platform for the voltammetric monitoring of epinephrine *Process Biochem.* **92** 476–85

[86] Joseph T and Thomas N 2020 A facile electrochemical sensor based on titanium oxide (TiO2)/reduced graphene oxide (RGO) nano composite modified carbon paste electrode for sensitive detection of epinephrine (EP) from ternary mixture *Mater. Today Proc.* **41** 606–9

[87] Immanuel S and Sivasubramanian R 2020 Fabrication of two-dimensional chemically reduced graphene oxide nanosheets for the electrochemical determination of epinephrine *Bull. Mater. Sci.* 43

[88] Ramu A G, Umar A, Ibrahim A A, Algadi H, Ibrahim Y S A, Wang Y, Hanafiah M M, Shanmugam P and Choi D 2021 Synthesis of porous 2D layered nickel oxide-reduced graphene oxide (NiO-rGO) hybrid composite for the efficient electrochemical detection of epinephrine in biological fluid *Environ. Res.* **200** 111366

[89] Ahmed T A, Ghandour M A and Khalil M M 2020 Stripping voltammetric determination of Epinephrine applying sulfacetamide modified glassy carbon electrode *Int. J. Electrochem. Sci.* **15** 3143–56

[90] Emran M Y, Shenashen M A, El-Safty S A, Reda A and Selim M M 2021 Microporous P-doped carbon spheres sensory electrode for voltammetry and amperometry adrenaline screening in human fluids *Microchim. Acta* **188** 7–11

[91] Luk H N, Chou T Y, Huang B H, Lin Y S, Li H and Wu R J 2021 Promotion effect of palladium on bivo4 sensing material for epinephrine detection *Catalysts.* **11** 1083

[92] Mattioli I A, Cervini P and Cavalheiro É T G 2020 Screen-printed disposable electrodes using graphite-polyurethane composites modified with magnetite and chitosan-coated magnetite nanoparticles for voltammetric epinephrine sensing: a comparative study *Microchim. Acta* **187** 318

[93] Kim S H, Lee J W and Yeo I H 2000 Spectroelectrochemical and electrochemical behavior of epinephrine at a gold electrode *Electrochim. Acta* **45** 2889–95

[94] Pichai E and Lakshmanan M 2021 Neurotransmitters and neurotransmission ed A Paul, N Anandabaskar, J Mathaiyan and G M Raj *Introduction to Basics of Pharmacology and Toxicology: Volume 2 : Essentials of Systemic Pharmacology : From Principles to Practice* (Singapore: Springer Singapore) pp 69–95

[95] Ou Y, Buchanan A M, Witt C E and Hashemi P 2019 Frontiers in electrochemical sensors for neurotransmitter detection: towards measuring neurotransmitters as chemical diagnostics for brain disorders *Anal. Methods* **11** 2738–55
[96] Hejazi M, Tong W, Ibbotson M R, Prawer S and Garrett D J 2021 Advances in carbon-based microfiber electrodes for neural interfacing *Front. Neurosci.* **15** 658703
[97] Lee Y, Kong C, Chang J W and Jun S B 2019 Carbon-fiber based microelectrode array embedded with a biodegradable silk support for in vivo neural recording *J. Korean Med. Sci.* **34** 115
[98] Schmidt A C, Wang X, Zhu Y and Sombers L A 2013 Carbon nanotube yarn electrodes for enhanced detection of neurotransmitter dynamics in live brain tissue *ACS Nano.* **7** 7864–73
[99] Dankoski E and Wightman R 2013 Monitoring serotonin signaling on a subsecond time scale *Front. Int. Neurosci* **7** 44
[100] Roberts J G, Mitchell E C, Dunaway L E, McCarty G S and Sombers L A 2020 Carbon-fiber nanoelectrodes for real-time discrimination of vesicle cargo in the native cellular environment *ACS Nano.* **14** 2917–26
[101] Zhou T, Niu Y, Li Z, Li H, Yong Z, Wu K, Zhang Y and Li Q 2021 The synergetic relationship between the length and orientation of carbon nanotubes in direct spinning of high-strength carbon nanotube fibers *Mater. Des.* **203** 109557
[102] Shao Z, Puthongkham P, Hu K K, Jia R, Mirkin M V and Venton B J 2020 Thin layer cell behavior of CNT yarn and cavity carbon nanopipette electrodes: effect on catecholamine detection *Electrochim. Acta* **361** 137032
[103] Gupta P, Tsai K, Ruhunage C K, Gupta V K, Rahm C E, Jiang D and Alvarez N T 2020 True picomolar neurotransmitter sensor based on open-ended carbon nanotubes *Anal. Chem.* **92** 8536–45
[104] Puthongkham P and Venton B J 2019 Nanodiamond coating improves the sensitivity and antifouling properties of carbon fiber microelectrodes *ACS Sens.* **4** 2403–11
[105] Schrand A M, Dai L, Schlager J J, Hussain S M and Osawa E 2007 Differential biocompatibility of carbon nanotubes and nanodiamonds *Diam. Relat. Mater.* **16** 2118–23
[106] Hu M 2016 Application of electrochemical redox cycling: toward differentiation of dopamine and norepinephrine *Anal. Chem.* **88** 5574–8

Chapter 10

Electrochemical detection of amoxicillin as an antibiotic drug by using surface modified carbon based sensors

M M Charithra and J G Manjunatha

As a beta-lactam antibiotic, amoxicillin (AMX) is typically consumed to inhibit bacterial infection thus qualitative analysis of the AMX is very important. Hence, the construction of a sensing device based on poly aspartic acid modified carbon paste electrode (PAPMCPE) for the analysis of AMX was reported. The PAPMCPE exposed an enhanced electrocatalytic effect corresponding to the voltammetric signal of the AMX (pH 7.0) compared to the bare carbon paste electrode (BCPE). The morphology of the prepared electrodes was evaluated by field emission scanning electron microscopy (FE-SEM). The electrochemical characterization of the projected electrodes was done by using the redox probe. The charge transfer resistance of the proposed electrode was inspected through electrochemical impedance spectroscopy (EIS). The influence of pH and the impact of scan rate was examined through the linear sweep voltammetry (LSV) technique. The pH and scan rate study discloses the electro-oxidation of the AMX was pH-dependant and regulated by diffusion. The designed electrode yielded a low detection limit (0.2 μM) for AMX. Furthermore, the prepared electrode revealed good reproducibility, anti-interference, repeatability, selectivity, and anti-fouling property. The viability of the constructed sensor was examined by analyzing the AMX in tablet and blood serum samples.

10.1 Introduction

Antibiotics are the most essential medicines since they suppress or destroy microorganisms that cause infectious illnesses in people and animals. They are classified into different types based on their chemical structure and other factors [1]. Beta-lactam antibiotics are the usually utilized drugs comprised of a beta-lactam ring that is responsible for their anti-bacterial activity [2]. Amoxicillin (AMX) is one

doi:10.1088/978-0-7503-5127-0ch10

of the beta-lactam antibiotics mainly used against respiratory infections, otitis, pneumonia, genital infections, and urinary infections [3–6]. Despite the fact that AMX is very safe at therapeutic dosages, it is well-known that its abuse creates the ideal circumstances for bacteria to evolve resistance mechanisms, thus sensible use of this drug is necessary [7]. After oral intake, AMX is quickly absorbed and partially translated to metabolites. Approximately 60% of the AMX is eliminated unaltered in the urine [8]. Thus, its monitoring at low concentration level is very significant.

A great deal of attention has been paid to pharmaceutical analysis due to its relevance in the quality control of pharmaceutical goods. Therefore, various approaches which have been implemented in the quantitative analysis of AMX are chromatography techniques [9–12], spectrophotometric methods [13], capillary electrophoresis [14], chemiluminescence [15]. Although these approaches are sensitive, they have certain shortcomings such as the need for expensive instruments, vast quantities of reagents, time-consuming protocols, and require professional personnel to manage the devices. Thus, to overcome these drawbacks, electrochemical techniques are the right substitute for the analysis of the selected compounds [16]. Electrochemical approaches are the most noteworthy method in the field of analytical chemistry. Voltammetric techniques are among the simplest and most sensitive methods within the electrochemical methods [17, 18]. Therefore, voltammetric detection of AMX is crucial.

Graphite is the carbonaceous material that holds massive physicochemical, electrical, and mechanical properties due to this it is frequently used in the field of electrochemistry [19–21]. Graphite paste electrode employed as a voltammetric sensor due to its favorable large surface area, anti-fouling features, signal improvement, quick electron transferability, and displays of synergetic activity with numerous electro-active molecules used as a modifier [22–29]. One of the achievable routes for surface modification is electropolymerization that includes electrochemical synthesis of thin polymer films on electrode surfaces [30–36]. Hence, by employing the uncomplicated electropolymerization method it is important to propose the modified electrode for the electrochemical analysis of the selected analyte.

In this chapter, a voltammetric sensor was constructed by utilizing poly aspartic acid modified carbon paste electrode (PAPMCPE) to sense AMX through the cyclic voltammetry (CV) and linear sweep voltammetry (LSV) approach. The PAPMCPE acts as an electrocatalyst and displays an enriched electrocatalyst activity for AMX analysis. Surface analysis of the proposed electrodes reveals the satisfactory modification of aspartic acid (AP) by the electropolymerization process. The proposed sensor offers a low detection limit for AMX detection. Corresponding to reports accessible in the literature, bare carbon paste electrode (BCPE) modified with AP as a voltammetric sensor was not yet exploited for electroanalysis of AMX using the CV technique. In light of the above matter, we were inspired to propose a sensing device based on the PAPMCPE for the detection of AMX. Overall, the prepared electrode has been effectively employed in tablet and blood serums sample to detect the AMX.

10.2 Experimental section

10.2.1 Reagents and solutions

AMX (99%), paracetamol (PR, 99%) were acquired from the Tokyo Chemicals Industries Company Ltd (Japan). AP (99%) was purchased from Molychem, India. Disodium hydrogen phosphate ($Na_2HPO_4{\cdot}H_2O$, 99.5%), sodium dihydrogen phosphate ($NaH_2PO_4{\cdot}H_2O$, 99%), were bought from the Himedia Chemical, Bangalore. Silicone oil and graphite powder were bought from Nice Chemicals India. All chemicals of the existing purity are employed in this work. A stock solution of AMX ($25{\times}10^{-4}$ M) was made by dissolving the required mass of the AMX with distilled water. The phosphate supporting electrolyte solution was prepared by mixing the appropriate volume of the $Na_2HPO_4{\cdot}H_2O$ and $NaH_2PO_4{\cdot}H_2O$ solution. The whole voltammetric measurement was carried out at laboratory temperature (25±1 °C).

10.2.2 Instrumentation

The voltammetric and impedimetric measurement was taken by employing the electrochemical working platform CHI-6038E (CH-Instruments-USA) associated to the computer for the data attainment. The workstation was linked to a 50 mL electrochemical glass cell that encompasses the standard three-electrode system of $Hg/HgCl_2$ electrode as a reference electrode, platinum wire as the auxiliary electrode, and BCPE or PAPMCPE as working sensor. VITSIL-VBSD/VBDD water distillation system was engaged to obtain the distilled water. The proper pH of the supporting electrolyte solution was taken from the EQ-610 digital pH meter. Field emission scanning electron microscopy (FE-SEM) was obtained from the DST-PURSE Laboratory, Mangalagangotri, Mangalore University, India.

10.2.3 Fabrication of working sensor

Primarily BCPE was prepared by blending 70% graphite powder with 30% silicone oil in an agate mortar with a pestle for 20 min to get a consistent paste. The fine homogenized paste was packed on the crater of the Teflon tube and it was smoothed with tissue paper. The copper wire offers an electrical supply to be attached to the one end of the Teflon tube. Next the developed bare electrode was modified through the electropolymerization of AP in 0.2 M supporting electrolyte (detailed information is given in the results and interpretation section) and the obtained electrode was termed as PAPMCPE. The constructed PAPMCPE was used for the analysis of the AMX.

10.3 Results and interpretations

10.3.1 Electropolymerization of the AP at the surface of the bare electrode

The analytical signal for the targeted analyte is effected by electropolymerization, and hence a comprehensive study of the electropolymerization of the AP is essential in order to analyze the AMX. Thus, electropolymerization of AP was conducted by documenting the 10 cycles of CV for the AP ($1{\times}10^{-4}$ M) in buffer solution (0.2 M) at the bare electrode, and the recorded CVs are illustrated in figure 10.1. As inferred from the CVs, CV response of the AP escalated with intensifying the CV cycles assuring that a satisfactorily thin layer of the AP was electrodeposited at the bare electrode. Next, the electropolymerization cycles were optimized by scanning the electropolymerization cycle of 5, 10, 15, and 20 (inset figure 10.1). The maximum response was gained at 10 electropolymerization cycles of AP and hence 10 cycles of the electropolymerization was expended in the whole analysis.

10.3.2 Morphological studies of the prepared electrodes

FE-SEM analysis is a helpful tool for overcoming the morphological characteristics of a thin sensor film. Hence, FE-SEM photos of the modified and unmodified electrode were captured at a magnification of 10 μm and are displayed in figure 10.2. Desquamate of the graphite was conspicuous from the FE-SEM images of the bare electrode and it is highlighted in figure 10.2(a). The FE-SEM images of the

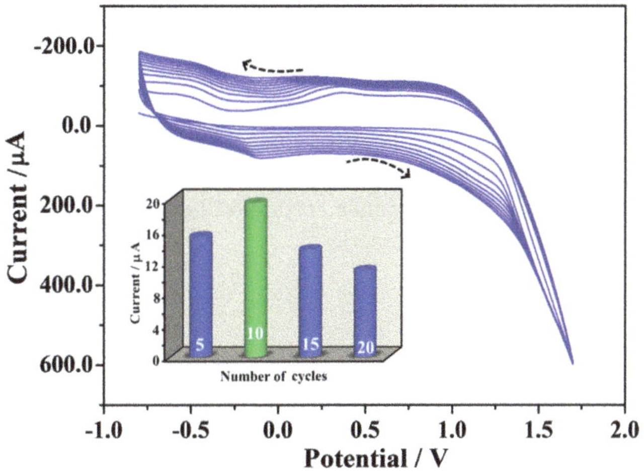

Figure 10.1. CVs documented for 1×10^{-4} M AP in 0.2 M supporting electrolyte (pH 7.0) at a scan rate of 0.1 V s^{-1} on the surface of the BCPE.

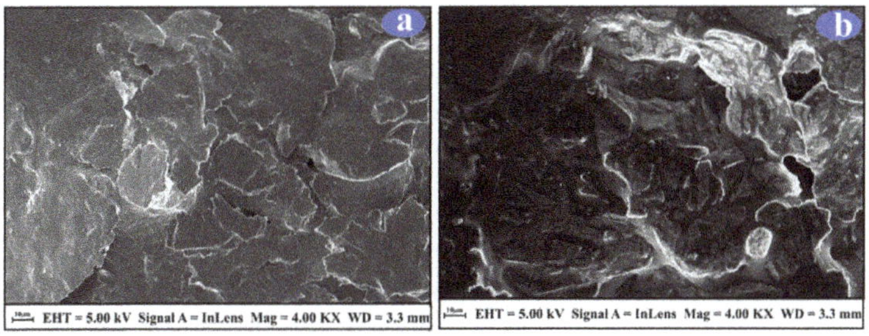

Figure 10.2. FE-SEM images of the BCPE (a), and PAPMCPE (b).

PAPMCPE exposing the layer covering the desquamate of the graphite point out that the electrodeposition of the AP at the unmodified electrode was effectively accomplished. The aforementioned FE-SEM images attest that the morphological features of the prepared electrodes and the sufficient deposition of the AP at the bare electrode were achieved by the electropolymerization procedures.

10.3.3 Electrochemical characterization of the modified and unmodified electrode

$K_4[Fe(CN_6)]$ is employed as a redox probe in order to promulgate sufficient modification to enrich the analytical response of the analyte. Hence, the voltammetric response of the 1×10^{-3} M $K_4[Fe(CN_6)]$ in 0.1 M KCl as buffer solution at the scan rate of 0.1 V s^{-1} on the surface-active area of the electrode was recorded and CVs are described in figure 10.3. The bare electrode displayed a low response whereas the modified electrode exhibited a boosted electro catalytic property. Thus, we can clarify that the modified electrode possesses the capability to enhance redox reaction of the analyte. Therefore, the surface area of the designed electrode can be determined by means of the Randles–Sevcik equation (10.1)

$$I_{pa} = 2.69 \times 10^5 n^{3/2} A D^{1/2} C v^{1/2} \tag{10.1}$$

Here, A is the surface-active area, n is the number of electrons, v the scan rate, I_{pa} the anodic peak current, C the concentration of the solution, and D the diffusion co-efficient [37]. The calculated surface-active area of the modified sensor is 0.06 cm^2 and 0.04 cm^2 for bare electrode. Hence, the modified sensor has a superior active surface area to the unmodified electrode.

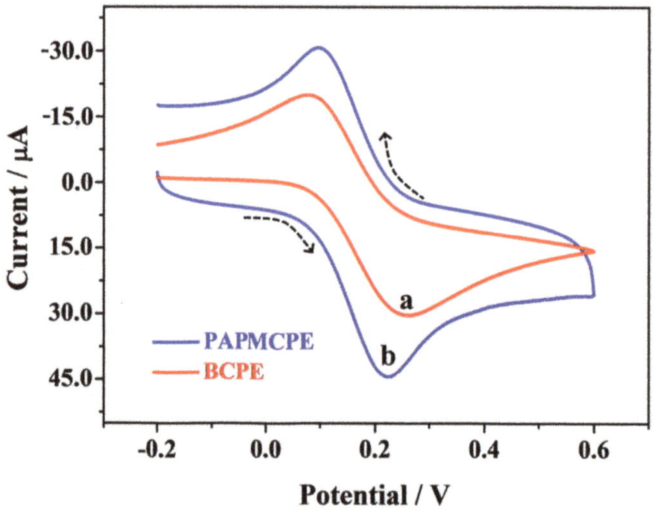

Figure 10.3. CVs documented for 1×10^{-3} $K_4[Fe(CN_6)]$ in 0.2 M KCl as supporting electrolyte at a scan rate of 0.1 V s^{-1} on the surface of the BCPE (curve a) and PAPMCPE (curve b).

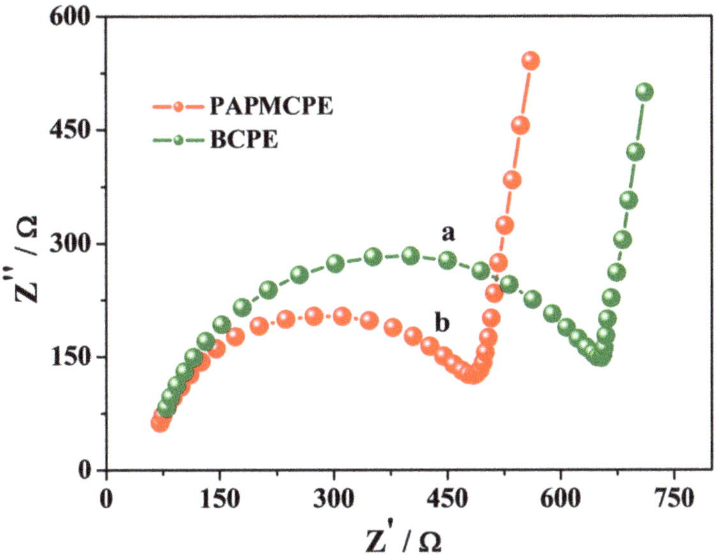

Figure 10.4. Nyquist plot for modified (curve b) and bare electrode (curve a).

10.3.4 Impedance study

EIS is a relevant system to convey the charge transfer resistance of surface of the prepared electrode. Therefore, an impedance study of the prepared sensing layer was performed by documenting the Nyquist plot of the redox probe $K_4[Fe(CN_6)]$ in 0.1 M KCl as a buffer solution at the surface of the unmodified and the modified electrode (figure 10.4). As inferred from the Nyquist plot, the larger diameter of the semicircle appeared at the bare electrode and the smaller diameter of the semicircle existed at the modified electrode. Furthermore, by fitting the Randles equivalent circuit the charge transfer resistance (R_{ct}) of the electrode was assessed. The modified electrode (3357 Ω) exhibited a lower R_{ct} value than the unmodified electrode (4734 Ω). Finally, we can conclude from the EIS study that the modified electrode has higher conductivity that the bare electrode.

10.3.5 Voltammetric sensing of AMX at the bare and modified electrode

CV was explored to verify the sensing performance of the modified and bare electrodes for the analysis of the substance of interest. Hence, figure 10.5 describes the CVs recorded using the PAPMCPE (curve c) and BCPE (curve a) for AMX (1×10^{-4} M) in 0.2 M buffer solutions at the scan rate of 0.1 V s^{-1} and also for the blank determination (without AMX, curve b). As is apparent from the CVs, the voltammetric response of AMX was frail in the absence of a modified electrode. However, in the presence of a modified electrode the analytical signal of AMX was boosted with high current response and low potential. Furthermore, blank analysis at the modified electrode does not yielded any peak response for AMX. Therefore, the aforementioned outcomes point out that the modified sensor enriches the

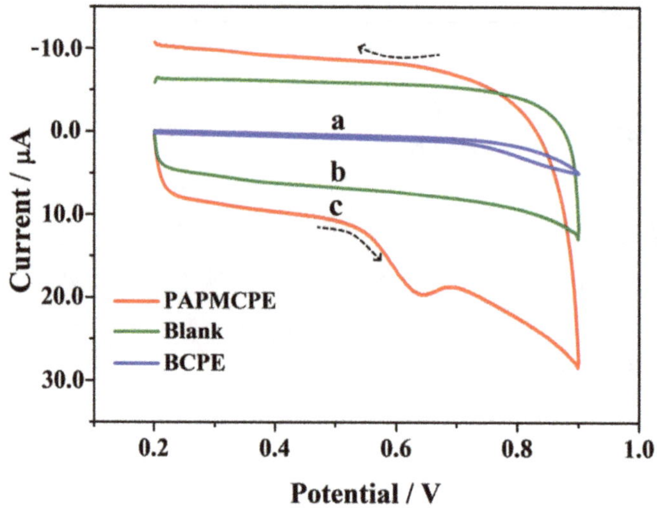

Figure 10.5. CVs for electrochemical response of the 1×10^{-4} M AMX in 0.2 M supporting electrolyte (pH 7.0) at a sweep rate of 0.1 V s^{-1} on the surface of the BCPE (curve a), PAPMCPE (curve c), and without analyte (curve b) at the modified electrode.

oxidation reaction of the AMX. In all cases, the oxidation processes was identified during the anodic potential evaluation, with the absence of an equivalent reduction process during the cathodic potential evaluation verifying that the oxidation process of AMX at the modified electrode was irreversible.

10.3.6 Influence of supporting electrolyte pH

In any biological system, pH takes a major role. Hence, it is important to study the impact of the pH on the oxidation reaction of the AMX on the surface of the proposed electrode. Thus, the voltammetric detection of the AMX in 0.2 M buffer solution having various pH 6.0–7.5 on the PAPMCPE was executed by utilizing the LSV method and the corresponding LSVs are demonstrated in figure 10.6(a). The plot of E_{pa} with respect to pH was constructed. As viewed from the graph, the potential was shifted a little to the negative side which suggests that the process is pH-dependant and the linear fit equation is represented as follows (10.2)

$$E_{pa}(V) = -0.0691(pH) + 14295\ (R^2 = -0.9967) \tag{10.2}$$

The obtained slope from the above fitted linear equation is near to the theoretical nerstian value. Hence, we can conclude that same number of electrons and protons are engaged during the oxidation reaction of AMX at the prepared sensing layer. A well-shaped highly intensified peak was identified at pH 7.0. All of the other pH portray the low response that was revealed in the graph of the I_{pa} with respect to pH. Therefore, pH 7.0 seems to be the most appropriate option for the whole experiment. The proposed electro-oxidations are displayed in the Scheme 10.1.

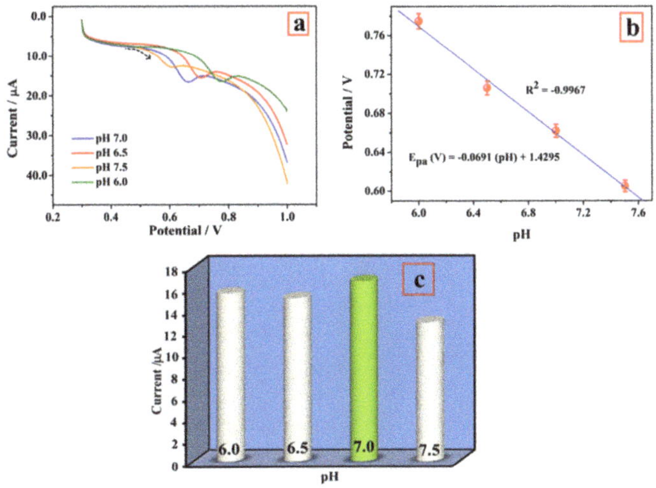

Figure 10.6. (a) LSVs recorded for the voltammetric response of 1×10^{-4} M AMX in 0.2 M supporting electrolyte with varying pH 6.0–7.5 at a scan rate of 0.1 V s^{-1}, (b) graph of potential against pH, (c) plot of I_{pa} vs. pH.

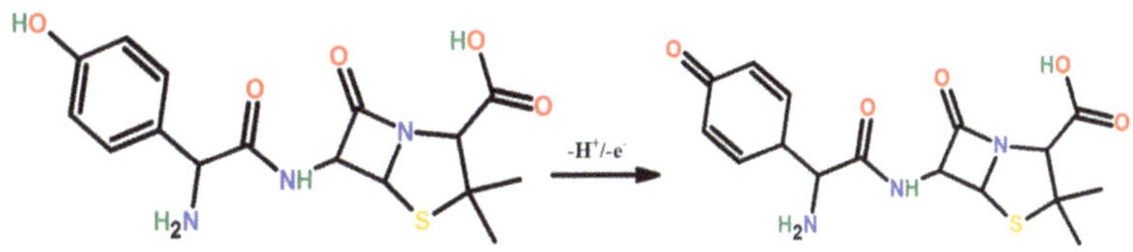

Scheme 10.1. The electrochemical oxidation reaction of AMX.

10.3.7 Effect of the potential sweep rate

The additional aspects of the electrochemical reactivity between the surface electrode and electro-active solution are gained from the scan rate studies. Thus, the effect of the scan rate on the redox reaction of the AMX at the prepared sensing film was scrutinized by recording the LSVs for AMX (1×10^{-4} M) having 0.2 M buffer solution (pH 7.0) when the scan rate ranges were varied from 0.1–0.2 V s^{-1} (figure 10.7(a)). As is apparent from the figure, the voltammetric signal is intensified and potential is moved a little towards the positive side by rising the scan rate. Next, the linear relationship between the square root of the scan rate and the current response is depicted in figure 10.7(b) which states that the process is regulated by diffusion and the linear regression equation is represented below in equation (10.3)

$$I_{pa}(\mu A) = 62.214\upsilon^{1/2}/(Vs^{-1})^{1/2} - 320.48 \ (R^2 = 0.9975) \qquad (10.3)$$

In addition, the slope value 0.4 demonstrated by the graph of the logarithm of scan rate versus logarithm of the current response (figure 10.7(c)) is close to the speculative value 0.5 which proves that the oxidation process is controlled by the diffusion. The fitted linear data is illustrated as the following equation (10.4)

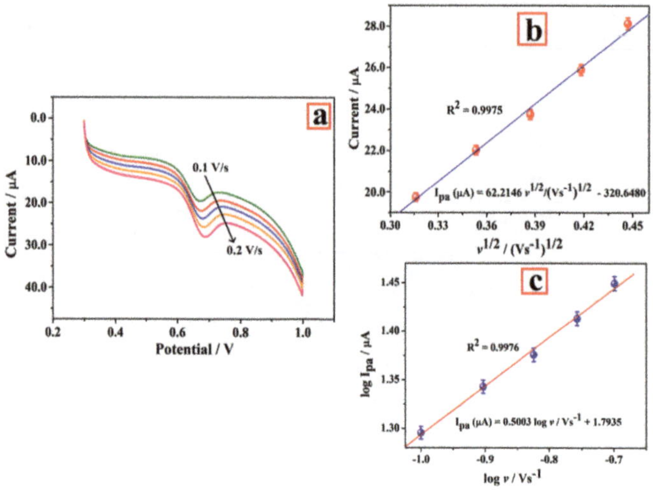

Figure 10.7. (a) LSVs attained for the voltammetric behavior of 1×10^{-4} M AMX in 0.2 M supporting electrolyte when the scan rate range is varied from 0.1–0.2 V s^{-1}, (b) graph of square root of scan rate with respect to peak current, (c) plot of logarithm of scan rate versus logarithm of peak current.

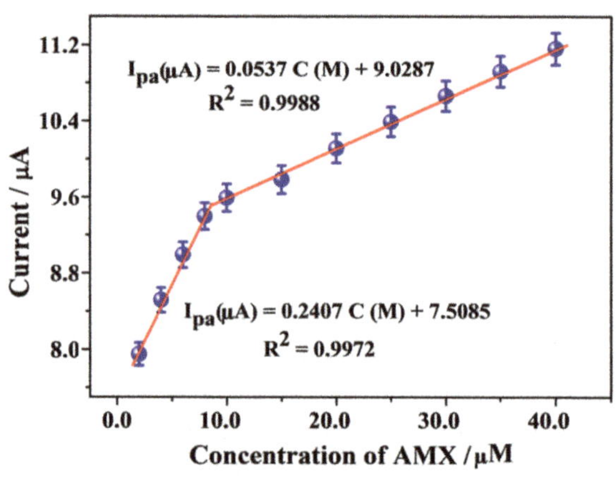

Figure 10.8. The graph of peak current versus concentration of AMX.

$$I_{pa}(\mu A) = 0.5003 \log v/(\mathrm{Vs}^{-1}) + 1.7935 \; (R^2 = 0.9976) \tag{10.4}$$

10.3.8 Analytical curve and detection limit

The validation of analytical data is a key component of expressing the efficiency of the developed sensing device and hence the detection limit (DL) and quantification limit (QL) are the mathematical terms to describe the sensing capability of the prepared electrode. Therefore, the analytical plot (figure 10.8) was created by recording the response of current with respect to the concentration of the AMX solution from 2.0–40 μM in 0.2 M buffer solution at the scan rate of 0.1 V s^{-1} on the surface of the sensing layer. As can be inferred from the calibration plot, the analytical signal of the AMX is intensifying by escalating the concentration of the AMX solution. The analytical curve response is linear in the range 2.0–8.0 μM and 10–40 μM. The data fitted linear segment relation is portrayed as in equations (10.5) and (10.6)

$$I_{pa}(\mu A) = 0.2407\ C(\mathrm{M}) + 7.5085\ (R^2 = 0.9972) \tag{10.5}$$

$$I_{pa}(\mu A) = 0.0537\ C\ (\mathrm{M}) + 9.0287\ (R^2 = 0.9988) \tag{10.6}$$

DL and QL were calculated using equations (10.7) and (10.8)

$$\mathrm{DL} = 3S_d/b \tag{10.7}$$

$$\mathrm{QL} = 10S_d/b \tag{10.8}$$

Here, S_d is the standard deviation of the blank detection and b is the gradient value of the calibration curve. Therefore, the assessed DL and QL are 0.20 and 0.67 μM. The obtained analytical parameters of the proposed method and the previous method for the detection of the AMX are listed in table 10.1. Data from table 10.1 indicates that the analytical features are good and have DL values as low as those attained by the other voltammetric methods [38–44]. Finally, we underline that the prepared sensing film provided the best DL value for the voltammetric analysis of the AMX (table 10.1).

TiO_2/CMK/AuNPs/Nafion/GCE: titanium dioxide sol modified with gold nanoparticles 3-type mesoporous carbon and Nafion, AuNPs/PdNPs/ErGO/: A gold-palladium nanoparticle decorated electrochemically reduced graphene oxide, Ni/CR/CPE: MWCNT/GCE: multiwalled carbon nanotube and glassy carbon electrode, DMBQ/ZnO/CNTs/CPE, 8,9-dihydroxy-7-methyl-12H-benzothiazolo [2,3-b] quinazolin-12-one (DMBQ) and ZnO/CNTs nanocomposite carbon paste electrode, GPE: graphite pencil electrode, EPPG: edge plane pyrolytic graphite.

Table 10.1. Analytical features of the proposed method using the adopted sensor compared to the data of other voltammetric methods applied to detect the AMX.

Electrodes	Method	Linear range (μM)	DL (μM)	References
TiO_2/CMK/AuNPs/Nafion/GCE	CV	2.5–133.0	0.3	[38]
AuNPs/PdNPs/ErGO/GCE	SWV	30.0–350.0	9.0	[39]
Ni/CR/CPE	Amperometry	8.0–100.0	5.0	[40]
MWCNT/GCE	CV	0.6–8.0	0.2	[41]
DMBQ/ZnO/CNTs/CPE	SWV	1.0–950	0.5	[42]
Treated GPE	SWV	1.0–80	0.22	[43]
Unmodified EPPG	SWV	5.0–500	0.84	[44]
PAPMCPE	CV	2.0–4.0	0.20	Present work

Table 10.2. Recovery studies for the electroanalysis of the AMX in tablet and blood serum samples using the proposed method.

Samples	Concentration added (μM)	Founded value (μM)	Recovery (%)
Tablet sample	10	10.2	102
	20	19.5	97.5
	30	30.8	102.6
Blood serum sample	10	9.8	98.0
	20	20.7	103.5
	30	29.9	99.6

10.3.9 Stability, reproducibility, repeatability

The basic quality of the prepared sensing film is its stability, repeatability, and reproducibility. Therefore, the stability of the designed sensor was inspected by an uninterrupted run of the CV response of 30 cycles for the working solution (1×10^{-4} M AMX) in buffer solution (0.2 M) and the corresponding result reveals that the electrochemical activity of the electrode continued even after 30 cycles. Then, the reproducibility of the projected electrode was probed by repeatedly measuring a separately prepared sample containing 1×10^{-4} M AMX in buffer solution with the constant sensor and resulting RSDs of AMX peak intensities were 3.8%. Next, keeping the sample solution unchanged with the individually prepared sensor taken for CV measurement and this reproducibility test gave a RSD value of 3.5%. Therefore, the RSD value below 4.0% guides that the reproducibility and the repeatability of the sensor is sufficient for the electroanalysis of the AMX. Finally, we can conclude that the prepared sensing layer possess an anti-fouling feature, ample repeatability, and reproducibility properties.

10.3.10 Selectivity of the modified electrode

Selectivity is a crucial factor to explain the capability of the prepared electrode to sense the targeted analyte in the presence of other biological compounds. Hence, the selectivity of the modified electrode was probed by documenting the CVs for AMX (1×10^{-4} M) and PR (1×10^{-4} M) in 0.2 M supporting electrolyte solution at a scan rate of 0.1 V s^{-1} in the potential window of 0.0 V–0.9 V (figure 10.9). As is apparent from the CVs, a well-separated analytical signal related to the boosted current response at the modified electrode (curve a) corresponding to the AMX and PR ensued in the potential of 0.64 and 0.39 V respectively. However, a low voltammetric signal was generated for AMX and PR at the non-modified electrode. From the above results, we can confirm that the modified electrode holds tremendous selectivity.

10.3.11 Interference study with some metal ions and the organic compounds

Anti-interference is the core characteristic of the modified electrode, hence it was evaluated in the presence of some metal ions and organic compounds because of

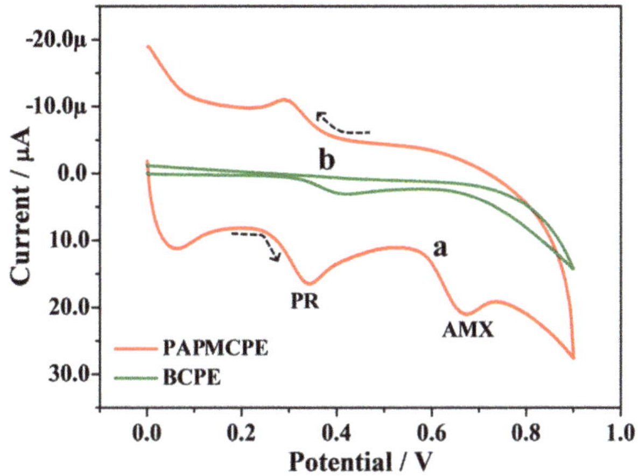

Figure 10.9. CVs documented for the simultaneous detection of 1×10^{-4} M AMX and PR in 0.2 M supporting electrolyte at the scan rate of 0.1 V s^{-1} on the surface of BCPE (curve b), PAPMCPE (curve a).

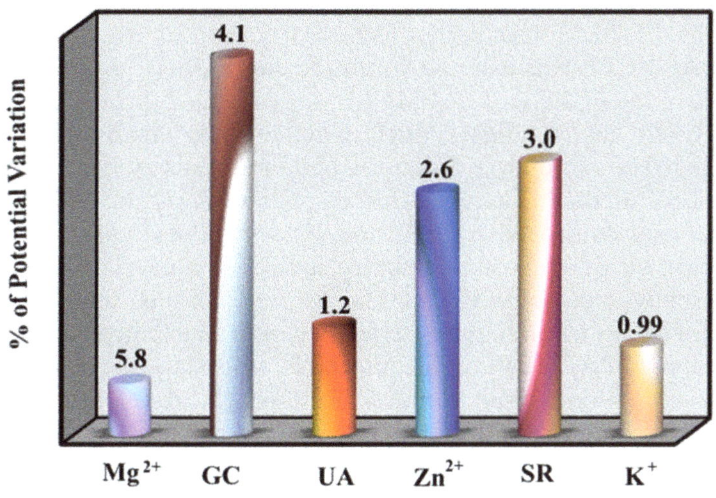

Figure 10.10. The plot of percentage of potential variation at different interferents.

their presence in biological fluids. A graph was plotted (figure 10.10) with the peak potential with respect to the percentage of the interfering compounds. From figure 10.10, it can be seen that the potential interference was less than 5% by the other metal ions (Na^+, Mg^{2+}, K^+) and the organic compounds such as glucose (GC), sucrose (SR), and urea (UA). Hence, the modified electrode possesses an admirable anti-interference feature in the presence of metal ions and organic compounds.

10.3.12 Analytical applications

The analytical applicability of the proposed sensor was inspected by detecting the AMX in tablet and blood serum samples. Therefore, a commercially available tablet with AMX content was acquired from a local pharmacy. Next, the tablet was ground to a fine powder and a solution of tablet was prepared with the distilled water. The solution was placed in an electrochemical cell having 0.2 M buffer solution and voltammetric measurements were performed. The attained recovery outcomes are given in table 10.2. Furthermore, the practicability of the prepared sensing film was evaluated with a blood serum sample. By invigorating the serum sample with the standard solution of AMX in 0.2 M supporting electrolyte solutions, recoveries between the 98%–101% were acquired and the recovery results are depicted in table 10.2. Thus, the aforementioned recovery outcome suggests that the modified electrode may have applications of pharmaceutical biological fluids analysis.

10.4 Conclusion

In summary, a voltammetric sensor based on PAPMCPE to sense AMX has been introduced in this chapter. The electrochemical reaction of AMX on the surface of PAPMCPE displays a better analytical signal than the bare electrode via CV method. The topographical characteristic of BCPE and PAPMCPE was accomplished by the FE-SEM technique. The electrochemical characterization of the BCPE and PAPMCPE was assessed by employing $K_4[Fe(CN_6)]$ as a redox probe. The resistivity of the BCPE and PAPMCPE was estimated through the EIS analysis. The PAPMCPE has a higher surface-active area than the non-modified electrode. The pH condition was optimized and the impact of scan rate was attained by using the LSV method. The oxidation reaction of AMX at the PAPMCPE was pH-dependent and controlled by diffusion. The analytical curve of AMX at the modified sensor yields the complimentary low DL value. The selectivity of the designed sensor was demonstrated in the existence of PR. The prepared sensor possesses anti-fouling feature, anti-interference, reproducibility, repeatability. As a result, the proposed PAPMCPE was appreciably utilized for the analysis of AMX in tablets as well as in blood serum samples. Therefore, the developed electrode can be used for usual assessment of AMX.

Acknowledgement

Charithra M M gratefully acknowledge the financial support from the SC/ST Fellowship No. MU/SCT RF/CR17/2017–18 Mangalore University.

References

[1] Wang J, MacNeil J D and Kay J F 2011 *Chemical Analysis of Antibiotic Residues in Food* 38 (New York: Wiley)

[2] Morin R B and Gorman M 2014 *The Biology of B-Lactam Antibiotics* 1st edn (New York: Elsevier)

[3] Wong A, Santos A M, Cincotto F H, Moraes F C, Fatibello-Filho O and Sotomayor M D P T A 2020 New electrochemical platform based on low cost nanomaterials for sensitive detection of the amoxicillin antibiotic in different matrices *Talanta* **206** 120252

[4] Isla A, Troconiz I F, Canut A, Labora A, Martin-Herrero J E, Pedraz J L and Gascon A R 2011 Pharmacokinetic/pharmacodynamic evaluation of amoxicillin, amoxicillin/clavulanate and ceftriaxone in the treatment of paediatric acute otitis media in Spain *Enferm. Infecc. Microbiol. Clin.* **29** 167–73

[5] Mathur S, Fuchs A, Bielicki J, Van Den Anker J and Sharland M 2018 Antibiotic use for community-acquired pneumonia in neonates and children: WHO evidence review *Paediatr. Int. Child Health* **38** S66–75

[6] Fernandez J, Ribeiro I A C, Martin V, Martija O L, Zuza E, Bettencourt A F and Sarasua J R 2018 Release mechanisms of urinary tract antibiotics when mixed with bioabsorbable polyesters *Mater. Sci. Eng.* C **93** 529–38

[7] Deroco P B, Rochafilho R C and Fatibellofilho O 2018 A new and simple method for the simultaneous determination of amoxicillin and nimesulide using carbon black within a dihexadecylphosphate film as electrochemical sensor *Talanta* **179** 115

[8] Kirby W M, Gordon R C and Regamey C 1974 The pharmacology of orally administered amoxicillin and ampicillin *J. Infect. Dis.* **129** 154–5

[9] Harshita M P, Aparna J T, Reddy L K, Suresh P V, Prabahar A E and Rao N R 2014 Simultaneous determination of residual NSAIDs and anti-biotic in raw milk by RPHPLC *Int. J. Pharm. Sci. Drug Res.* **5** 5498–502

[10] Zhang C, Zeng J, Xiong W and Zeng Z 2020 Rapid determination of amoxicillin in porcine tissues by UPLC-MS/MS with internal standard *J. Food Composit. Anal.* **92** 103578

[11] Gulfen M, Canbaz Y and Ozdemir A 2020 Simultaneous determination of amoxicillin, lansoprazole, and levofloxacin in pharmaceuticals by HPLC with UV–vis detector *J. Anal. Test.* **4** 45–53

[12] Batrawi N, Wahdan S and Alrimawi F 2017 A validated stability-indicating HPLC method for simultaneous determination of amoxicillin and enrofloxacin combination in an injectable suspension *Sci. Pharm.* **85** 6

[13] Khayata W and Zakri D A 2017 Two simple spectrophotometric methods for the simultaneous determination of amoxicillin trihydrates and flucloxacillin sodium *Res. J. Pharm. Technol.* **10** 1327

[14] Marra M C, Cunha R R, Munoz R A, Batista A D and Richter E M 2017 Single-run capillary electrophoresis method for the fast simultaneous determination of amoxicillin, clavulanate, and potassium *J. Sep. Sci.* **40** 3557–62

[15] Xie X and Song Z 2006 Ultrasensitive determination of amoxicillin using chemiluminescence with flow injection analysis *Spectroscopy* **20** 270417

[16] Charithra M M and Manjunatha J G 2021 Electrochemical sensing of adrenaline using surface modified carbon nanotube paste electrode *Mater. Chem. Phys.* **262** 124293

[17] Prinith N S, Manjunatha J G and Hareesha N 2021 Electrochemical validation of L-tyrosine with dopamine using composite surfactant modified carbon nanotube electrode *J. Iran. Chem. Soc.* **18** 3493–503

[18] Prinith N S, Manjunatha J G, Tigari G, Alothman Z A, Alanazi A M and Pandith A 2021 Mechanistic insights into the voltammetric determination of Riboflavin at Poly (Serine) modified graphite and carbon nanotube composite paste electrode *Chemistryselect.* **6** 10746–57

[19] Tigari G and Manjunatha J G 2020 Poly(glutamine) film-coated carbon nanotube paste electrode for the determination of curcumin with vanillin: an electroanalytical approach *Monatsh. Chem.* **151** 1681–688

[20] Tigari G and Manjunatha J G 2020 Optimized voltammetric experiment for the determination of phloroglucinol at surfactant modified carbon nanotube paste electrode *Instrum. Exp. Tech.* **63** 750–7

[21] Hareesha N and Manjunatha J G 2021 Electro-oxidation of formoterol fumarate on the surface of novel poly(thiazole yellow-G) layered multi-walled carbon nanotube paste electrode *Sci. Rep.* **11** 12797

[22] Hareesha N, Manjunatha J G, Amrutha B M, Sreeharsha N, Basheeruddin A, Md and Khalid A 2021 A fast and selective electrochemical detection of vanillin in food samples on the surface of poly(glutamic acid) functionalized multiwalled carbon nanotubes and graphite composite paste sensor *Colloids Surf. A: Physicochem. Eng.* **626** 127042

[23] Pushpanjali P A, Manjunatha J G, Hareesha N, D'Souza E S, Charithra M M and Prinith N S 2021 Voltammetric analysis of antihistamine drug cetirizine and paracetamol at poly (l-leucine) layered carbon nanotube paste electrode *Surf. Int.* **23** 101154

[24] Pushpanjali P A, Manjunatha J G and Srinivas M T 2020 Highly sensitive platform utilizing poly (L-methionine) layered carbon nanotube paste sensor for the determination of voltaren *FlatChem* **24** 100207

[25] Charithra M M, Manjunatha J G, Sreeharsha N, Basheeruddin Asdaq S M and Anwer Md K 2021 Polymerized carbon nanotube paste electrode as a sensing material for the detection of adrenaline with folic acid *Monatsh. Chem.* **152** 411–20

[26] Charithra M M and Manjunatha J G 2021 Electroanalytical determination of acetaminophen using polymerized carbon nanocomposite based sensor *Chem. Data Collect.* **33** 100718

[27] Charithra M M and Manjunatha J G 2021 Fabrication of poly (evans blue) modified graphite paste electrode as an electrochemical sensor for sensitive and instant riboflavin detection *Mor. J. Chem.* **9** 007–17

[28] Charithra M M, Manjunatha J G and Raril C 2020 Surfactant modified graphite paste electrode as an electrochemical sensor for the enhanced voltammetric detection of estriol with dopamine and uric acid *Adv. Pharm. Bull.* **10** 247–53

[29] Charithra M M and Manjunatha J G 2021 Electroanalytical determination of acetaminophen using a polymerised carbon nanotube based sensor *J. Electron. Mater.* **50** 6929–40

[30] Charithra M M and Manjunatha J G 2018 Sensitive detection of anthrone using poly (l-arginine) modified carbon paste electrode by voltammetric method *Open Access J. Chem.* **3** 8–14

[31] Manjunatha J G, Deraman M and Basri N H 2015 Electrocatalytic detection of dopamine and uric acid at poly (basic blue b) modified carbon nanotube paste electrode *Asian J. Pharm. Clin. Res.* **8** 48–53

[32] Manjunatha J G 2016 Poly (nigrosine) modified electrochemical sensor for the determination of dopamine and uric acid: a cyclic voltammetric study *Int. J. ChemTech Res.* **6** 136–46

[33] Beitollahi H, Khalilzadeh M A, Tajik S, Safaei M, Zhang K, Jang H W and Shokouhimehr M 2020 Recent advances in applications of voltammetric sensors modified with ferrocene and its derivatives *ACS Omega* **5** 2049–59

[34] Karimi-Maleh H, Shojaei A F, Karimi F, Tabatabaeian K, Shakeri S and Moradi R 2016 Simultaneous determination of 6-mercaptopruine, 6-thioguanine and dasatinib as three important anticancer drugs using nanostructure voltammetric sensor employing pt/mwcnts and 1-butyl-3-methylimidazolium hexafluoro Phosphate *Biosen. Bioelectron.* **86** 879

[35] Deroco P B, Fatibello-Filho O, Arduini F and Moscone D 2019 Effect of different carbon blacks on the simultaneous electroanalysis of drugs as water contaminants based on screen-printed sensors *Electroanalysis* **31** 2145–54
[36] Alhedabi T, Cattey H, Roussel C, Blondeau-Patissier V, Gharbi T and Herlem G 2017 Experimental and theoretical studies on electropolymerization of polar amino acids on platinum electrode *Mater. Chem. Phys.* **185** 183–94
[37] Charithra M M and Manjunatha J G 2020 Electrochemical sensing of paracetamol using electropolymerised and Sodium lauryl sulfate modified carbon nanotube paste electrode *Chemistryselect* **5** 9323
[38] Pollap A, Knihnicki P, Kuśtrowski P, Kozak J, Gołda M, Kotarba A and Kochana J 2018 Sensitive voltammetric amoxicillin sensor based on TiO_2 sol modified by CMK-3-type mesoporous carbon and gold ganoparticles *Electroanalysis* **30** 2386–96
[39] Kumar N and Rosy G R N 2017 Gold-palladium nanoparticles aided electrochemically reduced graphene oxide sensor for the simultaneous estimation of lomefloxacin and amoxicillin *Sens. Actuators B Chem.* **243** 658–68
[40] Ojani R, Raoof J-B and Zamani S 2012 A novel voltammetric sensor for amoxicillin based on nickel–curcumin complex modified carbon paste electrode *Bioelectrochem.* **85** 44–9
[41] Rezaei B and Damiri S 2009 Electrochemistry and adsorptive stripping voltammetric determination of amoxicillin on a multiwalled carbon nanotubes modified glassy carbon electrode *Electroanalysis* **21** 1577–86
[42] Karimi-Maleh H, Tahernejad-Javazmi F, Gupta V K, Ahmar H and Asadi M H 2014 A novel biosensor for liquid phase determination of glutathione and amoxicillin in biological and pharmaceutical samples using a ZnO/CNTs nanocomposite/catechol derivative modified electrode *J. Mol. Liq.* **196** 258–63
[43] Hai Yen P T, Anh N H, Thu Ha V T, Hung L Q, Phong P H and Thu Hien C T 2020 Electrochemical properties of amoxicillin on an economical, simple graphite pencil electrode and the ability of the electrode in amoxicillin detection *Vietnam J. Chem.* **58** 201–5
[44] Sharma R 2015 Estimation of amoxicillin in presence of high concentration of uric acid and other urinary metabolites using an unmodified pyrolytic graphite sensor *J. Electrochem. Soc.* **162** 8–13

Chapter 11

Chemically modified carbon electrodes for metal ions and organic molecule sensing applications

Shajahan Shanavas, Mohammad Abu Haija, Arun Thirumurugan, Roberto Acevedo and R Ramesh

11.1 Introduction

Developing a new range of electrochemical sensors is a rapidly developing area in electrochemistry due to its wide range of applications in industries [1–3]. Electrochemical sensors are used in several day-to-day routines for medical purposes, environmental monitoring, research facilities, amd chemical and pharmaceutical industries (figure 11.1). Additionally, the utilization of electrochemical sensors in ecological monitoring especially to analyze gas pollutants in the atmosphere and liquid phase pollutants in water resources will provide a great way to control hazards for living things [4, 5]. Electrochemical sensor frameworks are more well-known in contrast to other instrumental strategies. This might be because of their field compact capacities, and less difficult instrumentation necessities, which thus brings about less expense [6]. The advancement of a decent electrochemical sensor requires information on major electrochemistry like a mass exchange, thermodynamics and energy of electron movement, and primarily, the surface properties of the cathode [7, 8]. The exhibition of electrochemical sensors also profoundly relies upon the kind of terminal materials utilized during the investigation. Some of the traditional electrodes used during electroanalysis are Pt, Hg, Au, and carbon. However, these electrodes still have some limitations on their sensitivity and the inherent properties which restrict sensing and various electrochemical properties are some of their limitations [9–13]. The sensitivity of the electrodes will be affected due to the precipitation process or the absorption on the surface of electrodes. The very slow reaction rates during some electrochemical reactions is also a huge disadvantage due to the requirement of high overpotentials [14, 15]. Another major drawback of conventional electrodes is their basic oxidation and reduction behavior where these electrodes will act as electron transfer sites. The

doi:10.1088/978-0-7503-5127-0ch11

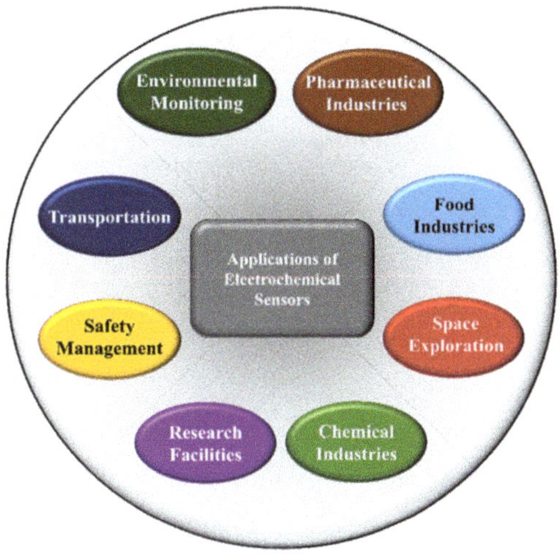

Figure 11.1. Applications of electrochemical sensors.

sensitivity, selectivity, and stability of the electrodes will be more important when facing the new complex challenges in the highly complicated medical and environmental matrix [16, 17].

The surface modification of electrodes is one of the very highly efficient techniques used to achieve the desired goals. The surface modification of electrodes enhances its specific redox process, decreases the surface oxidation process, reduces the required overpotential, accelerates electrochemical reaction rates, charge transfer, and transport properties which results in improved electrochemical sensing properties [18–20]. In addition, the significant improvement in the modified electrodes can be observed when compared to the unmodified electrodes which is due to the intrinsic properties of the surface-modified electrodes to perform specific electrochemical sensing reactions. The species bound on the surface of the electrode will exhibit a synergistic effect with the electrode to improve the biocompatibility, electrical conductivity, electrochemical reactivity, and catalytic properties and these properties will greatly help with the signal transduction in electrochemical biosensors [21–23]. These improved properties of modified electrodes will greatly enhance the efficiency of studying electrochemical behavior in electro-organic synthesis, photoelectrochemical energy conversion, electrocatalysis, semiconductor stabilization, and as well as sensing properties [24, 25]. Several different techniques have been previously studied to understand the electrochemical behavior of modified electrodes namely (i) non-specific adsorption, (ii) non-covalent approach and (iii) electrostatic approach [26–28]. Better selectivity and reproducibility are observed in the covalent surface modification and it also avoids unspecific adsorption. The schematic diagram of a modified substrate is shown in figure 11.2. In this chapter, we discuss the different techniques of surface modifications in electrodes and their pros and cons. When compared to other electrode materials, carbon-based electrodes currently have widespread use in electroanalysis because

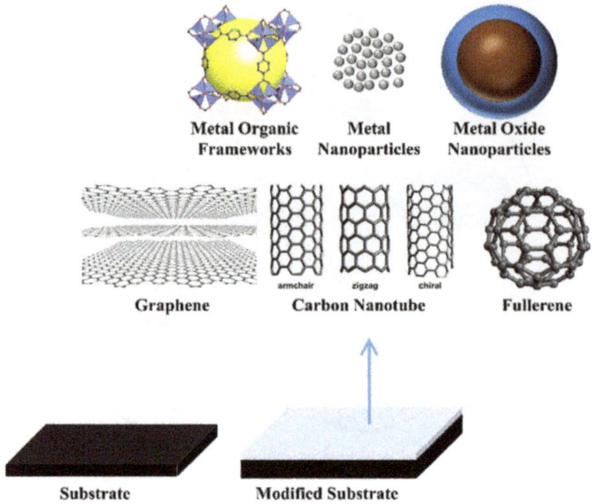

Figure 11.2. Modified substrates for electrochemical applications.

they are cheap, they have low background current, broad potential window, are chemically inert, and their high ability to sense different molecules [29, 30]. As one of the most abundant non-metallic elements in the world, carbon can be found uniformly distributed in every part of the Earth. Carbon has different allotropic forms with different physical and chemical properties. The carbon atoms are arranged to form crystalline forms especially diamond (sp3), graphite (sp2), fullerene (distorted sp2), and carbine (sp1). Diamond and graphite are the only naturally obtained crystalline carbon allotropes, the others have been synthesized synthetically. Each different allotropic form of carbon atom exhibits different physical and chemical properties. As a result of the specific physical and chemical properties, these carbon-based materials can be tailored for different applications [31–33].

11.2 Different types of electrodes in the modification process

11.2.1 Metal electrodes

The attractive optical, thermal, catalytic, and electric properties of nanometal particles have attracted huge interest within the scientific community due to their potential application in various fields of biology, chemistry, physics, material science, and medicine [34]. The interdisciplinary fields in physical and life sciences have a huge interest in electrode materials for electrocatalytic applications. The attractive physical and chemical properties of metallic nanoparticles have drawn huge attention when compared to bulk materials [35]. The quantum confinement effect of the nanoparticles has led nanostructures materials into widespread applications due to their improved magnetic, optic, electronic, physical, and chemical properties. For example, the bulk form of gold is highly inactive whilst gold nanoparticles are a very good catalyst for the carbon monoxide and hydrocarbons oxidation process [36]. The bulk form of metals like gold, platinum, and

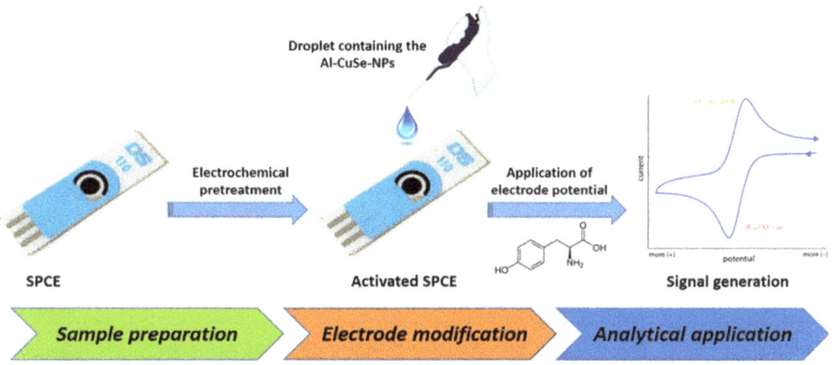

Figure 11.3. Electrode modification using metal nanoparticles (source: Elsevier).

palladium are non-magnetic but when the size of these metals is decreased to the nano level they behave like magnetic particles. The ferromagnetic gold particles can be made by capping gold nanoparticles with suitable molecules to alter the surface charge of the particles [37]. Khaled *et al* [38] fabricated the modified carbon electrode using aluminum doped copper selenide nanoparticles for L-tyrosine determination in pharmaceutical samples, see figure 11.3.

11.2.2 Electrodes from glassy carbon

Among the other forms of carbon electrodes, glassy carbon electrodes have been more widely used electrodes due to their highly stable nature. Glassy carbon electrodes were prepared by charring polymeric resins like polyacrylonitrile at high temperatures in an inert atmosphere. The huge advantage of glassy carbon electrodes is that they have ribbon-like structures where the graphite sheets are cross-linked to each other and hence the glassy carbon electrodes are harder and more robust than the graphite electrodes [39–41]. The glassy carbon electrodes have been the most widely used in electrochemical applications due to their excellent mechanical and electrical properties with very wide working potentials in both anodic and cathodic reactions. The electron transfer reaction in the glassy carbon electrodes is mainly dependent on the surface structure of the electrodes and it can be adjusted by the surface pretreatment process. Several surface pretreatment techniques have been used to obtain highly efficient glassy carbon electrodes including chemical, mechanical, electrochemical, heat, and laser irradiation. The most common and popular method for modifying the surface of glassy carbon electrodes is simply mechanically polishing the surface to a mirror-like appearance [42–44].

11.2.3 Electrodes from carbon paste

Carbon paste electrodes are made by blending organic solvents like silicone grease, mineral oil, paraffin oil, Nujiol, and bromonaphthalene with graphite powder. The huge advantage of carbon paste electrodes is that they are highly cost-effective and have easily renewable surfaces. Carbon paste electrodes can be also prepared by mixing modifier material with the binder and graphite powder. These carbon paste electrodes are enhanced by the surface renewability but the stability of these electrodes is still challenging due to the etching of modified molecules while performing the electrochemical process. In addition, the binder molecules in the carbon paste significantly slow the kinetics of electrodes and as a result, the long-term operational and storage stability of the electrodes will be affected. Examples of some carbon paste electrodes and molecular imprinted polymer electrodes with their modifying materials as well as binders are given in table 11.1.

Table 11.1. Modified electrode-based electrochemical sensors and their sensing studies.

S.no.	Modifier	Binder	Analyte	Real samples	References
1.	Cd-IIP/CPE	Mineral oil	Cd(II)	Water samples	[45]
2.	$CoFe_2O_4$ NPs/IL/CPE	Paraffin	Epirubicin	Injection and serum samples 14	[46]
3.	IIP/MWCNT-CPE	Paraffin	Pb(II)	NA	[47]
4.	IIP/CNT-CPE	n-Eicosane	Bi(III)	Environmental, pharmaceutical, and biological samples	[48]
5.	Ce-IIP/MWCNT/CPE	n-Eicosane	Cerium(III)	Spiked samples of drinking water and sea water	[49]
6.	MOF/CPE (TMU–16-NH_2)	Mineral oil	Cd(II)	NA	[50]
7.	$Zn_4O(BDC)_3$/CPE (MOF-5)	Mineral oil	Pb(II)	Real water samples and standard reference materials	[51]
8.	In-CeO_2/GCPE	Mineral oil	Uric acid	Human blood serum and urine samples	[52]
9.	MIP-CPE	Paraffin	Cu(II)	Wastewater of Yellow River	[53]
10.	IIP-CPE	Mineral oil	Cd(II)	Aqueduct water, tap water, dam water, waste water, and river water	[45]
11.	PME PVC-GE, PVC-PA-GE, PVC-CNTs-GE, PVC-CNTs-PAGE	n-Eicosane	Ag^+	Voltammetric sensor: tap water, well water. Potentiometric sensor: tap water, well water, mineral water, river water	[54]
12.	SCMNP, isonicotinic acid	Paraffin	Mo(IV)	Tap water, tomato, tea	[55]
13.	4-(2-pyridylazo) resorcinol as a lead binding ligand	Paraffin	Pb(II)	Distilled water, tap water, Caspian Sea water, and waste water	[56]

(*Continued*)

Table 11.1. (*Continued*)

S.no.	Modifier	Binder	Analyte	Real samples	References
14.	Br-PADAP	Paraffin	UO(II)	Tap, sea, Persian Gulf, and river water.	[57]
15.	Magnetic nanoparticle MWCNT	Paraffin	Pb(II)	River water, aqueduct water, copper factory waste water and coal processing waste water	[58]
16.	Eriocrome cyanin R	Paraffin	Pd(II)	Tap water, soil	[59]
17.	benzo-15-crown-5-CPE	Mineral oil	Tl^+	Tap, well, and waste water, human hair	[60]
18.	IIP-CPE-MWCNTs	Paraffin	Ce(III)	Drinking water and sea water	[61]

*IIP = Ion imprinted polymer; CPE = Carbon paste electrode; IL = Ionic liquid; MOF = Metal organic framework; GCPE = Glassy carbon paste electrode; MWCNT = Multi walled carbon nanotube.

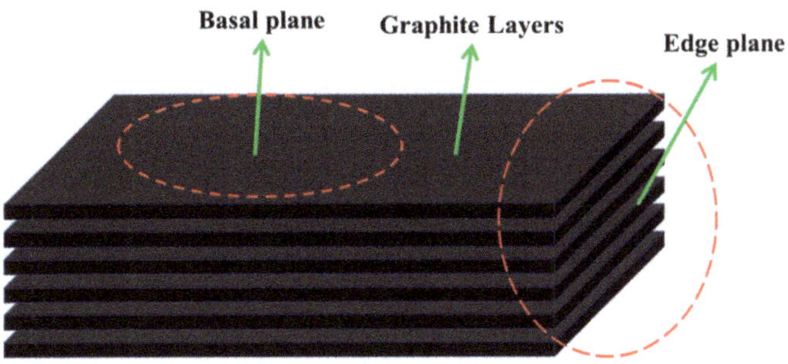

Figure 11.4. Schematic representation of basal plane and edge plane in pyrolytic graphite sheets.

11.2.4 Pyrolytic graphite electrodes

On the surface of pyrolytic graphite, both basal plane and edge plane will be present. The several layers of graphite sheets parallel to the surface with an interlayer spacing of 0.335 nm will be present in the highly ordered pyrolytic graphite electrodes. The exposed edges of graphite layers will be enhanced by the surface defects [62]. The different electrochemical properties can be obtained in pyrolytic graphite electrodes due to the nature of chemical bonding. In several redox reactions, the rate constant for the electron transfer process is more than 103 times faster when compared to the basal plane graphite electrodes. The low background currents and improved electrocatalytic signals can be obtained while using the edge plane pyrolytic graphite electrode [63, 64].

11.3 Modification of electrodes with chemical techniques

The rapid expansion of industries related to several fields like energy storage, energy conversion, chemical sensing, synthesis of new materials, corrosion protection, electrochromic displays, chemical sensing, genetic engineering, and microelectronics has drastically increased the demand for electrochemical devices with more efficient performance and accuracy. Although several electrochemical sensors are available with different types of electrodes, the chemically modified electrodes are a modern approach to obtain highly efficient electrochemical electrodes with a wide region of electrochemical investigations. The direct surface properties controllable ability in the chemical modified electrodes makes it efficient for various electrochemical applications.

The concept of chemically modified electrodes was first introduced by Murray by modifying SnO_2 electrodes with amine groups. He did several coupling reactions on the functionalization of electroactive organic moieties by using amine-modified electrodes [65, 66].

The chemically modified electrodes were produced by anchoring the modifier molecules on the surface of the electrode through covalent or non-covalent bonding. The pathway for the interaction of electrode and analyte is achieved by anchoring modifiers on the surface of the substrate. Chemically modified electrodes consist of two components namely substrate and modifier. The surface properties of electrodes play a major role in the electrochemical reaction process, the electrochemical reaction will occur on the surface of electrode materials. In the electrode reactions, the surface structure of the electrodes plays a major role in an interface that promotes electrode kinetics. Usually, carbon-based substrate materials-based electrodes were used for electrochemical reactions. Still, these carbon electrodes lag behind the chemically modified electrode materials on high overpotentials, sensitivity, stability, and selectivity [67]. These chemically modified electrodes will have some specific physical and chemical properties which are not available in carbon-based electrodes. Hence, the suitable modifier molecules with specific functional groups and with a structural framework of chemical reactions will be used to modify carbon electrodes to enhance their electrochemical performance. The reactions on chelation will enhance the electrocatalytic properties of electrodes to obtain a facile electron charge transfer process [68].

Chemically modified electrodes have been widely used in metal ion quantification applications due to their adjustable properties like sensitivity and selectivity towards the target molecules. The most widely investigated electrodes to quantify metal ions using a graphene oxide modified electrode is the best example for chemically modified electrodes. The required overpotential for the lead ions on the surface of carbon particles can be reduced by using GO on the surface of electrodes. Similarly, the thiol-containing functional groups on the surface of carbon particles can be used for mercury quantification due to their very strong binding interaction. The electrode modification consists of two primare modification processes including chemical and electrochemical modifications. Another popular modifying technique is electrografting/ electropolymerization [69].

11.3.1 Chemical modification

In the chemical modification of electrodes, several strategies have been reported including ball milling and microwave methods, covalent modification of electrode substrate material through diazonium salt reduction in electrodes using hypo phosphorus acid and using chemical modifiers with carbon paste electrodes [70].

11.3.2 Electrochemical modification

In the area of the development of chemically modified electrodes, the most fascinating approach is electrochemically assisted covalent modification of carbon substrate material with specific modifier molecules. Two different techniques have been used, namely, the oxidative and reductive strategies. The amines and alcohols were used in the oxidative grafting techniques and the functionalization of electrodes with triarylsulfonium, diaryliodonium, and aryldiazonium salts were achieved by reductive grafting techniques [71]. The mechanism involved in this electrochemical modification is the generation of aryl radicasl at the solution and electrode interface through the reduction of diazonium salt and its subsequent attachment in the surface of carbon electrode substrates [72].

The attachment is due to the binding between the carbon atom from modifier molecules and the carbon atom from electrodes through covalent bonding. The electrochemical modification can be achieved through cyclic voltammetry or controlled potential electrolysis either in an aqueous or non-aqueous medium. The electrode kinetics of a particular redox reaction is primarily decided by the thickness of the modified molecule at the interface and plays an important role in the electroanalysis area. The successful modification of electrodes can be achieved by altering and optimizing the experimental parameters like concentration of modifier salt, applied current, or potential length of the modifier molecules. Jae-Wook Shin *et al*, [73] reported the fabrication of graphene oxide-covered silver nanoparticles on the surface of indium tin oxide-coated electrode substrate for an electrochemical dopamine biosensor (figure 11.5). Silver nanoparticles are a great alternative for gold and platinum nanoparticles due to their effective electrochemical performance in a lower concentration of dopamine and are, comparatively, cost-effective. In addition, the highly efficient nature of graphene oxide nanosheets makes them the best material for electrochemical performances. Still, the efficiency of pristine silver nanoparticles and pristine graphene oxide sheets is very low due to the instability of silver nanoparticles and loss due to electrical conductivity in graphene oxide sheets.

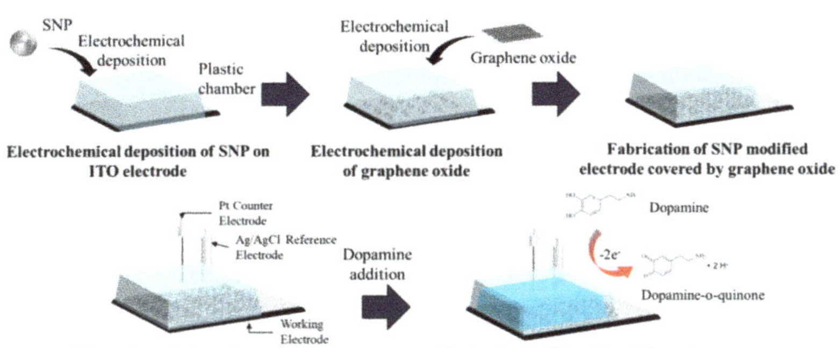

Figure 11.5. Schematic diagram of an electrochemical modification of electrode for dopamine detection [73].

Hence, it is necessary to modify the electrodes to overcome these disadvantages. The formation of graphene oxide-covered silver nanoparticles on the surface of indium tin oxide-coated electrode substrate showed promising results.

References

[1] Abdalla A and Patel B A 2020 *Curr. Opin. Electrochem.* **20** 78–81

[2] Alizadeh N and Salimi A 2018 *Electroanalysis.* **30** 2803–40

[3] Privett B J, Shin J H and Schoenfisch M H 2010 *Anal. Chem.* **15** 4723–41

[4] Rassaei L, Marken F, Sillanpää M, Amiri M, Cirtiu C M and Sillanpää M 2011 *TRAC-Trend Anal Chem.* **30** 1704–15

[5] Pang X, Shaw M D, Gillot S and Lewis A C 2018 *Sens. Actuators B: Chem.* **266** 674–84

[6] Liu L, Zhou Y, Liu S and Xu M 2018 *ChemElectroChem.* **5** 6–19

[7] Wang J and Zeng H 2021 *Adv. Colloid Interface Sci.* **288** 102346

[8] Shanavas S, Ahamad T, Alshehri S M, Acevedo R and Anbarasan P M 2021 *Optik.* **226** 165970

[9] Li S, Zhang C, Wang S, Liu Q, Feng H, Ma X and Guo J 2018 *Analyst* **143** 4230–46

[10] Deshmukh M A, Shirsat M D, Ramanaviciene A and Ramanavicius A 2018 *Crit. Rev. Anal. Chem.* **48** 293–304

[11] Zuo Y, Xu J, Zhu X, Duan X, Lu L and Yu Y 2019 *Microchim. Acta* **186** 171

[12] Shanavas S, Ahamad T, Alshehri S M, Sultan A, Acevedo R and Anbarasan P M 2020 *Opt. Laser Technol.* **123** 105902

[13] Yao Y, Wu H and Ping J 2019 *Food Chem.* **274** 8–15

[14] Perumal K, Shanavas S, Karthigeyan A, Ahamad T, Alshehri S M and Murugakoothan P 2020 *Diam. Relat. Mater.* **110** 108091

[15] Varun D N, Manjunatha J G, Hareesha N, Sandeep S, Mallu P, Karthik C S, Prinith N S, Sreeharsha N and Asdaq S M 2021 *Monatsh. Chem.* **152** 1183–91

[16] Elumalai N, Prabhu S, Selvaraj M, Shanavas S, Navaneethan M, Harish S, Ramu P and Ramesh R 2021 *J. Mater. Sci., Mater. Electron.* **32** 28633–47

[17] Rao H, Zhao X, Liu X, Zhong J, Zhang Z, Zou P, Jiang Y, Wang X and Wang Y 2018 *Biosens. Bioelectron.* **100** 341–7

[18] Zhou Y, Li Y, Han P, Dang Y, Zhu M, Li Q and Fu Y 2019 *New J. Chem.* **35** 14009–19

[19] Hareesha N, Manjunatha J G, Amrutha B M, Sreeharsha N, Asdaq S B and Anwer M K 2021 *Colloids Surf. A: Physicochem. Eng. Asp.* **626** 127042

[20] Ranjith K S, Vilian A E, Ghoreishian S M, Umapathi R, Hwang S K, Oh C W, Huh Y S and Han Y K 2022 *J. Hazard. Mater.* **421** 126775

[21] Cesewski E and Johnson B N 2020 *Biosens. Bioelectron.* **159** 112214

[22] Charithra M M and Manjunatha J G 2021 *Chem. Data Collect.* **33** 100718

[23] Moon J M, Thapliyal N, Hussain K K, Goyal R N and Shim Y B 2018 *Biosens. Bioelectron.* **102** 540–52

[24] Charithra M M, Manjunatha J G, Sreeharsha N, Asdaq S M and Anwer M K 2021 *Monatsh. Chem.* **152** 411–20

[25] Martínez N P, Isaacs M and Nanda K K 2020 *New J. Chem.* **15** 5617–37

[26] Vonasek E, Lu P, Hsieh Y L and Nitin N 2017 *Cellulose.* **10** 4581–9

[27] Jeong S, Pinals R L, Dharmadhikari B, Song H, Kalluri A, Debnath D, Wu Q, Ham M H, Patra P and Landry M P 2020 *Sci Rep.* **10** 1

[28] Razavi M, Kompany-Zareh M and Khoshkam M 2021 *Spectrochim. Acta—A: Mol. Biomol. Spectrosc.* **246** 119016
[29] Contreras-Naranjo J E, Perez-Gonzalez V H, Mata-Gómez M A and Aguilar O 2021 *Electrochem. Commun.* **130** 107098
[30] Souza E S, Manjunatha J G, Chenthatti R, Tigari G and Ravishankar D K 2021 *Biointerface Res. Appl. Chem.* **11** 14661–72
[31] Dunlap R A 2018 *Novel Microstructures for Solids* (Bristol: IOP Publishing) ch 8
[32] Khalaj Z, Monajjemi M and Diudea M V 2017 *Sustainable Nanosystems Development, Properties, and Applications* (IGI Global) pp 185–213
[33] Okwundu O S, Aniekwe E U and Nwanno C E 2018 *Metall. Mater. Eng.* **24** 145–71
[34] Wang F and Hu S 2009 *Microchim. Acta* **165** 1–22
[35] Han T, Jin J, Wang C, Sun Y, Zhang Y and Liu Y 2017 *Nanomaterials.* **2** 40
[36] Singh J A, Overbury S H, Dudney N J, Li M and Veith G M 2012 *ACS Catal.* **6** 1138–46
[37] Nealon G L, Donnio B, Greget R, Kappler J P, Terazzi E and Gallani J L 2012 *Nanoscale.* **4** 5244–58
[38] Murtada K, Salghi R, Ríos A and Zougagh M 2020 *J. Electroanal. Chem.* **87** 114466
[39] Wang Z, Guo H, Gui R, Jin H, Xia J and Zhang F 2018 *Sens. Actuators B: Chem.* **255** 2069–77
[40] Zilberg R A, Sidelnikov A V, Maistrenko V N, Yarkaeva Y A, Khamitov E M, Kornilov V M and Maksutova E I 2018 *Electroanalysis.* **4** 619–25
[41] Bartolome J P and Fragoso A 2017 *Inorg. Chim. Acta* **468** 223–31
[42] Tian Y, Deng P, Wu Y, Li J, Liu J, Li G and He Q 2020 *J. Electrochem. Soc.* **167** 046514
[43] Yan Q, Zhi N, Yang L, Xu G, Feng Q, Zhang Q and Sun S 2020 *Sci Rep.* **10** 1
[44] Amare M and Admassie S 2020 *J. Mater. Res. Technol.* **9** 11484–96
[45] Coelho M K, De Oliveira H L, De Almeida F G, Borges K B, Tarley C R and Pereira A C 2017 *Int. J. Environ. Anal. Chem.* **97** 1378–92
[46] Karimi F, Shojaei A F, Tabatabaeian K and Shakeri S 2017 *J. Mol. Liq.* **242** 685–9
[47] Alizadeh T, Hamidi N, Ganjali M R and Rafiei F 2017 *J. Environ. Chem. Eng.* **5** 4327–36
[48] Alizadeh T, Hamidi N, Ganjali M R and Nourozi P 2017 *Sens. Actuators B: Chem.* **245** 605–14
[49] Alizadeh T, Ganjali M R, Akhoundian M and Norouzi P 2016 *Microchim. Acta* **183** 1123–30
[50] Roushani M, Valipour A and Saedi Z 2016 *Sens. Actuators B: Chem.* **233** 419–25
[51] Wang Y, Wu Y, Xie J and Hu X 2013 *Sens. Actuators B: Chem.* **177** 1161–6
[52] Temerk Y and Ibrahim H 2016 *Sens. Actuators B: Chem.* **224** 868–77
[53] Zhihua W, Xiaole L, Jianming Y, Yaxin Q and Xiaoquan L 2011 *Electrochim. Acta* **58** 750–6
[54] Shamsipur M, Hashemi B, Dehdashtian S, Mohammadi M, Gholivand M B, Garau A and Lippolis V 2014 *Anal. Chim. Acta* **852** 223–35
[55] Afkhami A, Moradi M, Bahiraei A and Madrakian T 2018 *Anal. Bioanal. Chem. Res.* **5** 41–53
[56] Bojdi M K, Mashhadizadeh M H, Behbahani M, Farahani A, Davarani S S and Bagheri A 2014 *Electrochim. Acta* **136** 59–65
[57] KalateBojdi M, Behbahani M, Najafi M, Bagheri A, Omidi F and Salimi S 2015 *Electroanalysis.* **10** 2458–67
[58] Ghanei-Motlagh M and Taher M A 1999 *Anal. Bioanal. Chem. Res* **4** 295–306

[59] Bojdi M K, Behbahani M, Sahragard A, Amin B G, Fakhari A and Bagheri A 2014 *Electrochim. Acta* **149** 108–16

[60] Güney S and Güney O 2016 *Sens. Actuators B: Chem.* **231** 45–53

[61] Chen J, Bai H, Xia J, Liu X, Liu Y and Cao Q 2018 *J. Rare Earths* **36** 1121–6

[62] Goyal R N, Chatterjee S and Rana A R 2010 *Talanta* **83** 149–55

[63] Iamprasertkun P, Hirunpinyopas W, Keerthi A, Wang B, Radha B, Bissett M A and Dryfe R A 2019 *J. Phys. Chem. Lett.* **10** 617–23

[64] Shen A, Zou Y, Wang Q, Dryfe R A, Huang X, Dou S, Dai L and Wang S 2014 *Angew. Chem.* **126** 10980–4

[65] Untereker D F, Lennox J C, Wier L M, Moses P R and Murray R W 1977 *J. Electroanal. Chem. Interf. Electrochem.* **81** 309–18

[66] Wier L M and Murray R W 1979 *J. Electrochem. Soc.* **126** 617

[67] Luong J H, Male K B and Glennon J D 2009 *Analyst* **134** 1965–79

[68] Wang Y, Chen S, Zhao S, Chen Q and Zhang J 2020 *J. Mater. Chem.* A **8** 15845–52

[69] Ullah W, Herzog G, Vilà N and Walcarius A 2021 *Electrochem. Commun.* **122** 106896

[70] Kempahanumakkagari S K, Adarakatti P S and Malingappa P 2018 *J. Environ. Chem. Eng.* **6** 2674–83

[71] Koefoed L, Pedersen S U and Daasbjerg K 2017 *Langmuir* **33** 3217–22

[72] Yáñez-Sedeño P, Campuzano S and Pingarrón J M 2018 *J. Sens.* **18** 675

[73] Shin J W, Kim K J, Yoon J, Jo J, El-Said W A and Choi J W 2017 *J. Sens.* **12** 2771

Chapter 12

Electrochemical sensors based on carbon nanomaterial using Langmuir–Blodgett and layer-by-layer thin films for chemical and biological analyses

Anerise de Barros, Amanda de S M de Freitas, Cristiane Carla Maciel and Marystela Ferreira

The fabrication of electrochemical sensors requires the electrodes coating essential to integrate functionality, chemical and physical stability, and high sensitivity in devices. The layer-by-layer (LbL) and Langmuir–Blodgett (LB) techniques are two of the several coating strategies for modification of the surfaces and, used for sensors fabrication. Both methods play a significant and efficient role in electrode modification using different materials and on distinct surfaces with molecular-level control thickness on the composition of the structures formed. This chapter focuses on recent electrochemical sensors based on carbon nanomaterials using the LB and LbL techniques to modify electrodes in detecting especially chemical and biological species.

12.1 Introduction

The development of electrochemical sensors is still a hot field for new technologies that confer benefits to society (Manjunatha *et al* 2013, Hareesha and Manjunatha 2020b). The fabrication of devices displays high performance with miniaturized futures, allowing the field application. Beyond the fast response analysis and easy acquisition are the key futures for the electrochemical sensor classes (Manjunatha *et al* 2013, Hareesha *et al* 2019, Hareesha and Manjunatha 2020a, 2020b, Hareesha *et al* 2020, Lahcen *et al* 2020, Nemčeková and Labuda 2021). One of the essential factors to confer excellent performance of the electrochemical sensors is associated with the homogeneity coating of these devices, integrating functionality, chemical and physical stability, and high sensitivity. In this sense, methodologies of surface

doi:10.1088/978-0-7503-5127-0ch12

coating are essential to achieve these characteristics and, techniques as Langmuir–Blodgett (LB) and layer-by-layer (LbL) are some of the several coating strategies for sensors fabrication (Zhao *et al* 2019). Briefly, the LB technique requires using Langmuir trough equipment, in which the material is spread on the aqueous interface and compressed using a mechanical barrier to form a monolayer of material interest (De Oliveira *et al* 2017). This technique has as the main advantage that the high control of thickness film and uniformity electrode coating confers high sensitivity and reproducibility response, which is desirable for biosensor applications.

In contrast to the LB technique, the LbL assembly is a simple method for material deposition and has the main advantage of large-scale fabrication (Richardson *et al* 2015, Zhao *et al* 2019). This technique was adapted for industrial scale by the roll-to-roll deposition for flexible electrode coating method. Thus, the LB technique as LbL provides reasonable control of film thickness; however, a lower control of uniformity deposition can be present on a large scale. In summary, both methods play a significant and efficient role in electrode modification by using different materials and to distinct surfaces with molecular-level control thickness on the composition formed. The intercalated layers with other materials provide unique properties for developing the electrochemical sensors, guaranteeing high sensitivity and selective response. Combining these methods with carbon-based materials is very attractive for the development of electrochemical sensors due to their electron transfer and conduction mechanisms to improve the sensitivity of chemical and biosensors devices (Brett 2018, Bezzon *et al* 2019, Hareesha and Manjunatha 2020a, 2020b, Nemčeková and Labuda 2021). The most significant advantages of developing electrochemical sensors are associated with the direct analysis of the real sample, eliminating the pre-treatment steps as used in other analytical techniques (Bard and Faulkner 2003, Manjunatha *et al* 2013, Asadzadeh-Firouzabadi and Zare 2017, Raril and Manjunatha 2018, Hareesha *et al* 2019, Tigari *et al* 2019, Hareesha and Manjunatha 2020a, 2020b, Hareesha *et al* 2020, Tigari and Manjunatha 2020, Kokab *et al* 2021). The electrochemical sensors are classified according to the recognition/specificity mechanism (chemical or biological) and the method used, whose the most common are potentiometric, amperometric, voltammetric, and electrochemical impedance. Figure 12.1 illustrates a schematic representation of carbon-based nanomaterials in mechanism oxidation for the sensing and electrochemical methods.

Briefly, carbon-based nanomaterials such as carbon nanotubes (CNTs), fullerenes, carbon and graphene quantum dots (CQDs and GQDs), graphene oxides (GO), and reduced graphene oxides (rGO), are attractive materials in the building of electrochemical sensors mainly by acting as efficient electron acceptors and donors (Brett 2018, Bezzon *et al* 2019, Kamran *et al* 2019) For instance, the chemical bonding of CNTs consists of a fully sp^2 hybridization orbital, similar to the bonding of graphene with the free orbital to establish π-bonds. This promotes electron delocalization over long distances, making CNTs exhibit elevated electron conductivity (Bezzon *et al* 2019, Speranza 2021). Single-walled carbon nanotubes (SWCNTs) are most likely the best candidates for electronics due to their

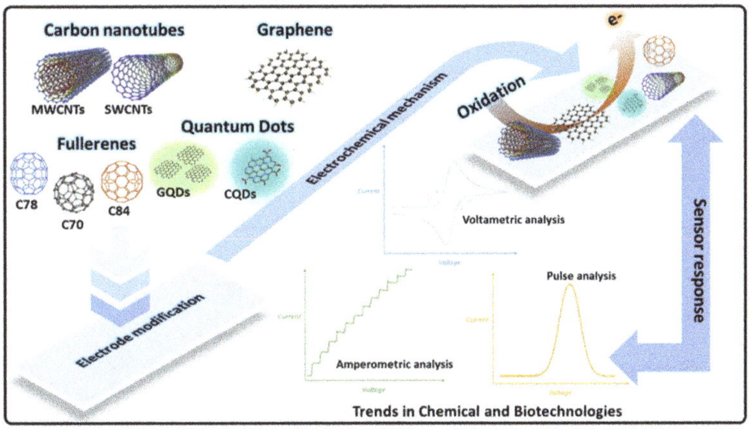

Figure 12.1. Schematic representation of carbon-based nanomaterials for the mechanism of oxidation in sensing platforms and the electrochemical responses. Carbon and graphene quantum dots (CQDs and GQDs, respectively). Single-walled carbon nanotubes (SWCNTs) and multiple-walled carbon nanotubes (MWCNTs).

long-range semiconducting properties, while the multi-walled carbon nanotubes (MWCNTs) have improved resistance to the chemical reagents with respect to SWCNTs (Kamran *et al* 2019, Kumar *et al* 2019, Speranza 2021). On the other hand, graphene and its derivates consist of a large-area single layer of sp^2-bonded carbon atoms arranged in a 2D structure, promoting an even higher delocalization of charge carriers. Due to their large surface area, the non-covalent functionalization with molecules enables the adjustment of graphene properties and essential functionalities for sensing devices by incorporating *ad hoc* chemical receptors (Speranza 2019). In this sense, this chapter approaches the use of carbon-based nanomaterials to develop electrochemical sensing platforms based on the last ten years. We limited the chemical and biosensing application example by the specific techniques of electrode coating, such as LB and LbL modification, to demonstrate the importance of coating techniques combined with carbon-based nanomaterials in the fabrication of sensor devices.

12.2 Electrochemical sensing based on carbon nanomaterials: electrode coating by Langmuir–Blodgett and layer-by-layer techniques

12.2.1 Electrochemical sensor modified by layer-by-layer technique

Electrochemical sensors based on LbL thin films is one of the most common methods used in sensors fabrication. The facilities for manipulating the thickness and multi-layer composition using different materials are the key features to achieve the sensing/mediating/transducing devices (Brett 2018). In this way, sensors based on carbon nanomaterials have been extensively employed in several fields to detect various compounds such as drugs, foods, body fluids, waste water, disease diagnosis, forensic analysis, etc (Manjunatha *et al* 2013, Raril and Manjunatha 2018, Kim *et al* 2019).

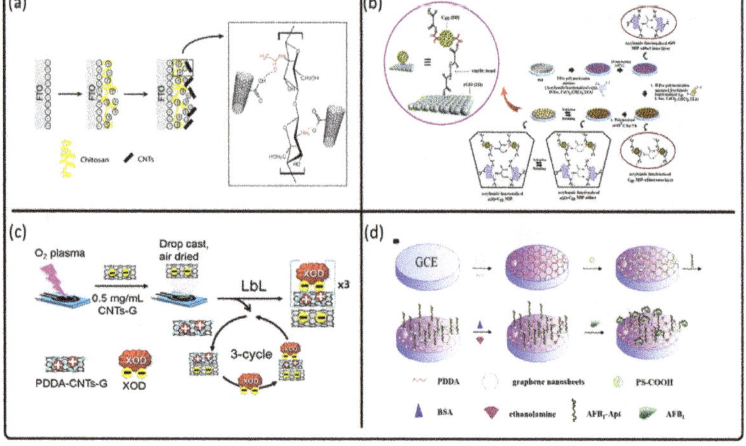

Figure 12.2. Schematic representations of the LbL methods modified different electrodes types: (a) electrode fluorine-doped tin oxide (FTO) with 1-bilayer chitosan/carbon nanotubes (Chi/CNTs) film. (b) A pencil graphite electrode (PGE) substrate, spin-coated with acrylamide functionalized reduced graphene oxide (rGO) D-serine and acrylamide functionalized C_{60}, L-serine, for development of acrylamide functionalized rGO-C_{60}MIP sensor. (c) Screen-printed carbon electrode surfaces (SPCE) xanthine oxidase charged (XOD) and poly (diallyldimethylammonium chloride) (PDDA) film wrapped in multi-walled oxidized carbon nanotubes and graphene complexes (CNTs-G), forming the material PDDA-CNTs-G. (d) GCE electrode modified with graphene nanosheets (GNs) + PDDA, carboxyl-functionalized polystyrene nanospheres (PS-COOH), aptâmero aflatoxin B1 (AFB_1-apt), bovine serum albumin (BSA), aflatoxin B1 (AFB_1). Reprinted (adapted) from Pavinatto *et al* (2015) (panel (a)) Jaiswal *et al* (2019) (panel (b)), Si *et al* (2018) (panel (c)), and Lin and Shen (2020) (panel (d)) with permission from Elsevier.

The use of carbon-based nanomaterials to develop electrochemical sensors has the main advantage of facile functionalization, allowing the interaction with different molecules for selective or non-selective detection. Figures 12.2(a)–(d) illustrate a schematic representation of LbL films built for carbon-based nanomaterials modifying electrodes interface.

From the electronic point of view, carbon-based nanomaterials are classified as excellent materials for the development of sensing platforms due to the facile properties of donors and accepted electrons for the redox process in the electrode interface (Manjunatha *et al* 2013, Raril and Manjunatha 2018, Hareesha *et al* 2019, Hareesha and Manjunatha 2020a, 2020b, Hareesha, *et al* 2020). For instance, Estevam *et al* (2015) verified the influence of GO on the functionalization of 3-*n*-propylpyrydinium silsesquioxane chloride polymer (SiPy+) in LbL thin films intercalated with nickel tetrasulfophthalocyanine ($NiTsPc^{4-}$) deposited on fluorine-doped tin oxide (FTO) electrode. As a result, the presence of GO has significantly improved in the LbL $(SiPy+/NiTsPc^{4-})_n$ films properties as to the electrochemical response, amount of mass deposition, and resistivity of the film. Moreover, the combination of $(SiPy+(GO)/NiTsPc^{4-})_n$ bilayers with the GO mixed SiPy+ revealed that the Q-band absorbance increases linearly with the amount of $NiTsPc^{4-}$ compared to architecture LbL film without GO. This effect is associated with the π–π interaction between the molecules of graphene and phthalocyanine.

Additionally, the VC and EIS measurements show that the GO nanosheets increase the conductivity substantially in $(SiPy+(GO)/NiTsPc^{4-})_n$ bilayers, consequently decreasing the resistivity of the LbL film and increasing the sensitivity of the electrode when applied as sensors and biosensors.

The application of MWCNTs was performed in a study by functionalization with the amine group (AMWNTs) to form alternating multilayers using gold nanoparticles (AuNPs) onto a gold electrode. Sequentially, the hairpin DNA probe 1 (H1) was immobilized by a thiol group at the 3′ end and a ferrocenyl moiety (Fc) at the 5′ end, in the AMWNTs/AuNPs multilayer films established by Au–S bond. The electrochemical DNA (E-DNA) sensor was evaluated for specific detection of the FMS-like tyrosine kinase 3 (FLT3) gene. The presence of AMWCNTs in the sensor architecture provides sensitive and enzyme-free detection of FLT3 due to the generation of dual-signal sensor amplification, leading to a limit of detection (LoD) of 0.1 pmol L^{-1}. Compared to the real sample, the electrochemical sensors agree with classical polymerase chain reaction (PCR), which has been significantly different to the simple operation and isothermal amplification without the use of enzymes or specialized instrumentation, with great potential in DNA diagnosis and clinical analysis (Sun *et al* 2016). The double-signal amplification effect was also studied by Wang *et al* (2018) to develop an ultrasensitive electrochemical biosensor to detect K562 cells. The synergistic interaction between CdS QDs functionalized with polystyrene microspheres (PS) as bio-probe, and LbL multilayers of GO and polyaniline (GO/PAni) increases the detection signal intensity of tumor cells. Here, the GO/PAni LbL films are responsible for increased electron transfer rate, and the multiple layers of CdS QDs increased the signal intensity, leading to the LoD of 3 cells mL^{-1} ($S/N = 3$) for a linear range from 10 to 1.0×10^7 cells mL^{-1}. Screen-printed carbon electrode surfaces (SPCE) were modified with multilayers of xanthine oxidase (XOD), and MWCNTs oxidized functionalized with poly(diallyldimethylammonium chloride) (PDDA) and graphene sheets (figure 12.2(c)) by LbL technique. The LbL architecture, called the acronym PDDA-CNTs-G, was used to determine Hypoxanthine (Hx) and Uric Acid (UA) concentrations in human serum and buffer sample solutions. The CNTs-G complexes used in the layers had the function of amplifying the biosensors signal and increasing the surface area for enzyme adsorption. The results showed a linear dynamic range from 5 to 50 μmol L^{-1} Hx with a LoD of 4.40 μmol L^{-1}. In addition, the sensor was evaluated for detection of Hx in typical human serum solutions in myocardial infarction (MI) patient serum sample solutions. The results showed a high potential application for this sensing platform in detecting Hx for clinical diagnosis with excellent detection and recovery results in human serum samples of the patients with MI (Si *et al* 2018).

Jaiswal *et al* (2019) modified a pencil graphite electrode (figure 12.2(b)) with rGO nanosheets functionalized by acrylamide printed D-serine and fullerene-C60 functionalized with acrylamide L-serine by the LbL technique. Thermal treatment was performed for both D- and L-serine to guarantee the effective functionalization, using heat treatment of 50 °C for 10 min and 60 °C for 3 h, respectively. Thus, double printing quantifies D- and L-serine at an ultra-trace level in aqueous and real samples. Furthermore, these D-serine and L-serine isomers have the same redox behavior, forming a double-layer printed molecular imprinted polymer (MIP) to solve the diffusion restriction compared to the single-layer printed MIP with better electronic and synergism properties. The modified electrodes with the configuration rGO D-serine/fullerene_L-serine displayed excellent results without limitations of regeneration, cross-reactivity, and false positives. Pulse anodic stripping and cyclic voltammetry analysis revealed a high sensitivity with a LoD 0.24 ng mL^{-1} for D- and L-serine isomers using a single electrode. This electrochemical sensor is very promising as a biomarker of various diseases at the ultra-trace level in clinical patients. A selective and sensitive electrochemical sensor based on LbL MXene-$Ti_3C_2T_x$ intercalated with amino CNTs (MXne/NH_2-CNTs) composite and molecularly imprinted polymer was developed for fisetin detection. The use of NH_2-CNTs as interlayer spacers in the LbL structure inhibited the aggregation of MXene flaxes and, as a result, increased the effective area and enhancement of the electrical conductivity and electrocatalytic activity for detecting fisetin. The electrochemical response showed an excellent linear relationship with the fisetin concentration with a LoD of 1.0 nmol L^{-1} (Ma *et al* 2020).

LbL films composed of kegging-type vanadium-substituted polyoxometalates (PMo_9V_3), graphene sheets, and $Ru(bpy)_3^{2+}$ were evaluated from electrochemical detection for ascorbic acid (AA) analyte in a concentration range of 7.5×10^{-7} to 2.07×10^{-4} mol L^{-1}. The amperometric and cyclic measurements reveal that introducing graphene and $Ru(bpy)_3^{2+}$ improves the electrocatalytic activity for AA catalysis leading to the LoD of 10 μmol L^{-1}. The sensor also displayed great potential for detecting ascorbic acid in real samples, such as commercial fruit juice (Guo *et al* 2020). Lin and Shen (2020) fabricated an electrochemical aptasensor for the detection of aflatoxin B1 (AFB1) by the LbL method using graphene nanosheets (GNs) functionalized with PDDA as positive charge and carboxylated polystyrene nanospheres (PS-COOH) as negative charge (figure 12.2(d)). The AFB1 detection occurs by an amide bond on the PS-COOH surface by a selective interaction with good stability, reproducibility, and selectivity with a LoD of 0.002 ng mL^{-1}. Recovery values were determined between 94.5%, and 103.3% detect AFB1 in oil and soy sauce. The high performance of this aptamer sensor is attributed to the pathway electron transport of GNs, and abundant sites to immobilize AFB1 on the PS-COOH surface. An electrochemical sensor for detecting a synthetic estrogen female hormone (17 α-ethinylestradiol, EE2) was fabricated using CNTs intercalated with chitosan from LbL technique (figure 12.2(a)). The electrochemical response shows a decrease of resistance in the LbL film with the increase of the number bilayer caused by the number of carbon nanotubes, and this effect improved significatively on the lower detection limit of 0.09 μmol L^{-1} (Pavinatto *et al* 2015).

The development of an efficient, selective, and highly sensitive electrochemical sensor for the simultaneous analysis of multiple drugs was reported by Kokab *et al* (2021). The authors fabricated an electrochemical sensing platform by acid- and base-functionalized CNTs with zinc oxide (ZnO) nanoparticles between the layers (COOH-CNTs/ZnO/NH_2-CNTs) to detect paracetamol, diclofenac, and orphenadrine drugs. The performance of COOH-CNTs/ZnO/NH_2-CNTs multilayers films onto a glassy carbon electrode (GCE) surface was probed by electrochemical impedance spectroscopy (EIS), cyclic voltammetry (CV), and square-wave anodic stripping voltammetry (SWASV). The functionalized porous architecture with a large effective surface area provided a conduit for efficient mass transport and active sites for anchoring adsorbate molecules, leading to substantially lower oxidation overpotentials and amplified current signals. The COOH-CNTs/ZnO/NH_2-CNTs exhibited a six-fold increase in active surface area over bare GCE. Under optimized SWASV conditions, the designed sensor demonstrated simultaneous detection of paracetamol, diclofenac, and orphenadrine with a LoD of 46.8, 78, and 60 fmol L^{-1}, respectively. Besides the specificity, stability, and reliability studies of the designed sensing platform in multiple biological and pharmaceutical samples with 96%–102% recoveries.

The mechanical properties of carbon-based nanomaterials have a significant attraction for obtaining flexible devices, and the LbL method is an excellent ally in the fabrication of flexible devices (Oh *et al* 2018, Dhanjai and Mugo, 2019, Park *et al* 2020). For instance, a flexible skin-attachable electrochemical sensor device for monitoring physiological and metabolic information was reported by Oh *et al* (2018). For this, a patterned stretchable was fabricated via LbL deposition of CNTs on top of Au electrodes to detect glucose and pH. $CoWO_4$/CNT and polyaniline/CNT nanocomposites were coated onto the CNT–Au electrodes, respectively. The electrochemical response showed high performance with sensitivities of 10.89 μA mmol L^{-1}–1 cm^{-2} and 71.44 mV pH −1 for glucose and pH, respectively, with mechanical stability up to 30% stretching and air stability for 10 days. The sensor also showed good adhesion even to wet skin, allowing the detection of glucose and pH in sweat from running while being attached to the skin. Dhanjai and Mugo (2019) fabricated a flexible electrochemical sensor for adrenaline detection in clinical analysis of sports medicine. For this, polydimethylsiloxane (PDMS) substrate was modified by LbL assembly of carbon nanotube-cellulose nanocrystals (CNT/CNC) as a first layer following the adrenaline imprinted of polyaniline-phenyboronic acid (PAni/PBA) as the second layer. The flexible electrochemical sensor shows selectivity adrenaline detection in real zebrafish brain samples with a LoD of 0.001 μmol L^{-1} and 15.7% reproducibility.

Park *et al* (2020) modified a gold electrode in a PET substrate with alternate multilayers of titanium carbide ($Ti_3C_2T_x$) and MWCNTs nanocomposites to detect copper and zinc (Cu and Zn) ions in sweat and urine samples as noninvasive smart monitoring for human health. LbL assembly allowed MWNTs into $Ti_3C_2T_x$ lamellar structure (figure 12.3), improved the aggregation issues of $Ti_3C_2T_x$ layers, and increased the interconnection between the $Ti_3C_2T_x$ layers. The authors interpreted that the combination of $Ti_3C_2T_x$ and MWNTs provides more sites for the efficient electron transfer and electrochemical reaction, whose $Ti_3C_2T_x$ has a primary function of adsorbed Cu and Zn at the same time MWNTs favors the electron transfer in the electrochemical reaction. The materials combination leads to a limit of detection (LoD) of 1.5 and 0.1 ppb for Cu and Zn, respectively. For real samples, the sensor's response has a detection limit of 20 and 300 ppb for Cu ions in urine and sweat samples and 200 and 500 ppb for Zn ions in urine and sweat samples.

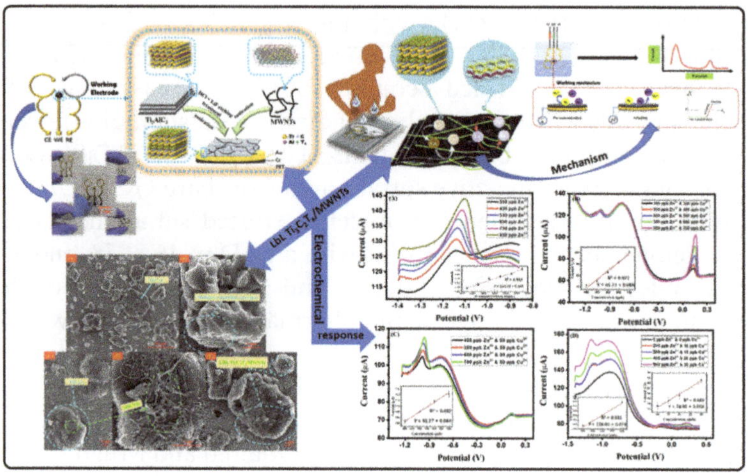

Figure 12.3. Flexible electrochemical sensor based on LbL thin film of $Ti_3C_2T_x$/MWNTs for simultaneous metals ions detection in human sweat. Reprinted (adapted) from Park *et al* (2020). Copyright (2020) American Chemical Society.

The detection of heavy metal ions and pesticides in drinking and tap water and soil is also an emergent field, where the electrochemical sensing platforms have an essential role in expanding the commercialization of this class of sensors. For instance, Jimenez *et al* (2021) developed a sensing platform for the detection of lead (II) (Pb (II)) in drinking water using polyethyleneimine-functionalized CNTs and GO onto gold interdigitated electrodes. As a result, the electrochemical response obtained by potentiometry stripping analysis shows a LoD of 36 pmol L^{-1} and a sensitivity of 4.3 μA μmol L^{-1}. The excellent performance of this sensor is associated with the high affinity of heavy metal ions with the functional groups present in the CNTs and GO. The determination of bromate ions (BrO_3^-) in tap and mineral water was achieved from structure blocks LbL assembly composed by oxidized MWCNT intercalated with poly(sodium 4-styrene sulfonate) (PSS). The amperometric response shows a LoD of 43 nmol L^{-1} (Lee *et al* 2017).

Rodrigues *et al* (2018) fabricated LbL thin films composed of GO reduced with PDDA and PSS (GPDDA and GPSS) in comparison with structure films composed of GPDDA/GPSS, including gold nanoparticles (AuNPs). The authors inserted AuNPs in GPDDA/GPSS multilayer structures to improve the catalytic property of the sensors, consequently the sensitivity and the detection limit decreased. As a result, the combination of GPDDA/GPSS used as polycation and polyanion promotes a similar or a few better detection limits than the LbL structure GPDDA/GPPS including AuNPs (figure 12.4). The sensors with AuNPs show a LOD of 16.52, 15.94, and 16.53 ppm of methyl parathion in tap water, soil, and cabbage, respectively. While the sensors without AuNPs have a LoD of 13.75, 16.91, 17.02 ppm for the same analytes. In conclusion, we can see that the rGO functionalized with different polyelectrolytes can reach a good multilayer combination and promote an excellent electrochemical property. In this way, LbL thin

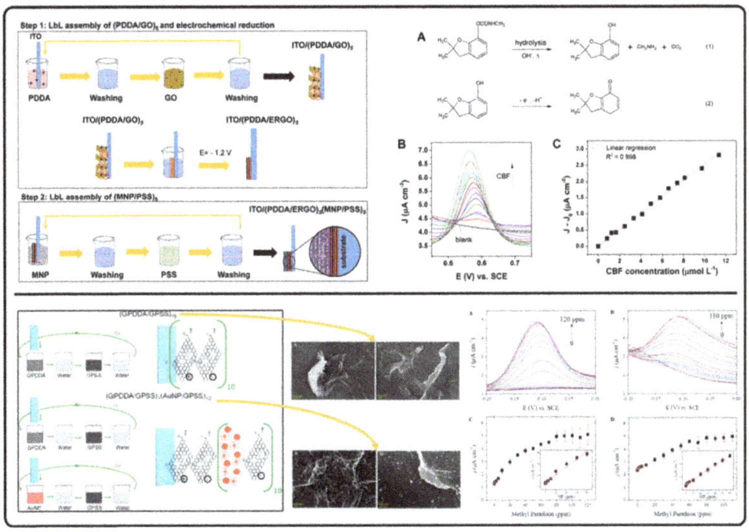

Figure 12.4. LbL graphene oxide films with magnetite nanoparticles for electrochemical detection of carbofuran pesticides in soil and tap water. Panel (a) reprinted (adapted) from Miyazaki *et al* (2020), with permission from Elsevier. Panel (b) reprinted (adapted) from Rodrigues *et al* (2019). Copyright (2019) American Chemical Society.

films composed of PDDA intercalated with GO and combined with magnetite nanoparticles were deposited onto ITO electrodes to detect carbofuran pesticides in soil and tap water. The GO was reduced in the substrate interface from electrochemical (ErGO, electrochemical reduced graphene oxide). For this, the authors deposited five bilayers of PDDA intercalated with GO, following the electrochemical reduction. In the second step, five additional bilayers of magnetite nanoparticles intercalated with poly(styrene sulfonate) (PSS) were deposited to detect carbofuran pesticides, as shown in figure 12.4. The electrochemical sensors offer a detection limit of 0.407 μmol L^{-1} for a laboratory sample. For the real samples, the sensors displayed a LOD of 5.51 and 5.76 and μmol L^{-1} for soil and tap water, respectively (Miyazaki *et al* 2020).

12.2.2 Electrochemical sensor modified by the Langmuir–Blodgett technique

The development of the electrochemical sensors by the LB technique has as its main advantage that the highly organized thin films and, consequently, has an excellent coating of modified electrodes (Siontorou *et al* 2017, Possarle *et al* 2020). The fabrication of monolayers composed of carbon nanomaterial using the LB technique is also interesting to promote an organization/orientation level of nanosheets or nanotubes (rGO, GO, SWCNTs, MWCNTs, GQDs, CQDs, and fullerenes) that lead to an increase in the conductivity layer (Wang *et al* 2016). From the point of view of biosensing application, this effect is a fascinating way to improve sensitivity and reproducibility and achieve a lower detection limit. The development of devices by the LB technique is still widely used for several fields' applications (De Barros *et al* 2015, 2016, Wang *et al* 2015, Kursunlu *et al* 2020). Here, we show the advances of electrochemical sensors modified with carbon-based nanomaterials. For instance, Liu *et al* (2006) produced monolayers of CNTs conjugates with cytochrome c (Cyt c) stabilized under the air–water interface, and CNTs-Cyt c monolayers were transferred to an ITO substrate by LB methods. The Langmuir isotherm reveals that the CNTs-Cyt c monolayer is highly stable, guaranteeing the excellent coating of the ITO electrode. The electrochemical results show three futures important for CNTs-Cyt modified ITO electrode: (i) a reversible redox process for Cyt c; (ii) a significative increase of current can be observed; (iii) a reduction potential for the Cyt c slightly increased compared to the pure Cyt c. The authors have associated these effects with the presence of CNTs that favor the electron transfer between the electrode surface and Cyt c. In this way, the electrochemical properties of SWCNTs conjugate and hydrogenase (H_{2ase}) monolayer are deposited onto ITO substrate from the LB technique. The results indicate the influence of SWCNTs directly related to H_{2ase} to promote electron transfer between the electrode surface and also active center of H_{2ase}, whose intensity current is increased and was proportional to the scan rate (υ) for the electrodes modified with the LB films of pure H_{2ase} and SWCNTs–H_{2ase} (Liu *et al* 2007).

Langmuir and LB ultrathin films using functionalized graphene with a carboxylate group (G-COOH) were studied to provide a better adsorption urea on monolayer composition film. The results indicate a strong interaction between the Langmuir monolayer of G-COOH and urea present in the aqueous subphase. The electrochemical sensing performance for the LB film of G-COOH with urea adsorbed was compared with a spin-coated film of G-COOH. As a result, the LB film could detect a urea concentration five times lower than spin-coated film, with a LoD of 8.3 μmol L^{-1} for LB G-COOH while the spin-coated film was of 41.6 μmol L^{-1}, as can be seen in figure 12.5(a–g). The sensing performance was also evaluated to detect urea in milk samples using the LB and spin-coating film of G-COOH. The sensitivity for these platforms in a real sample containing urea in milk was approximately five times better than spin-coating film (Poonia *et al* 2018). In this way, Rodrigues *et al* (2018) fabricated EIS devices for biosensing applications in the detection of urea. For this, the urease was incorporated in an algal polysaccharide and CNT matrix from LB deposition film onto p-Si-SiO_2 substrate, as illustrated in figure 12.5(h–k). The EIS results reveal that the presence of CNT tuned the catalytic activity of urease and also preserved the enzyme activity, exhibiting constant capacitance and distinctive output signal over all concentrations, with a LoD of 100 μmol L^{-1}.

The determination of daidzein compound is widely used in the treatment of cardio-cerebrovascular diseases, prevention of the onset of diabetes, and the alleviation of menopause symptoms was performed using an electrochemical sensor of GCE modified with MWCNTs and PAni. The MWCNTs functionalized with PAni were vertically arrayed on the electrode interface through the LB technique and compared by electrodes modified with randomly MWCNTs-PAni deposition. The voltammetric analysis indicates a lower detection limit for the MWCNTs-PAni LB film than random MWCNTs-PAni deposition, with a LoD of 8×10^{-8} mol L^{-1} for MWCNTs-PAni LB and stability of 97.5%. The authors attributed the efficiency

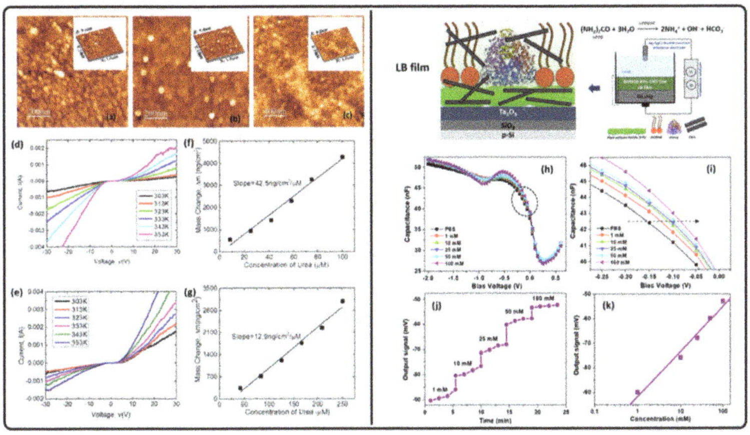

Figure 12.5. LB film of G-COOH with 4.7 μmol L^{-1} (a) 9.5 μmol L^{-1} (b) 95.1 μmol L^{-1} deposited onto Si/SiO_2. Current-voltage in different temperatures (d, e) and adsorption of urea (f, g) for LB and spin-coating film of G-COOH, respectively. LB film of CNTs, algal polysaccharides, and urease is used as EIS devices to determine urea in different concentrations, in (h, i) EIS measurements by capacitance versus voltage, and zoom for the specific region, respectively; (j, h) Curve for output signal for different concentrations and calibration curve, respectively. Reprinted (adapted) from Poonia *et al* (2018), with permission from Elsevier. Reprinted (adapted) with permission from Rodrigues *et al* (2018), copyright 2018 American Chemical Society.

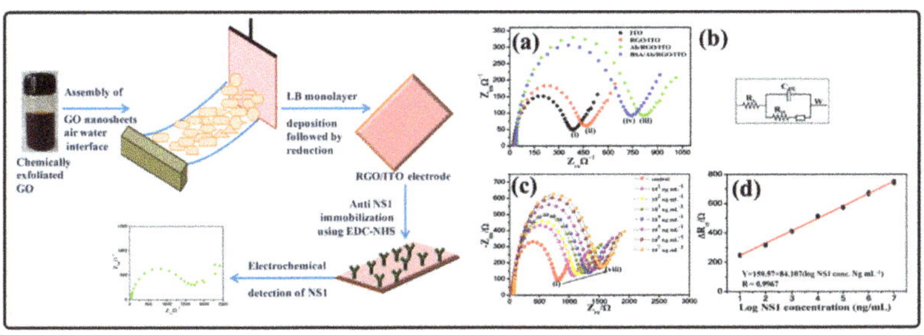

Figure 12.6. Schematic representation of ITO modification with rGO and NS1 antibodies through LB deposition film. (a–d) Electrochemical impedance spectroscopy for specific detection of NS1 antigen in standard and spiked sera sample comparison. Reprinted (adapted) with permission from Solanki *et al* (2021), copyright 2021 American Chemical Society.

of the sensor response to the highly ordered vertical alignment of MWCNTs-PAni, facilitating the electron transportation on the electrode interface (Wang *et al* 2016). An electrochemical sensor to diagnose dengue was developed by Solanki *et al* (2021) using rGO nanosheet coating ITO electrode through of LB technique. First, the performance of biosensing was evaluated by EIS measurements to NS1 protein quantification and subsequently for dengue detection, as shown in figure 12.6. For this, the LB film of GO was reduced by hydrazine vapor to obtain rGO-coated ITO electrode, and NS1 antibodies were grafted onto the ordered LB films using covalent linking. Then, the specific detection was performed by NS1 antigen in standard and spiked sera sample with a LoD of 0.069 ng mL^{-1} and 0.081 ng mL^{-1}, with a sensitivity of 8.41 and 36.75 Ω per ng mL, respectively.

In addition to the electrical properties for carbon-based nanomaterials favoring the electron transference in biosensor application, an attractive characteristic in the fabrication of sensors using biological materials is associated with the stability and temperature resistance provided by carbon-based nanomaterials. For instance, Abiko *et al* (2021) evaluated the temperature resistance for the Langmuir monolayer of stearyl phosphonate-modified SWCNTs (organo-SWCNTs) with different biomolecules adsorption, as trypsin enzyme, lysozyme enzyme, cytochrome C protein, and glucose oxidase, as summarized in figure 12.7. For this, the individual biomolecules were mixed in an aqueous subphase, and the SWCNTs spread on the air–water interface. The adsorption behavior on the SWCNTs surface was investigated in a high-temperature environment. The biomolecules presented in the aqueous subphase interact with the organo-SWCNT monolayers, forming a bioconjugated structure. These biomolecules adsorbed on organo-SWCNTs have an oriented design that maintains a secondary structure to 100°C–150°C. This characteristic can be fascinating to biosensor applications, where it is necessary to submit the sensors in a high-temperature environment. The stability and the enzymatic activity are also studied by Possarle *et al* (2020), incorporating

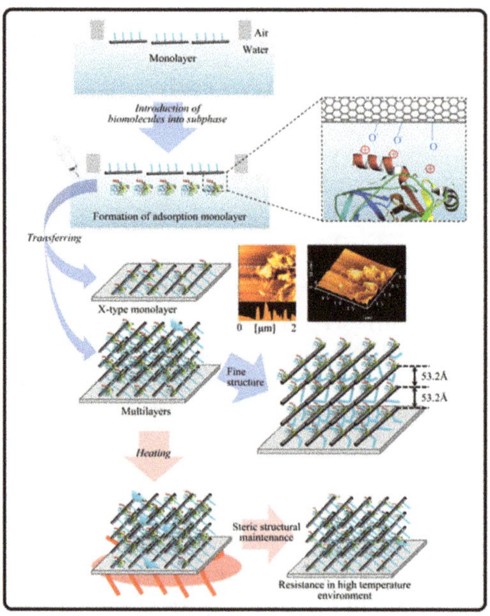

Figure 12.7. Summary of Langmuir organo-SWCNT monolayer to incorporate biomolecules presented in aqueous subphase and schematic diagram of the morphological formation of biomolecules adsorbed onto organo-SWMCT. A diagram showing the maintenance of the steric structure of SWCNT template adsorbed biomolecules during high-temperature exposure. Reprinted (adapted) from Abiiko *et al* (2021), with permission from Elsevier.

asparaginase in CNTs matrix mixed with stearic acid (SA) from the Langmuir monolayer method. The results reveal that asparaginase was adsorbed on SA-CNT monolayer and can be easily transferred to a solid support, and the enzymatic activity is dependent on the number of monolayers transferred to the solid support. These LB film designs are promising to act as a smart film based on lipid-CNT-enzyme hybrids, providing a favorable environment for enzyme conformation. The molecular arrangement facilitates the catalytic performance as specific units in optical devices and biosensors.

12.3 Future perspectives

The use of carbon-based nanomaterials in the development of sensors and biosensor technologies is very active due to their excellent electric and mechanical properties. This chapter has demonstrated the importance of the used methods for coating electrodes, such as LbL and LB techniques, to improve the acting of electric properties when the CNTs are highly aligned. In addition, the mechanical properties have are particularly attractive in the development of sensor technologies to apply flexible devices that can be easily extended to commercialization with benefit to society. Different synthesis processes and immobilization techniques have been explored to engineer devices with enhanced properties. In conclusion, the combination of carbon-based nanomaterials with these techniques guarantees an excellent electrode coating, promoting functionality, chemical and physical stability, and high sensitivity in devices.

References

Abiko Y *et al* 2021 Adsorption immobilization of biomolecules from subphase on Langmuir monolayers of organo-modified single-walled carbon nanotube *Colloids Surf., A* **621** 126559

Bard A J and Faulkner L R 2003 Electrochemical methods: fundamentals and applications *Anti-Corros. Methods Mater.* **50** 1–43

Asadzadeh-Firouzabadi A and Zare H R 2017 Application of cysteamine-capped gold nanoparticles for early detection of lung cancer-specific miRNA (miR-25) in human blood plasma *Anal. Methods* **9** 3852–61

De Barros A *et al* 2015 Synergy between polyaniline and OMt clay mineral in Langmuir–Blodgett films for the simultaneous detection of traces of metal ions *ACS Appl. Mater. Interfaces* **7** 6828–34

De Barros A *et al* 2016 Incorporation of gold nanoparticles into Langmuir–Blodgett films of polyaniline and montmorillonite for enhanced detection of metallic ions *Sens. Actuators* B **236** 408–17

Bezzon V D N *et al* 2019 Carbon nanostructure-based sensors: a brief review on recent advances *Adv. Mater. Sci. Eng.* **2019** 4293073

Brett C M A 2018 Perspectives and challenges for self-assembled layer-by-layer biosensor and biomaterial architectures *Curr. Opin. Electrochem.* **12** 21–6

Dhanjai Y N and Mugo S M 2019 A flexible-imprinted capacitive sensor for rapid detection of adrenaline *Talanta* **204** 602–6

Estevam R B *et al* 2015 Preparation and characterization of LbL films based on graphene oxide nanoparticles interacting with 3-n-propylpyridinium silsesquioxane chloride *Surf. Coat. Technol.* **275** 2–8

Guo D *et al* 2020 Electrochemical ascorbic acid sensor of composite film based on Keggin-type Vanadium-substituted Polyoxometalates decorated with graphene and Ru(bpy)32+ *Colloids Surf.,* A **592** 124550

Hareesha N *et al* 2019 Design of novel surfactant modified carbon nanotube paste electrochemical sensor for the sensitive investigation of tyrosine as a pharmaceutical drug *Adv. Pharm. Bull.* **9** 132

Hareesha N, Manjunatha J G and Manjunatha J G 2020 Elevated and rapid voltammetric sensing of riboflavin at poly(helianthin dye) blended carbon paste electrode with heterogeneous rate constant elucidation *J. Iran. Chem. Soc.* **17** 1507–19

Hareesha N and Manjunatha J G 2020a A simple and low-cost poly (dl-phenylalanine) modified carbon sensor for the improved electrochemical analysis of Riboflavin *J. Sci.: Adv. Mater. Devices* **5** 502–11

Hareesha N and Manjunatha J G 2020b Fast and enhanced electrochemical sensing of dopamine at cost-effective poly(DL-phenylalanine) based graphite electrode *J. Electroanal. Chem.* **878** 114533

Jaiswal S *et al* 2019 Enantioselective analysis of D- and L-serine on a layer-by-layer imprinted electrochemical sensor *Biosensors and Bioelectronics* (Elsevier B.V)

Jimenez M J M *et al* 2021 Polyethyleneimine-functionalized carbon nanotube/graphene oxide composite: a novel sensing platform for Pb(II) acetate in aqueous solution *ACS Omega* **6** 18190–9

Kamran U *et al* 2019 Functionalized carbon materials for electronic devices: a review *Micromachines* **10** 23410

Kim H S *et al* 2019 Carbon nanotubes immobilized on gold electrode as an electrochemical humidity sensor *Sens. Actuators, B: Chem.* **300** 127049

Kokab T *et al* 2021 Simultaneous femtomolar detection of paracetamol, diclofenac, and orphenadrine using a carbon nanotube/zinc oxide nanoparticle-based electrochemical sensor *ACS Appl. Nano Mater.* **2021** 4699–712

Kumar S *et al* 2019 Electrochemical sensors and biosensors based on graphene functionalized with metal oxide nanostructures for healthcare applications *ChemistrySelect* **4** 5322–37

Kursunlu A N *et al* 2020 Fabrication of LB thin film of pillar[5]arene-2-amino-3-hydroxypyridine for the sensing of vapors *Mater. Lett.* **267** 127538

Lahcen A A *et al* 2020 Electrochemical sensors and biosensors using laser-derived graphene: a comprehensive review *Biosens. Bioelectron.* **168** 112565

Lee Y G, Lee H J and Jang A 2017 Amperometric bromate-sensitive sensor via layer-by-layer assembling of metalloporphyrin and polyelectrolytes on carbon nanotubes modified surfaces *Sens. Actuators, B: Chem.* **244** 157–66

Lin T and Shen Y 2020 Fabricating electrochemical aptasensors for detecting aflatoxin B1 via layer-by-layer self-assembly *J. Electroanal. Chem.* **870** 114247

Liu A R *et al* 2006 Monolayers, Langmuir–Blodgett films of carbon nanotubes-cytochrome c conjugates and electrochemistry *Colloids Surf.,* A **284–285** 485–89

Liu A R *et al* 2007 Electrochemical properties of carbon nanotubes–hydrogenase conjugates Langmuir–Blodgett films *Electrochim. Acta* **52** 3222–8

Ma X *et al* 2020 Hierarchical porous MXene/amino carbon nanotubes-based molecular imprinting sensor for highly sensitive and selective sensing of fisetin *Sens. Actuators, B: Chem.* **309** 127815

Manjunatha J G, Kumara Swamy B E and Deraman M 2013 Electrochemical studies of dopamine, ascorbic acid and their simultaneous determination at a poly (rosaniline) modified carbon paste electrode: a cyclic voltammetric study *Anal. Bioanal. Electrochem* **5** 426–38

Miyazaki C M *et al* 2020 Combining electrochemically reduced graphene oxide and Layer-by-Layer films of magnetite nanoparticles for carbofuran detection *J. Environ. Chem. Eng.* **8** 104294

Nemčeková K and Labuda J 2021 Advanced materials-integrated electrochemical sensors as promising medical diagnostics tools: a review *Mater. Sci. Eng.,* C **120** 111751

Oh S Y *et al* 2018 Skin-attachable, stretchable electrochemical sweat sensor for glucose and pH detection *ACS Appl. Mater. Interfaces* **2018** 13729–40

De Oliveira R F, De Barros A and Ferreira M 2017 *Nanostructured Films: Langmuir–Blodgett (LB) and Layer-by-Layer (LbL) techniques* pp 105–23

Park J Y *et al* 2020 High-performance flexible electrochemical heavy metal sensor based on layer-by-layer assembly of Ti3C2Tx/MWNTs nanocomposites for noninvasive detection of copper and zinc ions in human biofluids *ACS Appl. Mater. Interfaces* **12** 48928–37

Pavinatto A *et al* 2015 Layer-by-layer assembled films of chitosan and multi-walled carbon nanotubes for the electrochemical detection of 17α-ethinylestradiol *J. Electroanal. Chem.* **755** 215–20

Poonia M, Manjuladevi V and Gupta R K 2018 Ultrathin film of carboxylated graphene at air-water and air-solid interfaces *Surf. Interfaces* **13** 37–45

Possarle L H R R, Siqueira Junior J R and Caseli L 2020 Insertion of carbon nanotubes in Langmuir–Blodgett films of stearic acid and asparaginase enhancing the catalytic performance *Colloids Surface* B **192** 111032

Raril C and Manjunatha J G 2018 Sensitive electrochemical analysis of resorcinol using polymer modified carbon paste electrode: a cyclic voltammetric study *Anal. Bioanal. Electrochem* **10** 488–98

Richardson J J, Björnmalm M and Caruso F 2015 Technology-driven layer-by-layer assembly of nanofilms *Science* **348** aa2491

Rodrigues G H S *et al* 2019 Layer-by-layer films of graphene nanoplatelets and gold nanoparticles for methyl parathion sensing *ACS Appl. Nano Mater.* **2** 1082–91

Rodrigues R T *et al* 2018 Carbon nanotubes and algal polysaccharides to enhance the enzymatic properties of urease in lipid Langmuir–Blodgett films *Langmuir* **2018** 3082–93

Si Y *et al* 2018 Layer-by-layer electrochemical biosensors configuring xanthine oxidase and carbon nanotubes/graphene complexes for hypoxanthine and uric acid in human serum solutions *Biosens. Bioelectron.* **121** 265–71

Siontorou C G *et al* 2017 Artificial lipid membranes: past, present, and future *Membranes* **7** 387

Solanki S *et al* 2021 Ultrasensitive immunosensor based on Langmuir–Blodgett deposited ordered graphene assemblies for dengue detection *Langmuir* **37** 8705–13

Speranza G 2019 The Role of functionalization in the applications of carbon materials: an overview *C–J. Carbon Res.* **5** 84

Speranza G 2021 Carbon nanomaterials: synthesis, functionalization and sensing applications *Nanomaterials 2021* **11** 967

Sun Y *et al* 2016 Enzyme-free and sensitive electrochemical determination of the FLT3 gene based on a dual signal amplified strategy: vontrolled nanomaterial multilayers and a target-catalyzed hairpin assembly *Biosens. Bioelectron.* **78** 7–13

Tigari G *et al* 2019 Determination of riboflavin at carbon nanotube paste electrodes modified with an anionic surfactant *ChemistrySelect* **4** 2168–173

Tigari G and Manjunatha J G 2020 A surfactant enhanced novel pencil graphite and carbon nanotube composite paste material as an effective electrochemical sensor for determination of riboflavin *J. Sci.: Adv. Mater. Devices* **5** 56–64

Wang J *et al* 2018 Ultrasensitive electrochemical detection of tumor cells based on multiple layer CdS quantum dots-functionalized polystyrene microspheres and graphene oxide–polyaniline composite *Biosens. Bioelectron.* **100** 1–7

Wang L *et al* 2015 A new strategy for enhancing electrochemical sensing from MWCNTs modified electrode with Langmuir–Blodgett film and used in determination of methylparaben *Sens. Actuators, B: Chem.* **211** 332–8

Wang L *et al* 2016 The construction of well-aligned MWCNTs-PANI Langmuir–Blodgett film modified glassy carbon electrode and its analytical application *Sens. Actuators, B: Chem.* **228** 214–20

Zhao S *et al* 2019 The future of layer-by-layer assembly: a tribute to ACS Nano Associate Editor Helmuth Mohwald *ACS Nano* **13** 6151–69

Chapter 13

Development of electrochemical sensors for the analysis of herbicides

Nida Aydogdu and Ersin Demir

In recent years, agricultural areas have been decreasing rapidly due to urbanization, industrialization, and for economic reasons. Therefore, pesticides have been widely used to obtain high yields of crops per hectare. Herbicides are usually organic substances used to control weeds. Reliable, fast, portable, and inexpensive analytical methods are needed for the analysis of herbicide residue in the harvest. Electrochemical methods have become popular in herbicide determinations, together with the newly discovered generation sensors in the last three decades. In particular, electrochemistry, which is a cheap, fast, portable, and reliable method, and the nanomaterial sensor platform, which allows the determination of numerous substances have gained great interest in analytical methods. In addition, the sensor platform in electrochemical methods such as voltammetry, potentiometry, amperometry, and polarography has been preferred for the analysis of maximum residue limits (MRL) values in herbicides. In this chapter, the electrochemical methods used for the determination of herbicides, the sensor platform, and the validation parameters were examined one by one. For this, herbicide analyses were examined in the last five years based on the 'web of science' database. Approximately 170 studies were summarized in detail according to the sensor type, working range, method, detection limit, and interference agent, etc.

13.1 Introduction

The most valid definition for pesticides has been made by the Food and Agriculture Organization of the United Nations (FAO). Briefly and clearly, pesticides can be defined as substances or mixtures used to prevent, control, or minimize harmful microorganisms [1]. Especially due to the acceleration of industrialization and urbanization, it causes a decrease in agricultural areas. This means that pesticide use is indispensable to obtain a more efficient product per unit of area. Moreover,

doi:10.1088/978-0-7503-5127-0ch13

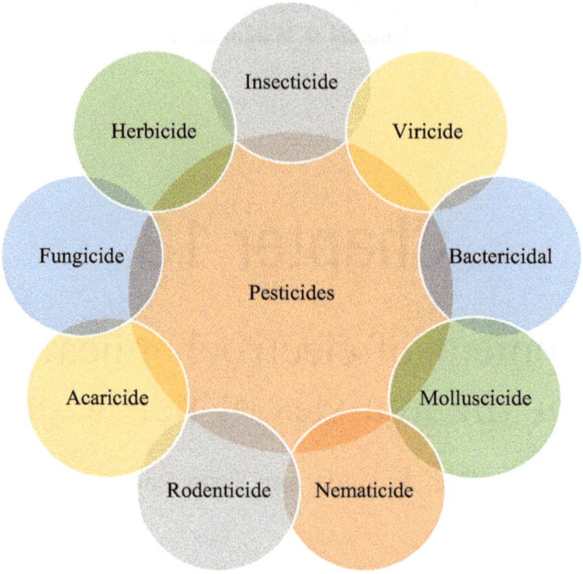

Figure 13.1. Classification of pesticides according to their intended use.

pesticides are essential chemicals to be used to destroy or minimize the damage of harmful weeds. By using these substances, the aim is to increase production, increase quality, and increase economic return. They have been used as the most important factor in increasing agricultural production since the 1940s. However, pesticides that enter the atmosphere, water, and soil, enter the food chain and affect species outside the area where they are used. Therefore, rapid, accurate, inexpensive, and reliable analytical methods are needed for the sensitive determination of pesticide residues. Pesticides can be classified as acaridae, fungicide, herbicide, insecticide, molluscicide, repellent rodenticide, and others in terms of usage areas (figure 13.1).

The term herbicide is derived from the Latin word for plant (herb-) and is the general name given to pesticides used to control weeds. Herbicides are substances that have the ability to kill or damage weeds. These substances can be produced both naturally and synthetically. Natural herbicides are known as allelochemicals. Recognition of herbicides goes back many years. It is known that the ashes of conquered places were used as a non-selective herbicide in 1200 BC [1]. The use of the first inorganic herbicides dates back to the end of the 1800s. The $CuSO_4$, $FeSO_4$, H_2SO_4, and $NaClO_3$ as inorganic herbicides played an important role in the control of weeds [2]. The first organic herbicide 'ammonium sulfamate' was developed by Dupond in the 1940s, during the Second World War. On the other hand, another widely used and early organic pesticide is 2,4-Dichlorophenoxy acetic acid (2,4-D) [3]. Especially in weed control, it has been the preferred herbicide since the 1940s [4]. Since the discovery of organic herbicides, numerous pesticides have been produced. It is only possible to classify these herbicides according to their functional groups (chemical structure) (figure 13.2). The best classification according to the chemical

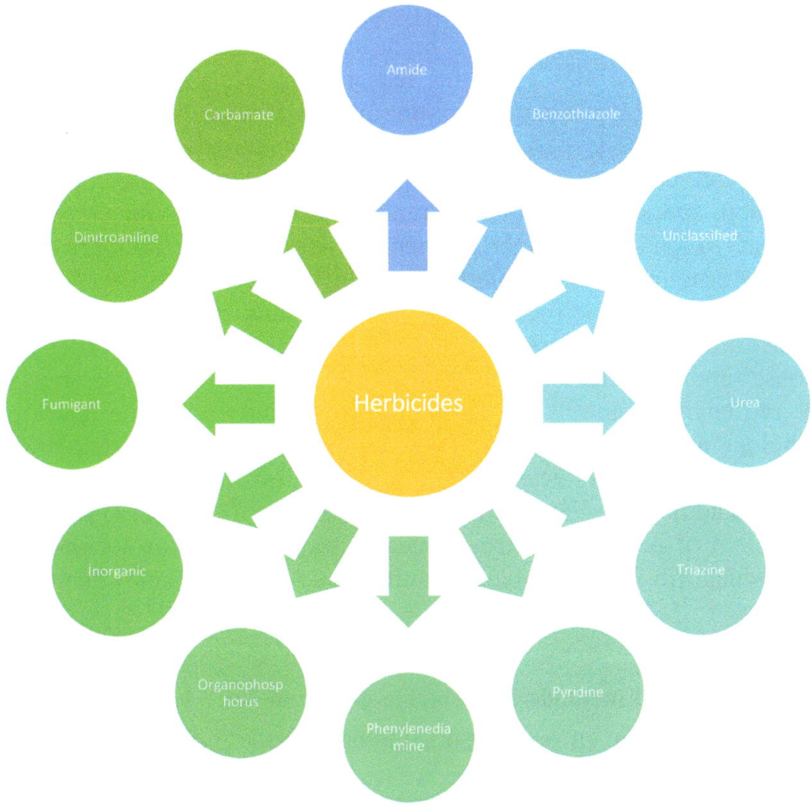

Figure 13.2. Classification of common herbicides according to their chemical structure.

structures of herbicides was made according to ISO/TC 81 and it is known that there are about 600 herbicides under 44 in the main functional groups [5]. In addition, there are some herbicides under the name of unclassified in this classification.

While most herbicides are not carcinogenic, excessive consumption and unconscious use play an important role in human health. Over and above, they have very important uses for both for crop yield and biological warfare and can be harmful due to their toxic properties. Moreover, as a result of various transports, they can be dangerous not only for living things but also in the environment. In addition, even very low amounts of herbicides, which are the most widely used pesticides, can be carcinogenic. Therefore, the maximum residue limit (MRL) values of herbicides are of great importance. For both people's access to reliable food sources and environmental health, the MRL value should be determined with accurate, reliable, and rapid analytical methods.

According to the literature review, modern analytical methods and advanced instrumental devices are widely used for herbicide residue quantification. These traditional methods are high-performance liquid chromatography (HPLC) and gas chromatography (GC), electrophoresis, mass spectroscopy (MS), spectrophotometric, and a combination of two or more techniques. Common features of these methods are low detection limits, high precision, and accuracy, wide operating range, and good repeatability. For this reason, they are the most used analytical methods and attract great attention in routine analysis and quality control laboratories. These advanced methods have some drawbacks as well as extraordinary features. These shortcomings can be given as long pre-processing, long analysis time, expensive instrumentation, specialist analysts, too much organic solvent consumption, and no green chemistry. The cheaper, faster, and reliable alternative methods are being sought for the residual amount analysis of herbicides, which have a significant effect on human health and access to reliable food sources for living things. Electrochemical analysis methods have attracted great interest from scientists in the last three decades due to their fast response time, extreme cheapness, no requirement for pre-processing, selective, accurate, and environmentally friendly [6–11]. Moreover, with the development of nanotechnology and the production of three-dimensional (3D) materials, the use of many nanosensors from the determination of numerous herbicides has gained great momentum. In addition to these extraordinary features, the design of microchips and the development of portable electrochemical detectors enable an on-site analysis of herbicides, which are indispensable for human and environmental health and food safety.

13.1.1 Electrochemical methods in herbicide detection

The literature review of all pesticides included in the classification of herbicides under approximately 40 main headings was carried out based on the last five years. Based on the recent studies for herbicide determination, voltammetric studies have been commonly used. Other electrochemical methods can be listed as potentiometry, amperometry, polarography, photoelectrochemical (PEC) biosensor. The most commonly used analytical methods in electrochemical sensor studies for herbicide determinations made so far are summarized in figure 13.3.

Among these electroanalytical methods, voltammetry techniques are the most preferred in herbicide analyses in natural and drug formulation samples. Especially since they are fast, accurate, reliable, and sensitive techniques, they allow the determination of not only herbicides but also medical drugs, metals, nucleic acids, vitamins, and numerous electro-active substances [12–19]. Among the voltammetric techniques, cyclic voltammetry (CV), differential pulse voltammetry (DPV), square wave voltammetry (SWV), and linear sweep voltammetry (LSV) are the most frequently used techniques.

13.1.1.1 Electrochemical sensor platform for the analysis of herbicides

In this section, the sensor platform in the voltammetric method which is the most widely used in herbicide determination with electroanalytical was first examined. The main indicator sensors most used in these studies are: gold electrode (AuE), glassy carbon electrode (GCE), carbon paste electrode (CPE), bismuth electrode (BiE), multi-walled carbon nanotube paste electrode (MWCNTPE), boron-doped diamond electrode (BDDE), pencil graphite electrode (PGE), screen-printed

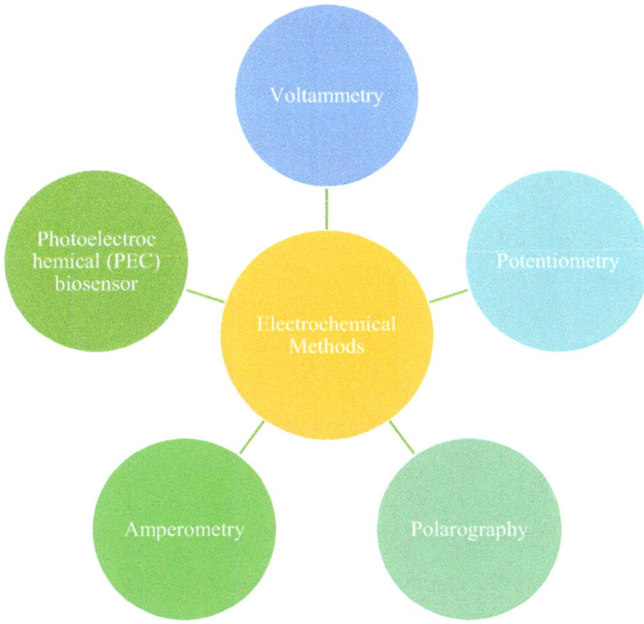

Figure 13.3. Common electrochemical methods used in herbicide detection.

electrode (SPE), and indium tin oxide (ITO). While these sensors are sometimes used alone, they are sometimes used in composites with nanoparticles, polymers, reduced graphene oxides (rGO), and metal-organic frameworks (MOF). In addition, it is understood that the mercury electrode is frequently used in reducible herbicide determinations. Supplementary to these sensors, thin films, bismuth, ITO, and antimony electrodes have been used rarely for the analysis of herbicides by voltammetric techniques. While working sensors with mercury, bismuth, thin-film, and carbonaceous are frequently used in herbicide determinations with cathodic peaks, BDDE, Au, ITO, SPE, carbonaceous, and modified composite sensors are preferred in the analysis of herbicides exhibiting anodic behavior (figure 13.4).

In voltammetric studies carried out with bare electrodes, the limits of quantification (LOQ) of herbicide were found to be approximately μM or μg L^{-1}. In particular, the LOQ values of composite electrodes developed for materials showing both conductivity and catalytic properties are reduced to lower (trace amount) levels. These studies are further evidence that voltammetric methods are suitable for residue quantification of herbicides by using bare or modified sensors. Also, the most widely used material from the sensor platform for the electrochemical determination of herbicides is carbon electrodes. Carbon sensors are preferred by scientists because of their excellent near conductivity, outstanding thermal and mechanical properties, and excellent catalytic effect. These materials include GCE, CPE, MWCNTPE, SPE, SWSVNT, BDDE, PGE, and graphene indicator sensors. In addition to carbonaceous sensors, hybrid sensors are of great interest due to the ease of preparation of composite sensors with many substances with catalytic and

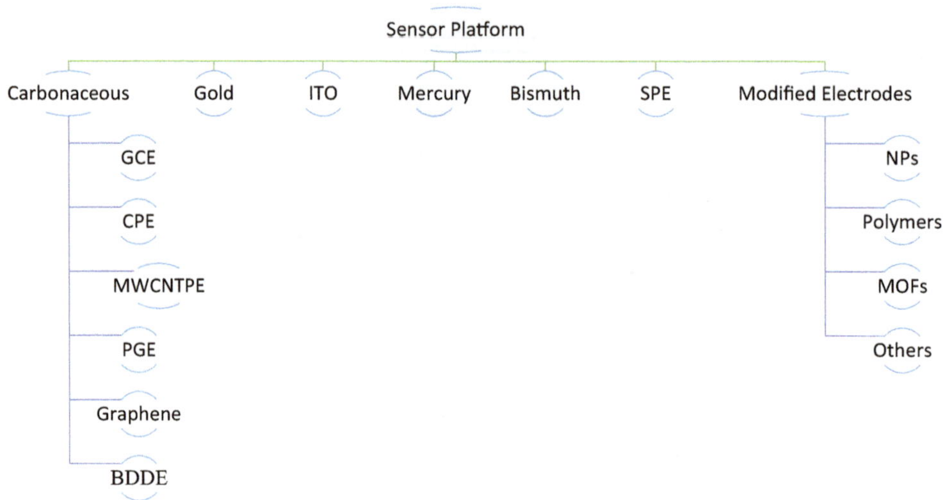

Figure 13.4. Sensor platform used in electrochemical methods for herbicide analysis.

conductive properties. While these materials are occasionally combined carbons, hybrid materials such as metal oxide nanoparticles, polymers, phthalocyanines, clays, and MOFs have also been manufactured. As a result, carbonaceous nanomaterial-based sensors with unique properties provide the opportunity for selective, reproducible, sensitive, and reliable herbicide determinations.

In the study in 2019 carried out by Demir, the carbon paste electrode (CPE), glassy carbon electrode (GCE), and composite sensor modified as α-Fe_2O_3 (hematite) nanoparticles/CPE was fabricated for desmedipham herbicide determination [20]. The α-Fe_2O_3/CPE sensor increases the desmedipham anodic peak current approximately 3.5 times compared to plain CPE, while this rate is 7.0 times compared to GCE. Also, using the composite sensor, desmedipham exhibited the separation of the two anodic peaks as well-defined. Then, the sensor platform performed the morphology illumination study with scanning electron microscopy (SEM) and energy dispersive x-ray analysis (EDX). With the new generation sensor developed, the working range was to be 0.15–4.50 mg L^{-1}, and the limits of detection (LOD) were found as 41.00 μg L^{-1}. The analysis of desmedipham at high recovery was carried in natural samples such as strawberry fruit juice with a spiked method using square wave stripping voltammetry (SWSV) on the α-Fe_2O_3/CPE sensor. Furthermore, for the detection of desmedipham herbicide, selectivity studies were performed in the presence of coexisting substances with both the developed sensor and the recommended electrochemical method (table 13.1). Consequently, the electrochemical behavior, mechanism elucidation, and analytical study of desmedipham herbicide have been accomplished successfully with a fast, reliable, and sensitive sensor [20].

Zhou *et al* in their recent study, created a synergistic effect by synthesizing oryzalin (ORZ) with the as prepared oxidized multi-walled carbon nanotubes

Table 13.1. Carbonaceous sensors used for herbicide analysis.

Analyte	Electrode	Method	Linear range	LOD	LOQ	Samples	Optimum pH	Peak potentials	İnterference	References
Amitrole	TiNP-fMWCNT-CPE	CV	0.1 to 1.5 μg mL^{-1}	0.727 ng mL^{-1}	2.42 ng mL^{-1}	Water, food, human blood plasma, serum, urine	pH 4.5 (Acetate buffer)	+0.786 V		[21]
Amitrole	TiNP-fMWCNT-CPE	DP-AdSV	0.01 to 1.3 μg mL^{-1}	0.267 ng mL^{-1}	0.891 ng mL^{-1}	Water, food, human blood plasma, serum, urine	pH 4.5 (Acetate buffer)	+0.775 V		[21]
Amitrole	TiNP-fMWCNT-CPE	LS-AdV	0.01 to 2.0 μg mL^{-1}	0.166 ng mL^{-1}	0.55 ng mL^{-1}	Water, food, human blood plasma, serum, urine	pH 4.5 (Acetate buffer)	+0.775 V	İons, organic compound	[21]
Amitrole	Fe. WO3/CTAB-CPE	SWV	0.8 to 500 μM	0.82 nM	2.82 nM	Soil, water	pH 5.0 (PBS)	+1.06 V	$FeSO_4$, $CaCl_2$, $ZnCl_2$, $MgCl_2KCl$, $MnSO_4$urea, starch, sscorbic acid, diuron, 2,4-D	[22]
Amitrole	WO3·0.33H_2O/CPE	SWV	1.0×10^{-8} M to 24 × 10^{-5} M	2.33 nM	7.8 nM	Soil, water	pH 5.0 (PBS)	+1.08 V	$CaCl_2$, $NaClCuSO_4$, $FeSO_4KNO_3$	[23]
Atrazine	CuONPs/ILs/CPE	SWV	0 to 2 nM	2.0×10^{-12} M	1.0×10^{-11} M	Waste water	pH 7.0 (PBS)	+0.1 V		[24]
Atrazine	Cellulose/PGE	SWV	5.0 ng mL^{-1} to 320 ng mL^{-1}	0.008 ng mL^{-1}	7.6 ng mL^{-1}	Water	pH 4.3 (BRB)	−0.9 V	K^+, Na^+, $Zn^{2+}$$Mg^{2+}$, Fe^{3+} trichlopyr, alachlor, diuron	[25]
Bromacil	Thylakoid-Modified Carbon paper Electrode	Amperometry	0 to 16 mg L^{-1}	0.21 mg L^{-1}		Water	pH 7.4 (PBS)			[26]
Chloridazon	MIP-CNT/CPE	DPV	5.0×10^{-7} to 4.0×10^{-4} M	6.2×10^{-8} M		Water	Ph 3.0 (BRB)	−0.6 V	Carbendazi, metamitron	[27]
Cyanazine	MIP-CPE	DP-CSV	5.0 to 1000 nM	3.2 nM	10.7 nM	Onion, tomato, lettuce, rice	pH 2.7	−0.6 to −0.8 V	Propazine	[28]
Cyanazine	MIP–CPE	DPSV	0.05 to 9.0 μM	0.010 μM		Foodstuffs, River water	pH 2.40	−0.45 V		[29]
Desmedipham	α-Fe_2O_3/CPE	SWSV	0.15 to 1.20 mg L^{-1} and 1.20 to 4.50 mg L^{-1}	41.00 μg L^{-1} and 50.00 μg L^{-1}	137.00 μg L^{-1} and 167.00 μg L^{-1}	Strawberry juices	pH 1.4 (H_2SO_4)	+1.1 and +1.2 V	Rizolex, fluometuron, teflubenzuron, Fe (III), Cr(III)Cu (II), Ni(II)Zn (II)	[20]
Dinitramine	PLC-AuNPs/PGE	DP-AdSV	1.0×10^{-9} to 1.0×10^{-8} M and 1.0×10^{-8} to 5.0×10^{-6} M	2.0×10^{-10} M		Tap (drinking) water, River water	pH 7.0 (BRB)	+1.18 V	Ba^{2+}, Fe^{3+}, $Hg^{2+}$$P^{2+}$, Mg^{2+}, $Zn^{2+}$$Na^+$, Ca^{2+} pymetrozin, pirimicarb, propanil, vhloridazon2, 4-dichlorophenoxy, acetic acid, carboxin, trifluralin	[30]

(Continued)

Table 13.1. (*Continued*)

Analyte	Electrode	Method	Linear range	LOD	LOQ	Samples	Optimum pH	Peak potentials	Interference	References
Diquat	CPE-C	CV	1×10^{-5} to 5×10^{-5} M	5.33×10^{-8} M	1.77×10^{-7} M	River water	pH 3.0 (K_2SO_4)			[31]
Fomesafen	CNTPE	SWV	0.30 to 40 mg L^{-1}	0.089 mg L^{-1}	0.297 mg L^{-1}	Apricot juice, cherry juice, lake water	pH 3.0 (BRB)	−0.5 V	Anilazine, pymetrozine, triflumizole, cobalt (III), lead (II), magnesium (II), nickel (II)	[32]
Fluometuron	FePc/MWCNTPE	DPSV	0.4 to 15.0 mg L^{-1}	69.8 μg L^{-1}	233.0 μg L^{-1}	Tap water, commercial formulation	pH 6.0 (BRB)	+0.98 V	Captan, halosulfuron methyl, monocrotophos, pencycuron, tolclofos-methyl, teflubenzuron, Cu (II), Fe (II)Pb (II), Zn (II)	[33]
Fluometuron	FePc/MWCNTPE	SWSV	0.4 to 7.5 mg L^{-1}	101.0 μg L^{-1}	337.0 μg L^{-1}		pH 6.0 (BRB)			[33]
Glyphosate	Sg-OS/EG/CPE	SWV	10 to 100 μM	0.98 μM		Soil	pH 6.8 (PBS)		Ca^{2+}, $Mg^{2+}Cu^{2+}$	[34]
Glyphosate	CPE	SW-AdSV	2×10^{-6} to 2×10^{-4} M	1.46 μM	4.88 μM	Commercial formulation	pH 6.5 (PBS)	−0.8 V	Isopropylamine	[35]
Glyphosate	CuO/MWCNTs–IL/HF-PGE	DPV	5 nM to 1.1 μM	1.3 nM	4.3 nM	Soil, water	pH 7.0 (PBS)	+0.6 V	Zn^{2+}, Cd^{2+}, $Ca^{2+}Mg^{2+}$, Na^{+}, $NH_4^{+}Br^{-}$, $NO_3^{2-}SO_4^{2-}$, PO_4^{2-}, glufosinate, bialaphos, tridemorph, chlorpyrifos, cypermethrin, phosphonic acid	[36]
Hexazinone	MIP/CPE	DPAdCSV	1.9×10^{-11} to 1.1×10^{-10} M	2.6×10^{-12} M	8.7×10^{-12} M	River water	pH 2. 5 (HCl)	−1.3 V		[37]
Imazethapyr	AgNP-CuNP-fMWCNT-PGE	CV	0.1 to 2.0 μg mL^{-1}	0.527 ng mL^{-1}	1.42 ng mL^{-1}		pH 4.5 (acetate buffer)			[38]
Imazethapyr	AgNP-CuNP-fMWCNT-PGE	SW-AdV	0.01 to 5.0 μg mL^{-1}	0.159 ng mL^{-1}	0.534 ng mL^{-1}	Water, soybean, peas, groundnut, tomato, human blood plasma, serum, urine	pH 4.5 (acetate buffer)		Ions and organic compounds	[38]
Molinate	ZnO/CPE	SWV	0.002 to 0.25 μM	1.03 $\times 10^{-8}$M	3.35 $\times 10^{-8}$M	Soil, water	pH 3.0 (PBS)	+1.2 V	Metal ions	[39]
Naptalam	CPE	SWV	0.6 to 1.0 μM and 1.0 to 10.0 μM	110 nM	360 nM		0.1 M H_2SO_4			[40]
Naptalam	TRGOPE	SWV	0.1 to 1.0 μM and 1.0 to 6.0 μM	15 nM	51 nM	River water	0.1 M H_2SO_4	0.95 V	Boscalid, fenfuram, metobromuron, novaluron, tebufenpyrad, dicamba, Zn^{2+}, Pb^{2+}, Cu^{2+}	[40]

Oxadiargyl	ADPE/MGRPE	DPV	0.03 to 1.4 mg L^{-1}	1.3 mg L^{-1}		River water, soil, rice	pH 3.0 (PBS)			[41]
Phenmedipham	MWCNT/CPE	SWV	0.02 to 2.0 mg L^{-1}	6.96 μg L^{-1}	23.2 μg L^{-1}	Saturated tea sugar	pH 2.0 (BRB)	+1.3 V	Carbendazim, benomyl, aclonifen, ethofumesat, metamitron, P-acetanisidide (methacetin)	[42]
Propazine	MIP–CPE	DPSV	0.01 to 1.00 μM	0.001 μM		Foodstuffs, River water	pH 2.34	−0.44 V		[29]
Propazine	MIP/CPE	CSV	0.01 to 1.0 μM and 1.0 to 55.0 μM	0.001 μM	0.003 μM	Onion, tomato, lettuce, rice, ater	pH 3.0 (HCl)	−0.5 V	Cyanazine	[43]
Propham	12.5FS/MH@FS/CPE	DPV	0.01 to 5.0 μM	4.4 nM		Potato, urine, River water	pH 6.0 (PBS)	+0.85 V and + 1.04 V	Carbofuran, carbaryl, silverdiethyl-dithiocarbamate, fenoxaprop-p-ethyl, glucose, ascorbic acid, citric acid, uric acid, urea, calcium chloride, phenol, nitrophenol	[44]
Propham	p(Thp)-Ox/PGE	DPV	0.005 to 1.0 μM and 2.0 to 15.0 μM	1.0 nM		Potato, urine, River water	pH 2.0 (PBS)	+1.14 V	Carbofuran, carbaryl, phenol	[45]
Quinclorac	Zn/Al-LDH-quinclorac/MWCNT/CPE	SWV	0.03 to 0.70 and 1.00 to 10.0 and30.0 to 300 μM	4.4 nM		Baby milk, mineral water	0.1 mM BPA	+0.4 V	Al^{3+}, Cu^{2+}, $K^{+}Mg^{2+}$, Na^{+}, $Ni^{2+}Zn^{2+}$, $NO_3^{-}SO_4^{2-}$ Dopamine, acetaminophenethanol	[46]
Thiobencarb	EDPE/CPE	DPV	0.96 to 106 μg L^{-1}	0.025 μg L^{-1}		Water	pH 3.0 (acetate buffer)	+0.15 V		[47]
Trifluralin	Graphite-MWCNT/CPE	SWV	$1.0 \times 10^{-8}$to 8.0×10^{-6} M	3 nM	10 nM	Water	pH 3.0 (BRB)	−0.1 V to +0.5 V	Fe^{2+}, Al^{3+}, $Pb^{2+}Zn^{2+}$, Co^{2+}, $Ni^{2+}NO_3^{-}$, $CO_3^{2-}S^{2-}$, SO_4^{2-}	[48]
Trifluralin	PGE	DPV	1.0 to 75.0 μM (ox) and 0.50 to 100.0 μM (red)	0.39 μM (ox) 0.20 μM (red)		Water	pH 4.0 (BRB)	0.1 V (ox) and 0.6 V (red)	Ca^{2+}, Mg^{2+}, $S^{2-}Na^{+}$, $NO_3^{-}CO_3^{2-}$, SO_4^{2-}, carbendazim, nitrobenzen, atrazine, phenol, hydroquinone, monolinuron,, thiamethoxam	[49]

(O-MWNTs) with CoMoS hydrothermal method for herbicide analysis. Then they built a CoMoS/O-MWNTs/GCE hybrid sensor by coating the GCE with the drop-dry technique. They applied to scan electron microscopy (SEM) and transmission electron microscope (TEM) analyses to elucidate the sensor surface morphology. Moreover, they showed that the composite sensor has a fast charge transfer rate, large active surface area, and sensitive peak signal. They determined a wide linear operating range of 0.48–85 μM for ORZ herbicide detection. They performed the analytical application of the method and the new generation sensor in fruit juice samples and analyzed the herbicide of oryzalin with satisfactory percentage recovery. To summarize, the CoMoS/O-MWNTs hybrid sensor enables rapid, practical, sensitive, and reliable analysis of oryzalin herbicide (table 13.2). In addition, they show that this composited sensor can be an alternative to replace the mercury electrode [50].

Diamond is located on the sensor platform with its excellent mechanical and extremely durable properties. However, it is rarely used directly as a working electrode due to its low conductivity. Boron-doped (BD) was developed to increase the conductivity of diamond and took its place in the literature as a BDDE. With its superior mechanical and durability properties, the use of BDDE in herbicide determinations has gained great momentum in the last decade. Moreover, these sensors are in high demand compared to carbonaceous and metal electrodes due to their usability in high anodic voltages, very acidic environments. In addition, BDDE allows for a more precise analysis of electro-active herbicides due to its low residual current, high thermal conductivity, and current density.

Jevtić Sonja *et al* in their study in 2018, developed an electroanalytical method on a boron-doped diamond electrode (BDDE) based on the first oxidation of pethoxamid herbicide [114]. They used the square wave voltammetric (SWV) technique for this, and pethoxamid exhibited an anodic at 1.35 V in a pH 4 Britton–Robinson (BR) buffer solution. With the developed SWV technique on the BDDE, the working range was determined as 3–100 μM and the limit of detection (LOD) as 1.37 μM. The analytical application of the method was performed by standard addition method in River water and commercial formulation samples on BDDE by SWV in pH 4.0 BR buffer solution. As a result, they presented a new, fast, inexpensive, and reliable alternative analytical method to the scientific world based on the oxidation of the first pethoxamid herbicide [114].

In the study published by Talay Pınar *et al* they performed the first anodic voltammetry study of dinoterb herbicide from the dinitrofolic group. For this purpose, they used a cathodically pretreated boron-doped diamond electrode (BDDE) sensor. They developed a square wave adsorptive stripping voltammetric (SW-AdSV) voltammetric method for the oxidation peak of the –OH group in the dinoterb molecule. They determined that dinoterb herbicide has irreversible electrode reaction and adsorption-controlled substance transport. In the presence of cetyltrimethylammonium bromide, which is a cationic surfactant, they found the working range 0.0075–0.75 mg L^{-1} and a detection limit of 2.2 μg L^{-1} in pH 7.0 Britton–Robinson (BR) buffer solution (table 13.3). They performed the application of the proposed methodology on spiked River water and soil matrix samples and

Table 13.2. Glassy carbon electrode and its modified derivatives sensor in herbicide analysis.

Analyte	Electrode	Method	Linear range	LOD	LOQ	Samples	Optimum pH	Peak Potentials	İnterference	Ref
Acetochlor	GCE	CV								[51]
Aclonifen	GCE	SWV	0.25 to 3.70 μg mL^{-1}	0.073 μg mL^{-1}	0.25 μg mL^{-1}	Soil, dam water	pH 4.0 (BRB)	+1.2 V	Methiocarb, bendiocarb, difenoconazole, carboxine, benomyl	[52]
Aclonifen	g-C3N4/GCE	SWV	0.01 to 1.2 μM	1.28 nM	4.27 nM	Water, soil	pH 8.0 (PBS)	−0.7 V	$CaCl_2$, $ZnCl_2$ $MgCl_2$, $CuSO_4FeSO_4$, KNO_3	[53]
Aclonifen	PNIPAM/PANI-Cu/GCE	DPV	0.01 to 10 μM and 18 to 76 μM	9 nM		Lake water	pH 7.0 (PBS)	−0.7 V	Chloramphenicol, clothianidin, paraoxon-ethyl, flutamide,, imidacloprid, metronidazole, metribuzin	[54]
Aclonifen	GdNbO4/GCE	DPV	0.02 to 78 μM	1.15 nM		River water, soil	0.1 M acetate buffer	−0.6 V	Clothianidin, imidacloprid, thiamethoxam, metronidazole, chloramphenicol, metribuzin	[55]
Alachlor	MIP/GCE	CV	1 nM to 1 μM	0.78 nM		Milk, eggs, meat	pH 7.0 (PBS)		Propachlor, metalaxyl	[56]
Amitrole	GO-MWCNTs/GCE	Amperometry	5 to 3545 μM	0.45 μM		Lake water, tap water	pH 7.0 (PBS)		Na^+, K^+, $Ca^{2+}Ba^{2+}$, Mg^{2+}, $Fe^{2+}Co^{2+}$, Ni^{2+}, F^-Cl^-, $I^-(COO)_2{}^{2-}$	[57]
Amitrole	CoTCPc-PANI-Co2O3NPs-GCE	Amperometry		66 nM	221 nM	Waste water	pH 4.0 (PBS)		Nitrates	[58]
Amitrole	N,S-OMC/GCE	DPV	3 to 750 μM	0.7 μM		River	(PBS)	+0.6 to +0.9 V	Fe^{2+}, Ni^{2+}, $Ba^{2+}Mg^{2+}$, Na^+, $Ca^{2+}K^+$, Br^-, Cl^- NO_3^-, glucose, hydroquinone, fructose, sucrose	[59]
Amitrole	CoTCPc-PANI-Co2O3NPs-GCE	LSV	10 to 50 mM				pH 7.0 (PBS)	+0.8 to +1.0 V		[58]
Amitrole	MWCNTs/Ca-ZnO/GCE	SWV	0.01 to 0.6 μM	3.58 nM	1.26×10^{-8} M	Soil, water	pH 3.0 (PBS)	+1.188 V		[60]
Amitrole	N-CMK-5/GCE	DPV	0.02 to 0.25 mM	0.007 mM		Water	pH 7.4 (PBS)	+0.6 to +0.75 V	Fe^{2+}, Mg^{2+}, $Ni^{2+}Ba^{2+}$, Ca^{2+}, Na^+K^+, Cl^-, $Br^-NO_3^-$, hydroquinone, fructose, sucrose, glucose	[61]
Asulam	CuNPs/MWCNTs-IL-Chit/GCE	Amperometry	1 to 11 μM and 11 to 200 μM	0.33 nM		Water	pH 9.0 (PBS)	+0.9 V		[62]
Atrazine	Au NPs/NiHCF NPs/ERGO/GCE	DPV	0.25 to 250 pM	0.1 pM	0.23 pM	Lake water, River water	pH 7.41 (PBS)		Simazine, propanil, malathion, 2,4-D*p*-nitrochlorobenzene2,2-binaphthol, bisphenol A	[63]

(Continued)

Table 13.2. (*Continued*)

Analyte	Electrode	Method	Linear range	LOD	LOQ	Samples	Optimum pH	Peak Potentials	İnterference	Ref
Atrazine	Al_2NiCoO_5/GCE	DPV	0.5 ppb to 100 ppb	0.3 ppb		Human blood serum	pH 7.0	+0.21 V	Na^+, Ca^{2+}, bovine serum albumin (BSA), human serum albumin (HSA)	[64]
Aziprotryne	GCE	Amperometry	0.6 to 135.0 μg mL^{-1}	0.2 μg mL^{-1}		Environmental, tap water	0.1 M HCl		Ametryne, atrazine, atrazine-2-hydroxy, atrazine-desethyl, cyanazine, hexazinone, simazine, asulam, diuron, terbumetone, Cu(II)	[65]
Bentazone	SDS/NiO nanoseeds/GCE	DPV	0.1 to 60 μM	1.2 nM	0.012 μM	Ridge gourd, potato	pH 7.0 (PBS)	+0.6 to +0.8 V	K^+, mexacarbate, carbaryl, methyl parathion	[66]
Bentazone	CB-CTS-ECH/GCE	SWV	1.99 to 65.4 μM	1.4 μM		Water	pH 3.0 (PBS)	+1.1 V	Mg^{2+}, Ba^{2+}, humic acid, vermicompost	[67]
Bentazone	CuO/GCE	DPV	5 to 60 μM	8 nM	26 nM	Water	pH 7.0 (PBS)		Carbofuran, endosulfan, carbaryl, Mg^{2+}, K^+	[68]
Bifenox	MWCNT/GCE	SWV	0.2 to 60 μM	0.08 μM		River water	pH 5.0 (acetate buffer)		Cl^-, SO_4^{2-}, $NO_3^-Co^{2+}$, Ni^{2+}, $Zn^{2+}Pb^{2+}$, Mn^{2+}, Cd^{2+}, uric acid, glucose, folic acid, aniline, diazinon, malathion, triton X-100	[69]
Bifenox	MnFe2O4@CTS/GCE	SWV	0.3 to 4.4 μM	0.09 μM	0.3 μM	Tap water, River water	pH 3.0 (BRB)	+0.3 V	Isoproturon, diuron, pirimicarb, 4-nitrophenol, pendimethalin	[70]
Butralin	Co-Ag BMNPs-PVP/GCE	SWV	0.1 μM to 1.0 μM	32 nM	106 nM	Honey, apple jam	pH 2.0 (BRB)	−0.4 V	İnorganic ions, nitro compounds	[71]
Butylate	GCE	SWV	10 to 50 μM	4.96 μM	16.5 μM		pH 4.43 (BRB)	+1.7 V		[72]
Clomazone	Pt-MWCNT/GCE	DPSV	0.61 to 20.56 ng mL^{-1}	0.38 ng cm^{-3}	0.61 ng cm^{-3}		pH 7.0 (PBS)	−0.25 V		[73]
Clopyralid	A-MWCNT-FS/GCE	DPV	0.005 to 10.0 μM	0.8 nM		Urine, River water, sugar beet, wheat	pH 2.0 (PBS)	−0.8 V		[74]
Cyanazine	Grafene/GCE	DPV	0.55 to 1500 nM	0.25 nM	0.83 nM	Water	pH 1.5	−0.45 V	Ca^{2+}, CO_3^{2-}, Na^+F^-, NO_3^-, Ag^+Mg^{2+}, SO_4^{2-}, Fe^{2+}	[75]
Dimethenamid-P	DSA Ti/PtOx/GCE	CV					5.70 mM NaCl	−0.2 V		[76]
Dimethenamid-P	DSA Ti/RuO_2/GCE	CV					5.70 mM NaCl	−0.2 V		[76]
Dinitramine	MWCNT/GCE	CV	4.0×10^{-7} to 5.5×10^{-5} M	0.1×10^{-7} M		Tap water, waste water	pH 2.0 (BRB)	−0.27 V		[77]
Dinitramine	MWCNT/GCE	SWV	4.0×10^{-8} to 1.4×10^{-6} and 1.4×10^{-6} to 2×10^{-5} M	0.8×10^{-8} M		Tap water, waste water	pH 2.0 (BRB)		Trifluralin, 2-methyl, 4,6-dinitropheolNa^+, K^+, $Mg^{2+}Ca^{2+}$, Pb^{2+}, $Al^{3+}Fe^{3+}$, Cl^-, Br^-I^-, NO_3^-, $CO_3^{2-}SO_4^{2-}$	[77]

Diquat	ND-MS/GCE	SWV	5.0×10^{-7} to 4.6×10^{-5} M	1.1×10^{-7} M		Tap water, River water	pH 5.0 (acetate buffer)		$C_6H_8O_6$, $NH_4ClCaCl_2$, $NaClNa_2SO_4C_5H_4N_4O_3$ C4H7$N_3ONH_2CONH_2$	[78]
Diquat	β-CDe/ hydroxyapatite/ GCE	DP-AdsV	5×10^{-8} to 5×10^{-7} M	4.6×10^{-10} M		Spring water	pH 6.0 (aqueous solution)	−0.5 V	$Cu^{2+,}$ $Ni^{2+,}$ $Cd^{2+}Al^{3+}$, Mg^{2+}, K^+Na^+, Cl^-, NO_3^-, paranitrophenol, methyl, parathion, methylene blu	[79]
Diuron	rGO–AuNPs/ Nafion/GCE	DP-AdSV	1.0×10^{-9} to 1.0×10^{-7} M	0.3 nM		Orange juice, mineral water, tap water, malathion, amitraz	0.02 M $HClO_4$	+0.5 V	Na^+, Ca^{2+}, Ba^{2+} K^+, Al^{3+}, $Co^{2+}Cl^-$, NO_3^-, BrO_3^-, Mg^{2+}, Cd^{2+}, $Fe^{2+}Fe^{3+}$, F^-, $C_2O_4^{2-}Ni^{2+}$, Cu^{2+}, Mn^{2+}, biotin, pyrocatechol violet, 1-nitroso-2-naphtol phenol, aminopyrene, nitropyrene, bendiocarb, malathion, amitraz, carbaryl	[80]
Ethyl parathion	AgNP/GCE	SWV	40 to 8000 nM	40 nM	210 nM	Lettuce, honey	pH 4.0 (BRB)	−0.58 V		[81]
Glyphosate	Cu/GCE	DPV	6 to 30 μM	0.186 μM	0.62 μM	Drinking water, bi-distilled water	pH 6.0 (ABS)	−0.015 V	Al^{3+}, Fe^{3+},Mg^{2+}, Ca^{2+}, cation free	[82]
Halosulfuron	GCE	SWSV	4.1 to 50 μg mL^{-1}	1.23 μg mL^{-1}	4.10 μg mL^{-1}	Dam water, soil	pH 1.0 (H_2SO_4)		Ni^{2+}, Co^{2+}, $Zn^{2+}Pb^{2+}$, Cu^{2+}, $Mg^{2+}Na^+$, NO_3^-, PO_4^{3-}, anilazine, cyromazin	[83]
Imazamethabenz	GCE	CV								[84]
Lactofen	GCE	SWV	50 to 250 nM	9.5 nM	32 nM		pH 8.5 (Borate buffer)	−0.4 V		[85]
Lenacil	Thin-layer with GCE	Amperometry	0.1 to 20 μg mL^{-1}	0.014 μg mL^{-1}		Tap water, beetroot juice				[86]
Lenacil	Wall–jet with GCE	Amperometry	0.1 to 20 μg mL^{-1}	0.10 μg mL^{-1}		Tap water, beetroot juice				[86]
MCPA	PANI-β-CD/ fMWCNT/GCE	CV	10 to 50 μM	1.06 μM		Water	pH 6.0 (PBS)	+0.661 V and+1.181 V	Ca^{2+}, Mg^{2+}, NO_3^-, PO_4^{3-}2,4-D, dicamba, atrazine, glyphosate	[87]
Mefenacet	GCE	SWV	1.13×10^{-5} to 6.17×10^{-5} M			Phitopharmaceutical preparation	pH 1.9 (BRB)	+ 1.6 V		[88]
Mesotrione	CB/GCE	SWV	0.040 to 7.2 μM	0.026 μM		Water, sugarcane juice	pH 2.0 (PBS)	−0.2 V	Mg^{2+}, Ca^{2+}, $Ba^{2+}Cu^{2+}$, Al^{3+}, $Pb^{2+}Cd^{2+}$, Ni^{2+}, $NO_3^-Cl^-$, SO_4^{2-}, CO_3^{2-}, humic acid, vermicompost	[89]
Mesotrione	GCE	DPV	0.5 to 5.0 μM	0.17 μM	0.51 μM	Drinking and surface water	pH 6.0 (BRB)	0 to −1.5 V	Cd^{2+}, Cu^{2+}, $Pb^{2+}Zn^{2+}$, sulcotrione	[90]
Mesotrione	NfrA2@MgAl-NPs/ GCE	Amperometry	5 to 60 μM	3 μM		Water	pH 6.5 (PBS)			[91]
Metam	GCE	SWV	17 to 29 mg L^{-1}			Water	pH 2.5 (BRB)	+1.46 V		[92]

(Continued)

Table 13.2. (*Continued*)

Analyte	Electrode	Method	Linear range	LOD	LOQ	Samples	Optimum pH	Peak Potentials	İnterference	Ref
Metamitron	GCE	Potentiometry	1 to 30 mg L^{-1}	92.91 µg L^{-1}	281.54 µg L^{-1}	Water	pH 2.0 (BRB)		Na^+, K^+, $Mg^{2+}Ca^{2+}$, SO_4^{2-}, $Cl^-NO_3^-$, $HCO_3^-Pb^{2+}$, Zn^{2+}, $Fe^{2+}Cd^{2+}$, metribuzin, imidacloprid	[93]
Metazachlor	GO-AgNp/GCE	SWV	0.1 to 10 mM	79 µM		Tap water	pH 7.0 (PBS)	0.2 V	Cl^-, I^-, Br^-F^-, Cu^{2+}, CO_3^{2-} NH_4^+, CN^-	[94]
Metazachlor	ErGO_AgNP/GCE	CV	37 to 1123 µM	27 µM		Forest soil	0.020 M NaOH		Humic acids	[95]
Metazachlor	MX_AgNP/GCE	CV	37 to 375 µM	40 µM		Park soil	0.020 M NaOH			[95]
Monalide	NpNiOx/GCE	DPV	1.0 to 340 µM	0.21 µM		Water	pH 2.0(PBS)		Zn^{2+}, Ca^{2+}, $Cu^{2+}Pb^{2+}$, Mg^{2+}, $Na^+SO_4^{2-}$, Cl^-, NO_3^-, 4-amino, 4-chlorophenol, 3-nitrophenol, 4-aminophenol, 4-chlorophenol	[96]
Oryzalin	Ag-Au BMNPs-CTS/GCE	SWV	0.1 to 7.0 µM	30 nM	100 nM	Grape juice	pH 2.0 (PBS)	−0.365 V	Pb^{2+}, Cd^{2+}, $Ni^{2+}Co^{2+}$, Zn^{2+}, o-nitrophenol, m-nitrophenol, p-nitrophenol, pendimethalin, butralin, profluralin, parathion	[97]
Oryzalin	GCE	SWV	0.4 to 1.0 µM	116 nM	386 nM		pH 2.0 (PBS)	−0.371 V		[97]
Oryzalin	CoMoS/O-MWNTs/GCE	DPV	0.48 to 85 µM	50 nM		Juice	pH 1.0 (PBS)	−0.1 V	NH_4^+, Na^+, $Ba^{2+}Zn^{2+}$ $CH3COO^-NO_3^-$, Cl^-, organic dyes, Congo red, Sudan I, rhodamine B	[50]
Oxadiazon	GCE	SWV	1.3×10^{-4} to 2.7×10^{-4} M	1×10^{-4} M			pH 12.8(Aqueous Electrolyte Buffer)	−0.1 V+0.6 V		[98]
Oxyfluorfen	MoO_2-MWCNT/GCE	DPSV	2.5 to 34.5 ng mL^{-1}	1.5 ng mL^{-1}	2.5 ng mL^{-1}	River water	pH 6.0 (BRB)	+0.3 V		[99]
Paraquat	Ag-GO/CuO/GCE	DPV	0.40 to 650.0 µM	0.07 µM		River water	pH 10.0 (PBS)	−0.6 V	K^+, Mg^{2+}, $Ca^{2+}Ag^+$, Na^+, $Cr^{3+}Cl^-$, NO_3^-, ethion, metribuzin	[100]
Pendimethalin	AgNP/GCE	SWV	70 to 2000 nM	36 nM	119 nM	Water	pH 4.0 (BRB)	−0.6 V		[81]
Picloram	CIT/MIL-101-GCE	SWV	12.5 to 40 µM	0.06 µM		River water	0.1 M H_2SO_4	+1.36 V	Ca^{2+}, Mg^{2+}, K^+Na^+, Ba^{2+}, $Cu^{2+}Bi^{3+}$, Al^{3+}, $As^{3+}Zn^{2+}$, Pb^{2+}, $Cd^{2+}Cl^-$, NO_3^-, PO_4^{3-}, CO_3^{2-}, clopyralid	[101]
Picloram	Au@MWCNT-PVCL/GCE	DPV	0.02 to 183 µM	1.5 nM		Agricultural and environmental	pH 7.0 (PBS)	+1.3 V	Cu^{2+}, Pb^{2+}, $Zn^{2+}Ni^{2+}$, Cd^{2+}, glyphosate, triclopyr, clothianidin, clopyralid, triasulfuron, paraoxon-ethyl	[102]

Profluralin	GCE	SWV	1.0×10^{-7} to 1.0×10^{-6} M	2.6×10^{-8} M	8.6×10^{-8} M	Tap water, River water	pH 2.0 (BRB)	−0.43 V	Cd^{2+}, $Zn^{2+}Co^{3+}$, Pb^{2+}, basticidine S, metam sodium, dinotefuran, aclonifen and nitrothal-isopropyl, minooctadine, methidathion, thiophanat methyl	[103]
Propanil	NpNiOx/GCE	DPV	0.1 to 80.0 μM and 80.0 to 290.0 μM	0.05 μM		Water	pH 2.0(PBS)		Zn^{2+}, Ca^{2+}, $Cu^{2+}Pb^{2+}$, Mg^{2+}, $Na^{+}SO_4^{2-}$, Cl^{-} NO_3^{-}, 4-amino, 4-chlorophenol, 3-nitrophenol, 4-aminophenol, 4-chlorophenol	[96]
Propham	MWCNT/GCE	SWV	3.0 to 38.6 μM	759 nM	2.28 nM		0.5 M H_2SO_4	+1.33 V		[104]
Propham	MWCNT/GCE	SW-AdSV	2.0 to 47.8 μM	365 nM	1.09 nM	River water	0.5 M H_2SO_4	−1.49 V		[104]
Pyridate	GCE	Amperometry				Water				[105]
R-Dinoseb	R-CSA-PANI/GCE	SWV	8×10^{-6} to 1.0×10^{-4} M	2.28×10^{-6} M	7.61×10^{-6} M		pH 1.8 (BRB)	+0.3 V and +0.7 V		[106]
S-Dinoseb	S-CSA-PANI/GCE	SWV		1.89×10^{-6} M	6.31×10^{-6} M		pH 1.8 (BRB)	+0.3 and +0.7 V		[106]
Sulcotrione	GCE	DPV	0.2 to 2 μM and 2 to 50 μM	0.05 μM		River water	pH 3.0 (BRB)	−0.84 V	Ca^{2+}, Mg^{2+}, $K^{+}Na^{+}$, Hg^{2+}, $Cu^{2+}Bi^{3+}$, Cr^{6+}, $As^{3+}Ni^{2+}$, Pb^{2+}, $Cl^{-}NO_3^{-}$, $SO_4^{2-}CO_3^{2-}$, PO_4^{3-} Cd^{2+}	[107]
Sulfentrazone	MWCNT/GCE	DPV	1.0×10^{-6} to 2.3×10^{-4} M	2.6×10^{-6} M	8.8×10^{-6} M	ss-DNA	pH 7.0 (acetate buffer,)	+1.0 V	Guanine, adenine	[108]
Sulfentrazone	MWCNT/GCE	DPV	1.0×10^{-6} to 2.3×10^{-4} M	2.6 μM	8.8 μM	ss-DNA	pH 7.0 (acetate BS)		Adenine, guanine	[108]
Sulfentrazone	GCE	DPV	5 to 100 mg L^{-1}	1.94 and 2.19 mM	6.46 and 7.31 mM	Soil	0.1 M KOH	+ 0.936 V		[109]
Triclopyr	PANI/C70/GCE	SWV	10 to 320 ng mL^{-1}	1.9 ng mL^{-1}	5.0 ng mL^{-1}	Tomatoes		+172 V	Common metabolites	[110]
Trifluralin	Fe3O4/SiO_2 ~ FLU/ MWCNT/GCE	SWV	1.0×10^{-9} to 2.0×10^{-7} M	1.0 nM		Water	pH 3.0 (BRB)	+0.3 V	Dinitramine, 2-methyl,4,6-dinitrophenol, Fe^{2+}, Al^{3+}, $Pb^{2+}Zn^{2+}$, Co^{2+}, $Ni^{2+}NO_3^{-}$, CO_3^{2-}, S^{2-}, SO_4^{2-}	[111]
Trifluralin	SDS/GCE	SWV	4.80×10^{-7} to 3.22×10^{-5} M	31 nM		Tap water, orange juice, green pepper, carrot, tomato, onion	0.02 M H_2SO_4	−0.3 V	Na^{+}, NH_4^{+}, $Ca^{2+}Mg^{2+}$, Al^{3+}, $Cu^{2+}Cl^{-}$, OH^{-}, $NO_3^{-}SO_4^{2-}$, atrazine, endosulfan, linuron, thiamethoxam, glyphosate, carbendazim	[112]
Trifluralin	rGO-PEI-AgNPs/ GCE	DPV and SWV	1 mM to 1 nM	1 nM		Human plasma	pH 7.4 (PBS)	+0.1 V		[113]

Table 13.3. Boron-doped diamond electrode (BDDE) used in herbicide analysis.

Analyte	Electrode	Method	Linear range	LOD	LOQ	Samples	Optimum pH	Peak Potentials	İnterference	References
2,4-dichlorophenoxy acetic acid	BDDE	DPV	1.0 to 9.0 μM	3.2 μM	9.8 μM	Water	0.1 M (H_2SO_4)		Ametryn, hexazinone, imazethapyr, glyphosate, imazethapyr, picoxystrobin, humic acid	[116]
Bentazone	BDDE	DPV	2.0 to 100.0 μM	0.5 μM		River water	pH 4.0 (BRB)	+1.0 V	K^+, Na^+, $Mg^{2+}Ca^{2+}$, Ni^{2+}, $Cu^{2+}Zn^{2+}$, Co^{2+}, Pb^{2+} NO_3^-, SO_4^{2-}, CO_3^{2-} $CH3COO^-$, sulcotrione, mesotrione	[117]
Bromacil	BDDE	DPV	5 to 75 μM	1.27 μM	4.22 μM	River water	pH 2.0 (BRB)	+1.65 V	NaClNaCl + Bro	[118]
Clomazone	BDDE	SWV	1 to 100 μM	0.21 μM		River water	pH 2 (BRB)	+ 1.6 V	Mezotrione, sulcotrione, pethoxamid	[119]
Dinoterb	BDDE	SW-AdSV	3.12×10^{-8} to 3.12×10^{-6} M	9.16×10^{-9} M	3.05×10^{-8} M	Water, soil	pH 7.0 (BRB)	+1.22V	Cl^-, NO_3^-, CO_3^{2-}, SO_4^{2-}, PO_4^{3-}, Na^+, K^+, Mg^{2+}, Ca^{2+}, Zn^{2+}, Fe^{3+}	[115]
Diuron	BDDE	DPV	1.0 to 9.0 μM	1.5 μM	4.4 μM	Water	0.1 M (H_2SO_4)		Ametryn, hexazinone, imazethapyr, glyphosate, imazethapyr, picoxystrobin, humic acid	[116]
Metolachlor	BDDE	PEF								[120]
Pethoxamid	BDDE	SWV	3.0 to 100.0 μM	1.37 μM		River water	pH 4.0 (BRB)	1.35 V	K^+, Na^+, $Ca^{2+}Mg^{2+}$, Cu^{2+}, $Ni^{2+}NO_3^-$, $SO_4^{2-}CO_3^{2-}$, Cl^- $HCO_3^-CH3COO^-$, clomazone, terbuthylazine, mesotrione, sulcotrione	[114]
Picloram	BDDE	SWV	6.70×10^{-6} to 6.70×10^{-5} M	390 nM		Amazon water	0.5 M H_2SO_4			[121]
Sulfonylurea herbicides	BDDE	H_2O_2								[122]
Terbutryn	BDDE	DPV	5×10^{-7} to 5×10^{-5} M	1.8×10^{-7} M	5.8×10^{-7} M	River water, drinking water	pH 2.0 (BRB)			[123]
Triclopyr	BDDE	DPV	1.0 to 108.8 μM	0.82 μM	2.72 μM	Water, urine	pH 2.0 (BRB)	+1.9 V	Cu^{2+}, Cd^{2+}, $Pb^{2+}Zn^{2+}$, Tl^+, ascorbic acid, barbituric acid, folic acid, glucose, creatinine, sucrose, urea, uric acid, clopyralid, picloram, glyphosate, triasulfuron, linuron	[124]
Triclopyr	BDDE	SWV	2.5 to 99.0 μM	1.85 μM	6.16 μM		pH 2.0 (BRB)	+1.9 V		[124]

analyzed the dinoterb herbicide with high recoveries using a BDDE sensor by SW-AdSV, fast, reliable, and accurate [115].

Although mercury electrodes having a toxic effect are outdated, they are indispensable sensors due to their extraordinary repeatability, high sensitivity, and accuracy. Mercury is the first indicator sensor that comes to mind, especially for the determination of herbicides in cathodic potential regions. The polygraph and the mercury electrode, which are the basis of voltammetry methods, became the most important sensor of the 19th century. The determination of many drugs, herbicides, and metals was carried out on this sensor by electrochemical methods. Mercury or mercury-coated film sensors are the first electrodes to be used in cases where it is still not able to detect solid electrodes or is potentially insufficient. Moreover, the hydrogen overvoltage of mercury sensors is high, thus providing opportunities to operate at more negative potentials.

In their study published by Mohan *et al* [125], the determination of chlornidine, dipropalin, and prodiamine herbicides containing dinitro functional group was performed by differential pulse adsorptive stripping voltammetry (DPAdSV) on the mercury sensor. They determined the optimum pH value as pH 2.0, 4.0, and 6.0 for chlornidine, dipropalin, and prodiamine, respectively. For each herbicide, some parameters that have a reasonable effect on peak signal and potentials, such as deposition time, deposition potential, scan rate, and pulse amplitude were optimized in detail. Then, using the hanging mercury drop electrode (HMDE) with DPAdSV under optimum modules; the detection limits for chlornidine, dipropalin, and prodiamine were found as 1.94, 1.82, and 4.52 μM, respectively. To speak well of the method and the sensor, they performed the analytical application in real samples such as soil and water and determined three herbicides with recoveries such as 99.1%–99.9% (table 13.4). As a result, they developed an accurate, reliable, and fast analytical method on the mercury sensor for three herbicides that gave a cathodic peak due to the reduction of nitro groups [125].

One of the sensors used in herbicide determinations is the indium tin oxide (ITO) electrode. ITO sensors are used in the electroanalytical analysis of numerous substances due to their low residual current, infinite conductivity, wide operating range, and stable structure. Moreover, due to the unique substrate affinity, excellent optical transparency, and stable physical properties of ITO electrodes, they have attracted great interest in the scientific community in the last two decades in sensor application. It also has the ability to be easily modified. Thus, it is possible to produce more sensitive new generation electrochemical sensors. Considering all these wonderful properties, it is impossible for it to not be used as a sensor in herbicide analysis.

In their scientific study published in 2017, Jinmeng Xu and colleagues developed the indium tin oxide (ITO) electrode based on a molecularly imprinted polymer for the voltammetric determination of glyphosate (GLY), which is used as a non-selective, broad-spectrum metal chelating herbicide. For this, they first coated the gold nanoparticles and Prussian Blue on the ITO sensor surface. Then, they permeated the glyphosate composite electrode in the presence of poly pyrrole. Voltammetric determination of such glyphosate herbicide produced a Ppy-MIP/

Table 13.4. Mercury sensor used in herbicide analysis.

Analyte	Electrode	Method	Linear range	LOD	LOQ	Samples	Optimum pH	Peak potentials	İnterference	References
Ametryn	ME	DPP					pH 6.5 (H_2SO_4)			[126]
Atraton	ME	DPP	2.0×10^{-8} to 1.0×10^{-5}M	1.5×10^{-8} M		Agricultural formulation, water, soil	pH 4.0 (BRB)	−1.0 V	Zn^{2+}, Pb^{2+}, $Cd^{2+}Cr^{2+}$, Co^{2+}, $NO^{-}SO_4^{2-}$, I^{-}, Cl^{-}	[127]
Atrazine	HMDE	DPV	0.05 to 0.60 mg L^{-1}	0.0347 mg L^{-1}		Water	pH 2.56 (BRB)	−0.936 V		[128]
Aziprotryne	HMDE	DPP	1×10^{-6} to 1×10^{-5} M	2.1×10^{-7} M		Water	pH 3.0 (BRB)		simazine, methoprotryne, terbutryne, propazine	[129]
Benfluralin	DME	DPP	1.25×10^{-5} to 2.0×10^{-9} M	1.02 $\times 10^{-9}$M		Soils, grains	pH 4.0	−0.16 V	other herbicides	[130]
Butralin	DME	DPP	1.2×10^{-5} M to 7×10^{-8} M and1.0×10^{-5} M to 2×10^{-8} M	6×10^{-8} M		Grains, soils, water	pH 4.0 (BRB)	−0.4 V–0.5 V	Atrazine, simizine, linuron, Zn^{2+}, Pb^{2+}, $Cd^{2+}Cr^{3+}$, Co^{3+}, $NO_3^{-}SO_4^{2-}$, I^{-}, Cl^{-}	[131]
Butylate	HMDE	SWV	2.0 to 9.0 μM	0.29 μM	0.96 μM	Water	pH 3.3 (BRB)	+1.1 V	Molinate, bentazone, MCPA, propanyl, humic acid, lead, calcium, iron, copper, manganese, aluminum, silicon	[72]
Chlornidine	DP-AdSV	HMDE		1.94×10^{-6} M		Soil, water, green grain	pH 2.0 (universal buffer)			[125]
Dinitramine	HDME	CV				ct-DNA	pH 7.3 (HCl Buffer)	−0.44 V–0.571 V		[132]
Dinitramine	HDME	DPV				ct-DNA	pH 7.3 (HCl Buffer)	−0.66 V–0.71 V		[132]
Dinoseb	HDME	DPV	0.01 to 0.25 mg L^{-1}	0.005 mg L^{-1} and 0.004 mg L^{-1}		Spinach, celery, cabbage, water	pH 5.72 (BRB)			[133]
Dipropalin	DP-AdSV	HMDE		1.821 $\times 10^{-6}$M		Soil, water, green grain	pH 4.0 (universal buffer)			[125]
DNOC	HDME	SSWV	0.01×10^{-6} to 0.55×10^{-6} M	2×10^{-8} M	4×10^{-8} M	Water	pH 6.0 (BRB)	−0.3 and−0.5 V		[134]
Ethalfluralin	HMDE	DP-AdSV	1.30×10^{-9} to 1.32×10^{-5} M	1.08×10^{-9} M		Formulations (sonalan, curbit), vegetable (cucurbits, pumpkins), real soil	pH 6.0	−0.4 V–0.8 V	Cr(III), Cd(II)Zn(II), Ni (II)Co(II), Cr(VI)Cu (II), Pb(II)F^{-}, Br^{-}, I^{-}	[135]
Fenoxaprop-p-ethyl	HMDE	SWV	0.30 to 15.0 μM	0.08 μM and 0.07 μM	0.25 μM and	0.24 μM	Tap water, river water, plant	pH 4.8 (BRB)	-0.7 and -0.9 V	[136]
Fluchloralin	DMEHMDE	DPP	2×10^{-5} M to 2.62×10^{-8} M	1.5×10^{-8} M		Grain, soil, water	pH 4.0	−0.24 V		[137]
Imazamethabenz	ME	Polarography								[84]
Imazamethabenz	ME	CV								[84]

Imazapyr	MDE	CV								[138]
İsopropalin	DME	DPP	1.2×10^{-5} M to 7×10^{-8} M and 1.0×10^{-5} M to 2×10^{-8} M	1.5×10^{-8} M		Grains, soils, water	pH 4.0 (BRB)	−0.3 V–0.4 V	Atrazine, simizine, linuron, Zn^{2+}, Pb^{2+}, $Cd^{2+}Cr^{3+}$, Co^{3+}, $NO_3^{-}SO_4^{2-}$, I^-, Cl^-	[131]
Lenacil	HMDE	DP-AdSV	1.5×10^{-5} to 2.5×10^{-8} M	2.0×10^{-8} M		Fruit, water	pH 4.0 (BRB)	+ 0.2 V		[139]
Metam	HMDE	SWV	0.25 to 1.0 mg L^{-1}			Water	pH 10.1 (BRB)	−0.72 V		[92]
Metamitron	Thin-Film ME (TFME)	Potentiometry	0.8 to 30 mg L^{-1}	68.53 μg L^{-1}	207.68 μg L^{-1}	Water	pH 7.0 (BRB)		Na^+, K^+, $Mg^{2+}Ca^{2+}$, SO_4^{2-}, $Cl^-NO_3^-$, $HCO_3^-Pb^{2+}$, Zn^{2+}, Fe^{2+} Cd^{2+}, metribuzin, omidacloprid	[93]
Methalpropalin	HMDE	DP-AdSV	13×10^{-5} to 2.0×10^{-8} M	1.87×10^{-8} M		Water	pH 6.0	−0.4 V–0.9 V	Cr(III), Cd(II)Zn(II), Ni (II)Co(II), Cr(VI)Cu (II), Pb(II)F^-, Br^-, I^-	[135]
Nitralin	HMDE	DP-AdSV	5×10^{-10} to 4.4×10^{-8} M	$3{,}5 \times 10^{-11}$ M		Ground water	pH 11.5 (BRB)	−0.5 V–0.6 V	Isomethiozine, chloridazone, metamitron	[140]
Nitralin	HMDE	SW-AdSV	4×10^{-10} to 1×10^{-8} M and 1×10^{-8} M to 1.1×10^{-7} M	4.4×10^{-10} M			pH 11.5 (BRB)	−0.6 V–0.7 V		[140]
Nitralin	HMDE	DP-AdSV	8.3×10^{-8} M to 1.5×10^{-6} M	2.47×10^{-8} M		Agricultural formulations, vegetables, grape juice	pH 6.0	−0.5 V	Ethalfluralin, nitralin, oryzalin	[141]
Oryzalin	HMDE	DP-AdSV	8.3×10^{-8} M to 1.5×10^{-6} M	2.47×10^{-8} M		Agricultural formulations, vegetable, grape juice	pH 6.0	−0.4 V	Ethalfluralin, nitralin, oryzalin	[141]
Pendimethalin	HMDE	SWV	5.0×10^{-7} to 4.5×10^{-6} M	1.92×10^{-8} M	6.42×10^{-8} M	Natural water, river sediment, baby food	pH 8.0 (BRB)	−0.60 V–0.71 V	Humic acid	[142]
Pendimethalin	HMDE	SWV					pH=8.0 (0.1 M Na_2SO_4)			[143]
Picloram	HMDE	DPV		3.9×10^{-8} M	8.25×10^{-8} M		0.1 M H_2SO_4			[144]
Picloram	HMDE	DP-AdSV		1.5×10^{-9} M	2.71×10^{-9} M		0.1 M H_2SO_4			[144]
Prodiamine	DP-AdSV	HMDE		$4.522 \times 10{-}6$ M		Soil, water, green grain	pH 6.0 (universal buffer)			[125]
Prometon	ME	DPP	3.5×10^{-8} to 1.13×10^{-5} M	2.0×10^{-8} M		Agricultural formulation, water, soil	pH 4.0 (BRB)	$^{-1}.06$ V	Zn^{2+}, Pb^{2+}, $Cd^{2+}Cr^{2+}$, Co^{2+} NO_3^-, $SO_4^{2-}I^-$, Cl^-	[127]
Propazine	HMDE	DPV	0.05 to 0.80 mg L^{-1}	0.0356 mg L^{-1}		Water	Ph 2.56 (BRB)	−0.946 V		[128]

(*Continued*)

Table 13.4. (*Continued*)

Analyte	Electrode	Method	Linear range	LOD	LOQ	Samples	Optimum pH	Peak potentials	İnterference	References
Quinmerac	ME	Amperometry					pH 1.8			[145]
Quinmerac	ME	CV					pH 1.78			[145]
Secbumeton	ME	DPP	3.0×10^{-8} to 1.0×10^{-5} M	2.8×10^{-8} M		Agricultural formulation, water, soil	pH 4.0 (BRB)	−1.15 V	Zn^{2+}, Pb^{2+}, $Cd^{2+}Cr^{2+}$, Co^{2+}, $NO_3^{-}SO_4^{2-}$, I^-, Cl^-	[127]
Simazine	HMDE	DPV	0.05 to 0.80 mg L^{-1}	0.0281 mg L^{-1}		Water	Ph 2.56 (BRB)	−930 V		[128]
Simetryn	ME	DPP					pH 6.5 (H_2SO_4)			[126]
Terbacil	HMDE	DP-AdSV	1.5×10^{-5} to 1.2×10^{-9} M	1.22×10^{-9} M		Fruit, water	pH 4.0 (BRB)	+ 0.3 V		[139]
Terbumeton	ME	DPP	2.0×10^{-8} to 1.12×10^{-5} M	1.0×10^{-8} M		Agricultural formulation, water, soil	pH 4.0 (BRB)	−1.2 V	Zn^{2+}, Pb^{2+}, $Cd^{2+}Cr^{2+}$, Co^{2+}, $NO_3^{-}SO_4^{2-}$, I^-, Cl^-	[127]
Terbuthylazine	HMDE	DPV	0.05 to 0.80 mg L^{-1}	0.0182 mg L^{-1}		Water	Ph 2.56 (BRB)	−0.968 V		[128]
Terbutryn	HMDE	DPV	5×10^{-9} to 1×10^{-6} M	8.9×10^{-10} M	2.9×10^{-9} M	River water, drinking water	pH 2.0 (BRB)			[123]
Trifluralin	DME	DPP	1.4 ×10–5 to 2.85 ×10–9 M	1.05 ×10–9 M		Soils, grains	pH 4.0	−0.2 1 V	other herbicides	[130]

AuCl4 NPs-PB/ITO hybrid sensor. By differential pulse voltammetry, on the Ppy-MIP/$AuCl_4$ NPs-PB/ITO hybrid sensor, they found the linear range to be 400–1200 μg L^{-1} and the detection limit was calculated as 92 μg L^{-1}. Finally, the analytical application of the developed hybrid ITO sensor performed glyphosate analysis on spiked samples with high recoveries (table 13.5). To sum up, they have created a powerful sensor for glyphosate herbicide determination that is simple, inexpensive, and fast to detect [146].

Screen-printed electrodes (SPE), which are very promising for both portable and miniature electrochemical analyzer systems, are also the most used sensor platform for substances such as herbicides, drugs, vitamins, and cations, etc. In particular, these sensors have been neglected by the scientific world, as they can be produced from many different materials such as glass, ceramics, and paper, and the electrode surface can be easily modified. The development of these unique sensors is extremely important in terms of increasing the importance of field analyses and making herbicide residue analyses online in terms of food quality safety. Moreover, they are used in the determination of numerous substances from herbicides to heavy metals due to their low cost, easy preparation, production in different geometries and shapes, and being included in the mass production line. In addition, SPE sensors, as multi-channel sensors, enable simultaneous determinations to take place. Therefore, SPE sensors are among the portable, inexpensive, simple, and reliable sensors in the defense industry, clinical and academic research. In addition, very reproducible results are obtained by being disposable. Considering all these features, it can be said that SPE is the biggest sensor candidate for online analysis of herbicides.

Raquel de Oliveira Silva *et al* in their study in 2019, developed a new nanosensor for the determination of sulfentrazone herbicide by modifying the screen-printed electrode (SPE) with a multi-walled carbon nanotube (MWCNT) [148]. They investigated the electrochemical behavior of sulfentrazone by square wave voltammetry (SWV) in pH 5.0 Britton–Robinson (BR) buffer solution on electrochemically activated MWCNT/SPE. They obtained a linear calibration plot with a high Pearson correlation coefficient ($r = 0.9995$) between 1.0–25 μM concentration of sulfentrazone herbicide. They found that the limit of detection in pure water was lower than 0.8 μM. The repeatability of the disposable sensor was calculated as 0.94% relative standard deviation (RSD). They performed sulfentrazone herbicide analysis with high recovery in soy milk and groundwater samples with the proposed MWCNT/SPE sensor (table 13.6). According to all these results, it can be said that disposable SPE sensor is reliable, inexpensive, reproducible, and sensitive [148].

Determination of analytes in electrochemistry is based on electrical change processes occurring on the surface of the working sensors. Therefore, the structure, conductivity, affinity, and stability of the sensor surfaces are of great importance. In the modification processes, there is the improvement and development of the sensor surface. This process can be done sometimes by physical interaction, by forming multi-layered thin films, and occasionally by chemical reaction. The newly developed sensors can be named differently such as composite, hybrid, or modified electrodes. It can be said that these sensors basically increase the catalytic, conductivity, active surface, and stability of the indicator electrodes. Modified

Table 13.5. Indium tin oxide (ITO) sensor in herbicide analysis.

Analyte	Electrode	Method	Linear range	LOD	LOQ	Samples	Optimum pH	Peak potentials	İnterference	References
Acetochlor	HTNs-PLA/ITO electrodes	DPV	2.0×10^{-7} M to 1.5×10^{-5} M	0.15×10^{-9} M		Strawberry	0.5 M (H_2SO_4)	0.1 V	NH_4^+, K^+, Na^+Ca^{2+}, $Mg^{2+}Al^{3+}$, SO_4^{2-} NO_3^-, urea, ascorbic acid	[147]
Glyphosate	Ppy-MIP/ HAuCl4-PB/ ITO	DPV	400 to 1200 ng mL^{-1}	92 ng mL^{-1}		Corn	5 mM H_2O_2	−0.3 to +0.8 V	Glufosinate, gramoxone, pendimethalin, acetochlor	[146]

Table 13.6. Screen-printed electrodes (SPE) in herbicide analysis.

Analyte	Electrode	Method	Linear range	LOD	LOQ	Samples	Optimum pH	Peak potentials	İnterference	References
2,4,5-T	SPE/CD/CNO/HRP biosensor	Amperometry		0.010 μM	2.6 μg L^{-1}	River water	pH 6.5 (PBS)			[149]
2,4-D	SPE/CD/CNO/HRP biosensor	Amperometry		0.023 μM	5.1 μg L^{-1}	Soil	pH 6.5 (PBS)			[149]
Acifluorfen	FeS_2@Ag NL/SPCE	DPV	0.05 to 1126.45 μM	0.0025 μM	26.48 μM	Human urine, blood serum	pH 7.0 (PBS)	−0.6 V	Nithiazine, imidazole, omeprazole, fenitrothion, nitrofuran, dimetroimidazole, cholramphenicol, acetamiprid, thiamethoxam, omidacloprid, clothianidin, dinotefuran	[150]
Atrazine	AIRGOC/SPE	SWV	1.2 to 23.0 nM	0.4 nM	1.2 nM	River water, sugar cane juice	pH 3.0 (NaH_2PO_4)	−0.9 V		[151]
Glyphosate	HRP-biotin-streptavidin/SPE	SWV	0.08 to 11.0 μg L^{-1}	45 ng L^{-1}	80 ng L^{-1}	Groundwater, superficial water	pH 5. 0 (citrate/phosphate buffer)			[152]
Mesotrione	YIG/GCN/SPE	DPV	0.005 to 751.8 μM	0.95 nM		Fruit extracts	pH 7.0 (PBS)		Metronidazole, sulfamethoxazole, nilutamide, 4-nitrophenol, 2-nitrophenol, chloramphenicol, trimethoprim, amoxicillin, flutamide, nitrobenzene	[153]
Metam	PVA-SbQ/SPE	Amperometry								[154]
Paraquat	PtNPs@MIP/SPGrE/PVC-SPE	ASV	0.05 to 1000.0 μM	0.02 μM	0.05 μM	Yardlong bean, cabbage, lettuce, cucumber, kale, onion, tomato	0.1 M K_2SO_4	−0.71 V	Phenol, caffeine, profenofos, ascorbic acid, glyphosate, carbofuran, chlorpyrifos, diuron, chlorotoluron	[155]
Sulcotrione	HPPD/SPE	Amperometry	5 $\times10^{-10}$ M to 5 ×10–6 M	1.4 ×10–10 M		A. thaliana	pH 8.0 (PBS)			[156]
Sulfentrazone	MWCNT/SPE	SWV	1.0 to 30.0 μM	1.5 × 10^{-7} M	5.1 × 10^{-7} M	Water, soy milk	pH 5.0 (BRB)	+0.8 V	Na^+, Ca^{2+}, $Ba^{2+}Cu^{2+}$, Co^{2+}, $Ni^{2+}Al^{3+}$, Cl^-, NO_3^-, humic acid, linuron, clomazone, emamectin, nenzoate, propanil, tebuthiuron, trifluralin,	[148]

(*Continued*)

Table 13.6. (*Continued*)

Analyte	Electrode	Method	Linear range	LOD	LOQ	Samples	Optimum pH	Peak potentials	İnterference	References
									atrazine, carbaryl, glyphosate	
Sulfentrazone	MWCNT/SPE	SWV	1.0 to 25 μM	1.5×10^{-7} M	5.1×10^{-7} M	Pure water	pH 5.0 (BRB)			[148]
Sulfentrazone	MWCNT/SPE	SWV	1.0 to 20 μM	8.2×10^{-7} M	2.7 μM	Groundwater	pH 5.0 (BRB)			[148]
Sulfentrazone	MWCNT/SPE	SWV	1.0 to 25 μM	1.7×10^{-7} M	5.8×10^{-7} M	Soy milk	pH 5.0 (BRB)			[148]

electrodes are attracting attention as important nanomaterials in the development of portable precision sensors. The modified sensors thus developed allow sensitive, reliable, and accurate analysis of lower amounts of analytes. These composite sensors are also frequently used in numerous herbicide determinations. In particular in the last two decades, modified sensors have been constructed by using metal nanoparticles, polymers, phthalocyanines, and many different materials in herbicide analysis.

After voltammetric and polarography techniques, the most commonly used electrochemical method in herbicide determination is amperometric. In this method, the production of biosensors gains importance. For herbicide determinations, hybrid sensors are prepared using carbon nanomaterials. Signal amplitude, sensitivity and stability, and cost are key factors in the development of these sensor strategies. In basic sensor production, organic materials such as carbon nanotubes, basic electrodes, nanoparticles, polymers, or chitosan form an effective top layer for reactions on the electrode surface. This increases the efficiency of the sensors and enables more sensitive determination of the analytes. In potentiometric sensors, the base electrode is generally composed of carbonaceous materials such as graphite, screen-printed, and graphene. It has general production in the solid contact form using solid contact material such as polyaniline (PANI). An ion-selective membrane is coated on this layer. Finally, polyvinylchloride (PVC) is used as a plasticizer (table 13.7). Herbicide determinations were carried out by chronopotentiometric measurements with these developed sensors.

Due to the toxic feature of the mercury electrode, which is the first and most important of the sensors, scientists have sought alternative sensors. Sensor studies that are sensitive, reliable, and inexpensive, which can give good results especially in negative potential regions, have gained importance. Bismuth sensor has become an interesting material for electrochemical sensors due to its surprising properties such as wide negative potential range, stability, and ability to form alloys with heavy metals. Moreover, due to its low toxicity and being an environmentally friendly element, the bismuth working electrode was the best candidate to replace mercury. Bismuth and its film sensors have been involved in the determination of heavy metals, organic compounds, and herbicides in recent years. It is seen that bismuth sensors have been widely used in herbicide determinations exhibited cathodic behavior.

In their study published in 2017, de Figueiredo-Filho *et al* developed a bismuth-film-based mini and a disposable sensor for herbicide determinations of diquat (DQ) and paraquat (PQ) [159]. While they determined diquat with SWV, they analyzed paraquat herbicide with DPV using minisensor. They found the linear operating range and detection limit for DQ on the bismuth-film sensor to be 0.19–9.3 μM and 89 nM, respectively. This value was determined as 0.12–4.2 μM and 12 nM for PQ, respectively. Together with the proposed new generation sensor, they successfully performed the accuracy test of the electroanalytical method and the determinations of two herbicides in natural water samples (tables 13.8 and 13.9). Finally, they showed that the results obtained using high-performance liquid chromatography (HPLC) for the validation studies were not significantly different [159].

Table 13.7. Sensor platform in herbicide analysis by potentiometry.

Analyte	Electrode	Method	Linear range	LOD	LOQ	Samples	Optimum pH	Peak potential	İnterference	References
Bispyribac	SC/ISE	Potentiometry	1.0×10^{-2} to 8.6×10^{-6} M	1.33 μg mL^{-1}		Herbicide formulations, soil			Diquate, acetamiprid, dinotefuran, imidachloprid, cyromazine, flucarbazon, Cl^{-}, Br^{-}, $SO_4{}^{2-}$ $CH3COO^{-}$, $NO_3{}^{-}$	[157]
Flucarbazone	GC/PANI-MIP/FLU/ISE	Potentiometry	10^{-2} to 10^{-5} M	5.8×10^{-6} M		Soil	pH 7 (PBS)		Diquate, cyromazine, dinotefuran, bispyribac, $SO_4{}^{2-}$ $NO_3{}^{-}$, $CH3COO^{-}$ CN^{-}	[158]
Flucarbazone	GC/PANI-Aliquat/FLU/ISE	Potentiometry	10^{-2}–3×10^{-5} M	8.5×10^{-6} M		Soil	pH 7 (PBS)		Diquate, cyromazine, dinotefuran, bispyribac, $SO_4{}^{2}$ $^{-}NO_3{}^{-}$, $CH3COO^{-}$ CN^{-}	[158]
Glyphosate	Urease-GNPs/ISE	Potentiometry	0.5 ppm to 50 ppm	0.5 ppm		Drinking water			Dichlorvos, dimethoate, 2–4D, hexaconazole, paraquat, dichloride	[141]

Table 13.8. Bismuth sensor in herbicide analysis.

Analyte	Electrode	Method	Linear range	LOD	LOQ	Samples	Optimum pH	Peak potentials	İnterference	References
Alachlor	BiE	CV					pH 7.0 (Na_2SO_4)			[160]
Diquat	BiFE	SWV	0.19×10^{-6} to 9.30×10^{-6} M	8.7×10^{-8} M		Water	pH 4.5 (acetate buffer)	−0.6 V	Alkalines, earth alkalines, Hg^{2+}, Pb^{2+}, $Co^{2+}Ni^{2+}$, Cd^{2+}, Cu^{2+}, humic acid	[159]
Paraquat	BiFE	SWV	0.12×10^{-6} to 4.20×10^{-6} M	1.2×10^{-8} M		Water	pH 4.5 (acetate buffer)	−0.6 V	Alkalines, earth alkalines, Hg^{2+}, Pb^{2+}, $Co^{2+}Ni^{2+}$, Cd^{2+}, Cu^{2+}, humic acid	[159]
Pendimethalin	BiFE	SWCV	0.3×10^{-6} M to 1.0×10^{-6} M	37 nM	112.2 nM	Tap water, sea water	pH 3.0 (HCl)	−0.585 V		[161]

Table 13.9. Metal sensor platform in herbicide analysis.

Analyte	Electrode	Method	Linear range	LOD	LOQ	Samples	Optimum pH	Peak Potentials	Interference	References
2,4-DB	MIP/PtE	Amperometry	200 μM to 2 mM						2,4-DP, DAA, PAA, POAA, 4-chlorocathecol, 2-dhlorophenol	[162]
Acetochlor	AgE	CV								[51]
Aclonifen	m-AgSAE	DPV	2 ×10–10to 1 ×10–6 M	1.6 ×10–10 M		Drinking water	pH 9.0 (BRB)	−0.8 V		[163]
Aclonifen	m-AgSAE	DPV	2 ×10–9 to 1 ×10–6 M	1.9 ×10–9 M		River water	pH 12.0 (BRB)	−0.8 V		[163]
Ametryn	AuE	SWV	4.60 $\times 10^{-8}$to 7.20 $\times 10^{-7}$ M	0.021 μ g g^{-1}		Soil	pH 3.3 (BRB)	+0.5 V	Atrazine	[164]
Chlornitrofen	AgAFE	SW-AdSV	1.0 × 10−7 M to 1.5 × 10−6 M	3.0 × 10−8 M	9.9 × 10−8 M	River water	pH 8.8 (BRB)	−0.6 V	Oxycarboxine, fenfuram, bromacil, cadmium, copper, zinc	[165]
Difenzoquat	m-AgSAE	DPV	0 to 4 μM		0.4 μM	Deionised water, river water	pH 12.0 (BRB)	−1.39 V		[166]
Dimethenamid-P	AuE	CV					0.05 M $NaHCO_3$	−0.2 V		[76]
Dinitramine	Au NPs–S–$(CH_2)_3$–Si–SiO_2/Fe_3O_4/AuE	SWV	5.0 $\times 10^{-9}$ to −7.0 $\times 10^{-6}$M	0.3 $\times 10^{-9}$ M		Water	pH 7.0 (BRB)		Trifluralin, 2-methyl, 4,6-dinitropheol, Pb^{2+}, Al^{3+}, Fe^{3+} Br^-, I^-, $NO_3^-$$CO_3^{2-}$, SO_4^{2-}	[167]
Diquat	Thiol/GNP/CA DNA/AuE	DPV	1.0 $\times 10^{-9}$ M to 1.2 $\times 10^{-6}$ M	2.0 $\times 10^{-10}$ M		Human urine, grain	pH 8.0 (PBS)	−0.6 V	Na^+, K^+, $Ca^{2+}$$Mg^{2+,}$ $Zn^{2+,}$ Fe^{3+} Cl^-, SO_4^{2-}, $CO_3^{2-}$$PO_4^{3-}$, lysine, glucose, acetic acid, dopamine, ascorbic acid, glutamic acid, L-cystine, uric acid	[168]
Diquat	NP/PtE	SWV	9.5 × 10^{-8} to 1.6 × 10^{-4} M	5.27 M and 3.16 M	10.54 and 17.57 M	Water	1.0 × 10^{-4} M K_2SO_4	−0.7 V and −1.0 V	Zn^{2+}, Cu^{2+}, $Fe^{2+}$$Hg^{2+}$, MnO_4^-	[169]
Glyphosate	MIP-MOF/AuE	LSV	5.9 fM to 5.9 nM	5 fM		Water	pH 7.2 (PBS)			[170]

Glyphosate	Ppy-MIP/AuE	Amperometry	1 pM to 1 nM	1 pM			Methanol + Acetic acid (1:1)		Gluphosinate, ammonium, simetryn, omethoate, methidathion, terbuthylazine-2-hydroxy	[171]
Glyphosate	Ppy-MIP/AuE	DPV	5 to 800 ng mL^{-1}	0,27 mg mL^{-1}		Cucumber, water	Formic acid buffer	+0.1 V	AMPA, chlorpyrifos, aldicarb	[172]
Glyphosate	rGO/DWCNTs/ Oct-Fe_3O_4/Cs/ SPAuE	SWV	0.1 to 1000 ppb	0.08 ppb		River water		+0.2 V		[173]
Glyphosate	AuE	Amperometry	0 to 50 μM			Water	0.5 M H_2SO_4			[174]
Lactofen	AgAFE	SWSV	10 to 100 nM	2.0 nM	6.5 nM	Water	pH 8.5 (Borate buffer)	−0.5 V	dsDNA, nickel, copper, cobalt, zinc, cimetidine, dinotefuran, blasticidine, aclonifen, metam sodium	[85]
Molinate	AgSAE	SWV	9.36 to 243.49 μg L^{-1}	2.34 μg L^{-1}	7.81 μg L^{-1}	RiceRiver water	pH 4.0 (BRB)	−0.37 V	K^+, Na^+, $NH4^+OH^-$, CO_3^{2-}, SO_4^{2-}, $PO_4^{3-}Cl^-$, Fe^{2+}, Cd^{2+} Cu^{2+}, terbutryn, thiodicarb, atrazine, carbendazim, glyphosate, linuron	[175]
Picloram	m-AgSAE	DPV	1 $\times 10^{-7}$ M to 8 $\times 10^{-5}$ M	8.8 $\times 10^{-8}$ M	1.37 $\times 10^{-7}$ M	Water	0.1 M H_2SO_4		Glyphosate, triclopyr, metamitron, triasulfuron, clopyralid, Cu^{2+}, Pb^{2+}, Zn^{2+}, Tl^+, Cd^{2+}	[144]
Profluralin	AgAFE	SW-AdSV	6.0 $\times 10^{-8}$ to 8.0 $\times 10^{-7}$ M	1.7 × 10^{-8} M	5.7 × 10^{-8} M	Tap water, river water, soil	pH 8.5 (BRB)		DNA	[176]
Profluralin	AgAFE	SWV	0.4 × 10^{-7} to 7.0 × 10^{-6} M and	2.9 × 10^{-8} M	9.7 × 10^{-8} M	Tap water, river water, soil	pH 8.5 (BRB)	−0.3 V −0.5 V	DNA	[176]
Prometon	CPTS andHPMo ormosil deposited AuE	DPV	3,6 $\times 10^{-6}$ to 18 $\times 10^{-6}$ M				pH 1.0 (H_2SO_4)	+ 0.75 and −1.0 V		[177]
Simazine	ATP@AuNPs/AuE	Amperometry	0.03 to 140 μM	0.013 μM		Water, soil	pH 1.7 (H_2SO_4)		Picloram, acetochlor, terbutylazine	[178]

(Continued)

Table 13.9. (*Continued*)

Analyte	Electrode	Method	Linear range	LOD	LOQ	Samples	Optimum pH	Peak Potentials	İnterference	References
Simazine	PtE	SPEC-SPEF					pH 3.0 (K2SO4)		$NO_3^-NH_4^+$	[179]
Terbuthylazine	MIP/AuE	DPV	2.5×10^{-7} to 1.2×10^{-4} M	2.0×10^{-8} M		Tomato, soil	pH 7.0	+0.1 V		[180]
Terbutryn	m-AgSAE	DPV	1×10^{-7} to 1×10^{-5} M	2.9×10^{-8} M	9.8×10^{-8} M	River water, drinking water	pH 2.0 (BRB)		Cu^{2+}, Pb^{2+}, Zn^{2+}, Tl^{+}, Cd^{2+}, triclopyr, glyphosate, clopyralid, triasulfuron, atrazine, imidacloprid, chlorpyrifos, bentazon, chloridazon, picloram, metamitron, tribenuron, metribuzin	[123]
Terbutryn	p-AgSAE	DPV	2.9×10^{-8} to 5.9×10^{-5} M	4.3 ×10-9 M	1.4×10^{-8} M	River water, drinking water	pH 2.0 (BRB)			[123]
Trifluralin	p-AgSAE	SWV	0.25 to 1.75 μM	4.56 and 4.16 nM	14.6 nM and 13.30 nM	Eggplant, carrot, tomato, grape juice, orange juice	pH 7.0 (BRB)	−0.5 and −0.6 V	Phenolic compounds, organic acids, sugars	[181]

Demir *et al* in their scientific paper published in 2021, developed a modified sensor based on iron (III) phthalocyanine (FePc) for the first time for the determination of fluometuron herbicide [33]. For this purpose, FePc/MWCNT hybrid sensor was produced by chemically reacting FePc material and amine multi-walled carbon nanotube powders (NH_2-MWCNTP). Next, they coated this composite material on the GCE and MWCNTPE electrode surface by the drop-drying technique. Then, they elucidated the electrode surface morphology with scanning electron microscopy (SEM), Fourier transform infrared spectroscopy (FTIR) and, x-ray diffraction (XRD). They proved that the developed hybrid sensor is six times more sensitive to the anodic peak signal of fluometuron herbicide than the bare electrodes. The working range with DPSV and FePc/MWCNTPE sensor was found to be 0.4–15.0 mg L^{-1}. Furthermore, the limit of detection (LOD) and limit of quantification (LOQ) values were calculated as 69.8 and 233.0 µg L^{-1}, respectively. They carried out studies to elucidate the electrode mechanism of fluometuron herbicide by cyclic voltammetry (CV). Finally, they performed the analytical application of Cottonex 500 SC®, a commercial formulation, and tap water, on the proposed DPSV method and the recommended FePc/MWCNTPE (table 13.10). As a result, for the first time, they have developed a novel hybrid sensor by electrochemical method, and they have succeeded in analyzing fluometuron herbicide quickly, reliably, and accurately [33].

13.2 Conclusion

Herbicide analyses are of great importance for human health and food safety. These analyses should enable reliable, fast, inexpensive, portable, and online determinations. In this respect, the electrochemical sensor is extremely promising. Especially with the developing technology, it has become possible to develop new generation hybrid sensors by using three-dimensional materials. Due to their portability, herbicide analyses can be carried out in the field. In addition, the production of nanosensors from many different materials, the improvement of the electrode surface, the increase of conductivity, and the creation of stable structures have started to take place in the routine analysis of herbicides. In the last two decades, these sensors have opened new horizons in the scientific world due to their unique physicochemical properties such as good conductivity, excellent biocompatibility, and excellent catalytic. Moreover, this fast, simple, reliable, and inexpensive sensor platform has played a key role in the determination of numerous analytes compared to the existing classical analytical methods. Modified sensors have been the sensors that received the most attention.

Some ambiguous contractions

ADPE/MGRPE:	Graphene-paste electrode modified with an azo dye, 2-(4-((4-acetylphenyl)diazenyl)phenylamino)ethanol
AgAFE:	Silver amalgam film e lectrode
Ag-Au BMNPs:	Silver-gold core-partial shell bimetallic nanoparticles
AIRGOC:	Atrazine (ATZ)-imprinted reduced graphene oxide composite

Table 13.10. Others sensors in herbicide analysis.

Analyte	Electrode	Method	Linear range	LOD	LOQ	Samples	Optimum pH	Peak potentials	İnterference	References
Acetochlor	Ab-Pd@PtNPs 3D-printed sensor	NEMEIS	1 to 1000 ppb	3.2 ppb						[182]
Asulam	TNA/HRP biosensor	PEC biosensor	0.02 to 2.0 ng mL^{-1}	4.1 pg mL^{-1}		Water	pH 7.4 (PBS)		2, 4, 6-tri-chlorophenol, phenol, paraquat, atrazine, amitrole2, 4, 5-T	[183]
Benzobicyclon	FTO electrodes	SWV	10.00 to 582.96 μM	0.38 μM		Brown rice, rice straw	pH 7.0 (PBS)	+0.3 V	K^+, Na^+, Ca^{2+}, Mg^{2+}, rutin, vitamin B1, vitamin B2, p-hydroxybenzoic acid, gentisic acid, p-coumaric acid, vanillic acid, ellagic acid, ferulic acid, gallic acid	[184]
Bispyribac	PVC membrane	Potentiometry	1.0×10^{-2} to 9.0×10^{-6} M	1.81 μg mL^{-1}						[157]
Cacodylic acid	AgDDTC/Membrane sensor films	Potentiometry	1.7×10^{-5} to 1.0×10^{-2} M	1.0×10^{-5} M		Human urine	pH 4.0 (BRB)		I^-, NO_3^-, $NO_2^-ClO_4^-$, SO_3^{2-}, $N_3^-Cl^-$, PO_4^{3-}, $Br^-S_2O_3^{3-}$ SCN^- $CH3COO^-$ $C_2O_4^{2-}$, citrate, benzoate, salicylate	[185]
Dichlorprop	Chiral molecularly imprinted TiO_2 photoelectrodes	PEC oxidation							S-DCPP, R-DCPP, humic acid, glyphosate, omethoate, metolachlor, atranex	[186]
DNOC	Antimony Film Electrodes	SWV	1.0×10^{-6} to 15×10^{-6} M	1.12×10^{-6} M	1.42×10^{-6} M	Water	pH 2.0 (BRB)	−0.3 and −0.5 V		[187]
Glyphosate	Cu-Ppy/Activated carbon fiber	DPV	0.02 to 12 mg L^{-1}	0.01 mg L^{-1}	0.032 mg L^{-1}	Soil, fruit	(PBS)	−0.1 V	SO_4^{2-}, NH_4^+Tcf, AMPA	[188]
Methyl iodide	SCE	CV					pH 4.5		(Na2SO4)	[189]
Oxadiargyl	Sol–Gel Based Fibers	CV								[190]
Paraquat	Metal-catex composite electrode	DP-AdCSV	3.8×10^{-8} to 7.7×10^{-7} M	2.3 nM	7.78 nM	Juices, potato, onion	pH 6.5		Cl^-, PO_4^{3-}, SO_4^{2-}, NO_3^-, Ca^{2+}, Mg^{2+}	[191]
Paraquat	Paper-based silver sensor	SWV	3.0 to 100 μM	0.80 μM		Water, human serum, orange juice	0.1 M Na_2SO_4		Cu^{2+}, Fe^{2+}, Fe^{3+}, glucose, albumin, ascorbic acid	[192]
Simetryn	SCE	CV				Water	pH 2.3			[193]
Thiazopyr	SCE	DPP	0.43 to 38.6 mg mL^{-1}	0.127 mg mL^{-1}	0.430 mg mL^{-1}	Fruit juice, soil	pH 7.0 (BRB)	−1.2 V	Atrazine, acifluorfen, azinphos methyl	[194]
Trifluralin	Ag-citrate/GQDs/leaf electrode	CV	0.005 to 1 mM		5 μM	Apple skin, surface of leaves	pH 7.4 (PBS)			[195]
Trifluralin	Ag-citrate/GQDs/leaf electrode	DPV	0.001 to 0.08 mM		1 μM	Apple skin, surface of leaves	pH 7.4 (PBS)			[195]
Trifluralin	Ag-citrate/GQDs/leaf electrode	SWV	0.01 to 1 mM		10 μM	Apple skin, surface of leaves	pH 7.4 (PBS)			[195]
Trifluralin	Au@Ag cores-shell/GQDs/wearable sensing systems	Amperometry	10 nM to 1 mM		10 nM	Water, real leaves, apple	pH 7.0			[196]

Au NPs:	Gold nanoparticles
BDDE:	Boron-doped diamond electrode
BiFE:	*In situ* plated bismuth-film electrode
CB:	Carbon black
CD/CNO/HRP biosensor:	Carbon nano-onion/horseradish peroxidase conjugate on cyclodextrin polymer
CPE:	Carbon paste electrode
CTS:	Chitosan
DME:	Dropping mercury electrode
DPP:	Differential pulse polarography
ECH:	Epichlorohydrin
ERGO:	Electrochemically reduced graphene oxide
GCE:	Glassy carbon electrode
HMDE:	Hanging mercury drop electrode
Ionic liquid:	IL
ISE:	Ion selective electrodes
M-AgSAE:	Meniscus modified silver solid amalgam electrode
MWCNT:	Multi-walled carbon nanotube
NEMEIS:	Nanomaterial-enhanced multiplex electrochemical immunosensing
NiHCF NPs:	Nickel hexacyanoferrate nanoparticles
PANI-Aliquat:	Polyanline and aliquat
PGE:	Pencil graphite electrode
PLC:	Poly-l-cystein
PVA-SbQ:	Photocrosslinkablepoly-(vinylalcohol) bearing styrylpyridinium groups
SCE:	Saturated calomel electrodes
SPE:	Screen-printed electrode
SWCV:	Square wave cathodic voltammetry
TRGOPE:	Thermally reduced graphene oxide paste electrode

References

[1] Agar S, Aydinoglu H, Temel O, Ikizunal K and Ece H 1991 Pestisit kullanımının tarihçesi, bugünü ve geleceği *Turkish J. Entomol.* **15** 247–56

[2] Manno M 1996 Herbicides *Human Toxicology* (Amsterdam: Elsevier) pp 551–60

[3] Karanth S 2014 Chlorophenoxy Herbicides *Encyclopedia of Toxicology* 3rd edn vol 1 (Amsterdam: Elsevier) pp 900–2

[4] Xia Z-Y *et al* 2017 Biodegradation of the herbicide 2,4-dichlorophenoxyacetic acid by a new isolated strain of *Achromobacter* sp. LZ35 *Curr. Microbiol.* **74** 193–202

[5] BCPC 2021 https://pesticidecompendium.bcpc.org/class_herbicides.html

[6] Manjunatha J G, Kumara Swamy B E, Mamatha G P, Gilbert O and Sherigara B S 2011 Poly(maleic acid) modified carbon paste electrode for simultaneous detection of dopamine in the presence of uric acid: a cyclic voltammetric study *Anal. Bioanal. Electrochem.* **3** 146–59 https://researchgate.net/publication/279612676_Polymaleic_acid_modified_carbon_paste_electrode_for_simultaneous_detection_of_dopamine_in_the_presence_of_uric_acid_A_cyclic_-voltammetric_study

[7] Manjunatha J G, Kumara Swamy B E and Deraman M 2013 Electrochemical studies of dopamine, ascorbic acid and their simultaneous determination at a poly (rosaniline)

modified carbon paste electrode: a cyclic voltammetric study *Anal. Bioanal. Electrochem.* **5** 426–38

[8] Manjunatha J G, Swamy B E K, Deraman M and Mamatha G 2013 Simultaneous determination of ascorbic acid, dopamine and uric acid at poly (aniline blue) modified carbon paste electrode: a cyclic voltammetric study *Int. J. Pharm. Pharm. Sci.* **5** 355–61

[9] Hareesha N and Manjunatha J G 2020 A simple and low-cost poly (dl-phenylalanine) modified carbon sensor for the improved electrochemical analysis of Riboflavin *J. Sci. Adv. Mater. Devices* **5** 502–11

[10] Manjunatha J G 2020 A surfactant enhanced graphene paste electrode as an effective electrochemical sensor for the sensitive and simultaneous determination of catechol and resorcinol *Chem. Data Collect.* **25** 100331

[11] Tigari G and Manjunatha J G 2020 A surfactant enhanced novel pencil graphite and carbon nanotube composite paste material as an effective electrochemical sensor for determination of riboflavin *J. Sci. Adv. Mater. Devices* **5** 56–64

[12] Yaghoubi A, Ghojazadeh M, Abolhasani S, Alikhah H and Khaki-Khatibi F 2015 Correlation of serum levels of vitronectin, malondialdehyde and HS-CRP with disease severity in coronary artery disease *J. Cardiovasc. Thorac. Res.* **7** 113–17

[13] Tigari G and Manjunatha J G 2019 Electrochemical preparation of poly(arginine)-modified carbon nanotube paste electrode and its application for the determination of pyridoxine in the presence of riboflavin: an electroanalytical approach *J. Anal. Test.* **3** 331–40

[14] Hareesha N and Manjunatha J G 2020 Elevated and rapid voltammetric sensing of riboflavin at poly(helianthin dye) blended carbon paste electrode with heterogeneous rate constant elucidation J. Iran. Chem. *Soc.* **17** 1507–19

[15] Hareesha N and Manjunatha J G 2018 Sensitive electrochemical analysis of resorcinol using polymer modified carbon paste electrode: a cyclic voltammetric study *Anal. Bioanal. Electrochem.* **10** 488–98

[16] Hareesha N and Manjunatha J G 2020 Fast and enhanced electrochemical sensing of dopamine at cost-effective poly(DL-phenylalanine) based graphite electrode *J. Electroanal. Chem.* **878** 114533

[17] Hareesha N, Manjunatha J G, Raril C and Tigari G 2019 Sensitive and selective electrochemical resolution of tyrosine with ascorbic acid through the development of electropolymerized alizarin sodium sulfonate modified carbon nanotube paste electrodes *ChemistrySelect* **4** 4559–67

[18] Demir E and Silah H 2020 Development of a new analytical method for determination of veterinary drug oxyclozanide by electrochemical sensor and its application to pharmaceutical formulation *Chemosensors* **8** 25

[19] Demir E 2019 Sensitive and selective pathway of total antioxidant capacity in commercially lemon, watermelon and mango-pineapple cold teas by square wave adsorptive stripping voltammetry *Gazi Univ. J. Sci.* **32** 1123–36

[20] Demir E 2019 A simple and sensitive square wave stripping pathway for the analysis of desmedipham herbicide by modified carbon paste electrode based on hematite (α-Fe2O3 nanoparticles) *Electroanalysis* **31** 1545–53

[21] Khan I, Pandit U J, Wankar S and Limaye S N 2017 Design of electrochemical sensor based on fMWCNT-CPE decorated with ti nanofilm and its electrocatalytic behavior towards Aminotriazole *Electrocatalysis* **8** 196–213

[22] Ilager D, Seo H, Shetti N P and Kalanur S S 2020 CTAB modified Fe-WO3 as an electrochemical detector of amitrole by catalytic oxidation *J. Environ. Chem. Eng.* **8** 104580

[23] Ilager D, Seo H, Shetti N P, Kalanur S S and Aminabhavi T M 2020 Electrocatalytic detection of herbicide, amitrole at WO3·0.33H2O modified carbon paste electrode for environmental applications *Sci. Total Environ.* **743** 140691

[24] Kardaş F, Beytur M, Akyıldırım O, Yüksek H, Yola M L and Atar N 2017 Electrochemical detection of atrazine in wastewater samples by copper oxide (CuO) nanoparticles ionic liquid modified electrode *J. Mol. Liq.* **248** 360–63

[25] Annu S, Sharma A, Nitin and Jain R 2020 Cellulose fabricated pencil graphite sensor for the quantification of hazardous herbicide atrazine *Diam. Relat. Mater.* **105** 107788

[26] Rasmussen M and Minteer S D 2013 Self-powered herbicide biosensor utilizing thylakoid membranes *Anal. Methods* **5** 1140

[27] Ghorbani A, Ganjali M R, Ojani R and Raoof J 2020 Detection of chloridazon in aqueous matrices using a nano-sized chloridazon-imprinted polymer-based voltammetric sensor *Int. J. Electrochem. Sci.* **15** 2913–22

[28] Gholivand M B, Torkashvand M and Malekzadeh G 2012 Fabrication of an electrochemical sensor based on computationally designed molecularly imprinted polymers for determination of cyanazine in food samples *Anal. Chim. Acta* **713** 36–44

[29] Gholivand M B, Shariati-Rad M, Karimian N and Torkashvand M 2012 A chemometrics approach for simultaneous determination of cyanazine and propazine based on a carbon paste electrode modified by a molecularly imprinted polymer *Analyst* **137** 1190–8

[30] Amouzad F and Zarei K 2021 Electrochemical determination of dinitramine in water samples using a pencil graphite electrode modified with poly-L-cystein-gold nanoparticle *Chem. Pap.* **75** 493–501

[31] Echajia M, Haimouti A E L, Mbarki M and Oulfajrite H 2020 Electroanalysis of the herbicide diquat by electrochemical detector containing particles of clay in environmental *Anal. Bioanal. Electrochem.* **12** 168–79

[32] Demir E and İnam R 2017 Square wave voltammetric determination of fomesafen herbicide using modified nanostructure carbon paste electrode as a sensor and application to food samples *Food Anal. Methods* **10** 74–82

[33] Demir E, Göktug Ö, İnam R and Doyduk D 2021 Development and characterization of iron (III) phthalocyanine modified carbon nanotube paste electrodes and application for determination of fluometuron herbicide as an electrochemical sensor *J. Electroanal. Chem.* **895** 115389

[34] Mbokana J G Y, Dedzo G K and Ngameni E 2019 Grafting of organophilic silane in the interlayer space of acid-treated smectite: application to the direct electrochemical detection of glyphosate *Appl. Clay Sci.* **188** 105513

[35] Youssef Ibrahim Moharram I S, El-Hallag and Selim S M 2021 Quantitative Evaluation of Generic Glyphosate Using Carbon Paste Electrode Electrochemically Modified with Copper Ions *Russ. J. Electrochem.* **57** 644–53

[36] Gholivand M-B, Akbari A and Norouzi L 2018 Development of a novel hollow fiber-pencil graphite modified electrochemical sensor for the ultra-trace analysis of glyphosate *Sens. Actuators B Chem.* **272** 415–24

[37] Toro M J U, Marestoni L D and Sotomayor M D P T 2015 A new biomimetic sensor based on molecularly imprinted polymers for highly sensitive and selective determination of hexazinone herbicide *Sens. Actuators B Chem.* **208** 299–306

[38] Khan I, Pandit U J and Limaye S N 2017 Design of electrochemically modified fMWCNT-pencil graphite electrode decorated with Cu and Ag nanofilm and its electrocatalytic behavior towards imazethapyr *Electroanalysis* **29** 2423–36
[39] Shetti N P, Malode S J, Ilager D, Raghava Reddy K, Shukla S S and Aminabhavi T M 2019 A novel electrochemical sensor for detection of molinate using ZnO nanoparticles loaded carbon electrode *Electroanalysis* **31** 1040–9
[40] Brycht M *et al* 2018 Synthesis and characterization of the thermally reduced graphene oxide in argon atmosphere, and its application to construct graphene paste electrode as a naptalam electrochemical sensor *Anal. Chim. Acta* **1035** 22–31
[41] Saber-Tehrani M, Pourhabib A, Husain S W and Arvand M 2012 Enhanced electrocatalytic reduction of oxadiargyl and its determination on 2-(4-((4-acetylphenyl)diazenyl) phenylamino)-ethanol modified graphene-paste electrode *Electroanalysis* **24** 2395–404
[42] Demir E and Inam R 2018 Voltammetric determination of phenmedipham herbicide using a multiwalled carbon nanotube paste electrode *Turkish J. Chem.* **42** 997–1007
[43] Gholivand M B, Karimian N and Malekzadeh G 2012 Computational design and synthesis of a high selective molecularly imprinted polymer for voltammetric sensing of propazine in food samples *Talanta* **89** 513–20
[44] Özcan A, Gürbüz M and Özbal A 2018 Preparation of a double-step modified carbon paste electrode for the voltammetric determination of propham via bulk modification with fumed silica and drop-casting of maghemite-modified fumed silica nanocomposite *Sens. Actuators B Chem.* **255** 1517–24
[45] Özcan A, Gürbüz M and Özcan A A 2018 Preparation of a disposable and low-cost electrochemical sensor for propham detection based on over-oxidized poly(thiophene) modified pencil graphite electrode *Talanta* **187** 125–32
[46] Zainul R *et al* 2019 Zinc/aluminium–quinclorac layered nanocomposite modified multiwalled carbon nanotube paste electrode for electrochemical determination of bisphenol A *Sensors* **19** 941
[47] Arvand M, Vaziri M and Zanjanchi M A 2012 Electrochemical behavior and differential pulse voltammetric detection of thiobencarb on 2-(4-((4-ethoxyphenyl)diazenyl) phenylamino)ethanol-modified carbon paste electrode *J. Solid State Electrochem.* **16** 1151–59
[48] Haghighi M, Irandoust M and Shariati-rad M 2019 Simultaneous determination of antinonin and trifluralin by electrochemical method and net analyte signal interferent modeling *Microchem. J.* **146** 34–40
[49] Gege Ü, Karakaya S and Dilgin Y 2021 Sensitive electrochemical determination of trifluralin at a disposable pencil graphite electrode *Electroanalysis* **33** 1547–56
[50] Zhou X *et al* 2020 An effective strategy for developing the CoMoS nanosheets wrapped by oxidized multi-walled carbon nanotubes as an electrosensor of oryzalin *J. Electroanal. Chem.* **878** 114710
[51] Couto Petro A G, Thapa B, Karty J A, Raghavachari K, Baker L A and Peters D G 2020 Direct electrochemical reduction of acetochlor at carbon and silver cathodes in dimethylformamide *J. Electrochem. Soc.* **167** 155517
[52] İnam R and Çakmak Z 2013 A simple square wave voltammetric method for the determination of aclonifen herbicide *Anal. Methods* **5** 3314
[53] Shetti N P *et al* 2019 Electro-sensing base for herbicide aclonifen at graphitic carbon nitride modified carbon electrodeWater and soil sample analysis *Microchem. J.* **149** 103976

[54] Mutharani B, Ranganathan P, Chen S-M and Deepak Vishnu S K 2020 Stimuli-enabled reversible switched aclonifen electrochemical sensor based on smart PNIPAM/PANI-Cu hybrid conducting microgel *Sens. Actuators B Chem.* **304** 127232
[55] Gopi P K *et al* 2021 Electrochemical sensing base for hazardous herbicide aclonifen using gadolinium niobate ($GdNbO_4$) nanoparticles-actual River water and soil sample analysis *Ecotoxicol. Environ. Saf.* **207** 111285
[56] Elshafey R and Radi A E 2018 Electrochemical impedance sensor for herbicide alachlor based on imprinted polymer receptor *J. Electroanal. Chem.* **813** 171–7
[57] Devasenathipathy R 2017 Fabrication of graphene oxide-MWCNTs nanocomposite modified glassy carbon electrode for the sensitive determination of amitrole *Int. J. Electrochem. Sci.* **12** 5888–97
[58] Mafuwe P T, Moyo M, Mugadza T, Shumba M and Nyoni S 2019 Cobalt oxide nanoparticles anchored polyaniline-appended cobalt tetracarboxy phthalocyanine, modified glassy carbon electrode for facile electrocatalysis of amitrole *J. Solid State Electrochem.* **23** 285–94
[59] Zhou S *et al* 2019 Platelet nitrogen and sulfur Co-doped ordered mesoporous carbon with inexpensive methylene blue as a single precursor for electrochemical detection of herbicide amitrole *Nano* **14** 1950104
[60] Malode S J, Keerthi Prabhu K and Shetti N P 2020 Electrocatalytic behavior of a heterostructured nanocomposite sensor for aminotriazole *New J. Chem.* **44** 19376–84
[61] Zhou S *et al* 2021 Simplified synthesis of N-doped carbon nanotube arrayed mesoporous carbon for electrochemical detection of amitrole *Chem. Phys.* **542** 111074
[62] Roushani M, Mohammadi F and Valipour A 2020 Electroanalytical sensing of asulam based on nanocomposite modified glassy carbon electrode *J. Nanostructures* **10** 128–39
[63] Fan L *et al* 2019 Design of a facile and label-free electrochemical aptasensor for detection of atrazine *Talanta* **201** 156–64
[64] Durai L and Badhulika S 2020 Highly selective trace level detection of Atrazine in human blood samples using lead-free double perovskite Al_2NiCoO_5 modified electrode via differential pulse voltammetry *Sens. Actuators B Chem.* **325** 128792
[65] Zapardiel A, Bermejo E, Pérez J A and Chicharro M 2000 Determination of s-triazines with copper and glassy carbon electrodes. Flow injection analysis of aziprotryne in water samples *Fresenius J. Anal. Chem.* **367** 461–66
[66] Bakhsh H *et al* 2021 Sodium dodecyl sulfate stabilized NiO nanoseeds: a potential procedure for ultra-sensitive determination of bentazone in vegetables *J. Mater. Sci., Mater. Electron.* **32** 15917–29
[67] Vaz F C, Silva T A, Fatibello-Filho O, Assumpção M H M T and Vicentini F C 2021 A novel carbon nanosphere-based sensor used for herbicide detection *Environ. Technol. Innov.* **22** 101529
[68] Buledi J A *et al* 2021 Exploring electrocatalytic proficiencies of CuO nanostructure for simultaneous determination of bentazone and mexacarbate pesticides *Appl. Nanosci.* **11** 1889–902
[69] Salehzadeh H, Ebrahimi M, Nematollahi D and Salarian A A 2016 Electrochemical study of fenitrothion and bifenox and their simultaneous determination using multiwalled carbon nanotube modified glassy carbon electrode *J. Electroanal. Chem.* **767** 188–94

[70] de F, Morawski M, Caon N B, Sousa K A P, Faita F L, Parize A L and Jost C L 2021 Hybrid chitosan-coated manganese ferrite nanoparticles for electrochemical sensing of bifenox herbicide *J. Environ. Chem. Eng.* **9** 106298

[71] Gerent G G, Santana E R, Martins E C and Spinelli A 2021 A non-mercury electrode for the voltammetric determination of butralin in foods *Food Chem.* **343** 128419

[72] Fátima Barroso M, João Ramalhosa M, Delerue-Matos C, Vaz M C V F and Sales M G F 2008 Electrochemical study of butylate: application to the analysis of water *Int. J. Environ. Anal. Chem.* **88** 1049–62

[73] Ranđelović M S, Momčilović M Z, Milićević J S, Đurović-Pejčev R D, Mofarah S S and Sorrel C C 2019 Voltammetric sensor based on Pt nanoparticles suported MWCNT for determination of pesticide clomazone in water samples *J. Taiwan Inst. Chem. Eng.* **105** 115–23

[74] Özcan A and Gürbüz M 2018 Development of a modified electrode by using a nanocomposite containing acid-activated multi-walled carbon nanotube and fumed silica for the voltammetric determination of clopyralid *Sens. Actuators B Chem.* **255** 262–67

[75] Gholivand M B, Karimian N and Torkashvand M 2015 A graphene-based electrochemical sensor for sensitive determination of cyanazine *J. Anal. Chem.* **70** 384–91

[76] Glavaški O S *et al* 2014 Electrochemical degradation of the pesticide Dimethenamid-P at Gold, DSA platinum and ruthenium oxide electrodes in different electrolytes *Electroanalysis* **26** 1877–80

[77] Irandoust M and Haghighi M 2016 Electrochemical study and determination of dinitramine using glassy carbon electrodes modified with multi-walled carbon nanotubes *Electrochemistry* **84** 228–33

[78] Zambianco N A, Silva T A, Zanin H, Fatibello-Filho O and Janegitz B C 2019 Novel electrochemical sensor based on nanodiamonds and manioc starch for detection of diquat in environmental samples *Diam. Relat. Mater.* **98** 107512

[79] Tchoffo R, Ngassa G B P, Tonlé I K and Ngameni E 2021 Electroanalysis of diquat using a glassy carbon electrode modified with natural hydroxyapatite and β-cyclodextrin composite *Talanta* **222** 121550

[80] Zarei K and Khodadadi A 2017 Very sensitive electrochemical determination of diuron on glassy carbon electrode modified with reduced graphene oxide–gold nanoparticle–Nafion composite film *Ecotoxicol. Environ. Saf.* **144** 171–7

[81] de Lima C A, Santana E R, Piovesan J V and Spinelli A 2016 Silver nanoparticle-modified electrode for the determination of nitro compound-containing pesticides *Anal. Bioanal. Chem.* **408** 2595–606

[82] del Carmen Aguirre M, Urreta S E and Gomez C G 2019 A Cu^{2+}–Cu/glassy carbon system for glyphosate determination *Sens. Actuators B Chem.* **284** 675–83

[83] Sarigül T, Inam R and Aboul-Enein H Y 2010 Electro-oxidation of herbicide halosulfuron methyl on glassy carbon electrode and applications *Talanta* **82** 1814–9

[84] Montoya M R, Pintado S and Mellado J M R 2010 Electrochemical reduction of imazamethabenz methyl on mercury and carbon electrodes *Electrochim. Acta* **55** 3164–70

[85] Guziejewski D, Morawska K, Popławski T, Metelka R, Ciesielski W and Smarzewska S 2018 Lactofen—electrochemical sensing and interaction with dsDNA *Electroanalysis* **30** 94–100

[86] Shapovalova E N, Yaroslavtseva L N, Merkulova N L, Yashin A Y and Shpigun O A 2009 Separation of pesticides by high-performance liquid chromatography with amperometric detection *J. Anal. Chem.* **64** 164–70

[87] Rahemi V, Garrido J M P J, Borges F, Brett C M A and Garrido E M P J 2015 Electrochemical sensor for simultaneous determination of herbicide MCPA and its metabolite 4-chloro-2-methylphenol. Application to photodegradation environmental monitoring *Environ. Sci. Pollut. Res.* **22** 4491–9
[88] Garrido E M, Lima J C, Delerue-matos C M and Brett A M O 1999 Electrochemical behaviour and square wave voltammetry of the rice herbicides molinate, bensulfuron-methyl, mefenacet and thiobencarb *Int. J. Environ. Anal. Chem.* **75** 149–57
[89] Deroco P B, Lourencao B C and Fatibello-Filho O 2017 The use of modified electrode with carbon black as sensor to the electrochemical studies and voltammetric determination of pesticide mesotrione *Microchem. J.* **133** 188–94
[90] Barchanska H, Markowski P and Strzebin M 2018 Electrochemical determination of mesotrione and its degradation products on glassy carbon electrode *Int. J. Environ. Anal. Chem.* **98** 493–505
[91] Hdiouech S, Bruna F, Batisson I, Besse-Hoggan P, Prevot V and Mousty C 2019 Amperometric detection of the herbicide mesotrione based on competitive reactions at nitroreductase@layered double hydroxide bioelectrode *J. Electroanal. Chem.* **835** 324–8
[92] Barroso M F, Paíga P, Vaz M C V F and Delerue-Matos C 2005 Study of the voltammetric behaviour of metam and its application to an amperometric flow system *Anal. Bioanal. Chem.* **383** 880–5
[93] Đurović A, Stojanović Z, Kravić S, Rakić D and Grahovac N 2018 Novel electrochemical procedure for the determination of metamitron *Int. J. Environ. Anal. Chem.* **98** 369–85
[94] Filip J *et al* 2020 Simplified synthesis of silver nanoparticles on graphene oxide and their applications in electrocatalysis *Nanotechnology* **32** 025502
[95] Filip J, Vinter S, Skacelik P, Sotolarova J, Borska K and Osicka J 2021 Silver Integrated with Carbonaceous 2D Nanomaterials as an Electrocatalyst for Reductive Dechlorination of Chloroacetanilide Herbicide *J. Electrochem. Soc.* **168** 037504
[96] Zolgharnein J, Shariatmanesh T and Babaei A 2014 Simultaneous determination of propanil and monalide by modified glassy carbon electrode with nickel oxide nanoparticles, using partial least squares modified by orthogonal signal correction and wavelet packet transform *Sens. Actuators B Chem.* **197** 326–33
[97] Gerent G G and Spinelli A 2019 Ag–Au core-partial shell bimetallic nanoparticles applied in electrochemical determination of the potential endocrine disruptor oryzalin *J. Electroanal. Chem.* **855** 113484
[98] Garrido E M, Lima J L F C, Delerue-Matos C, Borges M F M and Brett A M O 2001 Electroanalytical determination of oxadiazon and characterization of its base-catalyzed ring-opening products *Electroanalysis* **13** 199–203
[99] Milićević J S *et al* 2020 Multiwalled carbon nanotubes modified with MoO_2 nanoparticles for voltammetric determination of the pesticide oxyfluorfen *Microchim. Acta* **187** 429
[100] Savary A and Yari A 2021 Determination of the herbicide paraquat using the new Ag-GO/CuO/GCE-modified glassy carbon electrode by differential pulse voltammetry *J. Chinese Chem. Soc.* **68** 1987–97
[101] Hadi M, Bayat M, Mostaanzadeh H, Ehsani A and Yeganeh-Faal A 2018 Sensitive electrochemical detection of picloram utilising a multi-walled carbon nanotube/Cr-based metal-organic framework composite-modified glassy carbon electrode *Int. J. Environ. Anal. Chem.* **98** 197–214

[102] Mutharani B *et al* 2021 Temperature-enabled reversible 'On/Off' switch-like hazardous herbicide picloram voltammetric sensor in agricultural and environmental samples based on thermo-responsive PVCL-tethered MWCNT@Au catalyst *J. Hazard. Mater.* **402** 123672

[103] Smarzewska S, Jasińska A, Ciesielski W and Guziejewski D 2017 First electroanalytical studies of profluralin with square wave voltammetry using glassy carbon electrode *Electroanalysis* **29** 244–8

[104] Leniart A, Brycht M, Burnat B and Skrzypek S 2016 Voltammetric determination of the herbicide propham on glassy carbon electrode modified with multi-walled carbon nanotubes *Sens. Actuators B Chem.* **231** 54–63

[105] Pachinger A, Eisner E, Begutter H and Klus H 1991 Determination of a metabolite of the herbicide pyridate in drinking and groundwater using high-performance liquid chromatography with amperometric detection *J. Chromatogr.* A **558** 369–73

[106] Basozabal I, Gómez-Caballero A, Unceta N, Aranzazu Goicolea M and Barrio R J 2011 Voltammetric sensors with chiral recognition capability: the use of a chiral inducing agent in polyaniline electrochemical synthesis for the specific recognition of the enantiomers of the pesticide dinoseb *Electrochim. Acta* **58** 729–35

[107] Stanković D M, Mehmeti E, Svorc L and Kalcher K 2015 Simple, rapid and sensitive electrochemical method for the determination of the triketone herbicide sulcotrione in River water using a glassy carbon electrode *Electroanalysis* **27** 1587–93

[108] de A C and Lima A *et al* 2010 Electroanalytical studies of sulfentrazone in protic medium, its degradation by the electro-Fenton process, and toxicity assessment using ss-DNA *Chemosphere* **81** 884–89

[109] Catrinck M N, Okumura L L, Silva A A, Saczk A A and Oliveira M F 2015 New and sensitive electroquantification of sulfentrazone in soil by differential-pulse voltammetry *J. Braz. Chem. Soc.* **26** 1751–59

[110] Pandey A, Sharma S and Jain R 2019 Voltammetric sensor for the monitoring of hazardous herbicide triclopyr (TCP) *J. Hazard. Mater.* **367** 246–55

[111] Irandoust M, Haghighi M, Taherpour A A and Jafarzadeh M 2018 Electrochemical sensing of trifluralin in water by fluconazole-immobilized Fe_3O_4/SiO_2 nanomagnetic core–shell linked to carbon nanotube modified glassy carbon electrode; an experimental and theoretical modeling *J. Iran. Chem. Soc.* **15** 719–32

[112] Ferro É C, Cardoso C A L and Arruda G J 2017 Voltammetric detection of trifluralin in tap water, fruit juice, and vegetable extracts in the presence of surfactants *J. Environ. Sci. Heal.—Part B Pestic. Food Contam. Agric. Wastes* **52** 762–9

[113] Jafari M, Hasanzadeh M, Karimian R and Shadjou N 2019 Sensitive detection of Trifluralin in untreated human plasma samples using reduced graphene oxide modified by polyethylene imine and silver nanoparticles: a new platform on the analysis of pesticides and chemical injuries *Microchem. J.* **147** 741–8

[114] Jevtić S *et al* 2018 First electrochemistry of herbicide pethoxamid and its quantification using electroanalytical approach from mixed commercial product *Electrochim. Acta* **277** 136–42

[115] Pınar P T, Allahverdiyeva S, Yardım Y and Şentürk Z 2020 Voltammetric sensing of dinitrophenolic herbicide dinoterb on cathodically pretreated boron-doped diamond electrode in the presence of cationic surfactant *Microchem. J.* **155** 104772

[116] Duarte E H, Casarin J, Sartori E R and Tarley C R T 2018 Highly improved simultaneous herbicides determination in water samples by differential pulse voltammetry using boron-doped

diamond electrode and solid phase extraction on cross-linked poly(vinylimidazole) *Sens. Actuators B Chem.* **255** 166–75

[117] Jevtić S *et al* 2018 Boron-doped diamond electrode—a prestigious unmodified carbon electrode for simple and fast determination of bentazone in River water samples *Diam. Relat. Mater.* **81** 133–7

[118] Brycht M *et al* 2016 Voltammetric behavior, quantitative determination, and corrosion investigation of herbicide bromacil *J. Electroanal. Chem.* **770** 6–13

[119] Djurdjić S, Vukojević V, Jevtić S, Pergal M V, Petković B B and Stanković D M 2018 Herbicide clomazone detection using electroanalytical approach using boron doped diamond electrode *Int. J. Electrochem. Sci.* **13** 2791–9

[120] Thiam A and Salazar R 2019 Fenton-based electrochemical degradation of metolachlor in aqueous solution by means of BDD and Pt electrodes: influencing factors and reaction pathways *Environ. Sci. Pollut. Res.* **26** 2580–91

[121] da Silva J I L, dos Santos V B, Neves C A and de Souza J P I 2021 An ecologically correct electroanalytical method to determine the herbicide Picloram in Amazon waters using a miniaturized boron-doped diamond electrode and a 3D compact electrochemical cell *Chem. Pap.* **75** 1055–67

[122] Guelfi D R V, Gozzi F, Machulek A, Sirés I, Brillas E and de Oliveira S C 2018 Degradation of herbicide S-metolachlor by electrochemical AOPs using a boron-doped diamond anode *Catal. Today* **313** 182–8

[123] Šelešovská R, Janíková L, Pithardtová K, Chýlková J and Tomášková M 2016 Sensitive voltammetric determination of herbicide terbutryn using solid electrodes based on silver amalgam and boron-doped diamond *Monatsh. Chem.—Chem. Mon.* **147** 207–17

[124] Janíková-Bandžuchová L, Šelešovská R, Schwarzová-Pecková K and Chýlková J 2015 Sensitive voltammetric method for rapid determination of pyridine herbicide triclopyr on bare boron-doped diamond electrode *Electrochim. Acta* **154** 421–9

[125] Mohan K C, Chandraiah T, Nagaraju M and Sreedhar N Y 2013 Elucidation of electrochemical reduction behavior of dinitro group containing pesticides Chlornidine, Dipropalin and Prodiamine residues in soil, water and agricultural formulations *J. Indian Chemical Soc.* **92** 467–71

[126] Higuera M J, Marín Galvín R, Rodríguez Mellado J M and Ruiz Montoya M 2002 EC (EE) processes in the reduction of some 2-methylthio-4,6-di(alkylamino)-1,3,5-triazines on mercury electrodes *Electrochem. Commun.* **4** 30–5

[127] Sreedhar M, Reddy T M, Balaji K and Reddy S J 2010 Electrochemical reduction behavior and polarographic determination of methoxy triazine herbicides in environmental samples *Anal. Lett.* **43** 674–86

[128] Qiu P, Ni Y and Kokot S 2014 Simultaneous voltammetric determination of four triazine herbicides in water samples with the aid of chemometrics *J. Environ. Sci. Heal. Part* B **49** 722–9

[129] Gálvez R, Pedrero M, Buyo F, De Villena F J M and Pingarrón J M 2000 Oil-in-water emulsions as suitable working media for the direct polarographic determination of aziprotryne and desmetryne from its organic extracts in water samples *Fresenius J. Anal. Chem.* **367** 454–60

[130] Sreedhar N Y, Samatha K R, Kumar Reddy P R and Reddy S J 1998 Differential pulse polarographic determination of benfluralin and trifluralin in formulations soils and grains *Int. J. Environ. Anal. Chem.* **72** 247–55

[131] Sreedhar M and Reddy S J 2002 Electrochemical reduction and differential pulse polarographic determination of butralin and isopropalin in environmental samples at a mercury electrode *Bull. Chem. Soc. Jpn.* **75** 2155–9

[132] Daneshmehr M-A, Ahmadi F, Ahmadi B and Shakiba E 2016 Deciphering the binding mode of dinitramine herbicide to ct-DNA, a thermodynamic discussion *Food Agric. Immunol.* **27** 23–39

[133] Ni Y, Wang L and Kokot S 2011 Simultaneous determination of three herbicides by differential pulse voltammetry and chemometrics *J. Environ. Sci. Heal. Part* B **46** 328–35

[134] Cuellar M, Moreno Betancourth J, Pfaffen V and Ortiz P I 2019 Response surface methodology for optimisation of 4,6-Dinitro-o-cresol quantification using hanging mercury drop electrode *Int. J. Environ. Anal. Chem.* **99** 74–86

[135] Thriveni T, Kumar J R, Lee J Y and Sreedhar N Y 2009 Study of the voltammetric behaviour of the ethalfluralin and methalpropalin and its determination in environmental matrices at hanging mercury drop electrode *Environ. Monit. Assess.* **151** 9–18

[136] Eş Z and Taşdemir I H 2015 Reduction behavior of fenoxaprop-p-ethyl and its voltammetric determination *Turkish J. Chem.* **39** 54–63

[137] Balaji K, Sridevi C, Ananda Kumar Reddy N, Reddy K M M S, Suresh C and Reddy 2010 Voltammetric behaviour and analysis of fluchloralin *E-J. Chem.* **7** 1605–11

[138] Mozo J D, López-López M, Olloqui-Sariego J L, Molina V M, Maraver J and Carbajo J 2010 Stereoselective electrochemical reduction of imazapyr in aqueous media without chiral auxiliaries *J. Electrochem. Soc.* **157** E149

[139] Thriveni T, Rajesh Kumar J, Sujatha D and Sreedhar N Y 2007 Behaviour and quantification studies of terbacil and lenacil in environmental samples using cyclic and adsorptive stripping voltammetry at hanging mercury drop electrode *Environ. Monit. Assess.* **128** 359–68

[140] Arranz A 1997 Study of the electro-adsorptive behaviour of the herbicide nitralin by means of voltammetric techniques *Talanta* **45** 417–24

[141] Thriveni T, Rajesh Kumar J, Sujatha D and Sreedhar N Y 2007 Voltammetric determination of the herbicides nitralin and oryzalin in agricultural formulations, vegetables and grape juice samples *Food Chem.* **104** 1304–9

[142] Galli A, De Souza D and Machado S A S 2011 Pendimethalin determination in natural water, baby food and river sediment samples using electroanalytical methods *Microchem. J.* **98** 135–43

[143] Galli A *et al* 2018 A mechanistic study of the electrochemical behavior of pendimethalin herbicide *J. Electroanal. Chem.* **826** 157–63

[144] Janíková-Bandžuchová L, Šelešovská R, Chýlková J and Nesnídalová V 2016 Voltammetric analysis of herbicide picloram on the silver solid amalgam electrode *Anal. Lett.* **49** 19–36

[145] Pintado S, Gómez L, Rodríguez-Amaro R, Montoya M R and Mellado J M R 2011 2D nucleation in the electroreduction of 8-quinolinecarboxylic acid, and the herbicide quinmerac on mercury electrodes *Electrochim. Acta* **58** 662–7

[146] Xu J, Zhang Y, Wu K, Zhang L, Ge S and Yu J 2017 A molecularly imprinted polypyrrole for ultrasensitive voltammetric determination of glyphosate *Microchim. Acta* **184** 1959–67

[147] Cheng H, Liu S, Hu C and Wang H 2018 Electro-copolymerization of symmetric flower-like luminol–aniline–hollow TiO_2–NH_2 nano-shell polymer in ionic liquid for Acetochlor detection *Int. J. Electrochem. Sci.* **13** 4182–97

[148] de Oliveira Silva R, da Silva É A, Fiorucci A R and Ferreira V S 2019 Electrochemically activated multi-walled carbon nanotubes modified screen-printed electrode for voltammetric determination of sulfentrazone *J. Electroanal. Chem.* **835** 220–6
[149] Sok V and Fragoso A 2021 Carbon nano-onion peroxidase composite biosensor for electrochemical detection of 2,4-D and 2,4,5-T *Appl. Sci.* **11** 6889
[150] Chen T-W, Rajaji U, Chen S-M and Jothi Ramalingam R 2019 Rapid sonochemical synthesis of silver nano-leaves encapsulated on iron pyrite nanocomposite: an excellent catalytic application in the electrochemical detection of herbicide (Acifluorfen) *Ultrason. Sonochem.* **54** 90–8
[151] Ahmed S *et al* 2020 A composite of imprinted polypyrrole beads and reduced graphene oxide for specific electrochemical sensing of atrazine in complex matrices *Monatsh. Chem.* **151** 1271–82
[152] Kergaravat S V, Fabiano S N, Soutullo A R and Hernández S R 2021 Comparison of the performance analytical of two glyphosate electrochemical screening methods based on peroxidase enzyme inhibition *Microchem. J.* **160** 105654
[153] Rajaji U *et al* 2021 Design and fabrication of yttrium ferrite garnet-embedded graphitic carbon nitride: a sensitive electrocatalyst for smartphone-enabled point-of-care pesticide (mesotrione) analysis in food samples *ACS Appl. Mater. Interfaces* **13** 24865–76
[154] Noguer T, Balasoiu A-M, Avramescu A and Marty J-L 2001 Development of a disposable biosensor for the detection of metam-sodium and its metabolite MITC *Anal. Lett.* **34** 513–28
[155] Somnet K, Thimoonnee S, Karuwan C, Kamsong W, Tuantranont A and Amatatongchai M 2021 Ready-to-use paraquat sensor using a graphene-screen printed electrode modified with a molecularly imprinted polymer coating on a platinum core *Analyst* **146** 6270–80
[156] Rocaboy-Faquet E, Barthelmebs L, Calas-Blanchard C and Noguer T 2016 A novel amperometric biosensor for ß-triketone herbicides based on hydroxyphenylpyruvate dioxygenase inhibition: a case study for sulcotrione *Talanta* **146** 510–6
[157] Abdalla N, Youssef M, Algarni H, Awwad N and Kamel A 2019 All solid-state poly (vinyl chloride) membrane potentiometric sensor integrated with nano-beads imprinted polymers for sensitive and rapid detection of bispyribac herbicide as organic pollutant *Molecules* **24** 712
[158] Kamel *et al* 2019 Novel solid-state potentiometric sensors using polyaniline (pani) as a solid-contact transducer for flucarbazone herbicide assessment *Polymers* **11** 1796
[159] de Figueiredo-Filho L C S, Baccarin M, Janegitz B C and Fatibello-Filho O 2017 A disposable and inexpensive bismuth film minisensor for a voltammetric determination of diquat and paraquat pesticides in natural water samples *Sens. Actuators B Chem.* **240** 749–56
[160] Lou Y-Y, Hapiot P, Floner D, Fourcade F, Amrane A and Geneste F 2020 Efficient dechlorination of α-halocarbonyl and α-haloallyl pollutants by electroreduction on bismuth *Environ. Sci. Technol.* **54** 559–67
[161] Gerent G G, Gonçalves C Q, da Silva P S and Spinelli A 2015 *In situ* bismuth-film electrode for square-wave cathodic voltammetric detection of pendimethalin at nanomolar level *Electrochim. Acta* **168** 379–85
[162] Mazzotta E and Malitesta C 2010 Electrochemical detection of the toxic organohalide 2,4-DB using a Co-porphyrin based electrosynthesized molecularly imprinted polymer *Sens. Actuators B Chem.* **148** 186–94
[163] Novotný V and Barek J 2017 Voltammetric determination of aclonifen at a silver amalgam electrode in drinking and River water *Ecol. Chem. Eng. S* **24** 277–84

[164] Tavares O, Morais S, Paíga P and Delerue-Matos C 2005 Determination of ametryn in soils via microwave-assisted solvent extraction coupled to anodic stripping voltammetry with a gold ultramicroelectrode *Anal. Bioanal. Chem.* **382** 477–84
[165] Brycht M, Burnat B, Nosal-Wiercińska A and Skrzypek S 2016 New sensitive square-wave adsorptive stripping voltammetric determination of pesticide chlornitrofen, and an evaluation of its corrosivity towards steel agricultural equipment *J. Electroanal. Chem.* **777** 8–18
[166] Gajdár J, Barek J and Fischer J 2019 Electrochemical microcell based on silver solid amalgam electrode for voltammetric determination of pesticide difenzoquat *Sens. Actuators B Chem.* **299** 126931
[167] Irandoust M and Haghighi M 2016 Fabrication of a new modified gold electrode based on gold nanoparticles and nanomagnetic $Fe_3O_4/SiO_2–(CH_2)_3–SH$ core–shell for electrochemical evaluation and determination of dinitramine herbicide in water *RSC Adv.* **6** 49798–805
[168] Niu L M, Liu Y, Lian K Q, Ma L and Kang W J 2018 Characterization of a sensitive biosensor based on an unmodified DNA and gold nanoparticle composite and its application in diquat determination *Arab. J. Chem.* **11** 655–61
[169] El Harmoudi H *et al* 2017 Square wave voltammetric determination of diquat using natural phosphate modified platinum electrode *Arab. J. Chem.* **10** S671–6
[170] Do M H *et al* 2015 Molecularly imprinted polymer-based electrochemical sensor for the sensitive detection of glyphosate herbicide *Int. J. Environ. Anal. Chem.* **95** 1489–501
[171] Mazouz Z *et al* 2017 Highly selective polypyrrole MIP-based gravimetric and electrochemical sensors for picomolar detection of glyphosate *Sensors* **17** 2586
[172] Zhang C *et al* 2017 A highly selective electrochemical sensor based on molecularly imprinted polypyrrole-modified gold electrode for the determination of glyphosate in cucumber and tap water *Anal. Bioanal. Chem.* **409** 7133–44
[173] Thanh C T *et al* 2021 Electrochemical sensor based on reduced graphene oxide/double-walled carbon nanotubes/octahedral Fe_3O_4/chitosan composite for glyphosate detection *Bull. Environ. Contam. Toxicol.* **106** 1017–23
[174] Uka B, Kieninger J, Urban G A and Weltin A 2021 Electrochemical microsensor for microfluidic glyphosate monitoring in water using MIP-based concentrators *ACS Sens.* **6** 2738–46
[175] de Jesus V O, Barbosa A M J, Petroni J M, Lucca B G and Ferreira V S 2017 Voltammetric determination of herbicide molinate in River water and rice samples using solid silver amalgam electrode fabricated with nanoparticles *Int. J. Environ. Anal. Chem.* **97** 468–83
[176] Morawska K, Jedlińska K, Smarzewska S, Metelka R, Ciesielski W and Guziejewski D 2019 Analysis and DNA interaction of the profluralin herbicide *Environ. Chem. Lett.* **17** 1359–65
[177] de Oliveira Margarido V, de Paula J H, Francisco K R and de Souza A L 2020 Atividade eletrocatalítica de um filme de ormosil híbrido contendo ácido fosfomolíbdico para o herbicida prometon *Matéria (Rio Janeiro)* **25** 1–8
[178] Zhang J, Wang C, Niu Y, Li S and Luo R 2017 Electrochemical sensor based on molecularly imprinted composite membrane of poly(*o*-aminothiophenol) with gold nanoparticles for sensitive determination of herbicide simazine in environmental samples *Sens. Actuators B Chem.* **249** 747–55
[179] Xu J, Olvera-Vargas H, Teo F Y H and Lefebvre O 2021 A comparison of visible-light photocatalysts for solar photoelectrocatalysis coupled to solar photoelectro-Fenton: application to the degradation of the pesticide simazine *Chemosphere* **276** 130138

[180] Chen J, Xu G R, Bai L Y, Tao J Z, Liu Y Q and Zhang Y P 2012 A high sensitive sensor for terbuthylazine determination based on molecularly imprinted electropolymer of 3-Thiophenemalonic acid *Int. J. Electrochem. Sci.* **7** 9825–34
[181] Gonçalves Filho D and de Souza D 2021 Electrochemical determination of trifluralin herbicide using silver solid amalgam electrode: application in fresh food samples *J. Braz. Chem. Soc.* **32** 1456–66
[182] Ruan X *et al* 2021 Nanomaterial-enhanced 3D-printed sensor platform for simultaneous detection of atrazine and acetochlor *Biosens. Bioelectron.* **184** 113238
[183] Tian J, Li Y, Dong J, Huang M and Lu J 2018 Photoelectrochemical TiO_2 nanotube arrays biosensor for asulam determination based on *in situ* generation of quantum dots *Biosens. Bioelectron.* **110** 1–7
[184] Puangjan A and Chaiyasith S 2018 A Co_3O_4 nano-octahedron modified fluorine doped tin oxide electrochemical sensor for detection of benzobicyclon *J. Electroanal. Chem.* **813** 20–30
[185] Abd-Rabboh H S M, Kamel A H and Alshehri F H A 2021 Cacodylate sensors and their application in the determination of amino acid levels in biological samples *J. AOAC Int.* **104** 113–21
[186] Zhang Y-N, Dai W, Wen Y and Zhao G 2017 Efficient enantioselective degradation of the inactive (S)-herbicide dichlorprop on chiral molecular-imprinted TiO_2 *Appl. Catal. B Environ.* **212** 185–92
[187] Betancourth J M, Cuellar M, Ortiz P I and Pfaffen V 2018 Multivariate cathodic square wave stripping voltammetry optimization for nitro group compounds determination using antimony film electrodes *Microchem. J.* **139** 139–49
[188] Shrivastava S, Kumar A, Verma N, Chen B and Chang C 2021 Voltammetric detection of aqueous glyphosate on a copper and poly(pyrrole)-electromodified activated carbon fiber *Electroanalysis* **33** 916–24
[189] Huang G, Huang L and Dong L 2019 Electrocatalytic deiodination of methyl iodide on a copper electrode *Res. Chem. Intermed.* **45** 3723–37
[190] Es-haghi A, Baghernejad M and Bagheri H 2014 Novel unbreakable solid-phase micro-extraction fibers on stainless steel wire and application for the determination of oxadiargyl in environmental and agricultural samples in combination with gas chromatography–mass spectrometry *Talanta* **128** 231–6
[191] Pourakbari Z, Aliakbar A and Sheykhan M 2020 A metal-catex composite electrode for determination of paraquat in various samples by Ad-differential pulse cathodic stripping voltammetry *Talanta* **212** 120793
[192] Deroco P B, Wachholz Junior D and Kubota L T 2021 Silver inkjet-printed electrode on paper for electrochemical sensing of paraquat *Chemosensors* **9** 61
[193] Velázquez-Manzanares M, Martínez H G, Yudi L M, Amador-Hernández J, de la Garza Rodríguez I M and Urbina E M C 2019 Electrochemical behavior of simetryn herbicide at water|1,2-dichloroethane interface *J. Electrochem. Soc.* **166** H286–90
[194] Mercan H and İnam R 2008 Differential pulse polarographic behaviour of thiazopyr herbicide and application to its determination in fruit juice and soil samples *Int. J. Environ. Anal. Chem.* **88** 879–90
[195] Saadati A, Hassanpour S and Hasanzadeh M 2020 Lab-on-fruit skin and lab-on-leaf towards recognition of trifluralin using Ag-citrate/GQDs nanocomposite stabilized on the flexible substrate: a new platform for the electroanalysis of herbicides using direct writing of nano-inks and pen-on paper technolo *Heliyon* **6** e05779

[196] Mahmoudpour M, Saadati A, Hasanzadeh M and Kholafazad-kordasht H 2021 A stretchable glove sensor toward rapid monitoring of trifluralin: a new platform for the on-site recognition of herbicides based on wearable flexible sensor technology using lab-on-glove *J. Mol. Recognit.* **34** e2923

Chapter 14

Application of electrochemical sensor for insulin detection

Amoolya Chandran, S Swathy, Goldamol S Pallam and K Girish Kumar

Human insulin, a pancreatic hormone has various functions in the human body. This dipeptide hormone plays a vital role in regulating carbohydrate metabolism to maintain the blood glucose level. Hence, it is considered to be a reliable biomarker in the management of diabetes. Meanwhile, the detection and determination of insulin is more complicated owing to its limited quantity in blood (picomolar), larger size (>MW 5.8 kDa), and interferences offered by endogenous molecules. In clinical applications, insulin quantification executed with high efficacy and potency by fully automated immunosensing strategies, encounter sophisticated instrumentation and longer analysis time. Compared to these conventional methods, electrochemical sensing platforms have been used recently due to their quick response, portability, and ease of operation with continuous real-time signals. Carbon materials generally used in electrochemical insulin sensors offer faster electron transfer rates thereby enhancing sensitivity and enabling miniaturization. Furthermore, incorporating the carbon composite materials with metal nanoparticles and insulin biorecognition elements offer promising stability and enhanced sensing features which can meet the clinical demand in existing and emerging insulin sensors. This chapter focuses on the transition of insulin determination from conventional analytical methods to electrochemical biosensors as a cost-effective and portable solution to real-time and point of care diagnostics.

14.1 Introduction

Insulin is a pancreatic hormone which performs a variety of functions in the human body, particularly in maintaining the blood glucose level by regulating the metabolism of carbohydrates and fats [1]. Structurally, this hormone consists of 51 amino acids as dipeptide chains connected by disulfide bonds. Insulin is secreted by the beta cells which is present in islets of Langerhans of the pancreas [2]. Insulin

doi:10.1088/978-0-7503-5127-0ch14

which is not attached to other proteins is called free insulin [3]. It is transported through body fluids (blood and plasma) and facilitates the storage of excess glucose as glycogen in fat cells, muscles, and the liver which results in weight gain [4]. Insulin plays a key regulator role in lipid and protein metabolism as well as a vital part in energy homeostasis and it accelerates cell growth via its mitogenic effect. The discovery of insulin has been a fascinating story, saving the lives of several million people worldwide.

The significant landmarks in the discovery of insulin demonstrate that it was not accidental but had been a long voyage over the course of history. After decades of research into the substance synthesized by the pancreas for glucose metabolism, Belgian researcher, Jean de Meyer made notable contributions but could not extract the substance (insulin) secreted by the islets of Langerhans successfully. Following him, Sir Edward Albert Sharpey-Schafer in 1910 and in 1916 confirmed that a substance is secreted by pancreatic islets which regulates glucose metabolism and he coined the name 'insulin', which was derived from the Latin word *insula*, meaning 'island' [5].

The detection of insulin is regarded as a breakthrough for the prognosis, systematic monitoring, adopting proper medication, and clinical advancements for diabetes. Diabetes is a long lasting disease that occurs due to low levels of insulin in blood. According to WHO statistics from 2014, more than 420 million people suffer from diabetes and it is estimated to increase to 570 million by 2030 which is approximately 4.4% of the global population. The acute illness which arises via diabetes can lead to dysfunctioning of, or even damage to, various organs in the human body including blood vessels, nerves, eyes, kidney, etc [6]. Two types of diabetes are known, diabetes mellitus and diabetes insipidus. Diabetes mellitus is the commonly known diabetes which is an abnormal condition caused by chronic hyperglycemia, that arises due to the synthesis of impaired insulin or inadequate insulin secretion or both. Diabetes insipidus is an insulin independent, antidiuretic hormonal disorder caused by vasopressin that assists in maintaining the fluid content of our body. In 1997 the American Diabetes Association (ADA) proposed a criteria to differentiate between various types of diabetes mellitus in different individuals which is identified as the established classification and is widely accepted [7]. Explicitly, they are type 1 diabetes, type 2 diabetes, other specific diabetes, and gestational diabetes mellitus. Gestational diabetes mellitus refers to a metabolic disorder during pregnancy, which manifests hyperglycemia during gestation due to insulin resistance caused by defective functioning of pancreatic β cells [8]. Other specific diabetes is also indicated by hyperglycemia caused by genetic disorders or by external factors, for example as a consequence of taking some medicines, drugs, etc [9]. Type 1 diabetes or insulin dependent diabetes is caused by the autoimmune destruction of β cells of the pancreas which results in absolute insulin deficiency whereas type 2 diabetes, or non-insulin dependent diabetes, is a non-communicable disease, where the individual cannot produce adequate amounts of insulin or the synthesized insulin cannot be utilized effectively [10, 11]. Thus it is obvious that, all types of diabetes mellitus lead to hyperglycemia. Insulin exhibits a significant role in maintaining glucose homeostasis and also manages a wide range of physiological processes, which means it is a potent biomarker for the early prognosis and diagnosis of many chronic diseases. Perceiving the role of insulin in the broad range of physiological processes, its quantification has a significant implication in the early detection of many pathological conditions [1].

The landmark revolution in diabetic detection and diagnosis has been achieved by studying physiological actions as well as the determination of insulin in body fluids. Mostly, the early detection of diabetes was done by the quantification of glucose in blood due to certain complications in insulin determination, owing to its limited quantity in blood (picomolar), larger size (>MW 5.8 kDa), and interferences offered by endogenous coexisting molecules [12, 13]. In order to tackle this shortcoming, insulin analogs were extracted and epitopes identified and employed to meet the clinical requirements [14]. Insulin was exogenously extracted to study the physical as well as chemical properties and to monitor consequential pathological conditions that emerge due to its variations in the normal blood levels.

Endogenous insulin is made by the human pancreas while exogenous insulin is derived from animals or through DNA recombinant technology [15]. Before the emergence of biotechnology, animal insulin was used to treat diabetes but encountered allergic effects. In the 1980s, with the advance of DNA recombinant technology, the use of animal insulin significantly dropped and human insulin was generated [2, 3]. Compared to animal derived insulin, artificially modified insulin molecules provide better pharmacokinetic and pharmacodynamics properties in terms of absorption, insulin receptor recognition, metabolic activity, etc. The insulin analogs are prepared synthetically by slightly modifying the native protein structure [4]. The recombinant insulins are generally classified into two categories: short acting and long acting insulin analogs. Short acting analogs are absorbed rapidly in the capillaries hence dissociated easily while long acting analogs show delayed absorption and extended duration of action [5]. The three short acting insulin analogs are insulin lispro, insulin aspart, and insulin glulisine, all are commercially available under the brand names of humalog, novolog, and apidra, respectively. Insulin lispro was the first introduced analog in 1996 by Eli Lilly.

Variegate analytical tools are available for the determination of insulin, which are mainly classified as non-immuno based and immune based. Immuno based assays cover radioimmunoassay (RIA) [16], enzyme immunoassay (EIA) [17, 18] and luminescent immunoassay (LIA) [19]. Non-immuno based assays cover high performance liquid chromatography (HPLC) [20, 21], capillary electrophoresis [22, 23], and electrochemical assays [24]. Enzyme-linked immunosorbent assay (ELISA) has long been used in the immunoassay category where a target antigen (insulin) has been detected, but its high cost and lack of specificity makes it less feasible. The modified ELISA techniques used nowadays are amplified luminescent proximity homogeneous assay (AlphaLISA) and homogeneous time-resolved fluorescence (HTRF). But these techniques are generally laboratory limited with expensive and inconvenient instrumentation, time-consuming procedures, and demand the use of highly skilled personnel. Electrochemical methods are considered to be the most convenient means of determination owing to their unique features such as simplicity, fast response, cost effectiveness, automation features, etc.

In addition the capability of miniaturization facilitates the construction of point of care devices.

The electrochemical properties of insulin was examined by various groups who explored different electroactive functionalities in its structure. In insulin, the electrically active sites are hydroxyl groups of tyrosine residues and disulfide bonds. The three active groups of tyrosine are vulnerable to electro-oxidation at the electrode surface. But the tyrosine residues are buried deeply inside the three-dimensional (3D) structure making it difficult for electro-oxidation at the surface [25]. Thus, widely examined electroactive entities are three disulfide bonds attributed to the property of facile reduction to its corresponding thiols [26]. Of the three disulfide bonds, one is deeply buried and another is shielded and is thus unavailable for electro-oxidation. However, another inter-chain S–S bond is exposed and takes part in electro-oxidation [27].

In 1964, the reduction of insulin was studied by analyzing the cathodic waves [27] obtained from polarographic studies. In 1977, by employing controlled potential coulometry the reduction of disulfide groups in insulin were studied using a mercury electrode [28]. All these studies paved the way to the electrochemical quantification experiments and facilitated the sensitive detection of insulin with promising selectivity [12, 28]. Even though a glucometer enables the determination of glucose at millimolar concentration, more efficient monitoring of hypo- and hyperglycaemic conditions could be achieved by coupling these strategies with ultra-sensitive analytical assays of diabetes biomarkers like insulin.

14.2 Carbon-based electrochemical determination of insulin

Electrochemical methods are regarded as user-friendly, robust, and reliable means of determination of biomolecules including insulin. The analytical behavior of electrochemical sensors is primarily tied to the material of which the working electrode is constructed [29]. Subsequently it affects the sensitivity, selectivity, reproducibility, repeatability, and linear range offered by the electroanalytical device. Materials which guarantee fast electron transfer, large surface-to-volume ratio, low background current, wide potential window, and satisfactory electrochemical as well as chemical stability are fundamental requirements for the working electrode material [12]. The most commonly used working electrode materials are platinum, gold, carbon, and mercury, in which carbon-based working electrodes have received greater attention nowadays for insulin determination.

14.2.1 Carbon-based electrodes

Carbon-based electrodes (CBE) have a great role in the electro-analysis of biomolecules [30]. The key characteristics of CBEs are stability, ease of preparation, reproducibility, wide potential range, large current density, surface renewability, surface area for uniformly placing catalysts in different specifications, etc [31]. These features vary depending on the specific material utilized for the fabrication of the electrode. All these factors prompted the research world to explore carbon-based electrodes in the electrochemical sensing of broad varieties of analytes [30]. Of the diverse types of CBE, glassy carbon electrodes (GCE), carbon paste electrodes (CPE), carbon ceramic electrodes (CCE), edge plane pyrolytic graphite electrodes (EPPG), screen printed electrodes (SPE), and graphite electrodes are widely used for insulin determination.

Glassy carbon (GC), which is usually called vitreous carbon, is a non-graphitized carbon-based inert material, extensively used as electrode material [32]. This material is generated by controlled pyrolysis of specific polymer resins. The gradual carbonization of the cleansed precursor resin is executed, followed by high temperature treatment [33]. A highly controlled and organized strategy is maintained, in terms of temperature and thickness specification (one-dimensional (1D) thickness) to ensure the complete removal of volatile by-products produced during the material formation. The endowed superior characteristics of GC are attributed to its preparation procedure. The prime features of GC include, minute pore size, small gas and liquid-permeability, oxidative resistance, superior electrical conductivity, and thermal, chemical as well as electrochemical stability [34]. When it is implemented for electrode fabrication, all these features permit the electro-analysis of a variety of biomolecules [35, 36]. Moreover, wide potential windows as well as rich surface characteristics of these electrode broaden the class of electro-sensitive GC detectable molecules. Ease of fabrication and maintenance added to its acceptance as a potent electrode with a wide range of applications [37].

Pencil graphite electrodes (PGEs) are also commonly used transducers for insulin determination. The electrode material employed for this type of electrode is pencil graphite lead [38]. These easily disposable electrodes exhibit interesting features such as cost effectiveness, ease to use, and are free from tedious surface cleaning procedures between measurements which reduce the overall sensor fabrication and analysis time [39]. Besides that, low background current, adjustable active surface area, high reproducibility and sensitivity with low detection limit and less sample volume analysis are the key features of PGE [40]. Hence, PGE is generally a preferred electrode for biomolecule analysis including insulin [41].

CPEs are widely employed in electrochemical determination of a variety of biomolecules owing to their affordability, simplicity of preparation, feasibility in renewable surface formation, less ohmic resistance and low background currents [42, 43]. These electrodes receive massive consideration owing to their versatility in terms of facileness in electrode modification by incorporating foreign materials of interest [44, 45]. This brought about an outstanding platform for characterization, detection, and determination of a number of biomolecules such as insulin and glucose [12, 46–48]. CPE formed by variety of carbonaceous materials like graphite, carbon nanotubes (CNTs) and so forth are broadly used, of which graphite based electrodes are the most explored for the determination of insulin [46, 49].

EPPG electrode, a special type of carbon-based electrode, is formed by cutting highly ordered pyrolytic graphite (HOPG) into the desired surface geometry so that the graphitic layers are perpendicular to the surface. Basal plane pyrolytic graphite electrodes (BPPG) are also formed from HOPG but EPPG have faster electrode kinetics, lower background current, and improved electrocatalytic signals compared to that of BPPG, CNTs-modified BPPG, boron doped diamond electrodes, and

GCE. This is attributed to the presence of a number of reactive functional groups formed by the facile reaction of oxo-groups with atmospheric oxygen which facilitates faster electron transfer between the electrode surface and insulin [50]. In addition, BPPG forms irreversible voltammograms whereas EPPG brings about reversible ones [51]. Owing to the aforementioned advantages, EPPG is preferred and is commonly utilized for the electro-analysis of insulin [51].

The suitability for miniaturization and automation for on-site as well as point of care analysis of compounds meant that SPEs propspered as a new era in electrochemical sensing scenario. More inherent features of the former exists in its versatility [52]. These electrodes exhibit high reproducibility, disposability, quick analysis time, broader potential window, robustness, material compatibility as well as flexibility in adopting apt electrode design. This flexibility facilitates small volume analysis which meets the need of on-site device fabrication with reliable sensing features [53]. Screen printing, photolithography, and ink-jet printing are the general techniques employed for SPE design. These methods have the potential to ensure its mass production and also offer control over the thickness and dimension of the electroactive surface. Screen printing technology is one of the significant methods used as a path for transition from lab to point-of-use [54].

The aforesaid factors prompted the research world to design carbon electrode based electrochemical sensors for the determination of insulin. Although the inherent characteristics of these electrodes show satisfactory sensing features, the bare form does not always meet all the electrochemical requirements for outstanding sensing performance. The challenges that are often encountered while using the bare electrodes are surface fouling, high over-potential, low sensitivity, and selectivity [54]. These factors forced researchers to modify the bare surface with suitable custom-made materials having peculiar structural and electrocatalytic properties onto the working electrode surface to meet their specific needs [55]. Electrode surface modification procedures include coating and/or depositing bonded materials in which the SPE prefers solvent-modifier based surface wrapping by coating methods such as spin-coating, drop-casting, and dip-coating. For GCE, electrodeposition, electropolymerization and drop-casting methods are commonly employed [35, 55]. In general, those entities which manifest high electrical conductivity, promising electron mobility, and large surface area with excellent electrocatalytic properties are used as electrode materials [56]. In the case of CPE, foreign species are added into the bulk of the electrode material during electrode fabrication.

14.2.2 Carbon-based electrode modifiers

The electrochemical performance of the working electrode can also be improved by adopting appropriate electrode modification [57, 58]. Carbon-based materials are widely used as electrode modifiers for insulin detection owing to its salient features like chemical inertness, low background current, high electrical conductivity, large surface-to-volume ratio, chemical stability, etc [59–61] which enhance the overall electroanalytical performance of the sensor. Electrochemical biosensors of insulin are customarily designed by using CNTs, graphene, and ordered mesoporous carbon (OMC) as the electrode modifiers [62].

14.2.2.1 Carbon nanotubes

These materials are cylindrical in structure with SP^2 hybridization, diameter in manometers and length in microns [63]. The advantages of employing carbon nanotubes (CNTs) in electrochemical sensing applications are mainly due to their large surface-to-volume ratio, high electron transfer capability, good sensitivity, and fast electrochemical reversible response in standard conditions. CNTs are better electrode modifiers than any other carbon-based material owing to their promising electrical conductivity, stability, large surface area, ultra-sensitivity, long-term stability, and reproducibility [64, 65]. These are extensively employed as electrode materials in amperometric enzyme electrodes, immunosensors, nucleic-acid sensing devices, etc [66–68]. In CNTs, the electron transfer happens due to the conducting sites that are available along with its structure, each producing a quantum conduction by transferring electrons or holes. CNTs are generally classified into two categories; single-walled carbon nanotubes (SWCNTs) and multi-walled carbon nanotubes (MWCNTs) [69]. SWCNTs are created using distinct graphite sheets which are rolled like a tube with diameters in the range 1–2 nm whereas MWCNTs are made by nesting multiple graphite cylinders having dimensions of 2–25 nm [60]. SWCNTs have been used in combination with materials like graphene oxide (GO) because of their fibrous morphology and fast electron transfer. In order to enhance the electrocatalytic feature of MWCNTs, composites are designed with chitosan, graphene quantum dots (GQDs), etc. These composites act as outstanding matrixes for bioreceptor element immobilization and are capable of forming new hybrid nanomaterials [70]. These composites have proved to be better for amplifying the current signal due to their large surface-to-volume ratio, excellent stability, conductivity, faster electron transfer and so forth [69]. Meanwhile, MWCNTs along with other metal nanomaterials have also been widely explored as functional materials for insulin detection because of their anti-fouling property, high electro-catalytic activity, sensitivity, etc [49, 64].

Another important class of carbonaceous materials which is used as an electrode modifier for insulin detection is graphene.

14.2.2.2 Graphene

Graphene is a two-dimensional (2D) single layered sheet of SP^2 hybridized carbon atoms and have electrical conductivity and mechanical properties similar to that of CNT [71]. They have been used for immobilizing biorecognition elements where π–π stacking interactions occur between aptamer and graphene [72]. Since the graphitic structures are inert, it is necessary to functionalize graphene to form its derivatives [73]. Surface functionalization of graphene generally occurs via physical adsorption, electrostatic interaction, oxidation treatment of carbonyl and carboxyl functional group, etc [74]. The graphitic materials such as GO, reduced graphene oxide (rGO), graphene nanoplates, and graphene nanoribbons are usually used as electrode

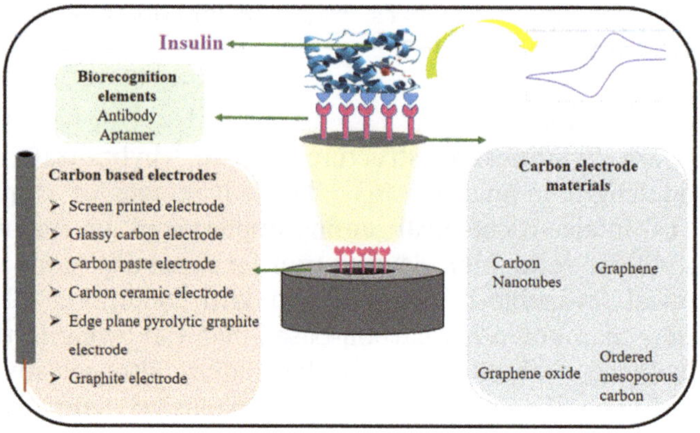

Figure 14.1. Schematic representation of carbon-based electrochemical determination of insulin. Different carbon-based electrodes and modifiers are portrayed in the representation.

modifiers [75]. Graphene along with metals, metal oxides, and semiconductor nanoparticles show good sensitivity for electrochemical insulin sensing due to its coupled effect resulting in enhanced sensing features especially electronic conductivity and large surface area [76].

14.2.2.3 Ordered mesoporous carbon

This is a subclass of 3D nanostructured porous carbon material which shows higher surface area than that of graphene and CNT. The specific surface area of ordered mesoporous carbon (OMC), graphene, and CNT are 1628 $m^2\ g^{-1}$, 457 $m^2\ g^{-1}$, and 145 $m^2\ g^{-1}$ respectively [77]. The main advantages of using OMC based electrochemical sensors for insulin detection include chemical inertness, commendable electrical conductivity, and large surface area to volume ratio. OMC with better chemical and electrical properties can be produced from soft-carbon materials and the properties of thus prepared OMCs can be improvised again by incorporating different components in or on the mesoporous walls [77].

A schematic representation of different carbon-based electrodes and modifiers used for the electrochemical determination of insulin is portrayed in figure 14.1.

14.3 Commonly employed electrochemical sensing methodologies

There are three modes by which electrochemical quantification of insulin can be executed. They are non-biorecognition element based, biorecognition element based, and molecularly imprinted polymer (MIP) based sensors [4].

14.3.1 Non-biorecognition element based electrochemical sensors

The direct oxidation of insulin on the working electrode encounters surface passivation at high potentials, slow reaction kinetics, poor reproducibility, and low sensitivity [41]. Thus electrode modification overcomes these hurdles. The nanocomposites are often used to achieve the same and are designed by affixing carbon materials in combination with metal or metal oxides and/or simple molecules or polymers to integrate the morphological, physical as well as chemical properties of the former with the latter. Thus, the electrode modifiers act as electro-catalysts to achieve better sensing features.

14.3.1.1 Carbon/metal/metal oxide composites for the determination of insulin

The electro-catalysts are compounds that accelerate electron transfer and offer good sensing features such as high reproducibility and sensitivity by reducing surface passivation [78]. The catalytic activity of different metals such as Ni, Cu, Co, and metal oxides such as NiO, CoO, RuO, have been extensively studied in the determination of insulin. The property of nanomaterials are different from bulk materials, especially their electrocatalytic activity [79]. Owing to their exceptionally high surface area, their electrocatalytic effects are promising and have accelerated its application as an electrode modifier in analytical devices [79]. Recently, the focus has been shifted to electro-catalysts based on transition metal oxide nanoparticles that are abundant and cheap. These catalysts have been identified as excellent in conductivity, long-term stability, and are cost-effective [80].

Ni is a widely used electrode modifier and its versatility is attributed to its ability to exhibit various oxidation state. It is cost-effective in comparison with noble metals such as Au, Ag, Pd, and has good electrical conductivity and electrocatalytic activity. The high catalytic activity of Ni based nanoparticle (metal nanoparticle/ metal oxide nanoparticle) accounts for the formation of nickel oxyhydroxide (NiO (OH)), a species of high electron affinity, produced in alkaline media during the electrodeposition of Ni. [81]. The electrocatalytic activity of Ni can also be credited to the presence of unpaired d electrons or empty d orbitals in oxidized Ni which is available for bonding with the analyte of interest.

To improve the electrocatalytic property of the electrode modifier, mixed oxides have been generally used which account for the magnified synergistic effect provided by the individual components in the composite. By the addition of Co to NiO, mixed oxides are produced which significantly enhance the electrode conductivity [82]. Generally, both Ni and Co show electrocatalytic activity in basic medium and their mixed oxides are used for oxygen evolution hence casted as electrocatalyst in amine and alcohol oxidation [83]. Carbon composite electrodes prepared by mixing metal oxide with carbon powder are an excellent source for improving electrocatalytic performance in the determination of insulin [84]. The electrocatalytic behavior of metal oxides can be improved by the presence of carbon materials which are innately porous and good electron mediators. Thus, supporting substrates such as carbon nanomaterials are important to hold these redox mediators for improving electrochemical process and electrocatalyzing insulin oxidation [50, 85, 86]. Carbon–metal oxide composite electrodes showed the best performance when subjected to electrochemical pretreatment in an alkaline medium (Ni-CoO) in which the metal oxides in the composite are converted to their respective hydroxides [50]. This can be done by a cyclic voltammetric scan over a potential range in this medium which leads to a significant improvement in the sensitivity during the amperometric detection of insulin.

The combination of inorganic redox mediators with carbon nanostructures like MWCNTs are known to increase surface area and catalytic site [87]. An additional coating of nafion, a negatively charged ionomer can prevent interference from the negatively charged uric acid, ascorbic acid, and acetaminophen. The utilization of polymer materials such as chitosan, nafion, and poly-ethylene glycol in the modification of electrode yields a polymer layer which prevents the attack of active sites from interfering species such as Cl^- and increases the stability by firmly fixing metal nanoparticles on the electrode surface [86]. The covering of chitosan polymer in NiO nanoparticle decorated MWCNTs increase the number of reactive sites of modified electrode, resulting in better current response. The significance of chitosan is in the fact that its absence results in a non-linear current response with the increase in insulin concentration [88].

The influence of NiO loading on polymer-carbon material (chitosan-MWCNT) composite for insulin determination was investigated by employing different compositions of NiO. It was observed that catalytic sites increase with increase in active surface area provided by NiO nanoparticles [49]. When considering the ability of metal nanoparticles and metal oxide nanoparticles as electrode material for the insulin determination, the metal oxides are superior by providing excellent electrochemical properties, better stability, wide linear ranges and low limit of detection [81].

The metal nanoparticles experience a serious problem of aggregation by the weak Van der Waals forces that exist between them which significantly affect their electrocatalytic activity by reducing the surface area. This can be tackled by dispersing metal nanoparticles in electronically conducting supports such as carbon nanofibres (CNFs), CNTs, graphene, and so forth. There are a number of premier works carried out on CNF as base material for metal nanostructures worldwide due to its promising electrocatalytic effects [89, 90]. Pyrolytic synthesis of CNF through ethylenediamine (EDA) assisted pathway forms electrocatalytic CNF decorated NiO nanoparticles which are excellent electrode material in insulin sensors. EDA provides excess amino groups for uniform loading of NiO nanoparticles over CNF base [91].

Graphene has a commendable dispersing property of forming robust and stable films by electrodeposition which leads to an enhancement in surface area for metal NP/metal oxide loading. Organic reagents/linkers are extensively used as binders for the synthesis of graphene-metal nanocomposites to attach metal nanostructures. But the design strategies demand more environmentally benign procedures. Moreover, these composites are generally used as electrode materials in super-capacitors but are less explored as electro-catalysts in the protein detection [92]. Using green chemical methodology, $Ni(OH)_2$ nanoparticles were directly synthesized over graphene which showed superior insulin oxidation compared to pure $Ni(OH)_2$ and graphene sheets. Graphene matrix has been used for adhering Ni^{2+} precursors and it is electrochemically converted to nickel hydroxide ($Ni(OH)_2$) to form a graphene-metal oxide hybrid. In an alkaline medium, $Ni(OH)_2$ is converted to more stable and highly electrocatalytic nickel oxyhydroxide–graphene (NiOOH–graphene) nanocomposite [76].

While constructing metal–graphene nanostructure assemblies, the uniformity and thickness of the deposited film are the factors of prime concern. Surface coating approaches such as electrophoretic deposition (EPD) ensure uniform thickness of coated $Ni(OH)_2$ nanoparticle film loaded over rGO modified electrodes for insulin detection [93]. EPD is a colloidal process where charged particles are deposited by passing an electric current [94]. The properties such as low cost, simplicity, and good control over thickness make this process advantageous in forming films of GO or rGO for insulin determination [95]. This process electrophoretically infuses insulin to the matrix which can be electrochemically released by applying a potential in the opposite direction.

A major step towards the emergence of electrochemical insulin sensing devices is the development of microelectrodes and screen printed carbon electrodes (SPCEs) as transducers. The small dimensions of microelectrodes offer properties including high signal to noise ratio, rapid mass and charge transfer, etc. A composite of CNT-Ni nanoparticles show synergetic effects of CNT and Ni nanoparticles on carbon fiber microelectrodes (CFMEs) which are prepared by electroless plating and chemical vapor deposition to first deposit Ni nanoparticles and then to grow CNT over the carbon fiber surface. The multicomponent assembly consists of Ni nanoparticles on the topmost layer, CNT in the middle layer, and Ni nanoparticles on the tip of CNT, showing excellent reproducibility and storage life up to eight weeks [96]. SPCE modified with Ni nanoparticles account for miniaturization along with reduced sample volume down to 50 μl from 50 ml needed for other kind of electrodes such as GCE and PGE [88]. Along with their chemical stability and low background current, these electrodes are efficient for electrochemical determination of insulin [97].

CCEs provide an easily renewable and reproducible electrochemical platform for insulin determination. CCE is an excellent alternative when catalysts and redox mediators are unstable and electrodes are easily passivated and poisoned by the oxidation of insulin. CCE electrodes have been doped with Ni powder or mixed with Ru complexes for sensitive determination of insulin and have potential in the development of miniaturized sensors [98].

Unlike Ni and Co, ruthenium oxide and iridium oxide based metal–carbon nanocomposite modified electrochemical transducers monitor insulin in physiological pH. A microsensor for insulin was fabricated by using the electrocatalyst ruthenium oxide electrochemically deposited on a CPE. A dual detector platform provides a real-time simultaneous monitoring of glucose and insulin using flow injection amperometry for the better management of diabetes and insulinoma [99]. The bilayer surface coating of RuO film on CNT modified electrodes provide excellent stability and sensitivity on insulin detection when compared to individually modified electrodes. The flow injection amperometry results showed a sensitivity of 1 nanomolar in the determination of insulin [86].

14.3.1.2 Carbon/monomer/polymer composites for the determination of insulin

Nowadays polymer modified electrodes are more extensively employed as functional materials for electro-analysis owing to their superior electrocatalytic properties [100, 101]. Generally, covalent linkage, physisorption, chemisorption, and so forth form monolayer coating of the material on the electrode surface, but in applicational point of view, multilayer coating of surface active material shows more demand and it can be fulfilled by executing polymers as electrode modifier [102]. Three types of polymer can be used as electrode materials, (1) redox polymers, which are accompanied by electroactive functionalities, assisting for specific recognition of analytes, (2) ion exchange polymers, which are capable of accommodating electroactive species via ion exchange means; and (3) electronically conductive polymers, whose oxidation resulted in the generation of delocalized cationic sites along the polymer chain [103]. Of these categories, electronically conducting polymers like polypyrroles, chitosan, etc, are extensively utilized as electrode modifiers in insulin sensors [41]. Polymer modified electrodes can be fabricated via many procedures like radiofrequency polymerization, dip casting, spin-coating, electropolymerization, etc [56]. Electropolymerization and electrodeposition procedures enable a controlled route to engineer the formation of polymeric coating of homogenous coverage [56, 102]. Electroactive and non-electroactive polymers can be generated, depending on the solvent and monomer used. Owing to outstanding selectivity, stability, sensitivity, larger surface area, and aromatic π–π conjugate bonds, polymers incorporated with different metal/metal oxide composites and nanomaterials are used as promising electrode materials for the facile electro-oxidation and/or electro-reduction of different biomolecules [104, 105].

Polypyrroles are conducting polymers, which show outstanding stability and biocompatibility [106]. They are extensively preferred due to their simplicity in polymerization, nontoxic nature, superior electrical conductivity, low cost, and high stability. If the carbon materials along with metal or metal oxide coated polymer nanocomposites are used as electrode modifiers in insulin sensors, the so formed transducer possess great advantages such as high surface area, excellent electroconductivity, biocompatibility, along with amplification of signals.

Chitosan biopolymer is one of the most appropriate candidates for the electroanalysis of insulin [88]. Chitosan, the deacetylated form of chitin, is a nitrogen containing polysaccharide which is capable of dissolving CNT [107]. This key feature intensified the wide applicability of chitosan based composites as a superior electrode modifier in insulin sensors. Decoration of chitosan with MWCNT has benefits like increased number of active sites resulting in the enhancement of the active surface area available for electro-redox reaction of the analytes. Thus the number of efficient electrochemical biosensors with plenty of combinations emerged based on chitosan-MWCNT along with metal/metal oxide film and polymer nanocomposites, utilizing the morphological as well as chemical features of the composite [108].

14.3.2 Biorecognition element based electrochemical sensors

To advance the sensor platform, sensitivity, selectivity, and biocompatibility, biomaterials such as antibodies, aptamers, proteins, and peptide composites were incorporated successfully. A number of electrochemical sensors have also been developed for insulin incorporating biorecognition elements which are capable of the ultra-sensitive sensing of insulin with promising selectivity and robustness.

14.3.2.1 Carbon-based aptasensors for insulin determination

Aptamers are single stranded synthetic oligonucleotides having high affinity and sensitivity towards analytes. They couple with target molecules covalently or non-covalently. The target molecules of aptamers generally include biomolecules (protein, lipid, DNA, RNA) and biological cells (blood cell, virus, bacteria, etc) [75]. Aptamers, whose name is derived from the term aptus, are synthetic antibodies which mimic the action of antibodies due to their specificity and affinity towards targets. They play a vital role in developing efficient and specific biosensors and are superior to antibodies because of their smaller size which makes them highly stable having the ability of reversible structure-switching [109]. The process called selection evolution of ligands by exponential enrichment (SELEX) has been primarily used to generate aptamer sequences but it provides sequences of low affinity and specificity towards the target [75]. Even though improvised methods like *in silico* sequence recombination and *in vitro* functional evaluation are available for selecting aptamers, there is a need to introduce techniques with high efficacy to create sequences with better binding ability [75].

Electrochemical insulin sensors which employ biorecognition element incorporated modified electrodes as transducers provide better selectivity, sensitivity, and electron transfer rates thereby an improvement in the reaction kinetics [60]. Currently, carbon nanomaterials-aptamers/carbon nanomaterials-metal nanocomposites-aptamers modified electrodes are frequently employed for insulin detection owing to their improved interfacial adsorption properties, better electrocatalytic activities, good biocompatibility, and increased electron transfer kinetics [110]. When the former is used as electrode material on GCE, integration of the specific advantages of individual components along with the benefits of GCE occurs, thereby current amplification takes place during insulin detection. Modification of GCE with aptamer immobilized mesoporous silica film using the electrochemically assisted self-assembly (EASA) method offers a large surface area for the electrode which can facilitate the grafting of aptamers due to the increased reactive sites [110]. When GCE is modified with nanocomposites of gold nanoparticles (AuNPs) and orange II functionalized graphene nanosheets (O-GNs), label free aptasensors for insulin detection will be possible with very low detection limits (10^{-15} M). AuNPs are able to enhance the electron flow by providing binding sites for aptamer immobilization whereas O-GNs can increase the electroactive surface area. The combined effect of AuNPs and O-GNs amplify the current signal to a greater extent [110].

SPE plays a vital role in aptamer modified insulin sensors compared to other carbon-based electrochemical sensors because of their ultra-sensitive analyte detection, easy fabrication, specificity, reproducibility, etc. Nanocomposites of OMC with 1,3,6,8-pyrenetetrasulfonate (OMC-TPS) modified SPE can impart reactive sites for immobilizing aptamers thereby increasing the specificity and sensitivity of the sensor for insulin detection. Even femtomolar range detection of insulin is possible due to the advantage of the OMC-TPS modification [77]. Another important nanomaterial, effective as electrode modifiers for insulin detection are carbon quantum dots (CQDs). The key features of CQDs such as low cost, easy synthesis, and facile functionalization accelerates its applicability as an efficient electrode material. The nanocomposites of CQDs along with chitosan can be employed due to their excellent stability, homogeneity, high selectivity, and large number of amino groups for insulin aptamer immobilization. Picomolar insulin detection is possible with the use of CQDs–chitosan nanocomposites.

Modification of aptamers at conductive Au-O-Phenylenediamine modified PGEs enhances the electroactive surface area thereby providing good sensitivity for insulin detection. It can offer a nanomolar concentration range for insulin quantification. During the course of electrochemical sensing, the advantages of pencil graphite modified electrodes like large electroactive surface area, reproducibility, low background current and high sensitivity assist the selective determination of insulin with ultra-low detection limit [38].

14.3.2.2 Carbon-based immunoassay (antibody)

An immunosensor is a biosensor involving specific interaction between antigens and antibodies by forming an immunocomplex between them. Electrochemical insulin immunosensors which detect the serum insulin level and hence distinguish between type 1 and type 2 diabetes in patients are of vital importance clinically. The functionalization of CNTs by the non-covalent incorporation of pyrene (Py) derivatives on the graphene sidewalls by Π–Π stacking result in the construction of carbon pyrenyl nanostructures for highly sensitive insulin immunosensing. The surface functionalization of carbon pyrenyl nanostructure assemblies by carbodiimide/N-hydroxysuccinimide (NHS) activation yield N-succinimidyl esters that are highly reactive to lysine residues in insulin antibody. This leads to covalent immobilization of insulin monoclonal antibody to carbon pyrenyl nanostructures such as MWCNT/Py composites. To reduce background interferences from the serum matrix, insulin can be extracted from the sample by conjugating to magnetic nanoparticles (MNPs) which can be separated easily by applying a magnetic field. The MNPs are modified with functionalities such as poly(acrylic acid) (PAA), poly (ethylene glycol) PEG and dextran for the bioconjugation of insulin [111]. The effect of surface carboxylic acid enrichment of carbon pyrenyl nanostructure linkers on the covalent attachment of monoclonal insulin antibody was also studied for the serum insulin detection. With greater numbers of carboxyl groups, MWCNT-COOH/Py-COOH shows better sensitivity by a greater amount of covalent anchoring of antibodies. A sandwich immunoassay, important for assaying large biomolecules such as insulin, consists of a capture antibody (anti-insulin antibody) and a detection antibody (peroxidase labeled anti-insulin antibodies). The sandwich assay format coupled with MNPs conjugated to detection antibody results in diminished serum matrix effect and hence amplified current signals [112]. The binding between insulin and surface functionalized antibody/sandwich immunoassay on carbon pyrenyl nanostructures is monitored by the indirect measurement of the charge transfer ions, e.g., ferri- or ferrocyanide redox couple.

A capacitive affinity biosensor based on interdigitated chain electrodes (ICE) first introduced by Newman in 1986, is an excellent sensor for direct detection of analytes without the use of redox couple and ensures label free, real-time detection of insulin.

The immobilization or adsorption of insulin antibodies on ICE generates an electric field on the electrodes which can be monitored by electrochemical impedance spectroscopy (EIS) [113]. The insulin binding on antibodies generates capacitance changes proportional to the concentration of insulin. The sensitivity of antibody–antigen binding can be enhanced by the electrodeposition of graphene on the electrode surface [114].

The sensitivity of electrochemical insulin immunosensors can be increased by employing nanomaterials as signal amplification labels. Carbon-based electrodes such as GCEs provide a highly conductive electrochemical platform for catalytic substrates such as molybdenum disulfide (MoS_2) which is extensively used in electrocatalysis and chemical sensors [115]. The conductivity of MoS_2 can be significantly improved by Pd nanoparticles which not only provides biocompatibility but also behaves as an effective anchoring substrate for antibodies by the formation of a Pd–N bond [116]. Similarly, the zinc silicate spheres loaded with Pd nanoparticles act as dual function labels to form a sandwich immunoassay by conjugating secondary antibodies and to enhance sensitivity [117]. The Pt–Co–Cu nanodendrites as signal labels exhibit synergistic catalytic effects of noble and non-noble metal nanocomposites for the construction of sandwich immunoassay [118]. The binding between insulin and nanomaterial labeled immunoassay is monitored by the decreased electrochemical response due to decrease in the electron transfer [119].

14.3.3 Carbon-based MIP sensors for insulin

As mentioned earlier, biorecognition element-carbon nanocomposite modified biosensors for insulin exhibit excellent features like enhanced selectivity, high specificity, etc. Meanwhile these biosensors encounter some loss of points: (i) difficulty in the adequate production of biorecognition elements and their high cost (ii) the demand of optimum storage conditions and expiration of the recognition element. In order to tackle these limitations and to promote a more advanced alternative to the former, MIP biosensors decorated with carbon nanocomposite were fabricated. MIP is a state of the art strategy to architecture polymers with three-dimensional (3D) imprints complementary to the structure and functionality of the analyte which facilitates specific interaction with the analyte. The MIP-target molecule interaction is similar to antibody–antigen interaction, results in the generation of the electric signal, proportional to the concentration of the analyte which facilitates its quantification. In addition to the inherent sensing features on par with the biorecognition element incorporated sensors, more pronounced reusability, repeatability, reproducibility, and long-term stability even under harsh laboratory conditions include the key features of MIP sensors.

MIPs have been more explored for small molecules attributed to their highly exposed imprint formation, temporary entrapment, structural consistency, facile rebinding, and stability even under severe external conditions [120]. Insulin, being a peptide hormone having large size, its direct MIP fabrication is tedious and inconvenient [13, 121].

Although, MIP sensor fabrication by bulk polymerization, surface polymerization, and epitope imprinting were reported to date, insulin MIPs via bulk polymerization is not preferred owing to the heterogeneous distribution of binding sites, partly embedded imprints, diffusion controlled template–polymer interactions, partial template removal, irregular morphology, etc. Surface polymerization, resulting in the formation of imprints at the surface or its vicinity would be a better choice. The molecularly imprinted nanomaterials, aggrandize the surface polymerization strategy by furnishing 3D binding cavities on the nanomaterial surface [79]. Carbon nanomaterials like MWCNTs act as a matrix for the development of MIPs owing to their peculiar physical and chemical features such as outstanding mechanical strength, promising electron transfer kinetics and high surface area [122, 123]. A MIP based electrochemical sensor for insulin was first reported in 2010, utilizing the coupled advantages of carbon nanomaterial integrated with MIP [124]. Pretreatment of pencil rod led to the formation of PGE and its surface was further modified with monomeric group functionalized MWCNT acting as a matrix on which the polymer was developed where the template, i.e., insulin was imprinted effectively. MIP was generated successfully on the modified electrode via free radical polymerization by adding a suitable functional monomer containing a phosphotidylcholine group, an initiator, cross-linker, and insulin as the template which was then eluted forming the imprint of the template. The sensor facilitated the nanomolar detection of insulin.

Epitope imprinting polymer (EIP) based sensors have the prime advantage of minimizing the structural complexity commonly confronted in MIP based macromolecular quantification, by implementing epitopes as the template molecule [125]. Thus the former can be effectively employed in insulin detection [125]. Epitopes are specific binding sites in proteins whose imprinting enables polypeptide chain entrapment and facile template removal. For insulin determination, this procedure of polymerization facilitates a very low limit of detection via sensitivity intensification. The carbon composite modified EIPs have not yet been reported for insulin quantification.

Meanwhile, designing insulin imprints with exceptional affinity towards the template is a challenging task. Magnetic MIPs (MMIP) are magnetic component integrated MIPs which are custom-made for the specific analytes. These magnetically receptive MIPs in which the template molecules are congregated and eluted magnetically, make the fabrication procedure more facile and controllable, and generate high affinity binding sites for analyte interaction [126]. In order to manipulate the analyte to be determined, MMIPs utilize their magnetic properties to form biomimetic imprints and renewable polymer films. The renewability of the MMIP membrane depends on the external magnetic composite immobilized on the electrode surface and varying MMIP composite concentration. While designing the MMIP sensor for insulin determination, incorporating nanocomposites onto the electrode surface, particularly carbon nanomaterials such as MWCNTs, GOs, OMCs, etc, fascinates the sensor research via outstanding selectivity with ultrasensitivity and excellent practical utility [127]. The electrochemical MMIP sensor for insulin was proposed in 2016 which enabled picomolar detection with high throughput in practical utility studies.

For insulin quantification, although MIP based sensors show excellent analytical features, all the MIP fabrication procedures, even EIP and MMIP are less assimilated due to the previously discussed inherent properties of insulin.

14.4 Conclusion and future aspects

The quantification of insulin can uplift the perceptions, diagnosis, real-time monitoring, and medication of hypo/hyperglycemia, insulin resistance, and type 2 diabetes. Thus, developing reliable sensors for insulin monitoring is of high demand. Employing electrochemical analytical techniques would be a superior option due to its ultra-sensitivity, exceptional stability, selectivity, simplicity, robustness, easy miniaturization, etc. Carbon-based electrodes modified with carbonaceous composites coupled with metal/metal oxide and/or polymer nanocomposites would be an excellent tool to achieve outstanding sensing features for insulin sensors. In addition to that, incorporating biorecognition elements such as aptamers, antibodies, etc during sensor fabrication brings about intensified selectivity for insulin determination. The most astonishing feature of the electrochemical method includes its versatility in terms of sensor design, electrode material compatibility, affordability etc. Current clinical methodologies are complex and are not suitable for on-site and point of care monitoring of insulin. In order to comply with rapid response and ultra-sensitive biosensor for insulin, the current analytical strategies have to be revolutionized with the new era of miniaturized sensing devices. This can be effectively achieved by implementing electrochemical techniques with SPEs as the working electrode. The aforementioned precedence of these electrodes in combination with electrochemical techniques facilitate the transition of conventional laboratorial methodologies of insulin quantification to economic, robust, ultra-sensitive, and fast response devices.

The unique physical, chemical, and electronic properties of carbon materials make them a most suitable electrode material for insulin determination. Coupling the advantages of carbon materials with that of metal or polymer composites would be a reliable choice to acquire sensors with outstanding features such as ultra-sensitivity, selectivity, stability with minimized surface fouling, low over-potential and enhanced peak current. In general, for the determination of insulin, carbon electrode integrated with carbon-based materials, metal oxides, metal nanoparticles, conducting polymers, and their composites are widely explored owing to their promising sensing features.

References

[1] Wilcox G 2005 Insulin and insulin resistance *Clin. Biochem. Rev.* **26** 19

[2] Landgraf W and Sandow J 2016 Recombinant human insulins - clinical efficacy and safety in diabetes *Therapy Eur. Endocrinol.* **12** 12–7

[3] Peterson M C and Shulman G I 2018 Mechanisms of insulin action and insulin resistance *Physiol. Rev.* **98** 2133–233

[4] Duttaroy A *et al* 2005 Development of a long-acting insulin analog using albumin fusion technology *Diabetes* **54** 251–8

[5] Hartman I 2008 Insulin analogs: Impact on treatment success, satisfaction, quality of life, and adherence *Clin. Med. Res.* **6** 54–67

[6] Qaid M M and Abdelrahman M M 2016 Role of insulin and other related hormones in energy metabolism—a review *Cogent Food Agric.* **2** 1267691

[7] De Vegt F, Dekker J M, Stehouwer C D A, Nijpels G, Bouter L M and Heine R J 1998 The 1997 American Diabetes Association criteria versus the 1985 World Health Organization criteria for the diagnosis of abnormal glucose tolerance: poor agreement in the Hoorn Study *Diabetes Care.* **21** 1686–90

[8] Plows J F, Stanley J L, Baker P N, Reynolds C M and Vickers M H 2018 The pathophysiology of gestational diabetes mellitus *Int. J. Mol. Sci.* **19** 3342

[9] American diabetes Association 2014 Diagnosis and classification of diabetes mellitus *Diabetes Care.* **28** S5–S10

[10] Kahanovitz L, Sluss P M and Russell S J 2017 Type 1 diabetes—a clinical perspective *Point of Care* **16** 37–40

[11] Asif M 2014 The prevention and control the type-2 diabetes by changing lifestyle and dietary pattern *J. Educ. Health Promot.* **3** 1

[12] Zhang M, Mullens C and Gorski W 2005 Insulin oxidation and determination at carbon electrodes *Anal. Chem.* **77** 6396–401

[13] Liu M, Weiss M A, Arunagiri A, Yong J, Rege N, Sun J, Haataja L, Kaufman R J and Arvan P 2018 Biosynthesis, structure, and folding of the insulin precursor protein *Diabetes, Obes. Metab.* **20** 28–50

[14] Nakayama M *et al* 2005 Prime role for an insulin epitope in the development of type 1 diabetes in NOD mice *Nature* **435** 220–3

[15] Guin D, Thakran S, Singh P, Ramachandran S, Hasija Y and Kukreti R 2021 Translational biotechnology: a transition from basic biology to evidence-based research *Transl. Biotechnol.* **2021** 3–24

[16] Vleurick L, Renaville R, VandeHaar M, Hornick J L, Istasse L, Parmentier I, Bertozzi C, Van Eenaeme C and Portetelle D 2000 A homologous radioimmunoassay for quantification of insulin-like growth factor-binding protein-2 in blood from cattle *J. Dairy Sci.* **83** 452–8

[17] Aronis K N and Mantzoros C S 2012 A brief history of insulin resistance: from the first insulin radioimmunoassay to selectively targeting protein kinase C pathways *Metab. Clin. Exp.* **61** 445–9

[18] Andersen L, Dinesen B, Jorgensen P N, Poulsen F and Roder M E 1993 Enzyme immunoassay for intact human insulin in serum or plasma *Clin. Chem.* **39** 578–82

[19] Murayama H, Matsuura N, Kawamura T, Maruyama T, Kikuchi N, Kobayashi T, Nishibe F and Nagata A 2006 A sensitive radioimmunoassay of insulin autoantibody: reduction of non-specific binding of [125I] insulin *J. Autoimmun.* **26** 127–32

[20] Määttä K R, Kamal-Eldin A and Törrönen R 2003 A. High-performance liquid chromatography (HPLC) analysis of phenolic compounds in berries with diode array and electrospray ionization mass spectrometric (MS) detection: ribes species *J. Agric. Food Chem.* **51** 6736–44

[21] Yilmaz B, Kadioğlu Y and Aksoy Y 2003 Simultaneous determination of gemcitabine and its metabolite in human plasma by high-performance liquid chromatography *J. Chromatogr. B Analyt Technol. Biomed. Life Sci.* **791** 103–9

[22] Ortner K, Buchberger W and Himmelsbach M 2009 Capillary electrokinetic chromatography of insulin and related synthetic analogues *J. Chromatogr.* A **1216** 2953–7
[23] Martin R S, Gawron A J, Lunte S M and Henry C S 2000 Dual-electrode electrochemical detection for poly(dimethylsiloxane)-fabricated capillary electrophoresis microchips *Anal. Chem.* **72** 3196–202
[24] Pu Y, Zhu Z, Han D, Liu H, Liu J, Liao J, Zhang K and Tan W 2011 Insulin-binding aptamer-conjugated graphene oxide for insulin detection *Analyst* **136** 4138–40
[25] Wang J, Liu C and Hua J 2021 Au–Ag nanoparticles-graphene quantum dots as sensor for highly sensitive electrochemical determination of insulin level in pharmaceutical samples *Int. J. Electrochem. Sci.* **16** 211016
[26] Zhan Z, Zhang H, Niu X, Yu X, Sun H, Sha X, Zhao Y, Wang Y and Li W J 2020 Microliter sample insulin detection using a screen-printed electrode modified by nickel hydroxide *ACS Omega.* **5** 6169–76
[27] Cecil R and Weitzman P D 1964 The electroreduction of the disulphide bonds of insulin and other proteins *Biochem. J.* **93** 1
[28] Stankovich M T and Bard A J 1977 The electrochemistry of proteins and related substances: Part II. *Insulin J. Electroanal. Chem. Interfacial Electrochem.* **85** 173–83
[29] Abraham P, Renjini S, Vijayan P, Nisha V, Sreevalsan K and Anithakumary V 2020 Review on the progress in electrochemical detection of morphine based on different modified electrodes *J. Electrochem. Soc.* **167** 037559
[30] Hasoň S, Daňhel A, Schwarzová-Pecková K and Fojta M 2018 Carbon electrodes in electrochemical analysis of biomolecules and bioactive substances: roles of surface structures and chemical groups *Nanotechnol. Biosens.* **2018** 51–111
[31] Tiwari J N, Vij V, Kemp K C and Kim K S 2016 Engineered carbon-nanomaterial-based electrochemical sensors for biomolecules *ACS Nano.* **10** 46–80
[32] Hassler M 2012 Other commonly used biomedical coatings: pyrolytic carbon coatings *Coat. Biomed. Appl.* **2012** 75–105
[33] Ranganathan S, Kuo T C and McCreery R L 1999 Facile preparation of active glassy carbon electrodes with activated carbon and organic solvents *Anal. Chem.* **71** 3574–80
[34] Yao Y, Liu Y and Yang Z 2016 A novel electrochemical sensor based on a glassy carbon electrode modified with Cu-MWCNT nanocomposites for determination of hydroquinone *Anal. Methods* **8** 2568–75
[35] Abdel-Aziz A M, Hassan H H and Badr I H A 2020 Glassy carbon electrode electromodification in the presence of organic monomers: electropolymerization versus activation *Anal. Chem.* **92** 7947–54
[36] Redivo L, Stredanský M, De Angelis E, Navarini L, Resmini M and Švorc Ĺ 2018 Bare carbon electrodes as simple and efficient sensors for the quantification of caffeine in commercial beverages *R Soc. Open Sci.* **5** 172146
[37] Swain G M 2007 Solid electrode materials: pretreatment and activation *Handbook of Electrochemistry* (Amsterdam: Elsevier) pp 111–53
[38] Annu S S, Jain R and Raja A N 2020 Review—pencil graphite electrode: an emerging sensing material *J. Electrochem. Soc.* **167** 037501
[39] David I G, Popa D E and Buleandra M 2017 Pencil graphite electrodes: a versatile tool in electroanalysis *J. Anal. Methods Chem.* **2017** 1905968

[40] Manjunatha J G 2020 A surfactant enhanced graphene paste electrode as an effective electrochemical sensor for the sensitive and simultaneous determination of catechol and resorcinol *Chem. Data Collect.* **25** 100331

[41] Ebrahimiasl S, Fathi E and Ahmad M 2018 Electrochemical detection of insulin in blood serum using ppy/gf nanocomposite modified pencil graphite electrode *Nanomed. Res. J.* **3** 219–28

[42] Apetrei C, Apetrei I M, de Saja J A and Rodriguez-Mendez M L 2011 Carbon paste electrodes made from different carbonaceous materials: application in the study of antioxidants *Sensors* **11** 1328–44

[43] Hareesha N and Manjunatha J G 2020 A simple and low-cost poly (DL-phenylalanine) modified carbon sensor for the improved electrochemical analysis of riboflavin *J. Sci. Adv. Mater. Devices.* **5** 502–11

[44] Vytřas K, Švancara I and Metelka R 2009 Carbon paste electrodes in electroanalytical chemistry *J. Serb. Chem. Soc.* **74** 1021–33

[45] Hareesha N, Manjunatha J G, Raril C and Tigari G 2019 Sensitive and selective electrochemical resolution of tyrosine with ascorbic acid through the development of electropolymerized alizarin sodium sulfonate modified carbon nanotube paste electrodes *ChemistrySelect.* **4** 4559–67

[46] Jaafariasl M, Shams E and Amini M K 2011 Silica gel modified carbon paste electrode for electrochemical detection of insulin *Electrochim. Acta* **56** 4390–5

[47] Comba F N, Rubianes M D, Herrasti P and Rivas G A 2010 Glucose biosensing at carbon paste electrodes containing iron nanoparticles *Sens. Actuators, B Chem.* **149** 306–9

[48] Erdem C, Zeybek D K, Aydoğdu G, Zeybek B, Pekyardimci Ş and Kiliç E 2014 Electrochemical glucose biosensor based on nickel oxide nanoparticle-modified carbon paste electrode *Artif. Cells. Nanomed. Biotechnol.* **42** 237–44

[49] Shepa J, Šišoláková I, Vojtko M, Trnková L, Nagy G, Maskaľová I, Oriňak A and Oriňaková R 2021 NiO nanoparticles for electrochemical insulin detection *Sensors* **21** 12–7

[50] Arvinte A, Westermann A C, Sesay A M and Virtanen V 2010 Electrocatalytic oxidation and determination of insulin at CNT–nickel–cobalt oxide modified electrode *Sens. Actuators, B Chem..* **150** 756–63

[51] Banks C E and Compton R G 2005 Edge plane pyrolytic graphite electrodes in electroanalysis: an overview *Anal. Sci.* **21** 1263–8

[52] Yamanaka K, Vestergaard M C and Tamiya E 2016 Printable electrochemical biosensors: a focus on screen-printed electrodes and their application *Sensors* **16** 1761

[53] Zavanelli N and Yeo W H 2021 Advances in screen printing of conductive nanomaterials for stretchable electronics *ACS Omega.* **6** 9344–51

[54] Tanner E E L and Compton R G 2018 How can electrode surface modification benefit electroanalysis? *Electroanalysis.* **30** 1336–41

[55] Zen J M, Kumar A S and Tsai D M 2003 Recent updates of chemically modified electrodes in analytical chemistry *Electroanalysis.* **15** 1073–87

[56] Alkire R C, Kolb D M, Lipkowski J and Ross P N 2011 Chemically modified electrodes *Chem. Modif. Electrodes.* **11** 1–267

[57] Jos T, Jose A R, Sivasankaran U and Girish Kumar K 2015 Electrochemical sensing of tinidazole on modified glassy carbon electrodes *J. Electrochem. Soc.* **162** B94–100

[58] Anand S K, Mathew M R, Radecki J, Radecka H and Girish Kumar K 2020 Individual and simultaneous voltammetric sensing of norepinephrine and tyramine based on poly

(L-arginine)/reduced graphene oxide composite film modified glassy carbon electrode *J. Electroanal. Chem.* **878** 114531
[59] Ghapanvari M, Madrakian T, Afkhami A and Ghoorchian A 2020 A modified carbon paste electrode based on Fe3O4@multi-walled carbon nanotubes@polyacrylonitrile nanofibers for determination of imatinib anticancer drug *J. Appl. Electrochem.* **50** 281–94
[60] Power A C, Gorey B, Chandra S and Chapman J 2018 Carbon nanomaterials and their application to electrochemical sensors: a review *Nanotechnol. Rev.* **7** 19–41
[61] Uslu B and Ozkan S A 2007 Electroanalytical application of carbon based electrodes to the pharmaceuticals *Anal. Lett.* **40** 817–53
[62] Tigari G, Manjunatha J G, Nagarajappa H and Prinith N S 2021 Research developments in carbon materials based sensors for determination of hormones *J. Electrochem. Sci. Eng.* **12** 3–23
[63] Anzar N, Hasan R, Tyagi M, Yadav N and Narang J 2020 Carbon nanotube—a review on synthesis, properties and plethora of applications in the field of biomedical science *Sensors Int.* **1** 100003
[64] Businova P, Prasek J, Chomoucka J, Drbohlavova J, Pekarek J, Hrdy R and Hubalek J 2012 Voltammetric sensor for direct insulin detection *Procedia Eng.* **47** 1235–8
[65] Mastrantonio F, Valgimigli F, Grassi L, Cappa P, De Micheli G and Carrara S 2013 Comparative performance of different nanostructured electrochemical sensors on insulin detection *Bionanoscience.* **3** 285–8
[66] Manjunatha J G 2017 A new electrochemical sensor based on modified carbon nanotube-graphite mixture paste electrode for voltammetric determination of resorcinol *Asian J. Pharm. Clin. Res.* **10** 295–300
[67] Gomes-Filho S L R, Dias A C M S, Silva M M S, Silva B V M and Dutra R F 2013 A carbon nanotube-based electrochemical immunosensor for cardiac troponin T *Microchem. J.* **109** 10–5
[68] Jacobs C B, Peairs M J and Venton B J 2010 Review: carbon nanotube based electrochemical sensors for biomolecules *Anal. Chim. Acta* **662** 105–27
[69] Peng L, You S J and Wang J Y 2010 Carbon nanotubes as electrode modifier promoting direct electron transfer from *Shewanella oneidensis Biosens. Bioelectron.* **25** 1248–51
[70] Zelada-Guillén G A, Blondeau P, Rius F X and Riu J 2013 Carbon nanotube-based aptasensors for the rapid and ultrasensitive detection of bacteria *Methods.* **63** 233–8
[71] Sur U K 2012 Graphene: a rising star on the horizon of materials science *Int. J. Electrochem.* **2012** 237689
[72] Rabchinskii M K *et al* 2021 Guiding graphene derivatization for covalent immobilization of aptamers *SSRN Electron. J.* http://dx.doi.org/10.2139/ssrn.3931621
[73] Banerjee A N 2018 Graphene and its derivatives as biomedical materials: future prospects and challenges *Interface Focus.* **8** 20170056
[74] Georgakilas V, Otyepka M, Bourlinos A B, Chandra V, Kim N, Kemp C, Hobza P, Zboril R and Kim K S 2012 Functionalization of graphene: covalent and non-covalent approaches, derivatives and applications *Chem. Rev.* **112** 6156–214
[75] Kurup C P, Tlili C, Zakaria S N A and Ahmed M U 2021 Recent trends in design and development of nanomaterial-based aptasensors *Biointerface Res. Appl. Chem.* **11** 14057–77
[76] Lin Y, Hu L, Li L and Wang K 2014 Facile synthesis of nickel hydroxide-graphene nanocomposites for insulin detection with enhanced electro-oxidation properties *RSC Adv.* **4** 46208–13

[77] Amouzadeh Tabrizi M, Shamsipur M, Saber R, Sarkar S and Besharati M 2018 An electrochemical aptamer-based assay for femtomolar determination of insulin using a screen printed electrode modified with mesoporous carbon and 1,3,6,8-pyrenetetrasulfonate *Microchim. Acta* **185** 59

[78] Testing P *Metal Nanoparticles for Electrochemical Sensing* pp 8–13

[79] Lahcen A A and Amine A 2019 Recent advances in electrochemical sensors based on molecularly imprinted polymers and nanomaterials *Electroanalysis.* **31** 188–201

[80] Habibi E, Omidinia E, Heidari H and Fazli M 2016 Flow injection amperometric detection of insulin at cobalt hydroxide nanoparticles modified carbon ceramic electrode *Anal. Biochem.* **495** 37–41

[81] Šišoláková I, Hovancová J, Oriňaková R, Oriňak A, Rueda Garcia D, Shylenko O and Radoňák J 2019 Comparison of insulin determination on NiNPs/chitosan-MWCNTs and NiONPs/chitosan-MWCNTs modified pencil graphite electrode *Electroanalysis.* **31** 103–12

[82] Armstrong R D, Briggs G W D and Charles E A 1988 Some effects of the addition of cobalt to the nickel hydroxide electrode *J. Appl. Electrochem.* **18** 215–9

[83] Singh R N, Koenig J-F, Poillerat G and Chartier P 1990 Electrochemical studies on protective thin Co_3O_4 and $NiCo_2O_4$ films prepared on titanium by spray pyrolysis for oxygen evolution *J. Electrochem. Soc.* **137** 1408–13

[84] Il Yoon Y and Myoun K J 2008 CoNi oxide/carbon-nanofiber composite electrodes for supercapacitors *Int. J. Electrochem. Sci.* **3** 1340–7

[85] Rafiee B and Fakhari A R 2013 Electrocatalytic oxidation and determination of insulin at nickel oxide nanoparticles-multiwalled carbon nanotube modified screen printed electrode *Biosens. Bioelectron.* **46** 130–5

[86] Wang J, Tangkuaram T, Loyprasert S, Vazquez-Alvarez T, Veerasai W, Kanatharana P and Thavarungkul P 2007 Electrocatalytic detection of insulin at RuOx/carbon nanotube-modified carbon electrodes *Anal. Chim. Acta* **581** 1–6

[87] Martínez-Periñán E, Revenga-Parra M, Gennari M, Pariente F, Mas-Ballesté R and Zamora F 2016 Insulin sensor based on nanoparticle-decorated multiwalled carbon nanotubes modified electrodes *Sens. Actuators, B Chem..* **222** 331–8

[88] Šišoláková I, Hovancová J, Oriňaková R, Oriňak A, Trnková L, García D R and Radonak J 2019 Influence of a polymer membrane on the electrochemical determination of insulin in nanomodified screen printed carbon electrodes *Bioelectrochemistry* **130** 107326

[89] Van Der Lee M K, Van Jos Dillen A, Bitter J H and De Jong K P 2005 Deposition precipitation for the preparation of carbon nanofiber supported nickel catalysts *J. Am. Chem. Soc.* **127** 13573–82

[90] Serp P, Corrias M and Kalck P 2003 Carbon nanotubes and nanofibers in catalysis *Appl. Catal. A Gen.* **253** 337–58

[91] Zhang L, Chu X, Yuan S M and Zhao G C 2015 Ethylenediamine-assisted preparation of carbon nanofiber supported nickel oxide electrocatalysts for sensitive and durable detection of insulin *RSC Adv.* **5** 41317–23

[92] Ren Y, Yu C, Chen Z and Xu Y 2020 Two-dimensional polymer nanosheets for efficient energy storage and conversion *Nano Res.* **14** 2023–36

[93] Belkhalfa H, Teodorescu F, Quéniat G, Coffinier Y, Dokhan N, Sam S, Abderrahmani A, Boukherroub R and Szunerits S 2016 Insulin impregnated reduced graphene oxide/Ni(OH)2 thin films for electrochemical insulin release and glucose sensing *Sens. Actuators, B Chem..* **237** 693–701

[94] Subramanian P, Niedziolka-Jonsson J, Lesniewski A, Wang Q, Li M, Boukherroub R and Szunerits S 2014 Preparation of reduced graphene oxide-Ni(OH)2 composites by electrophoretic deposition: application for non-enzymatic glucose sensing *J. Mater. Chem.* A **2** 5525–33
[95] Van Der Biest O O and Vandeperre L J 2003 Electrophoretic deposition of materials **29** 327–52
[96] Lu L, Liang L, Xie Y, Tang K, Wan Z and Chen S 2018 A nickel nanoparticle/carbon nanotube-modified carbon fiber microelectrode for sensitive insulin detection *J. Solid State Electrochem.* **22** 825–33
[97] Zhan Z, Zhang H, Niu X, Yu X, Sun H, Sha X, Zhao Y, Wang Y and Li W J 2020 Microliter sample insulin detection using a screen-printed electrode modified by nickel hydroxide *ACS Omega.* **5** 6169–76
[98] Salimi A, Roushani M, Soltanian S and Hallaj R 2007 Picomolar detection of insulin at renewable nickel powder-doped carbon composite electrode *Anal. Chem.* **79** 7431–8
[99] Wang J and Zhang X 2001 Needle-type dual microsensor for the simultaneous monitoring of glucose and insulin *Anal. Chem.* **73** 844–7
[100] Terán-Alcocer Á, Bravo-Plascencia F, Cevallos-Morillo C and Palma-Cando A 2021 Electrochemical sensors based on conducting polymers for the aqueous detection of biologically relevant molecules *Nanomaterials.* **11** 1–62
[101] Manjunatha J G 2020 A promising enhanced polymer modified voltammetric sensor for the quantification of catechol and phloroglucinol
[102] Aliofkhazraei M *et al* 2021 Development of electrodeposited multilayer coatings: a review of fabrication, microstructure, properties and applications *Appl. Surf. Sci. Adv.* **6** 100–41
[103] Murray R W 1984 Polymer modification of electrodes *Annu. Rev. Mater. Sci.* **14** 145–69
[104] Namsheer K and Rout C S 2021 Conducting polymers: a comprehensive review on recent advances in synthesis, properties and applications *RSC Adv.* **11** 5659–97
[105] Hareesha N and Manjunatha J G 2021 Electro-oxidation of formoterol fumarate on the surface of novel poly(thiazole yellow-G) layered multi-walled carbon nanotube paste electrode *Sci. Rep.* **11** 12797
[106] Guo T, Zhang C, Zhao J, Ma C, Li S and Li W 2019 Evaluation of polypyrrole-modified bioelectrodes in a chemical absorption-bioelectrochemical reduction integrated system for NO removal *Sci. Rep.* **9** 1–11
[107] Zhang M, Mullens C and Gorski W 2005 Insulin oxidation and determination at carbon electrodes *Anal. Chem.* **77** 6396–401
[108] Šišoláková I, Hovancová J, Oriňaková R, Oriňak A, Trnková L, Třísková I, Farka Z, Pastucha M and Radoňák J 2020 Electrochemical determination of insulin at CuNPs/chitosan-MWCNTs and CoNPs/chitosan-MWCNTs modified screen printed carbon electrodes *J. Electroanal. Chem.* **860** 113881
[109] Radi A-E 2011 Electrochemical aptamer-based biosensors: recent advances and perspectives *Int. J. Electrochem.* **2011** 863196
[110] Li T, Liu Z, Wang L and Guo Y 2016 Gold nanoparticles/Orange II functionalized graphene nanohybrid based electrochemical aptasensor for label-free determination of insulin *RSC Adv.* **6** 30732–8
[111] Singh V and Krishnan S 2015 Voltammetric immunosensor assembled on carbon-pyrenyl nanostructures for clinical diagnosis of type of diabetes *Anal. Chem.* **87** 2648–54

[112] Niroula J, Premaratne G, Ali Shojaee S, Lucca D A and Krishnan S 2016 Combined covalent and noncovalent carboxylation of carbon nanotubes for sensitivity enhancement of clinical immunosensors *Chem. Commun.* **52** 13039–42

[113] Mamishev A V, Sundara-Rajan K, Yang F, Du Y and Zahn M 2004 Interdigital sensors and transducers *Proc. IEEE* **92** 808–44

[114] Yagati A K, Park J and Cho S 2016 Reduced graphene oxide modified the interdigitated chain electrode for an insulin sensor *Sensors* **16** 109

[115] Khan M, Yousaf A B, Chen M, Wei C, Wu X, Huang N, Qi Z and Li L 2016 Molybdenum sulfide/graphene-carbon nanotube nanocomposite material for electrocatalytic applications in hydrogen evolution reactions *Nano Res.* **9** 837–48

[116] Gao Z, Li Y, Zhang C, Zhang S, Li F, Wang P, Wang H and Wei Q 2019 Label-free electrochemical immunosensor for insulin detection by high-efficiency synergy strategy of Pd NPs@3D MoS x towards H2O2 *Biosens. Bioelectron.* **126** 108–14

[117] Li Y, Tian L, Liu L, Khan M S, Zhao G, Fan D, Cao W and Wei Q 2018 Dual-responsive electrochemical immunosensor for detection of insulin based on dual-functional zinc silicate spheres-palladium nanoparticles *Talanta* **179** 420–5

[118] Shao F, Jiao L, Wei Q and Li H 2017 Ternary Pt-Co-Cu nanodendrites for ultrasensitive voltammetric determination of insulin at very low working potential *Microchim. Acta* **184** 2031–8

[119] Li S S, Hu Y Y, Feng J J, Lv Z Y, Chen J R and Wang A J 2014 Rapid room-temperature synthesis of Pd nanodendrites on reduced graphene oxide for catalytic oxidation of ethylene glycol and glycerol *Int. J. Hydrogen Energy* **39** 3730–8

[120] Ertürk G and Mattiasson B 2017 Molecular imprinting techniques used for the preparation of biosensors *Sensors* **17** 288

[121] Buijs J, Vera C C, Ayala E, Steensma E, Håkansson P and Oscarsson S 1999 Conformational stability of adsorbed insulin studied with mass spectrometry and hydrogen exchange *Anal. Chem.* **71** 3219–25

[122] Benmassaoud Y, Murtada K, Salghi R, Zougagh M and Ríos Á 2021 Surface polymers on multiwalled carbon nanotubes for selective extraction and electrochemical determination of Rhodamine B in food samples *Molecules* **26** 2670

[123] Khumsap T, Corpuz A and Nguyen L T 2021 Epitope-imprinted polymers: applications in protein recognition and separation *RSC Adv.* **11** 11403–14

[124] Prasad B B, Madhuri R, Tiwari M P and Sharma P S 2010 Imprinting molecular recognition sites on multiwalled carbon nanotubes surface for electrochemical detection of insulin in real samples *Electrochim. Acta.* **55** 9146–56

[125] Zhao C J, Ma X H and Li J P 2017 An insulin molecularly imprinted electrochemical sensor based on epitope imprinting *Chinese J. Anal. Chem.* **45** 1360–6

[126] Yang Y, Yan W, Guo C, Zhang J, Yu L, Zhang G, Wang X, Fang G and Sun D 2020 Magnetic molecularly imprinted electrochemical sensors: a review *Anal. Chim. Acta.* **1106** 1–21

[127] Zhu W, Xu L, Zhu C, Li B, Xiao H, Jiang H and Zhou X 2016 Magnetically controlled electrochemical sensing membrane based on multifunctional molecularly imprinted polymers for detection of insulin *Electrochim. Acta.* **218** 91–100

IOP Publishing

Electrochemical Sensors Based on Carbon Composite Materials
Fabrication, properties and applications
Jamballi G Manjunatha

Chapter 15

Carbon nanomaterial-based electrochemical sensors for biomedical applications

Monima Sarma

Carbon-based materials have been associated with human civilization since ancient days. As civilization progressed over the centuries, nanoscopic carbon-based materials evolved from the bulk. The most popular forms of carbon nanomaterials are graphene, carbon nanotubes, carbon dots, and fullerenes. The rapid growth of carbon nanomaterials is driven by their excellent thermal, chemical, mechanical, and electrical properties. Another integrated part of modern civilization is sensors. The demand for nanomaterials as the components of the sensors has increased since miniaturized electronic devices came along. The excellent surface properties of carbon nanomaterials allow easy functionalization through chemical modifications. Thus, target and task-specific materials are easily synthesized. The large surface area also promotes electron mobility. Carbon nanomaterial-based sensors have high selectivity and sensitivity, and they have a range of applications including biomedical applications. In this chapter, we intend to review and summarize the applicability of carbon nanomaterial-based electrochemical sensors in the detection, drug delivery, and treatment of various diseases. The electrochemical technique has many advantages over other spectrometry-based analytical techniques. The electrodes modified with appropriate modifiers outperform the bare electrodes. The superior physicochemical properties of the carbon-based nanomaterials make them perfect electrode modifiers for electrochemical sensing of important analytes. The field of carbon nanomaterial-based electrochemical sensors has rapidly expanded through the past decade when a large number of publications have demonstrated the various scopes of these materials. In this chapter, we highlight selected examples emphasizing their potentiality as electrochemical biosensors for clinically important biomolecules.

doi:10.1088/978-0-7503-5127-0ch15

15.1 Introduction

The present era is all about sensors. Sensors have become the central part of rapid technological evolution. The most complicated and efficient array of sensors is the human body where scores of sensors are connected and processed by the brain. An artificial analog of this gargantuan array of sensors is the Internet of Things (IoT). The science of sensors got an impetus with the unraveling of the potentiality of nanomaterials. As bulk materials are broken down into nanosized particles with dimensions in the range 1–100 nanometers, a different set of chemically and physically unique materials evolve. The huge increase in surface area as bulk materials are processed into nanomaterials makes a world of difference. The high surface area in nanomaterials results in a high surface-to-volume ratio, reactivity, and penetrability that enhance efficiency. The sensors of yesteryear were rather bulky but with sizeable progress in nanotechnology, miniaturized devices have come along [1].

Sensors are perturbation responsive mechanical devices. The stimulus that triggers the sensors mechanism could be a physical or chemical perturbation such as light, temperature, magnetism, pressure, electric field, movement, pH, chemical reactions, and so on. An analog response follows the trigger which is analyzed and processed by a built-in computer. The mechanism of each type of sensor is unique and task-specific. An important type of sensor is the biosensor that respond to a biological stimulus such as an antibody, receptor, nucleic acid, etc. Biomedical sensors are the key components of modern medical diagnostic equipment. The research in this area is a combination of electronic information technology and biomedicine. The progress in biomedical sensors is gradually leveling up medical diagnostic procedures. The incorporation of nanomaterials as the receptor or the transducer in biosensors has revamped biological and chemical analyses as the nanomaterial-based sensors have high sensitivity and specificity. As the number of chronic diseases is increasing across the globe, nanomaterial-based biosensors are found to be extremely useful for early detection due to their fast response time, high selectivity, and sensitivity. The choice of nanomaterial in biosensors is crucial. Generally, their selection as the receptors or the transducers in biosensors depends on their properties. Nanofabrication is an important step in biosensing where the manufacturing methods are mostly based on chemical bonding, photolithography, surface etching, and thin-film etching or growth [2].

There are three major types of biomedical sensors based on the information they process—physical, chemical, and biological. The first group of sensors is used to monitor respiration and respiratory rate, blood pressure and blood viscosity, flow rate, body temperature, etc. The second group of sensors, i.e., the chemical sensors, are found in the equipment that monitors various physiological chemical parameters such as pH, the concentration of ions, oxygen, and carbon dioxide concentration in human body fluid. The third category of biosensors is used for the detection of biological matters such as enzymes, antigens, antibodies, receptors, hormones, acetylcholine, serotonin, nucleic acid, among many others. The research in biomedical sensor technology aims at enhancing the efficiency of the relevant sensors to incorporate them into modern instruments [3].

An important modern application of biosensors is wearable devices such as smart glasses, smart watches, wristbands, armbands, running shoes, and so on. Since these devices make lives easier, they are in high demand in the developed and even in developing countries. These task-specific, wearable, mini-computers contain a sensor at their heart that feels a particular impulse such as motion, temperature, pressure, etc and they can be used to monitor blood sugar, blood pressure, heart rate, and so on. Most of these wearable devices are network-enabled and they could be linked with other sensor devices through the IoT to automate various actions. For instance, a heart rate monitor could be networked with emergency services such as ambulances for automatic action in case of abnormal patient conditions [4]. Google's project of monitoring blood glucose levels of diabetic patients by wearable smart contact lenses is worth mentioning. A tiny glucose sensor along with a controlling system was embedded into a contact lens that could determine glucose levels in tears [5]. Small molecules like nitric oxide, carbon monoxide, hydrogen sulfide, etc are important regulators of the nervous and cardiovascular systems. These molecules are also known as signaling molecules, and their detection in living systems is important in medical sciences for diagnosing various diseases and drug development. Biomedical sensors play an important role here. There are various other types of applications of biomedical sensors that we will discuss in the following sections.

Cost is an important determinant when it comes to the practical application of any device. Numerous sensors are known for detecting and analyzing gas, heavy metal ions, biomolecules, etc but their applicability is limited due to high manufacturing cost, the requirement of pretreatment, difficulty in operation, slow response time, low detection limit, and meager selectivity. Carbon-based nanomaterials are one of the most popular materials in nanotechnology due to their outstanding physicochemical properties, easy fabrication, and environmental friendliness. Apart from the pristine nanoparticles, they can also be combined into nanocomposites that have different chemical and physical properties. The most common carbon nanomaterials used in biosensors are based on pristine or composites of graphene, carbon nanotubes, and carbon dots.

Each carbon-based nanostructure has a unique crystal structure that makes each category of materials suitable for task-specific sensing. For instance, carbon nanodiamond structures usually contain nitrogen-vacancy defects that have an elegant quantum nature, responsible for the fluorescence of nanodiamonds, and make them promising in biosensing. Fullerenes, on the other hand, have zero-dimensional π-conjugated structures that are responsible for their phenomenal redox properties. Also, fullerenes can be chemically modified due to their attractive surface properties which expands their applications. The enhanced electron transfer kinetics, high surface-to-volume ratio, and biocompatibility of fullerenes are beneficial for their widespread applications in biosensing such as drug, nucleic acid, nucleobase detection, and so on. Fullerenes have a high electron affinity and are chemically reactive, especially toward free radicals. The tubular structure of carbon nanotubes and their interesting surface properties separate them from other types of carbon nanomaterials. The physicochemical properties of carbon nanotubes are dependent on the functional groups attached to their surface and they are highly suitable in biosensing because of their incredible electrical properties such as high carrier mobilities [6–11]. Each of these carbon nanomaterials has some advantages and limitations in their applications in biosensing. These are summarized in table 15.1.

The technique is another important factor in determining the cost of experimentation. Popular spectroscopic techniques such as atomic absorption spectroscopy, x-ray fluorescence, inductively coupled plasma mass spectrometry, neutron activation analysis, etc are costly, involve complicated sample preparation, intricate

Table 15.1. Summary of the advantages and limitations of carbon-based nanomaterials in biomedical applications. Adapted from [9]. Copyright 2021, MDPI.

Nanomaterial	Advantages	Limitations
Carbon nanotubes	• Excellent mechanical strength, high aspect ratio, good conductivity, high chemical stability. • Tunable physical properties through chemical modifications. • Biocompatibility.	• Poorly soluble in the aqueous medium. • Size inhomogeneity. • Metallic impurity contamination possibility. • Possible toxicity by respiration.
Graphenes	• Outstanding physicochemical properties. • Versatile electronic characteristics.	• Instability in the colloidal medium. • Reproducibility and stability issues in the biological media. • Sensitive in the oxidative environment.
Fullerenes	• Unique physicochemical properties. • Alteration of properties via chemical modification.	• Low solubility in the aqueous medium. • Sensitive to light and oxygen. • Accumulate in cell membranes.
Carbon nanodiamonds	• Exhibit luminescence. • Biocompatibility. • Amazing physicochemical properties.	• Synthetic difficulties. • Abnormal drug-releasing issues. • Aggregate formation.
Carbon nano onions	• Remarkable electronic, structural, and optical properties.	• Sparse solubility in the aqueous medium. • Low biocompatibility. • Susceptible to oxidation. • Less surface reactivity. • Aggregate formation.

instrumentation, and difficult data analysis. Using a cheaper technique such as electrochemical sample detection keeps the overall cost low, allows direct, specific, and real-time monitoring. Therefore considerable research attention has been given to developing highly selective and sensitive electrochemical sensors to analyze a diverse class of analytes. The biological analyte binds to the receptor fragment of the electrochemical sensor through various intermolecular interactions. Following this recognition process, signal transduction can occur through variations of potential difference, current, resistance, or capacitance.

The various electrochemical sensing methods are potentiometry, amperometry, voltammetry, conductometry, and so on [12]. The heart of potentiometry is the Nernst equation, i.e., variation of the cell potential based on the analyte concentration according to the equation below

$$E_{\text{cell}} = E^0_{\text{cell}} - \frac{RT}{nF}\ln\frac{a_{\text{ox}}}{a_{\text{red}}} \approx E^0_{\text{cell}} - \frac{RT}{nF}\ln\frac{[\text{ox}]}{[\text{red}]} \approx E^0_{\text{cell}} - \frac{0.0591}{n}\log\frac{[\text{ox}]}{[\text{red}]}$$

where, E_{cell} is the formal cell potential, E^0_{cell} is the standard cell potential, R is the universal gas constant (~8.314 J mol^{-1} K^{-1}), T is the temperature in Kelvin, n is the number of electrons involved, F is the Faraday constant (~96 485 C mol^{-1}) and terms in bracket are the concentration (more accurately, activity) of the oxidized and reduced species in solution respectively. The instrument that is used in potentiometry is called a potentiostat that has a three-electrode setup—a working electrode (e.g., glassy carbon etc), a reference electrode (e.g., saturated calomel electrode, silver-silver chloride electrode, etc), and a counter electrode (mostly, a platinum wire). The electrochemical reaction occurs at the working electrode surface. Therefore, modification of the working electrode surface to enhance selectivity and sensitivity is the central objective of research in electrochemical sensing.

In amperometry, on the other hand, the flow of current through the working electrode with time is used to determine the concentration of the redox-active species. Calculations are done according to the Cottrell equation:

$$I_t = \frac{nFAc^0_{\text{O}}D^{1/2}_{\text{O}}}{\pi^{1/2}t^{1/2}} = bt^{1/2}I_i$$

Here, I_t is current at time t (s), n is the number of electrons, F is the Faraday constant (96 485 C mol^{-1}), A is the area of the electrode (cm^2), c_{O} is the concentration of the oxidized species (mol L^{-1}), D_{O} is the diffusion coefficient of the oxidized species (cm^2 s^{-1}) [13].

The most popular electroanalytical technique is however voltammetry. In this technique, the flow of current (I) between two electrodes is measured as a function of the potential difference (E) between them [14]. Cyclic voltammetry (CV) and differential pulse voltammetry (DPV) are the commonly used voltammetry-based techniques. In CV, a potential difference between two electrodes dipped in a redox-active sample solution is scanned for a period, and the resultant flow to current is measured. Based on the mode of the potential sweep, oxidation (positive mode) or reduction (negative mode) of the analyte occurs at the working electrode surface that brings about a change in the current. The oxidation and reduction potentials are characteristics of the analytes. A cyclic voltammogram is then established by plotting the change of current with applied potential, and analysis of the voltammogram yields the performance of the sensor.

DPV, in contrast, is a differential technique. At first, a base potential is applied such that no electrochemical reactions occur. After this, the potential is slowly increased between the pulses, the current is measured at the start and the end of the pulse, and their difference is noted [13]. Both these techniques come with their own set of pros and cons. However, CV is more popular than DPV because of many advantages such as lower operating costs, faster analysis time, usability for both organic and inorganic analytes, outstanding sensitivity, and so on.

On balance, an electrochemical biosensor can be represented by the flow chart shown in scheme 15.1. A typical biosensor has three major components: (a) a bio-recognition element that specifically interacts with a target analyte, (b) a transducer to convert the bio-recognition process into a measurable signal, and (c) a machine to process the signal output. This chapter reviews the last two components of the flow chart, i.e., electrodes (transducer) and technique (electrical signal), for sensing of biological analytes. Detection of bio-analytes is especially tricky as the concentration of the target analytes in the biological medium could be significantly lower. Therefore, the selectivity and sensitivity of the biological sensors should be high enough for quantitative detection. Both the technique and electrode play an important role here. A wide range of electrode materials such as noble metals, conducting polymers, carbon-based nanomaterials is available for biological sensing applications [15–23]. However, in this chapter, we wish to review the carbon-based nanomaterials in the electrochemical detection of biological analytes. Since each carbon nanomaterial has unique physicochemical properties, we divided the chapter into various sections based on the type of nanomaterial used. The electron transport mechanism of the nanomaterials is influenced by their chemical composition, surface, solid-state structure, orientation crystal axis, etc. Breaking the chapter down into several subsections based on the structure of the carbon nanomaterials will help readers to correlate the nature of the nanomaterials with their applications.

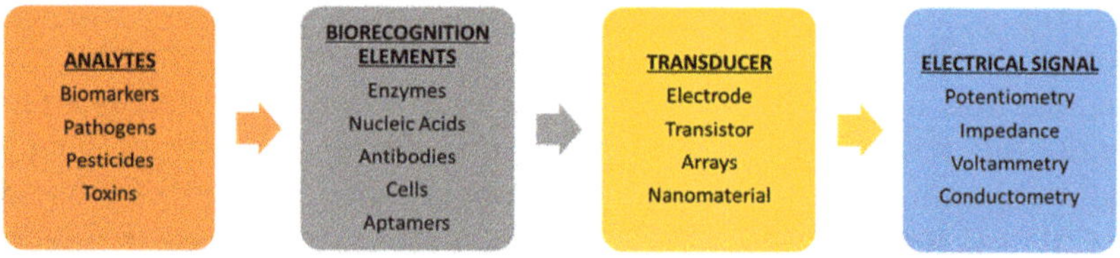

Scheme 15.1. Flow chart representing electrochemical biosensors. Adapted with permission from [10]. Copyright (2018) American Chemical Society. Further permission related to this figure should be directed to the ACS

15.2 Biosensors based on graphenes

Graphene is a single layer of sp^2 hybridized carbon atoms in a two-dimensional honeycomb crystalline lattice. This wonder carbon nanomaterial is viewed as an exfoliated sheet of a single atomic layer from graphite. Alternatively, graphite is a stacked layer of graphene sheets held together through $\pi\cdots\pi$ stacking interactions. Graphene is considered to be the precursor of all carbon nanostructures. It can be rolled into carbon nanotubes, stacked into graphite, and folded into fullerenes [24]. Examples of important graphene-based nanomaterials are pristine graphene, graphene oxide (GO), reduced graphene oxide (RGO), graphene quantum dots (GQDs), and graphene nanoribbons (figure 15.1).

Each carbon atom in a graphene sheet is bonded to three neighboring carbon atoms by a sigma bond and leaves an electron in the conduction band that is delocalized over the whole sheet (figure 15.1). The high electron mobility, large surface area, and high electrical conductivity make graphene highly important in practical applications. The graphene-based nanomaterials could be synthesized by top-down, bottom-up methods, chemical exfoliation, chemical or thermal reduction, or electron beam lithography. In the top-down synthesis of graphene, the dispersion forces between the sheets are distorted and segregated. The bottom-up method, on the other hand, involves a simpler synthesis but yields poorer product quality than the former. The most popular method of graphene synthesis is however exfoliation of single sheets from graphite. The graphene-based nanomaterials are generally hydrophobic whereas biomedical applications require hydrophilicity. This is achieved by attaching specific functional groups through appropriate chemical modifications which enhances the biocompatibility, dispersibility, and reactivity of graphene. Hydrogen bonding between polar functionalization on graphene sheets and water molecules in the biological environment further facilitates the biomedical applications of graphenes. Graphene-based materials are easily encapsulated onto a substrate due to their properties such as electrostatic interactions and stacking interactions. The encapsulated graphenes have diverse applications such as drug and gene delivery and bioimaging [25, 26]. In this section, we discuss the various biological applications of graphene-based nanomaterials.

Enzymes are proteins that behave as biological catalysts and bring down the activation energy of the biological reactions, similar to the chemical catalysts do on

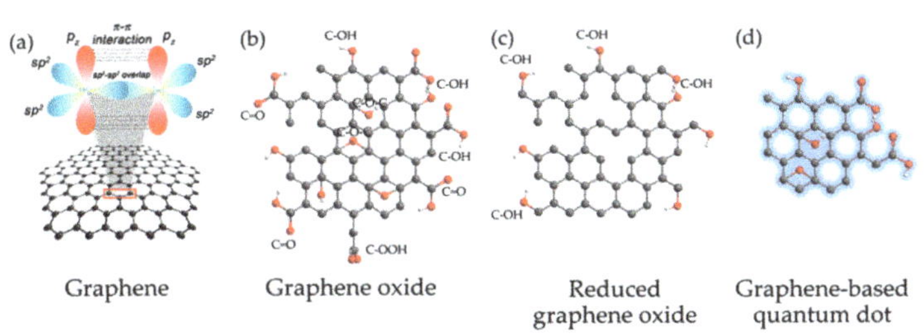

Figure 15.1. Structures of important graphene-based nanomaterials. Reproduced with permission from [25]. Copyright 2017. MDPI.

the chemical reactions. The target molecules are called substrates and the enzymes have a complementary shape to the substrates to form specific enzyme-substrate complexes. Sustenance of life requires faster biological reactions where enzymes play a central role [27, 28]. Since the enzymes have selective bio-recognition properties, they are suitable for the sensing of biological analytes. Enzymatic biosensors combine the substrate recognition process of enzymes with transducers and generate signals proportional to the substrate concentration with high selectivity and sensitivity. Due to the outstanding electrochemical properties of graphenes, interest grew to develop graphene-enzyme hybrid materials for electrochemical biosensing applications. Each carbon atom of graphene sheets is exposed to the surrounding. Therefore, any fluctuation of its environment could trigger a dramatic electrical response paving the way for highly sensitive biosensing by graphene-based materials. This section outlines some typical examples of such development.

The tricky part of electrochemical enzymatic biosensing is the electron transfer between enzymes and the electrodes. The most popular methods of electrical communication are direct electron transfer between the active site and the electrode, and mediated electron transfer between them using redox mediators. The former is modulated by the immobilization of the enzymes on the electrode surface, the orientation of the enzymes, and the localization of the active sites in the enzyme. On the other hand, the latter is affected by the redox properties of the mediator. The immobilization of enzymes on the electrode surface locks the bio-recognition element at a specific position thus increasing the concentration of the bio-recognition material at the suitable location, and preventing deactivation of the biomolecules. Both reversible (adsorption, metal chelation, disulfide bond formation) and irreversible (covalent attachment, entrapment, encapsulation, cross-linking) immobilization techniques were explored to obtain optimal performance from the relevant sensors. Adsorption is perhaps the most common reversible immobilization technique that makes use of van der Waals dispersion forces, hydrophilic and hydrophobic interactions, hydrogen bonding, etc. The disulfide bond formation, though is a type of covalent attachment, is considered a reversible immobilization method due to easy rupture of the bonds under mild conditions. The most common method for irreversible enzyme immobilization is covalent bonding. The stability of the bonds between the enzyme and the support curtails enzyme detachment. The concerned strategy incorporates activation of the support surface through some sort of chemical modifications followed by attachment of the biomolecule by different bonding mechanisms such as bonding using the side chain of lysine, thiol groups in cysteine, carboxylic acid groups of glutamic and aspartic acids [29].

15.2.1 Glucose sensing

The detection of glucose in body fluids is an important medical diagnosis, especially for diabetic patients. Diabetes mellitus destructively affects the pancreas to produce insulin resulting in an abnormal concentration of sugar in the blood. The glucose oxidase enzyme (GOx) is an oxidoreductase that catalyzes the oxidation of glucose to gluconolactone and hydrogen peroxide. This dimeric protein contains one flavin

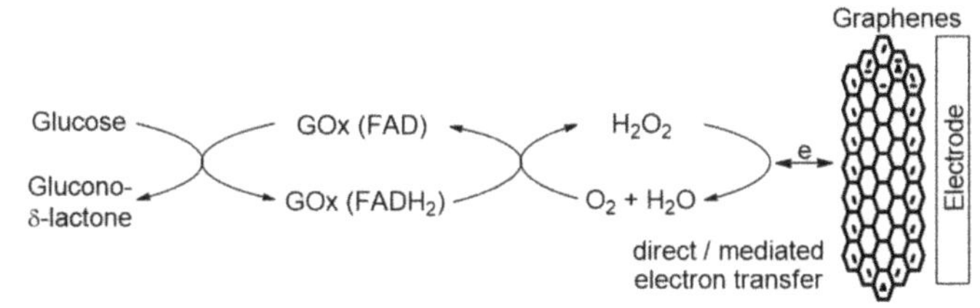

Scheme 15.2. Oxidation of glucose by glucose oxide enzyme on the graphene-electrode surface.

adenine dinucleotide (FAD) unit per monomer which is the redox-active site of the enzyme. Oxidation of glucose is accompanied by the reduction of FAD followed by the generation of hydrogen peroxide from oxygen and water and subsequent electron transfer (scheme 15.2). This typical reaction is the basis of electrochemical detection of free glucose in biological samples. Immobilization of glucose oxidase enzyme on graphene produces highly sensitive electrochemical sensors because of the plentiful active sites and protection from degradation of the biomolecules. The subsequent electron transfer was studied with and without using any mediator. Studies revealed that the mediator-free solutions had no response to glucose although direct electron transfer of the enzyme was observed. Therefore, additional nanomaterials were incorporated into the system to augment the catalytic activity and performance of the sensor [30].

Lin and coworkers reported glucose sensing by functional graphene sheet-based bio-nanocomposite film glucose oxidase/Pt/functional graphene sheets/chitosan. Here, platinum nanoparticles were used as the mediator to enhance the electrocatalytic activity and sensitivity of the biosensor. The relevant bio-nanocomposite film was obtained by chemical modification of graphene sheets by electrodeposition of platinum nanoparticles followed by physical modification of the thus obtained platinum-graphene nanocomposite film with enzymes. The glucose oxidase enzyme was then immobilized on the electrode surface (modified glassy carbon electrode). The electrochemical response of the relevant biosensor stems from the anodic oxidation of hydrogen peroxide. Although H_2O_2 undergoes electrochemical reactions at the solid electrodes, the concerned redox reactions take place at higher potentials that might cause interference from other electroactive materials in the solution. Platinum nanoparticles used in the study brings the potential down by acting as electrocatalyst. The pertinent biosensor exhibited excellent glucose sensing (linear range: 1.5×10^{-4}–4.2×10^{-3} M, limit of detection: 6×10^{-7} M) that was alluded to the fast electron transfer mediated by platinum. Remarkably, the sensor was able to detect glucose in the human sample with high sensitivity and fast response time [31].

Another classic example came from Chen and coworkers. They used RGO/zinc oxide (ZnO) composite on glassy carbon electrode (GCE) followed by immobilization of glucose oxidase on the electrode surface for glucose sensing. In this study, zinc oxide was used as the electron mediator due to its high specific surface area, nontoxicity, chemical stability, and high electron mobility. The GOx was immobilized on the electrode surface by drop-casting on the rGO/ZnO composite modified GCE. The immobilization of the enzyme on the electrode surface was facilitated by electrostatic attraction between the negatively charged GOx with the positively charged rGO/ZnO matrix. The relevant biosensor performed pretty well in glucose sensing with a linear range of 2.5 $\times 10^{-5}$–6.2 $\times 10^{-3}$ M and a limit of detection of 2.0 $\times 10^{-5}$ M [32].

As said previously, direct electron transfer is the most sought after process, although it is often hindered by the deep location of the prosthetic groups within electrically insulated pockets. Apart from blending metal-based nanoparticles with graphene nanosheets to uplift the device performance, efforts have been made to develop graphene-carbon nanomaterial hybrid three-dimensional architectures. The three-dimensional network along with the high surface area of such materials showed promising performance in the electrochemical biosensing of glucose. An example of this category of material is carbon nanotube—graphene hybrid from Chen *et al* Acid functionalized carbon nanotube and graphene oxide formed a homogeneous mixture upon sonication and the stability of the mixture was driven by dispersion interactions between them. The authors believed that graphene oxide could contain the irreversible aggregation of carbon nanotubes whereas the latter would prevent restacking of graphene oxide. The relevant three-dimensional hybrid exhibited a large surface area, excellent electronic conductivity, and an interconnected network of multi-walled carbon nanotubes with pores that were favorable for electron transport and subsequent glucose sensing. The relevant sensor exhibited good sensitivity of 1.27 μA mM L^{-1} cm^{-2} with a fast response time of fewer than four seconds. The response current of the relevant sensor decayed by only 10% after one week of storage at 4 °C [33].

15.2.2 Cholesterol sensing

Cholesterol is sterol or a modified steroid and is a long chain containing lipid molecules biosynthesized by all animal cells. It is an important structural component of animal cell membranes. Although cholesterol has an important role in biology, high accumulation of cholesterol in the body (>6.21 mM) might trigger multiple diseases such as arteriosclerosis (accumulation of cholesterol in the artery), cardiovascular disease, coronary artery disease, arterial hypertension, myocardial infarction, lipid metabolism dysfunction, and cerebral thrombosis. Thus, the determination of cholesterol is paramount in medical diagnosis because of the dangerous effects of cholesterol on the human body. Until now, various cholesterol sensors have been designed and developed. They include enzymatic, non-enzymatic, and redox mediator-based sensors. It was said in the previous section that enzymatic sensors offer high selectivity and sensitivity. Cholesterol oxidase (ChOx)-based electrochemical sensors are found to be effective in cholesterol sensing due to their simplicity, fast response time, and low cost. The ChOx enzyme belongs to the family of oxidoreductases and catalyzes the oxidation of cholesterol (scheme 15.3). Like the glucose oxidase enzyme, the redox-active group of ChOx is FAD. When the enzyme is immobilized on the electrode surface, the redox reaction is accompanied by

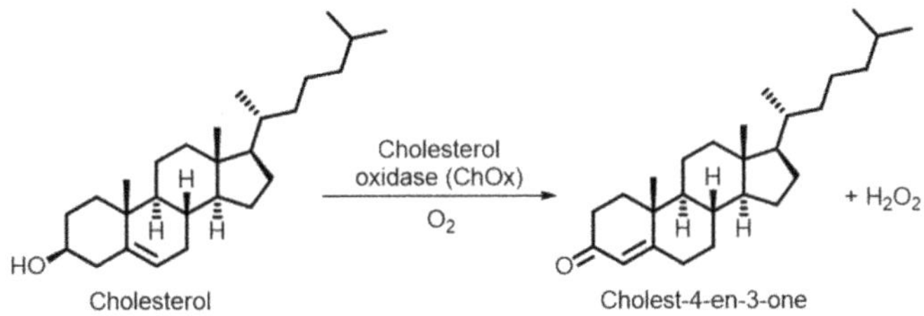

Scheme 15.3. Oxidation of cholesterol by cholesterol oxidase enzyme.

electron transfer and thus, electrochemical quantification of the relevant redox product, hydrogen peroxide, could be used as a means to determine cholesterol. The amount of the analyte could be calculated by measuring the electric current. However, the enzymes are short-lived and their immobilization on the electrode surface could often risk denaturing them. These limitations could be avoided by using non-enzymatic cholesterol sensors which mostly involve modification of the electrode surface with metal, metal oxide, or their composite nanomaterials. A somewhat hybrid device type is redox mediator-based sensors which use both enzymatic and chemical redox mediators. These types of devices might involve direct or indirect electron transfer between the electrode and the enzymes based on their architecture [34]. In this section, we discuss graphene-ChOx enzyme-based electrochemical sensors for cholesterol detection.

Chitosan is known as a biocompatible polymer and is used as an immobilizer because of its high permeability to water, outstanding membrane forming ability, and excellent adhesion. It is also known that chitosan can accumulate various metal ions depending on the nature of the ions and the pH of the solution. Yuan and coworkers developed a graphene–chitosan–Pt–Pd nanocomposite for electrochemical detection of cholesterol. In this study, chitosan played the dual role of solubilizing graphene nanosheets as well as accumulating platinum and palladium ions. The use of chitosan as a membrane forming component enhanced the porosity, surface area, and electron transfer rate of the relevant nanocomposite. Using cholesterol oxidase as the catalyst and amperometry as the technique, the relevant biosensor could detect cholesterol in a linear range of 2.2×10^{-6}–5.2×10^{-4} M and a detection limit of 7.5×10^{-7} M. The authors also examined the sensing ability of the sensor in real food samples such as meat, fish oil, margarine, and egg with satisfactory results [35].

A good-performing sensor requires a high signal-to-noise ratio, sufficient reproducibility, and a low limit of detection. Silver nanoparticles have a high catalytic activity for biochemical reactions and are often used to enhance the performance of biosensors. Huang *et al* reported ultrasensitive cholesterol sensing using graphene oxide and gold nanoparticles co-mediated silver deposition for signal amplification. In this study, cholesterol oxidase, cholesterol esterase, and graphene oxide were immobilized onto the surface of gold nanoparticles modified screen-printed carbon electrodes. The obtained sensor was applied to measure the different concentrations of cholesterol, and the following current response was recorded with linear sweep voltammetry. The electrochemical signals were amplified many-fold upon the addition of cholesterol. The current response of the sensor went up with analyte concentration maintaining good linearity. The sensitivity and limit of detection of the relevant sensor were calculated to be 0.084 $\mu A\ cm^{-2}\ \mu g\ mL^{-1}$ and 0.001 $\mu g\ mL^{-1}$ respectively [36].

15.2.3 Hydrogen peroxide sensing

Hydrogen peroxide is a reactive oxygen species and is a by-product that is hydrolyzed by oxidase enzymes. These reactive species are present in the biological systems playing important roles in physiological processes as a signaling molecule. Nevertheless, a high concentration of hydrogen peroxide in the body could be fatal. Therefore, detection of H_2O_2 is an important topic of research that is further complicated by its high reactivity and low physiological concentration. Various techniques such as titration, chromatography, light detection, and electrochemical sensing were tested for the detection of hydrogen peroxide in biological systems. Though titration and chromatography are cheap analytical techniques and can be used in the quantification of H_2O_2, they are unsuitable for *in vivo* and *in vitro* analyses. Electrochemical sensing of H_2O_2 based on proteins such as hemoglobin (Hb), myoglobin (Mb), horseradish peroxide (HRP), cytochrome c, catalase is found to be selective, sensitive, and effective. HRP is the most commonly used enzyme in H_2O_2 biosensors that contains a heme as a prosthetic group (Fe-protoporphyrin IX). Like the glucose and cholesterol sensors presented in the previous sections, the electron transfer for H_2O_2 sensing could be either direct or mediated. A simplified reduction cycle of H_2O_2 by HRP is shown in scheme 15.4 [37].

Graphene-based nanocomposites have been successfully employed in hydrogen peroxide sensing due to their excellent electron transfer property and bioactivity. Functionalized graphenes such as graphene oxide, carboxylate functionalized graphenes in conjunction with metallic nanoparticles or polymers are commonly used in electrochemical sensing of H_2O_2. Cai *et al* reported a simple and fast electrochemical synthesis of graphene-Polypyrrole (PPy) composite film using

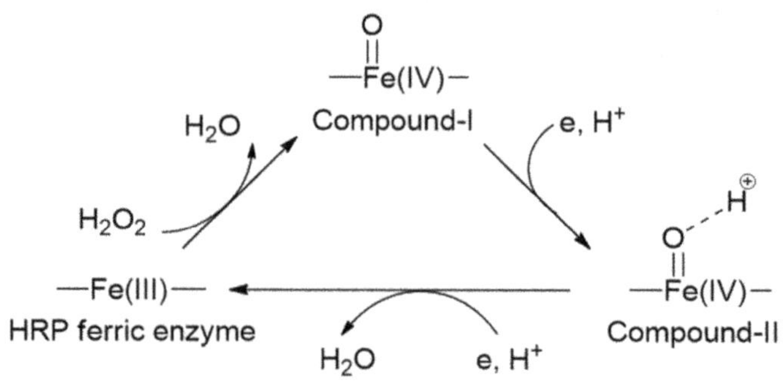

Scheme 15.4. A simplified reduction cycle of hydrogen peroxide by HRP enzyme.

graphene oxide and polypyrrole nanotube as the precursors without using any toxic reducing agents and subsequent application of the composite material in biosensing. According to the authors, the graphene nanosheets and PPy mutually benefitted each other as the former could improve the surface coverage of the PPy nanotubes on the electrode, and the latter could contain the aggregation of graphene. The hydrophilic graphene oxide has plenty of epoxy, carboxyl, hydroxyl functional groups which could participate in hydrogen bonding with the PPy nanotubes. The hydrogen-bonding interactions together with $\pi\cdots\pi$ stacking interactions between the components assisted homogeneous distribution and prevented precipitation of the PPy nanotubes. HRP was then immobilized onto the modified glassy carbon electrode. In the absence of H_2O_2, a pair of redox signals appear that were ascribed to the direct electron transfer between the enzyme and the modified electrode. The addition of hydrogen peroxide in the system caused enhancement of the cathodic current and reduction of the anodic current that is typical of an electrocatalytic reduction of H_2O_2. The fabricated biosensor exhibited a detection limit of 1.0×10^{-7} M at a signal-to-noise ratio of 3 [38].

In another report, Mercante *et al* demonstrated electrochemical biosensing of hydrogen peroxide by immobilizing HRP enzyme onto PEDOT:PSS-rGO-AuNPs nanocomposite film-modified electrode. Like the previous example, $\pi\cdots\pi$ stacking interactions between rGO surface and PEDOT: PSS polymer played an important role in realizing stable composite. The gold nanoparticles were grown on the rGO surface by *in situ* reductions. The HRP enzyme was subsequently attached to the PEDOT: PSS-rGO-AuNPs modified screen-printed gold electrodes by amide coupling reactions. As the authors saw it, the synergy between the polymer, rGO, and gold nanoparticles enhanced the electrocatalytic activity of the relevant biosensor. The analytical performance of the fabricated biosensor was evaluated by amperometry that exhibited a sensitivity of 677 $\mu A\ mM^{-1}\ cm^{-2}$ and a detection limit of 0.08 μM [39].

15.2.4 Biosensing of neurotransmitters

The development of devices for rapid and sensitive detection of neurotransmitters is an important and exciting field of research due to its major implications in clinical diagnosis. Diseases related to the neurotransmitters such as Alzheimer's disease, Parkinson's disease, schizophrenia, depression, and anxiety have become more prevalent these days. Early diagnosis of such diseases is thus paramount. The most concerning neurotransmitters and neuromodulators in the central nervous system are dopamine, serotonin, and adrenaline (also known as epinephrine) (scheme 15.5). These are monoamine neurotransmitters containing one amino acid attached to an aromatic ring by a carbon bridge and play an important role in the functioning of other organisms in the body. Diseases associated with neurological and psychiatric disorders such as Parkinson's disease, depression, and schizophrenia are usually treated by medicines that alter the levels of these neurotransmitters. Dopamine has an important role as a modulator of hormone secretions, cardiovascular and renal functions along with the regulation of various physiological functions. Serotonin is

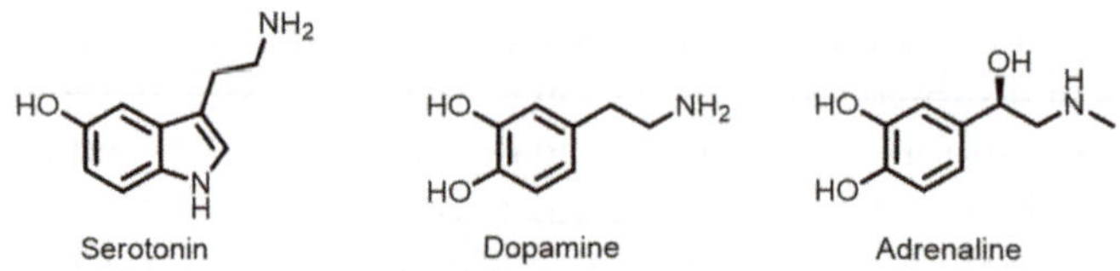

Scheme 15.5. Molecular structures of serotonin, dopamine, and adrenaline.

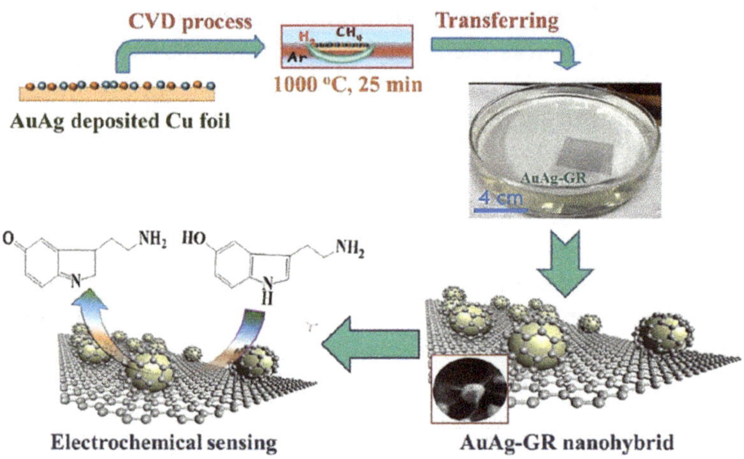

Figure 15.2. Synthesis of gold–silver–graphene nanohybrid and its application in electrochemical detection of serotonin. Reprinted from [41]. Copyright (2017), with permission from Elsevier.

responsible for swinging complex behaviors such as mood, appetite, sleep, cognition, perception, motor activity, temperature regulation, pain control, sexual behavior, and hormone secretion [40].

Hybrid nanocomposites based on graphene and metal alloy nanoparticles such as AuAg alloy was found to have interesting biosensing ability toward the neurotransmitters. The gold–silver alloy nanoparticles are known to have excellent conductivity, outstanding catalytic activity, and high chemical stability that make them suitable for the development of electrocatalysts. Hybridization of the gold–silver nanoparticles with graphene enhances the catalytic activity due to the mutual benefits of the two nanocomponents. The large surface area of graphene assists the adsorption of organic analytes and at the same time prevents aggregation and dissolution of the metal alloy nanoparticles. Lee and coworkers reported an efficient electrochemical biosensor based on AuAg-Graphene nanohybrid for the detection of serotonin. A schematic process for the synthesis of the relevant nanohybrid is shown in figure 15.2. Here, gold and silver were first deposited on a copper substrate which was then loaded into a quartz tube of a CVD system to form the AuAg-graphene nanohybrid film. The serotonin sensing performance was then evaluated using amperometry as the technique. The oxidation current of the system gradually increased with the increasing concentration of serotonin. The relevant electrochemical sensor exhibited a linear range of 2.7 nM–4.82 μM with a sensitivity of 0.766 μA μM^{-1} cm^{-2} and a detection limit of 1.6 nM. The fabricated electrochemical sensor showed good performance even with human serum samples [41].

The selectivity of the biosensors is often improved by incorporating biorecognition counterparts such as enzymes, nucleic acids, and antibodies. Using antibodies as a recognition element to develop immunosensors has many advantages such as direct detection of biomolecules, the possibility of miniaturization, and so on. The graphene oxide-chitosan combination has proved to be excellent in developing biosensors due to the various advantages of both components. One such benefit of the composite is the availability of more active sites for the immobilization of biomolecules such as antibodies. Adumitrăchioaie *et al* reported a GO-chitosan composite for the selective and sensitive sensing of serotonin using DPV as the electroanalytical technique. The relevant sensor exhibited a detection limit of 3.2 nM with a sensitivity of 0.05 mA mM^{-1} and excellent selectivity toward serotonin even in the presence of other neurotransmitters with similar structures. The authors further tested the efficacy of the sensor in biological samples such as urine, saliva, commercial serum, and tears [42].

Gelatin is a popular biopolymer that has excellent biomedical applications due to its biocompatibility and biodegradability. This colorless soluble polypeptide is usually isolated from animal skin and bone. Gelatin–graphene composite is found to be useful in the biosensing of dopamine. The hydrophobic and hydrophilic parts of gelatin make easier the exfoliation of graphite sheets. Chen and coworkers developed a graphene-gelatin composite modified screen-printed carbon electrode for the detection of dopamine using DPV as the technique. No electrical response was observed in the absence of dopamine, but the addition of the analyte in the system caused the appearance of an anodic signal due to the oxidation of dopamine to dopamine quinone. The relevant sensor exhibited a linear range of 0.05–79.5 μM with a sensitivity of 1.36±0.02 μA μM^{-1} cm^{-2} and a detection limit of 0.0045 μM [43]. A molybdenum sulfide-graphene composite was tested by Cheng *et al* for the electrochemical detection of dopamine using DPV as the technique. MoS_2 has a layered structure similar to graphene where the individual molybdenum sulfide layers are stacked together by van der Waals dispersion forces. The authors believed that graphene might undergo $\pi\cdots\pi$ stacking interactions with dopamine which might enhance the accumulation and detection efficacy of the graphene-MoS_2 modified GCE. The water-dispersible graphene was obtained by the reduction of graphene oxide in an alkaline condition and molybdenum sulfide was prepared using a solvothermal process. The GCE was then individually coated with graphene and MoS_2. The pertinent electrochemical sensor exhibited excellent sensing of dopamine in a linear range of 5.0×10^{-8}–1.0×10^{-5} M and a detection limit of 7.13×10^{-9} M. Performance of the sensor was further examined in a real sample of bovine serum with satisfactory results [44].

An interesting electrochemical biosensor for dopamine was reported by Vilian *et al* The research group prepared a three-dimensional RGO—polyurethane hybrid porous material using a directional melt crystallization process. RGO was

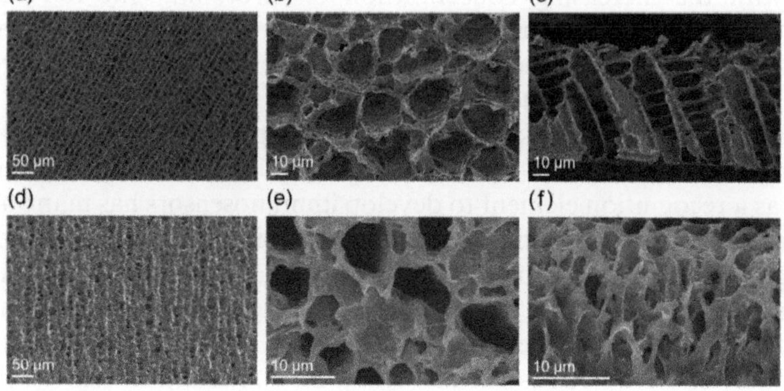

Figure 15.3. Scanning electron microscope (SEM) images of (a)–(c) polyurethane and (d)–(f) RGO—polyurethane hybrid showing the porous structure of the composite. Reproduced from [45], Copyright (2016) with permission from Elsevier.

incorporated into the pores by physical adsorption. Polyurethanes have gained immense attention lately as the support material for electrochemical applications due to their biocompatibility, excellent mechanical properties, hydrophilicity, and permeability. SEM images of polyurethane and the relevant RGO hybrid are shown in figure 15.3 that clearly show porous structures with pore sizes of 100–200 nm. Electrochemical biosensing of dopamine using the relevant composite was then examined using DPV as the analytical technique. The biosensor exhibited a linear range of 0.10–1.15 nM, a detection limit of 1 pM, and satisfactory performance in human urine samples with recoveries in the range 94.2%–97.4% [45].

15.3 Carbon nanotubes as electrochemical biosensors

Rolling the graphene sheets generates a new type of carbonaceous material with tubular morphology—carbon nanotubes (CNTs). These carbon nanomaterials with the nanosized diameter and micro-sized length were first reported in the 1990s and they could be either single-walled (single-walled carbon nanotubes, SWCNTs) or multi-walled (multi-walled carbon nanotubes, MWCNTs). They have amazing chemical, surface, electrical, thermal, and mechanical properties and a wide range of applications such as high-durability energy batteries, transparent conductive films for flexible solar cells, cancer therapy, biomedical imaging, drug delivery, gene therapy, tissue engineering, among many others. Easy chemical modification of the CNTs allows them to form composite materials with different physicochemical properties. For instance, pristine CNTs are sparsely soluble in most solvents. Chemical modification by attaching hydrophilic functional groups on their surface enhances solubility and biocompatibility. The durability and flexibility of carbon nanotubes make them good scaffolding materials in orthopedics [46]. The molecular dimension of CNTs makes them suitable for electrostatic screening during physiological procedures. Application of materials in biosensing demands appropriate surface charge density so that proteins can be adsorbed on the surface. The low charge carrier density of CNTs is equivalent to the surface charge density of protein molecules that renders the CNTs suitable materials for biosensing of bio-analytes [47]. In this section, we portray some typical applications of carbon nanotubes in electrochemical biosensing.

Enzymatic and non-enzymatic biosensing by graphene-based materials is discussed in section 15.2. The research community has a general tendency to test the various carbon nanomaterials for similar applications to establish a correlation between the device efficacy and the type of the nanomaterials. Like graphenes, carbon nanotubes also are found to be promising in enzymatic biosensing by using dehydrogenase enzymes such as alcohol dehydrogenase, phosphatase, fructose dehydrogenase, glucose dehydrogenase, or the other types of enzymatic biosensing (glucose, cholesterol, hydrogen peroxide) as discussed in section 15.2. Some typical examples of enzymatic and non-enzymatic biosensing by CNT-based materials are discussed in this section.

15.3.1 Enzymatic biosensing

An amperometric ethanol biosensor using alcohol dehydrogenase (ADH) enzyme immobilized in polyvinyl alcohol—MWCNT nanocomposite-modified glassy carbon electrode (PVA/MWCNT/ADH) was reported by Tsai *et al* The mechanism of ethanol sensing was based on the signal induced by β-nicotinamide adenine dinucleotide (NADH) generated in the enzymatic reaction. The relevant sensor exhibited a low electrochemical oxidation potential of NADH and lessened the electrode surface contamination by the oxidation product of NADH. The linear range, sensitivity, and detection limit were up to 1.5 mM, 196 nA mM^{-1}, and 13 μM respectively [48]. The individual semiconducting SWCNTs were also tested as versatile biosensors. In one such report, a predetermined number of glucose oxidase enzyme molecules were fixed with the nanotube and this process was found to alter the conductance of the system. The enzyme-coated carbon nanotube showed nice pH sensing with large and reversible changes in conductance with pH [49]. Intercalating carbon nanotube-platinum nanohybrid in waxed graphite electrode generated an electrode with high catalytic activity for the reduction of hydrogen peroxide. The cholesterol oxidase enzyme was immobilized with sol-gel on the electrode to fabricate a biosensor for cholesterol detection which exhibited a linear range of 1–100 μM with a detection limit of 1.4 μM [50]. A voltammetric biosensor for glutamate was developed by covalently attaching the glutamate dehydrogenase enzyme to the tip of carbon nanotubes in a vertically aligned CNT nanoelectrode array. The relevant biosensor exhibited a sensitivity of 0.976 mA mM^{-1} cm^{-2} in the linear range of 0.1–20 μM with a low detection limit of 10 nM [51].

15.3.2 Biosensing of dopamine

The detection of neurotransmitters by graphene-based electrochemical sensors is discussed in section 15.2.4. Biosensing of dopamine by CNT paste-based sensors was first reported by Britto *et al* in 1996 using bromoform as the binder. The team observed reversible two-electron oxidation of dopamine to dopamine quinone in cyclic voltammetry. The relevant biosensor performed better than other carbon-based sensors reported up until then. The authors presumed that enhanced surface area due to the presence of pores during packaging of the CNTs and the formation of the oxygen-containing functionalities on the surface of nanotubes during the oxidation of dopamine played important roles in the detection process [52]. Since then many other groups have come up with new device strategies for dopamine sensing by CNT-based electrochemical sensors.

A planar CNT-multi-electrode array (CNT-MEA) chip was developed by Suzuki *et al* for the detection of dopamine along with electrophysiological responses such as field postsynaptic potentials and action potentials. The multi-electrode arrays were found to be useful for noninvasive, real-time monitoring of electrophysiological activities and the alteration of the extracellular chemical microenvironment. The relevant chips were prepared by electroplating indium tin oxide (ITO) micro-electrodes and the CNT-plated ITO electrodes were found to have a higher current density as compared to the bare ITO electrodes. Chronoamperometric measurements using the fabricated CNT-MEA chips exhibited dopamine detection in a linear range of 1 nM–10 μM with a detection limit of 1 nM [53]. Pereira and the team proposed an enzymatic biosensor for dopamine based on functionalized MWCNTs and HRP extracted from zucchini. The enzyme was covalently attached with the carbon nanotubes to immobilize it on the nanotube surface. Dopamine is oxidized to dopaquinone in the presence of hydrogen peroxide at a low potential difference. The electrode was prepared by adding a solution of 1-ethyl-3-(3-dimethyl aminopropyl) carbodiimide (EDC) to oxidized MWCNTs to activate the carboxyl groups for amine bonding with primary amines, followed by the addition of the enzyme and subsequent homogenization. The team also studied the influence of various parameters such as concentrations of enzyme, hydrogen peroxide, phosphate buffer solution, along with the various electroanalytical techniques such as cyclic voltammetry, square wave voltammetry, and differential pulse voltammetry, and interference from foreign analytes such as uric acid and ascorbic acid on the sensing efficacy. Under the optimized condition, using square wave voltammetry as the technique, the relevant biosensor exhibited a linear range of 32–44 μM with a detection limit of 2 μM [54].

15.3.3 Non-enzymatic glucose sensing

Enzymatic glucose sensing by graphene-based electrochemical sensors is discussed in section 15.2.1. Although the method offers superior performance, enzymatic biosensing comes with some caveats such as stability issues, higher cost of enzymes, the requirement of enzyme immobilization, and critical operating conditions. Therefore, a section of researchers bet on the development of selective, sensitive, rapid, and cost-effective electrochemical non-enzymatic biosensors which is nevertheless welcome. They argue that the direct electrocatalytic oxidation of glucose at a non-enzymatic electrode would be beneficial over enzymatic electrodes. Lin *et al* demonstrated non-enzymatic glucose sensing by a nickel and copper nanoparticles decorated MWCNT-modified glassy carbon electrode. The metal nanoparticle-modified electrodes generally outperform the bare electrodes due to their large surface area and mass transport, and good biocompatibility. In the relevant work, carbon nanotubes were first deposited on the glassy carbon electrode by drop-casting followed by electrochemical deposition of nickel and copper nanoparticles on the MWCNT-modified glassy carbon electrode. The Ni/Cu/MWCNT nanocomposite formation was driven by electrostatic attraction between the negatively charged MWCNTs and the positively charged metal nanoparticles. Carboxylic functionalized carbon nanotubes were used to enhance dispersion and reduce mobility on the electrode surface. The relevant electrode exhibited good glucose sensing with a linear range of 25 nM to 800 μM, a low detection limit of 2.5×10^{-8} M, low overpotential, high sensitivity, fast response time, and good stability using amperometry as the electroanalytical technique. The sensing was also tested successfully in human serum samples without any interference [55].

15.3.4 Nucleic acid sensing

Electrochemical sensing of DNA and RNA at physiological levels has become a hot topic lately because of the growing interest in gene identification, molecular diagnostics, detection of pathogens, and forensic investigation. Both carbon nanotubes and graphenes are found to be promising in this area because of their ability to undergo π···π stacking interactions with the conjugated aromatic systems of the nucleobases. These bio-analytes exhibit inherently weak sensing signals which require amplification for successful sensing. The excellent electrical conductivity of graphenes and carbon nanotubes could assist amplify the sensing signal.

An impedimetric micro-fluidic DNA biosensor for quantification of the DNA sequence for chronic myelogenous leukemia (CML) was reported by Ghrera *et al* The sensor chip (pDNA/MWCNT/ITO) was fabricated by electrophoretic deposition of carboxyl-functionalized MWCNTs on ITO coated glass substrate using wet chemical etching technique. The CML-specific deoxyribonucleic acid probe was then immobilized on the carbon nanotube surface. The relevant sensor exhibited a linear range of 1 fM–1 μM with a detection limit of 1 fM and a response time of 1 min The researchers believed that modification of carbon nanotubes caused an enhanced surface area of the electrode for DNA loading that led to higher selectivity, sensitivity, and linear range [56]. A label-free micro-RNA sensor based on carbon nanotubes and the electroactive polymer 5-hydroxy-1,4-naphthoquinone-co-(3-(5-hydroxy-1,4-dioxo-1,4-dihydro naphthalene-2(3)-yl)-propanoic acid)-modified glassy carbon electrode was demonstrated by Tran *et al.* The fabricated electrode was tested to detect a miRNA biomarker miR-141. The electroactivity of the copolymer was found to be higher in the presence of the carbon nanotubes. The probe ODN-141-P was then immobilized on the fabricated electrode that showed selective voltammetric response to the target analyte (miR-141) without interference from the noncomplementary miRNA sequences molecules miR-103 and miR-29b-1. The concerned biosensor exhibited a low detection limit of 8 fM [57]. Another label-free electrochemical sensor was developed by Tang and the team for the detection of miRNA-24. The relevant sensor is composed of an MWCNT-modified glassy carbon electrode and performs the sensing operation by monitoring the oxidation signal of guanine. The complementary synthetic DNA probe for miRNA-24 was immobilized onto the modified electrode by covalent cross-linking. Differential pulse voltammetry was used as the electrochemical technique to monitor the changes in guanine oxidation signal as a result of the hybridization between the probes and miRNA-24. The relevant biosensor exhibited good selectivity and sensitivity, a low detection limit of 1 pM, and good reproducibility [58].

15.3.5 Detection of cancer biomarkers

Selective, sensitive, and rapid detection of cancer biomarkers is paramount to avoiding premature death of cancer patients. Although there are various methods for screening cancer biomarkers such as enzyme-linked immunosorbent assay, radioimmunoassay, fluorescence, mass spectrometric immunoassay, among many others, electrochemical sensing is gaining pace due to many obvious advantages. In this section, we discuss some representative examples of carbon nanotube-based electrochemical sensors for cancer biomarkers.

Signal amplification is a major issue for electrochemical immunosensors. One way to amplify the sensing signal is to form polymeric layers on the electrode surface through enzymatic reactions. Another simple approach is amplification through enzymatic precipitation of a nonconductive material that coats the electrode surface blocking the electron transfer process between the redox probes. The enzyme used in this strategy is usually HRP. This strategy was used to build a biosensor for the detection of prostate-specific antigen (PSA), a model protein for the screening and early diagnosis of prostate cancer. A high PSA level is associated with a higher risk of prostate cancer. This signal amplification strategy of bio-catalyzed precipitation by a multiple enzyme label method using CNTs, and gold nanoparticles (AuNPs) is shown in figure 15.5. The electrode has an Au/MWCNT/AuNP structure where MWCNTs were first deposited on the gold electrode surface followed by deposition of gold nanoparticles on the carbon nanotube surface. Finally, the monoclonal anti-PSA (Ab1) antibody was immobilized onto the electrode surface. As shown in figure 15.4, two methods were used to detect PSA: single and multiple HRP routes. In the second option, a higher number of HRP molecules produced more precipitate that led to higher signal amplification. The immunosensor showed good selectivity, long-term stability, a wide linear range of 1.0 pg mL^{-1} to 10.0 ng mL^{-1} with a low detection limit of 0.4 ± 0.03 pg mL^{-1} [59]. Zhu and the team developed a label-free method for the sensitive detection of HeLa and HL-60 cells with the electrochemical impedance technique using a polydopamine-coated carbon nanotubes-folic acid nanoprobe (CNT@PDA-FA/GCE) that could detect as few as 500 cells. The unique feature of this probe is that it does not require antibodies or enzymes and is not susceptible to denaturation or decomposition. The linear range and the detection limit were $5{\times}10^2$–$5{\times}10^6$ cells mL^{-1} and 500 cells mL^{-1} respectively [60].

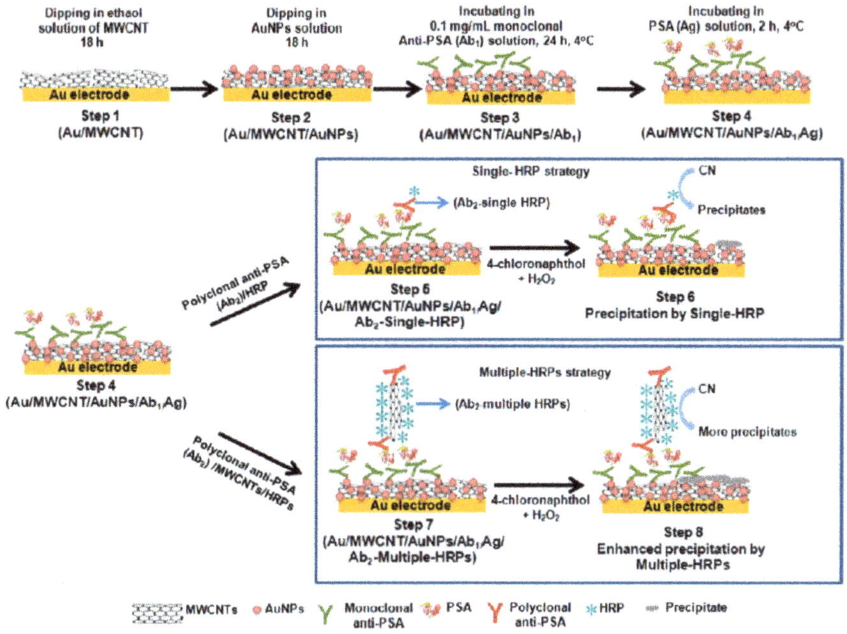

Figure 15.4. Signal amplification of MWCT-based PSA immunosensor using different strategies for enzymatic precipitation. Reprinted with permission from [59]. Copyright (2012) American Chemical Society.

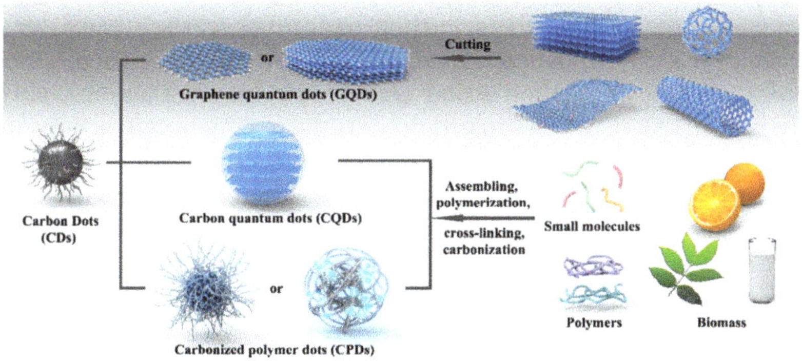

Figure 15.5. Various types of carbon dots and their preparation methods. Reprinted with permission from [64]. Copyright (2020) American Chemical Society.

15.4 Carbon dots as biosensors

Carbon dots are new entrants in the carbon nanomaterial family. These nanosized particles have a typical diameter in the order of 10 nm and are known for their unique photoluminescence properties. This class of carbon nanomaterials is further divided into three categories based on the nature of the carbon precursor, core structure, and quantum effect. They are graphene quantum dots (GQDs), carbon quantum dots (CQDs), and carbonized polymer dots (CPDs) (figure 15.5). Carbon dots are prepared by following either the top-down or the bottom-up synthetic methods and have many advantages over conventional organic probes in biosensing. There are many excellent review articles on this topic that readers might consider [61–65].

GQDs are carbon-based zero-dimensional anisotropic nanomaterials with a fabric-like structure. Interestingly, the structural features of graphene quantum dots have remarkable similarities with carbon dots as well as graphene, and they are found to be extremely useful in environmental, biological, electrochemical, and optoelectronic applications. GQDs have a large number of chemically active sites that are suitable for the chemical modification of the materials to alter their physicochemical properties (figure 15.1). For example, GQDs can be conjugated with other nanomaterials via $\pi\cdots\pi$ stacking interactions, or can form hybrids with antibodies, proteins, nucleic acids, etc. It is worth mentioning that GQDs have amazing fluorescence which makes them suitable for a variety of applications. GQDs are either obtained using a top-down synthetic methodology by the fragmentation of graphene oxide, graphene sheets, CNTs, graphite, or carbon fibers; or by a bottom-up synthetic method using small molecular precursor materials.

Although GQDs and carbon quantum dots (CQDs) fall under the category of carbon dots, their topological and preparation methods set them apart. GQDs have one- or multiple-layer graphite structures and are usually obtained by the top-down synthetic approach by cutting larger graphitized carbon materials such as graphite powder, carbon rods, carbon fibers, carbon nanotubes, carbon black, or graphene oxide into small pieces. CQDs, on the other hand, are spherical, usually obtained by the bottom-up synthesis method using small molecular precursors, polymers, or biomass (figure 15.5). Carbon quantum dots bear functional groups (carboxyl, hydroxyl, carbonyl, epoxide) at their edges that could be used for chemical modification to obtain composite materials with different physicochemical properties. Also, the presence of these functional groups allows other biomolecules such as antibodies, antigens, enzymes, proteins, nucleic acids to be immobilized on the surface for electrochemical sensing of biomolecules. Through the past decade, graphene quantum dots and carbon quantum dots have been heavily tested in sensing applications of various types of biomolecules using optical, electrochemical, photoelectrochemical processes as the transducer. However, in this section, we shall limit our discussion to electrochemical biosensors.

Huang and coworkers developed an ultrasensitive label-free electrochemical biosensor using GQD modified glassy carbon electrodes for detecting the hepatitis B virus (HBV) DNA. Hepatitis B virus is a deadly human pathogen that causes liver inflammation, cirrhosis, and even liver cancer. Therefore, the development of methods for sensitive detection of this pathogen is paramount. In the relevant study, the authors obtained graphene quantum dots by simple bottom-up synthesis and directly modified the electrode surface. Potassium ferricyanide was used as the indicator to monitor the electrochemical sensing process. DPV was used as the electroanalytical technique that revealed a limit of detection of 1 nM and a linear range of 10–500 nM [66]. Doong's group reported an interesting electrochemical biosensor for cholesterol based on β-cyclodextrin (β-CD)—nitrogen-doped graphene quantum dot (N-GQD) nanohybrid. Here, functionalized N-GQDs with a mean particle size of 8.5±0.5 nm were covalently attached with β-CD to obtain the desired nanoprobe. The cyclic polysaccharide, β-CD, contains a hydrophobic cavity surrounded by a hydrophilic outer surface and is known to form host–guest inclusion complexes with a wide range of guest molecules, cholesterol being one of them. The supramolecular interactions between β-CD and cholesterol are found to have positive implications in the extraction of the latter. In this work, the authors intended to use β-CD as an alternative to the cholesterol oxidase enzyme for selective electrochemical detection of cholesterol in human serum using N-GQD as a conductive matrix and ferrocene as the redox indicator. At first, a β-CD@N-GQD/ferrocene inclusion complex was synthesized by covalent bond formation between the amine-functionalized β-CD and the carboxylic acid groups of N-GQD followed by the inclusion of ferrocene in the cavity of β-CD. As cholesterol has a higher affinity toward the β-CD cavity than ferrocene, the addition of cholesterol into the system would release and replace ferrocene from the host. The redox process that follows the release of ferrocene was then analyzed using DPV as the transducer. The nitrogen-doped graphene quantum dots accelerate the electron transfer to the working electrode. The relevant probe exhibited a limit of detection of 80 nM with a linear range of 0.5–100 nM in human serum [67].

A concerning hazardous analyte is Bisphenol S which is used to produce plastic for various applications such as optical, electrical, automobile, packaging, appliances, electronic equipment, and so on. Be that as it may, Bisphenol S is found to be a toxic material due to its estrogenic activity, longer environmental persistence, and health hazards such as obesity, diabetes, genital malformation, hepatic disturbances, and carcinogenicity. Therefore, the discovery of selective and sensitive sensors for the pertinent analyte is crucial. Wang and coworkers developed a molecularly imprinted electrochemical sensor based on a GCE modified with GQDs coated on hollow nickel nanospheres (hNiNS) for rapid quantification of bisphenol S. The purpose of using the nickel nanospheres was to increase the active surface area and enhance conductivity which ultimately improved the electrochemical response of the sensor. Using DPV as the technique, the optimized sensor could detect Bisphenol S with a lower limit of 0.03 μM and exhibited a linearity range of 0.1–50 μM. The sensor was also successfully tested in real samples such as bottled mineral water [68]. Graphene quantum dots are found to be excellent candidates for the fabrication of electrochemical immunosensors which use antibodies as the bio-recognition elements. The term immunosensors stem from immunoreactions, i.e., specific recognition between antigens and antibodies. The electrochemical immunosensors could be based on different bioassay types such as direct, indirect, competitive, or sandwich of which the direct mode is the most common. The immunoassay processes have some commonalities such as the capture of the target analyte, obstruction of the non-reacted surface, and recognition of the analyte. The graphene quantum dots-based immunosensors are found to offer exciting results in cancer diagnosis, detection of cardiac biomarkers, detection of pathogens, etc [69].

15.4.1 Cancer diagnosis

Cancer is a menacing disease that has become prevalent in the past few decades and is a major cause of global mortality. Cancer is caused by the transformation of normal cells into tumor cells which could be triggered by various factors such as exposure to carcinogens, certain radiations, genetic factors, and so on. One of the major tumor markers for the diagnosis and containment of malignant tumors is the carcinoembryonic antigen (CEA). This oncofetal glycoprotein has a molecular weight of 180–200 kDa and its elevated level in human blood (>5 ng mL^{-1}) is fatal. Dong and coworkers came up with an ultrasensitive label-free electrochemical immunosensor based on nitrogen-doped graphene quantum dots for quantitative detection of CEA. The fabrication and sensing mechanism of the relevant electrode is schematically illustrated in figure 15.6. The overall electrode fabrication process consists of the preparation of graphene quantum dots supported PtPd bimetallic nanoparticles, immobilization onto gold nanoparticles through self-assembly and subsequent loading of the PtPd/NGQDs@Au nanocomposite on the glassy carbon electrode. The PtPd bimetallic nanoparticles provided exposed catalytic surface and enhanced conductivity which further accelerated the electrocatalytic hydrogen peroxide reduction. The choice of the gold nanoparticles for the fabrication of the relevant sensor was because of biocompatibility. Under the optimum amperometric condition, the immunosensor exhibited a wide linear range of 5 fg mL^{-1}–50 ng mL^{-1} with a low detection limit of 2 fg mL^{-1} and high sensitivity, selectivity, and stability [70].

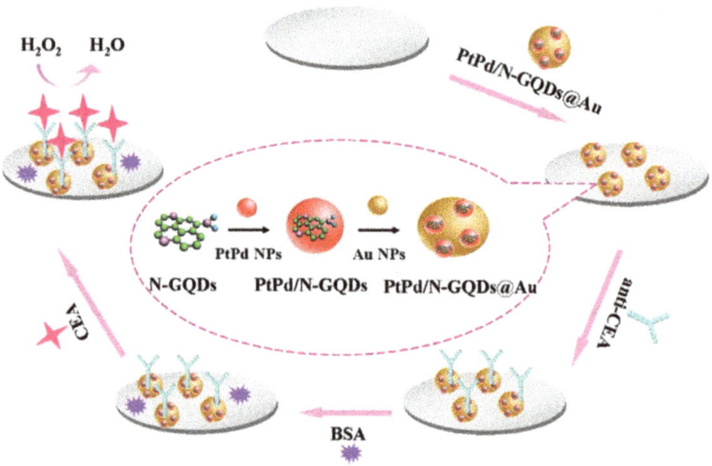

Figure 15.6. The schematic representation of the fabrication of the label-free electrochemical immunosensor PtPd/N-GQDs@Au for the detection of the carcinoembryonic antigen (CEA). Reprinted from [70], Copyright (2017), with permission from Elsevier.

Carcinoembryonic antigen (CEA) is a highly glycosylated protein and is one of the most used tumor markers for the diagnosis and monitoring of malignant tumors. The level of CEA in human blood increases upon formation of the cancer cells where a detectible concentration is about 5 ng mL^{-1}. Thus, the selective, sensitive, and label-free detection of CEA is imperative for early tumor diagnosis and screening as this process can yield information about antigen-antibody interactions. Doong and coworkers developed an electrochemically impedimetric N, S-codoped GQD-gold nanoparticle (AuNP)-nanostructured polyaniline (PANI) nanocomposite (N, S-GQD@Au-PANI) for the selective and sensitive detection of CEA. PANI is a low-dimensional organic conductor with a high surface area. Its capacity of adopting various protonated forms makes PANI a unique polymeric material. Also, PANI can be a good medium to immobilize nanoparticles to enhance their electrochemical performance. The authors justified that the use of an electrode modifier where gold nanoparticles are evenly distributed on the PANI surface would enhance the sensitivity of the relevant sensor by enhancing electron transferability and the number of immobilized antibodies. The detection mechanism of the relevant immunosensor was the change in impedance after the addition of CEA because of inhibition of electron transfer after the formation of antigen-antibody bioconjugate on the N, S-GQD@Au-PANI surface. A dynamic range of 0.5–1000 ng mL^{-1} with a detection limit of 0.01 ng mL^{-1} was obtained by the pertinent immunosensor [71].

15.4.2 Diagnosis and monitoring of cardiovascular diseases

Cardiovascular diseases occur due to disorders in the heart and blood vessels. Some of the common cardiovascular diseases are coronary heart disease, cerebrovascular disease, and rheumatic heart disease. Almost one-third of deaths worldwide are due to cardiovascular diseases, the most common causes being heart attacks and strokes. The common symptoms are elevated blood pressure, glucose, and lipid levels in the body where the chance increases with obesity. Therefore, the detection of these symptoms at an early stage would prevent premature deaths. The necessity of prognostic biomarker detection with minimal invasive methodologies is thus imperative for the successful management of cardiovascular diseases. The electrochemical immunosensors are found to be convenient for the quantification of cardiovascular biomarkers at extremely low concentrations in biological fluids. In this section, we discuss some typical developments in this area.

Bhatnagar *et al* demonstrated an ultrasensitive and rapid detection of heart attack (myocardial infarction) in humans based on a sensor comprising of troponin-I (cTnI) antibody conjugated with graphene quantum dots and polyamidoamine (PAMAM) nanohybrid-modified gold electrode. The cTnI is known as a gold standard marker because of its outstanding specificity and sensitivity for the detection of acute myocardial infarction. cTnI is released into the bloodstream just after the myocardial infarction when its level may dramatically vary from 0.04–1.4 ng mL^{-1} in a few hours. The relevant electrode fabrication and subsequent detection process are schematically illustrated in figure 15.7. Here, a screen-printed gold electrode was tagged with 4-amino thiophenol via gold–sulfur bond formations.

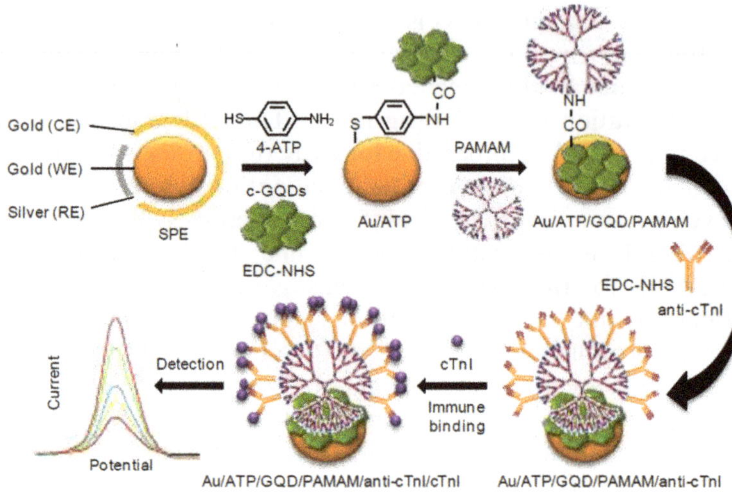

Figure 15.7. Schematic representation of the development of immunosensor for cardiac troponin-I (anti-cTnI). The relevant electrode was a gold nanoparticles-graphene quantum dots-PAMAM nanohybrid-based screen-printed electrode (CE: counter electrode, WE: working electrode, RE: reference electrode). Reprinted from [72], Copyright (2017), with permission from Elsevier.

The free amino groups were then coupled with the dangling carboxyl groups of graphene quantum dots. The PAMAM dendrimers were then fixed on graphene quantum dots via carbodiimide coupling that provided an ultra-high surface area for antibody immobilization and enhanced the sensitivity of the relevant sensor. This process was followed by the immobilization of the activated cardiac cTnI monoclonal antibody onto PAMAM. The authors measured a range of concentrations of cTnI in human blood serum by using cyclic voltammetry and differential pulse voltammetry which revealed high specificity of the sensor toward cTnI and negligible response to non-specific antigens. The sensor exhibited a good sensitivity of 109.23 $\mu A\ cm^{-2}\ \mu g^{-1}$, a lower detection limit of 20 fg mL^{-1}, and a linear range of 10^{-6}–10 ng mL^{-1} [72].

A label-free electrochemical immunosensor for the selective and sensitive detection of biomarker receptor tyrosine kinase (AXL) in human serum was demonstrated by Mollarasouli *et al* The relevant disposable sensor consisted of a screen-printed carbon electrode modified by amine-functionalized graphene quantum dots onto which anti-AXL antibody was immobilized. The affinity reactions were examined by measuring the differential pulse voltammetric response of the ferrocyanide/ferricyanide redox couple. The relevant immunosensor exhibited a wide linear range of 1.7–1000 pg mL^{-1} and a low detection limit of 0.5 pg mL^{-1} which were well below the recognized cut-off value for heart failure diagnosis in serum (71 ng mL^{-1}) [73]. Atrial fibrillation (AF) is a common complication during open-heart surgery which can be prevented by prophylactic beta-blockers (BB). Despite significant research efforts, successful protection of AF in patients with post-coronary artery bypass graft surgery (CABG) has not yet been achieved.

Although a correlation between high C-reactive protein (CRP) levels and AF occurrence has been presumed, ambiguity exists whether high plasma CRP levels are only indicative of AF occurrence, or whether high CRP levels cause the disorder. Bing *et al* reported an efficient electrochemical immunosensor for CRP detection in blood serum by using graphene quantum dots modified glassy carbon electrode that exhibited a low detection limit of 176 pM and a linear range of 0.5–70 nM [74].

15.4.3 Detection of pathogens and infectious diseases

The rapid growth of human society is accompanied by a growing concern about infections and diseases caused by food and waterborne pathogens such as bacteria, viruses, and fungi. Particularly in developing countries, the healthcare system is swamped by bacterial and viral infections which are mainly caused by poor sanitization, pollution, rampant use of pesticides, among many others. Uncontrolled release of the toxins into the environment has a serious consequence to society as these toxins travel through the food chain and come back to humans causing various types of chronic and acute diseases. Therefore, the development of rapid, sensitive, and selective biosensors for the detection and quantification of pathogens is important to prevent an outbreak. On-site monitoring is a special requirement. Even though many techniques have been tested for the detection of pathogens, they have many downsides such as high cost, intricate instrumentation, complicated data collection, and analysis. Graphene quantum dots-based electrochemical immunosensors are found to be promising in this area. Some typical examples are discussed in this section.

One of the major causes of acute viral hepatitis across the globe is the hepatitis E virus (HEV). There are four major genotypes of HEV that are endemic in many industrialized countries. The clinical diagnosis of HEV infection these days is mostly based on enzyme immunoassay and reverse transcription-polymerase chain reaction (RT-PCR). Though RT-PCR is a convenient method for the diagnosis of HEV there are some associated limitations such as reporting false results sometimes. A pulse-triggered ultrasensitive electrochemical sensor for HEV based on graphene quantum dots and gold-embedded polyaniline (PANI) nanowires was reported by Dutta Chowdhury *et al* The relevant sensing scheme is represented in figure 15.8. The N, S-graphene quantum dots@AuNPs nanocomposites were deposited on an electropolymerized polyaniline-coated glassy carbon electrode. The team used an impedimetric response to detect HEV. Under optimal conditions, the relevant biosensor exhibited a wide linear range of 1 $fg\,mL^{-1}$ to 100 $pg\,mL^{-1}$ with a low detection limit of 0.8 $fg\,mL^{-1}$. The specificity and sensitivity of the fabricated biosensor were evaluated in buffer and human serum media using HEV-like particles [75].

Hepatitis C is another infectious disease that affects the liver and is caused by the hepatitis C virus (HCV). HCV core antigen levels have been assumed to be potential markers for viral replication in the past decade. Valipour *et al* reported a label-free quantification method for hepatitis C virus core antigen based on silver nanoparticles-thiolated graphene quantum dots modified glassy carbon electrode. Riboflavin was used as the redox probe in the subsequent study. Silver nanoparticles

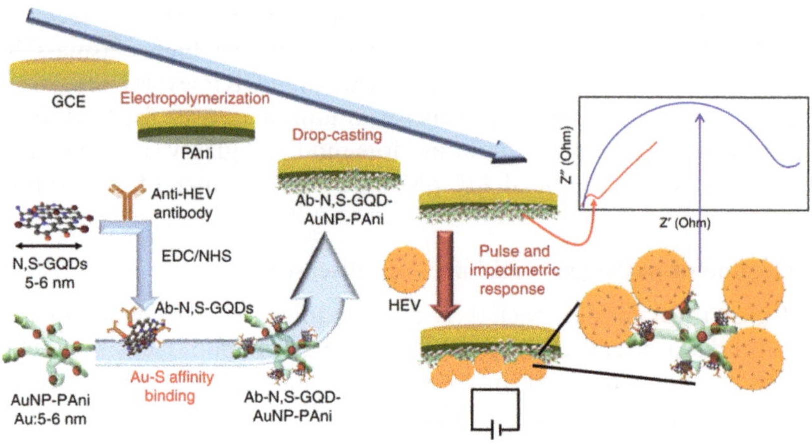

Figure 15.8. Schematic illustration of the Ab-N, S-GQDs@AuNP-PANI nanocomposite biosensor and its pulse-induced impedimetric sensing of HEV. Reproduced from [75]. (Copyright 2019. The author(s)). With permission of Springer.

were attached with graphene quantum dots through Ag–S bond formations. The anti-HCV molecules were then chemisorbed on the surface using the interactions between silver nanoparticles and amino groups of the antibody. DPV was used as the electroanalytical technique where the antigen-antibody recognition was evaluated based on the diminishing oxidation signal of riboflavin in a wide linear range of 0.05 pg mL^{-1} to 60 ng mL^{-1} with a limit of detection of 3 fg mL^{-1}. The sensor was also applied for the analysis of spiked human serum [76]. Tuberculosis (TB) is a dangerous airborne infectious disease caused by mycobacterium tuberculosis and is one of the most fatal diseases worldwide. TB usually infects the lungs (pulmonary tuberculosis) though other sites such as lymph nodes, abdomen, bones, etc could also be affected (extrapulmonary tuberculosis). A sandwich immunosensor for rapid and sensitive detection of mycobacterium tuberculosis antigen (culture filtrate protein, CFP-10) was designed and developed by Lee and coworkers. The relevant biosensor consisted of a glassy carbon electrode modified with graphene quantum dot coated Fe_3O_4@Ag core–shell nanohybrid. For the purpose of signal amplification, gold nanoparticles conjugated anti-CFP-10 antibodies were used as a label. The various nanocomponents in the pertinent biosensor played different roles. Fe_3O_4 enhanced the surface-to-volume ratio, silver nanoparticles increased the conductivity of the system, and graphene quantum dots assisted the loading of the antibody onto the electrode. The immunosensor excelled with a wide linear range of 0.005–500 μg mL^{-1} with a low limit of detection of 0.33 ng mL^{-1} [77].

15.4.4 Non-enzymatic biosensing of hydrogen peroxide and glucose

The enzymatic biosensing of hydrogen peroxide and glucose generally incorporates glucose oxidase enzyme (GOx) and HRP respectively immobilized onto an electrode surface. However, as mentioned in section 15.3.3, though the enzymatic biosensing offers high selectivity, sensitivity, and specificity, the process comes with some downsides such as high cost, issues with stability, and lifetime. Therefore non-enzymatic biosensing of these analytes has also gained equal popularity.

A silver nanoparticle–carbon quantum dots modified glassy carbon electrode was used for amperometric determination of hydrogen peroxide. The fabricated electrode was found to offer excellent performance toward the reduction of H_2O_2. The optimum electrode condition delivered the sensing operation with a long linear range of 0.2–27.0 μM with high sensitivity (1.5 μA μM^{-1}) and a low detection limit of 80 nM. The electrode was also successfully applied for the detection of the analyte in disinfected fetal bovine serum samples [78]. Another device was prepared by modifying a glassy carbon electrode with a CQD-MWCNT nanocomposite. The device performance was found to be better in the case of the nanocomposite-modified electrode than the electrode modified by CQD or MWCNT alone. The CDs/MWCNTs/GCE electrode performed the H_2O_2 sensing with a low detection limit of 0.25 μM, excellent stability, high selectivity, good reproducibility, and a long linear range (3.5 μM–30 mM) using amperometry as the technique. The electrode was also used for real-time tracking of hydrogen peroxide released from human cervical cancer cells with reasonable performance [79].

Metal oxide nanoparticles have also been tested as electrode modifiers for non-enzymatic electrochemical sensing of bio-analytes. For instance, a CQDs-octahedral cuprous oxide nanocomposite-modified glassy carbon electrode was used for non-enzymatic biosensing of glucose and hydrogen peroxide. The amperometric glucose sensing was performed with a low detection limit of 8.4 μM in a long linear range of 0.02–4.3 mM. The limit of detection and linear range for H_2O_2 sensing were 2.8 μM and 5 μM–5.3 mM respectively. The fabricated electrode was also tested in presence of interferents such as ascorbic acid, uric acid, dopamine, sodium chloride with negligible interference [80]. In a separate report, Maaoui *et al* demonstrated non-enzymatic glucose sensing using CQD/Cu_2O/GCE electrode and amperometry as the electroanalytical technique. The relevant electrochemical sensor performed with excellent device characteristics (linear range: 6 μM-6 mM, sensitivity: 2.9 ± 0.2 μA μM^{-1} cm^{-2}, LOD: 6 μM) [81]. Ionic liquids are also found to be a promising component for modifying electrodes for biosensing. One such example demonstrates amperometric hydrogen peroxide sensing by a glassy carbon electrode modified by PtNPs/CQDs/ionic liquid-graphene oxide nanocomposite with high selectivity, a wide linear range of 1–900 μM, and a low detection limit of 0.1 μM [82].

15.4.5 Electrochemical detection of dopamine

An electrochemical dopamine sensing by N-doped carbon quantum dots modified glassy carbon electrode (NCQD/GCE) using DPV as the electroanalytical technique. The concerned electrode exhibited a broad linear range of 0–1 mM, a low detection limit of 1 nM. The sensor performed well in presence of uric acid and ascorbic acid illustrating its great selectivity [83]. A Cu_2O/CQD/Nafion/GCE electrode was developed for electrochemical sensing of dopamine. The metal oxide nanoparticles were used to enhance the conductivity of the electrode. The authors assumed that the carboxy groups on CQDs and the sulfonyl groups on Nafion would selectively attract cations by ion exchange and repel the anions. The relevant biosensor exhibited a low detection limit of 1.1 nM in a wide linear range of 0.05–45.0 μM with high selectivity and sensitivity when combined with DPV as the electrochemical technique [84]. Another glassy carbon electrode modified by rGO-CQDs nanocomposite was employed for electrochemical sensing of dopamine using DPV as the technique. The authors believed that π···π stacking interactions between rGO and dopamine together with the carboxyl functionalities of CQDs would enhance the specificity of the sensing process. Under the optimal experimental condition, dopamine sensing by the relevant electrode proceeded with a low detection limit of 1.5 nM and a linear range of 0.01–450.00 μM. The electrode was able to suppress the interference from ascorbic acid and uric acid and lasted for at least a month [85].

15.4.6 Detection of other organic molecules

Apart from biosensing of conventional analytes such glucose, cholesterol, neurotransmitters, etc detection of other organic molecules such as ascorbic acid, uric acid, nicotinamide adenine dinucleotide (NADH), drugs, vitamins, amino acids are important in medical sciences. Therefore, considerable efforts have been given to develop sensors for the detection and quantification of these organic analytes of biological importance. We discuss a few representative examples in this section.

Chiral separation of L- and D-ascorbic acid in human cerebrospinal fluids and blood plasma samples by electro-polymerized molecularly imprinted polyaniline ferrocene sulfonic acid-carbon dots modified pencil graphite electrode was demonstrated by using DPV as the analytical technique. The molecularly imprinted film showed good chiral selectivity and sensitivity. The linear range and detection limit for the biosensing of L- and D-ascorbic acids were 6.0–165.0 nM, 1 pM, and 6.0–155.0 nM, 2 pM respectively [86]. A short-lived oxidant with potency to induce cell death is peroxynitrite which could be scavenged by uric acid. An abnormally low concentration of uric acid in the body could cause peroxynitrite toxicity. On the other hand, a high concentration of uric acid in the body could cause gout and hyperuricemia. Nitrogen-doped carbon dots modified glassy carbon electrodes were found to be promising for uric acid biosensing. For example, one such study reported uric acid biosensing by the N-CD/GCE electrode in a linear range of 20–260 nM with a low detection limit of 1.3 nM using amperometry as the technique [87].

NADH is an important coenzyme in many biological reactions where the ratio of its oxidized and reduced forms NAD/NADH points to an abnormality in the organisms such as the decline of energy production because of their direct influence on the ATP level, Parkinson's disease, insomnia. Analysis of NADH is thus an important research topic. Canevari *et al* came up with an interesting idea for the electrochemical detection of NADH. They demonstrated that superparamagnetic nanoparticles such as Fe_3O_4 nanoparticles-carbon quantum dots nanocomposite-modified screen-printed carbon electrodes could be beneficial for electrochemical sensing of NADH. The nanocomposite was held onto the screen-printed electrode by an external magnet. This is indeed an innovative way to modify electrodes as the magnetic modifiers could be released by taking the external magnet off. The pertinent magnetic biosensor exhibited good performance within a linear range of 0.2–5 μM with a detection limit of 20 nM using DPV as the technique [88]. A nitrogen-doped carbon quantum dots-SnO_2 nanocomposite-based screen-printed carbon electrode (N-CQD/SnO_2/SPCE) was developed for the sensitive determination of riboflavin (vitamin B2) in real samples such as riboflavin tablets and milk powder. The electrode was prepared by the sonochemical method as DPV was used as the electrochemical technique. The relevant electrode exhibited good performance with a wide linear range of 0.05–306 μM, a low detection limit of 8 nM, good sensitivity (2.496 μA $\mu M^{-1}\ cm^{-2}$), and excellent selectivity [89]. Another report demonstrated graphene quantum dots-poly aspartic acid-modified glassy carbon electrode (poly aspartic acid-GQDs/GCE) as the electrochemical biosensor for vitamins B2 and B9 using DPV as the signal transducer. The optimized experimental condition detected the relevant analytes with low detection limits of 0.22 and 0.1 μM respectively [90].

15.5 Conclusion

On balance, carbon-based nanomaterials offer versatile applications because of their superior electric, thermal, mechanical, and chemical properties. The electrochemical sensing of analytes including the molecules of biological significance is cheaper and handier than other analytical techniques and has gained immense research attention. Indeed, carbon nanomaterials have been one of the most widely used classes of nanomaterials in the past two decades. Each carbon nanomaterial has unique structural features that influence its physicochemical properties. The presence of chemically active functional groups on the surface allows the functionalization of the nanomaterials to alter their intrinsic properties and engineer goal-oriented nanocomposites. Outstanding electronic and surface properties of the carbon nanomaterials catapult conductivity of the electrodes and the carbon nanohybrid-modified electrodes outperform the bare electrodes. The π-electron system of the carbon nanomaterials can interact with the aromatic organic component of the biomolecules such as proteins, nucleic acids, and enzymes. Consequently, these electrodes are found to deliver superior performance for electrochemical biosensing of medically significant analytes that can be used for the early diagnosis of potential diseases. All things considered, carbonaceous materials are proving to be promising in various biomedical areas such as biosensing, bioimaging, drug delivery, among many others.

In this chapter, we attempted to emphasize electrochemical biosensing of clinically relevant biomolecules such as proteins, disease biomarkers, glucose, cholesterol, neurotransmitters, etc. We highlighted the most common carbon nanomaterials (graphene, graphene oxide, reduced graphene oxide, graphene quantum dots, carbon nanotubes, carbon quantum dots) as the electrode modifiers for electrochemical biosensing. Portable medical devices are becoming commonplace for their faster diagnosis and reduced cost. The superior electrochemical performance of carbon nanomaterials is beneficial for the miniaturization of analytical devices. Due to space limitations, we highlighted only specific examples to point out the potentiality of the relevant electrode modifiers.

All in all, carbon nanomaterials are major scientific breakthroughs in the modern age. The unique physicochemical properties of this important class of nanomaterials support their diverse applications. The amazing electronic properties of carbon nanomaterials make them promising electrode modifiers. The chapter so far discussed only the positive sides of these materials. It would not be complete without mentioning their negative outlooks, i.e., their toxicity to the human body and environment. The carbonaceous materials are constantly being engineered to fabricate goal-specific devices overlooking their toxic side effects. Though there is a lack of data, it has been found that the engineered carbon nanoparticles could be toxic if inhaled and ingested. During the production and testing stages, particles of the carbon nanomaterials could leach into the air exposing humans and other organisms to them. The directly affected group of people are those who handle them. This could complicate their health as the airborne carbon nanomaterials could enter their bodies through inhalation and ingestion. The consumers could be passively exposed to those nanoparticles too. One problem of assessing their safety is the lack of data. Whereas the available data in the literature is mostly about the toxicity of the pristine nanomaterials, it is the functionalized carbon nanoparticles that are mainly employed in the devices. Intrinsic and extrinsic behaviors of the nanomaterials further complicate their safety assessment. The intrinsic properties are much more predictable as they do not change much during the measurement. In contrast, the airborne nanoparticles could be transformed into aggregates and agglomerates through chemical reactions with other nanoparticles. The secondary nanomaterials generally have different toxic levels than their parents. In a nutshell, nanoparticles of different shapes, sizes, surface properties could coexist in the environment if used without care, and they could be damaging to the health of the people exposed to them. There are reports of cellular damages by the carbon nanomaterials which could be influenced by different mechanisms such as membrane impairment, production of reactive oxygen species, photo-catalytic activities, oxidative stress, DNA damage, among many others. The development of non-toxic and biocompatible carbon nanomaterials with low large-scale production costs is the need of the hour for the practical application of the materials. Although the laboratory results are often appealing, only a handful of instances went to clinical trials due to regulatory concerns because of their toxicity. We believe that the scientific community will be able to overcome this hurdle by discovering appropriate functional carbon nanocomposites with higher biocompatibility and low toxicity.

References

[1] Fraden J 2014 *Handbook of Modern Sensors: Physics, Designs, and Applications* (New York: Springer)
[2] Kumar H, Kuča K, Bhatia S K, Saini K, Kaushal A, Verma R, Bhalla T C and Kumar D 2020 *Sensors* **20** 1966
[3] Wan H, Zhuang L, Pan Y, Gao F, Tu J, Zhang B and Wang P 2020 *Biomedical Information Technology* 2nd edn, ed D D Feng (New York: Academic) pp 51–79
[4] Wearable Technology November 2021 https://en.wikipedia.org/wiki/Wearable_technology
[5] Google Contact Lens November 2021 https://en.wikipedia.org/wiki/Google_Contact_Lens
[6] Bezzon V D N, Montanheiro T L A, De Menezes B R C, Ribas R G, Righetti V A N, Rodrigues K F and Thim G P 2019 *Adv. Mater. Sci. Eng.* **2019** 4293073
[7] Peng Z, Zhao T, Zhou Y, Li S, Li J and Leblanc R M 2020 *Adv. Healthcare Mater.* **9** 1901495
[8] Maiti D, Tong X, Mou X and Yang K 2019 *Front. Pharmacol.* **9** 1401
[9] Mahor A, Singh P P, Bharadwaj P, Sharma N, Yadav S, Rosenholm J M and Bansal K K 2021 *J. Carbon Res.* **7** 19
[10] Wongkaew N, Simsek M, Griesche C and Baeumner A J 2019 *Chem. Rev.* **119** 120
[11] Chen A and Chatterjee S 2013 *Chem. Soc. Rev.* **42** 5425
[12] Christian G D, Dasgupta P K and Schug K A 2013 *Analytical Chemistry* 7th edn (New York: Wiley)
[13] Brett A M O and Brett C M 1994 *Electrochemistry Principles, Methods, and Applications* (Oxford: Oxford University Press)
[14] Holler F J, Skoog D A and Crouch S R 2018 *Principles of Instrumental Analysis* 7th edn (Boston, MA: Cengage Learning)
[15] Hareesha N, Manjunatha J G, Raril C and Tigari G 2019 *Adv Pharm Bull.* **9** 132
[16] Tigari G and Manjunatha J G 2020 *J. Sci.-Adv. Mater. Dev.* **5** 56
[17] Hareesha N and Manjunatha J G 2020 *J. Iran. Chem. Soc.* **17** 1507
[18] Hareesha N and Manjunatha J G 2020 *J. Electroanal. Chem.* **878** 114533
[19] Tigari G and Manjunatha J G 2019 *J. Anal. Test.* **3** 331
[20] Hareesha N, Manjunatha J G, Raril C and Tigari G 2019 *ChemistrySelect* **4** 4559
[21] Raril C and Manjunatha J G 2018 *Anal. Bioanal. Electrochem.* **10** 488
[22] Manjunatha J G, Kumara Swamy B E, Mamatha G P, Gilbert O and Sherigara B S 2011 *Anal. Bioanal. Electrochem.* **3** 146
Hareesha N and Manjunatha JG 2020 *J. Sci.-Adv. Mater. Dev.* **5** 502
[23] Manjunatha J G 2020 *Chem. Data Collect.* **25** 100331
[24] Geim A and Novoselov K 2007 *Nat. Mater.* **6** 183
[25] Suvarnaphaet P and Pechprasarn S 2017 *Sensors* **17** 2161
[26] Mathew T *et al* 2020 *FlatChem* **23** 100184
[27] Nelson D L and Cox M 2017 *Lehninger Principles of Biochemistry: International Edition* 7th edn (San Francisco, CA: WH Freeman)
[28] Voit D and Voit J G 2010 *Biochemistry* 4th edn (New York: Wiley)
[29] Fritea L, Tertis M, Sandulescu R and Cristea C 2018 *Enzyme Nanoarchitectures: Enzymes Armored with Graphene* ed C V Kumar (Amsterdam: Elsevier) pp 293–333
[30] Raymundo-Pereira P A, Baccarin M, Oliveira O N Jr and Janegitz B C 2018 *Electroanalysis* **30** 1888
[31] Wu H, Wang J, Kang K, Wang C, Wang D, Liu J, Aksay I A and Lin Y 2009 *Talanta* **80** 403

[32] Palanisamy S, Ezhil Vilian A T and Chen S M 2012 *Int. J. Electrochem. Sci.* **7** 2153
[33] Chen J, Zheng X, Miao F, Zhang J, Cui X and Zheng W 2012 *J. Appl. Electrochem.* **42** 875
[34] Yadav H M, Park J D, Kang H C and Lee J J 2021 *Chemosensors* **9** 98
[35] Cao S, Zhang L, Chai Y and Yuan R 2013 *Talanta* **109** 167
[36] Huang Y *et al* 2018 *Biosens. Bioelectron.* **102** 560
[37] Meier J, Hofferber E M, Stapleton J A and Iverson N M 2019 *Chemosensors* **7** 64
[38] Cai Z, Xiong H, Zhu Z, Huang H, Li L, Huang Y and Yu X 2017 *Synth. Met.* **227** 100
[39] Mercante L A, Facure M H M, Sanfelice R C, Migliorini F L, Mattoso L H C and Correa D S 2017 *Appl. Surf. Sci.* **407** 162
[40] Cernat A, Ştefan G, Tertis M, Cristea C and Simon I 2020 *Bioelectrochemistry* **136** 107620
[41] Thanh T D, Balamurugan J, Hien H V, Kim N H and Lee J H 2017 *Biosens. Bioelectron.* **96** 186
[42] Adumitrăchioaie A, Tertiş M, Suciu M, Graur F and Cristea C 2019 *Electrochim. Acta* **311** 50
[43] Rajkumar C, Thirumalraj B, Chen S-M and Chen H-A 2017 *J. Colloid Interface Sci.* **487** 149
[44] Cheng M, Zhang X, Wang M, Huang H and Ma J 2017 *J. Electroanal. Chem.* **786** 1
[45] Ezhil Vilian A T, An S, Choe S R, Kwak C H, Huh Y S, Lee J and Han Y K 2016 *Biosens. Bioelectron.* **86** 122
[46] Aoki K, Ogihara N, Tanaka M, Haniude H and Saito N 2020 *J. Mater. Chem.* B **8** 9227
[47] Deshmukh M A, Jeon J Y and Ha T J 2020 *Biosens. Bioelectron.* **150** 111919
[48] Tsai Y C, Huang J D and Chiu C C 2007 *Biosens. Bioelectron.* **22** 3051
[49] Besteman K, Lee J O, Wiertz F G M, Heering H A and Dekker C 2003 *Nano Lett.* **3** 6
[50] Shi Q, Peng T, Zhu Y and Yang C F 2005 *Electroanalysis* **17** 857
[51] Gholizadeh A, Shahrokhian S, Zad A I, Mohajerzadeh S, Vosoughi M, Darbari S, Koohsorkhi J and Mehran M 2012 *Anal. Chem.* **84** 5932
[52] Britto P J, Santhanam K S V and Ajayan P M 1996 *Bioelectrochem. Bioenerg.* **41** 121
[53] Suzuki I, Fukuda M, Shirakawa K, Jiko H and Gotoh M 2013 *Biosens. Bioelectron.* **49** 270
[54] Ribeiro F A D S, Tarley C R T, Borgesa K B and Pereira A C 2013 *Sens. Actuators* B **185** 743
[55] Lin K C, Lin Y C and Chen S M 2013 *Electrochim. Acta* **96** 164
[56] Ghrera A S, Pandey C M and Malhotra B D 2018 *Sens. Actuat.* B **266** 329
[57] Tran H V, Piro B, Reisberg S, Tran L D, Duc H T and Pham M C 2013 *Biosens. Bioelectron.* **49** 164
[58] Li F, Peng J, Wang J, Tang H, Tan L, Xie Q and Yao S 2014 *Biosens. Bioelectron.* **54** 158
[59] Akter R, Rahman M A and Rhee C K 2012 *Anal. Chem.* **84** 6407
[60] Zheng T T, Zhang R, Zou L and Zhu J J 2012 *Analyst* **137** 1316
[61] Ji C, Zhou Y, Leblanc R M and Peng Z 2020 *ACS Sens.* **5** 2724
[62] Khan S, Dunphy A, Anike M S, Belperain S, Patel K, Chiu N H L and Jia Z 2021 *Int. J. Mol. Sci.* **22** 6786
[63] Hassanvand Z, Jalali F, Nazari M, Parnianchi F and Santoro C 2021 *ChemElectroChem* **8** 15
[64] Liu J, Li R and Yang B 2020 *ACS Cent. Sci.* **6** 2179
[65] Su W, Wu H, Xu H, Zhang Y, Li Y, Li X and Fan L 2020 *Mater. Chem. Front.* **4** 821
[66] Xiang Q, Huang J, Huang H, Mao W and Ye Z 2018 *RSC Adv.* **8** 1820
[67] Ganganboina A B and Doong R A 2018 *Microchim. Acta* **185** 526

[68] Rao H, Zhao X, Liu X, Zhong J, Zhang Z, Zou P, Jiang Y, Wang X and Wang Y 2018 *Biosens. Bioelectron.* **100** 341
[69] Mansuriya B D and Altintas Z 2020 *Materials* **13** 96
[70] Yang Y, Liu Q, Liu Y, Cui J, Liu H, Wang P, Li Y, Chen L, Zhao Z and Dong Y 2017 *Biosens. Bioelectron.* **90** 31
[71] Ganganboina A B and Doong R A 2019 *Sci Rep.* **9** 7214
[72] Bhatnagar D, Kaur I and Kumar A 2017 *Int. J. Biol. Macromol.* **95** 505
[73] Mollarasouli F, Serafín V, Campuzano S, Yáñez-Sedeño P, Pingarrón J M and Asadpour-Zeynali K 2018 *Anal. Chim. Acta* **1011** 28
[74] Bing X and Wang G 2017 *Int. J. Electrochem. Sci.* **12** 6304
[75] Chowdhury A D, Takemura K, Li T C, Suzuki T and Park E Y 2019 *Nat. Commun.* **10** 3737
[76] Valipour A and Roushani M 2017 *Biosens. Bioelectron.* **89** 946
[77] Tufa L T, Oh S, Tran V T, Kim J, Jeong K J, Park T J, Kim H J and Lee J 2018 *Electrochim. Acta* **290** 369
[78] Jahanbakhshi M and Habibi B 2016 *Biosens. Bioelectron.* **81** 143
[79] Bai J, Sun C and Jiang X 2016 *Anal. Bioanal. Chem.* **408** 4705
[80] Li Y, Zhong Y, Zhang Y, Weng W and Li S 2015 *Sens. Actuators B Chem.* **206** 735
[81] Maaoui H, Teodoresu F, Wang Q, Pan G H, Addad A, Chtourou R, Szunerits S and Boukherroub R 2016 *Sensors* **16** 1720
[82] Chen D, Zhuang X, Zhai J, Zheng Y, Lu H and Chen L 2018 *Sens. Actuators B Chem.* **255** 1500
[83] Jiang G, Jiang T, Zhou H, Yao J and Kong X 2015 *RSC Adv.* **5** 9064
[84] Huang Q, Lin X, Lin C, Zhang Y, Hu S and Wei C 2015 *RSC Adv.* **5** 54102
[85] Hu S, Huang Q, Lin Y, Wei C, Zhang H, Zhang W, Guo Z, Bao X, Shi J and Hao A 2014 *Electrochim. Acta* **130** 805
[86] Pandey I and Jha S S 2015 *Electrochim. Acta* **182** 917
[87] Gowthaman N, Sinduja B, Karthikeyan R, Rubini K and John S A 2017 *Biosens. Bioelectron.* **94** 30
[88] Canevari T C, Cincotto F H, Gomes D, Landers R and Toma H E 2017 *Electroanalysis* **29** 1968
[89] Muthusankar G, Rajkumar C, Chen S M, Karkuzhali R, Gopu G, Sangili A, Sengottuvelan N and Sankar R 2019 *Sens. Actuators B Chem.* **281** 602
[90] Shadjou N, Hasanzadeh M and Omari A 2017 *Anal. Biochem.* **539** 70

Chapter 16

Fabrication of disposable sensors to test for environmental pollutants

Varsha Rani and Madan L Verma

Biosensors are an integral part of contemporary smart technology that are being employed in various application such as the biomedical and agri-food sectors. This chapter focuses on electrochemical biosensors, their characteristics, theoretical background, working principles, and advantages of screen-printed electrodes in the different fields, including healthcare, agriculture, and environment. The necessity of the detection of the analytes and the consequences of their presence in excess/absence in living systems and in the environment is also discussed. Engineering of screen-printed electrodes using different modifiers and nanomaterials in the detection of different contaminants in the environment has been critically discussed.

16.1 Introduction

A biosensor acts like a device for quantifying the concentration as well as the presence of the bioanalytes such as, microorganisms, biological structures, and biomolecules. Antibodies, enzymes, hormones, whole cell, organelles, and nucleic acids are the immobilized biomolecules present in biosensors [1]. The most commonly used biomolecules in biosensors are enzymes. Biosensors also contain transistor components and electronic systems like, displays, processors, and signal amplifiers (figure 16.1). Immobilized bioreceptors interact selectively with the bioanalyte and a reaction takes place which is measured using biotransducers. Biotransducers produce a measurable signal which is chemical, physical, or electric in nature. The electronic device, electronic amplifier increases the power of the signal. A number of biosensors are used for the detection of different types of analytes including, ethanol, cholesterol, pyruvate, lactate, glutamate, choline, creatinine, acetylcholine, and glucose [2]. Biosensors have applications in different sectors like food, healthcare, environment, and industry. One of the main

doi:10.1088/978-0-7503-5127-0ch16

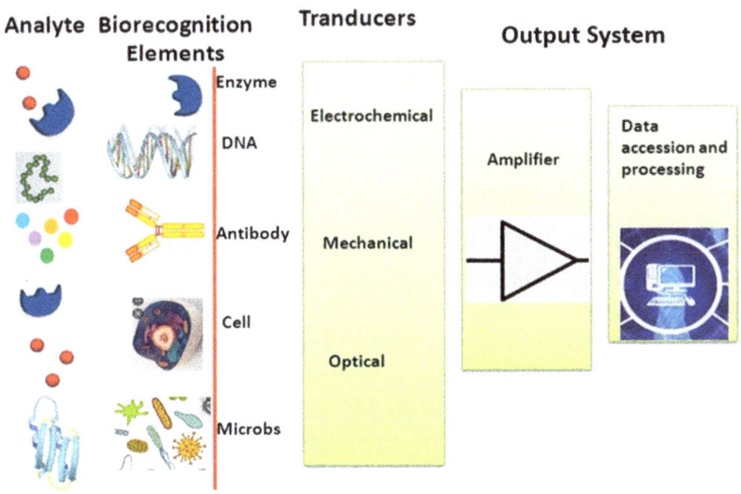

Figure 16.1. Schematic representation of a typical biosensor: This figure shows the parts of a typical biosensor: bioreceptor, transducer, and the output system consisting of an amplifier and the display.

requirements for biosensors is the immobilization of bioreceptors in the vicinity of the transducer [3].

There are two methods for the immobilization, chemical attachment and the physical entrapment method. A suitable reagent has to be covalently bonded with the transducer in the chemical attachment method. Bioreceptors are needed in very minute quantities and are used repeatedly for the measurements [1–3].

This chapter deals with biosensors, their characteristics, electrochemical biosensors, the principle by which they work, and the applications of biosensors.

16.2 Basic characteristics of a biosensor

There eight main parameters used to characterize biosensors. These parameters are selectivity, sensitivity, range, stability, detection limit, reproducibility, response time, and lifetime.

1. Selectivity is the ability of the biosensor to responding to a specific analyte. Selectivity is a desired feature of the biosensor to not respond or lack of response towards other interfering chemicals in the reaction.
2. Sensitivity is the response of the biosensor to the per unit change in the concentration of the analyte.
3. Range defines the concentration range, over which the biosensor sensitivity is good.
4. Stability of the biosensor, defines the change in the sensitivity over a fixed period of time.
5. Response time defines the time needed by the biosensor to indicate 63% of its final response because of a step change in the concentration of the analyte.
6. Reproducibility is the accuracy of obtaining the output of the biosensor.

7. Detection limit defines the analyte's lowest concentration at which the measurable response is being produced.
8. Lifetime of the biosensor defines the time period at which biosensor can be used without showing any significant deterioration in its performance [1].

16.3 Electrochemical biosensors and working principle

An electrochemical biosensor transduces the events which are biochemical in nature such as antigen–antibody interaction, enzyme substrate reactions into the electrical signals like, impedance, voltage, current, etc [2, 3]. The key component of the electrochemical biosensor is the electrode. This electrode is employed as a solid support for the movement of electrons as well as immobilizing the biomolecules such as, antibodies, enzymes, or nucleic acids [4, 5]. The presence of suitable enzymes is the most typical portion of the electrochemical biosensors in the recognizing layer which provides electroactive substances for detection. The biorecognition component can be a native enzyme and the analyte is equivalent to its substrate, which can also be an inhibitor. In affinity-based sensors, enzymes can be used like oligonucleotides with a specified sequence [6]. There is very limited processing of enzymes in biotechnology for the monitoring of clinical metabolites [7].

Low biocompatibility, loss of activity, and lower specificity are the results of inappropriate immobilization, though it is crucial to maintain the biological activity as well as the orientation of the biomolecules upon immobilization process.

A higher level of performance by the biosensors can be achieved by employing the appropriate functional material for the electrode being used. Different biosensing methods based on electrochemistry have been introduced for miniaturized and simple analytical devices for analyzing the on-site samples. It is still a big challenge to develop ideal devices for sensing the on-site samples with the required sensitivity as well as with higher reproducibility. High performance electrochemical analysis using nanomaterials as the supporting material which enhance the signal strength, have gained much attention [8]. Use of nanomaterials endow the larger surface area which enhances the mass transfer of the reactants and the loading capacity, thus producing synergic effects for the amplification of signal [9, 10].

16.4 Electrodes in electrochemical biosensors and fabrication of screen-printed electrodes

Analytical performance methods are mainly focused on the material of the electrode being used. Mercury is the material which has been used for many years as a suitable electrode material because of its high reproducibility, attractive behavior, smoother surface, and renewable nature. These distinct and very useful properties of mercury metal as an electrode in biosensors led to the Nobel prize in chemistry for Heyerovsky in 1959. Hanging and doping are the two types of electrode that have been widely used in different electrochemical and polarographic techniques [11]. Various non-mercury electrodes have been tried with the advancements in electroanalytical science. Carbon and bismuth electrodes have been used because of properties such as wider potential range, lower background current, chemically inert, and suitability for detection and sensing various compounds [12, 13].

Solid electrodes are miniaturized by reducing the volume of the sample which enhances its cost effectiveness as well as portability. There is a need to analyze the samples on-site, the previously used cumbersome cells and electrodes need to be replaced. New techniques of fabrication have allowed the replacement of old bulky electrodes and beaker type electrochemical cells. Fabricating the printed devices on substrates which are bendable has enabled new electrode systems with a wider range to be developed. Electrode systems which are portable, economic, and disposable can be developed via fabricating the electrodes with screen-printed technology. Counter, reference, and the working electrodes which are combined to represent the whole electrode system can be printed on the surface of the same substrate [14].

The glucose biosensor is an example of a screen-printed electrode which is being used for diabetes and represents a billion dollars annually in the global market [15, 16]. With the passage of time, society is constantly developing as well as growing and the demand for sensing devices is increasing day by day for the welfare of human beings and the environment. To achieve improved human health and a greener and safer environment, there is a need for rapid, portable, and accurate sensing devices. There is a need for decentralized analyses, thus new and advanced sensing devices are needed. The use of screen-printed electrodes in sensing devices is not only cost effective but they are also portable which is progress towards the decentralized analysis.

Screen-printed technology is a very important area of research through which electrodes can be modified using different types of inks commercially available for the counter, reference, and the working electrodes. This technique of screen-printed technology produces highly specific and calibrated electrodes for the analysis of specific analytes [14, 17, 18]. Various kinds of modifiers such as, inorganic nanomaterials, noble metals, enzymes, proteins, and DNA sequences are being used in these screen-printed electrodes for monitoring the environment [19]. *In situ* environmental monitoring can be easily achieved with improved performance using screen-printed electrodes [20, 21].

A screen-printed electrode is comprised of a chemically inert substrate on which three electrodes namely, the reference electrode, the working electrode, and the counter electrode are printed using the screen-printing methodology. The principal electrode is the working electrode which performs electrochemical reactions while the counter and reference electrodes are used to complete the electric circuit.

Biological and chemical events are converted into a detectable signal on the screen-printed electrode with transducer elements integration. Electrochemical detection methods include potentiometric, amperometric, and conductometic methods. The potentiometric method of detection is based on the potential differences or the voltage while the amperometric method of detection is based on the measurement of the current. The conductometric method of detection is based on the conductivity or the resistance. Among these three techniques for electrochemical sensing, amperometric detection is the methodology which has been widely adopted because of properties such as its higher applicability and higher sensitivity.

The fabrication technique for screen-printed electrodes consists of three main steps, screen-printed electrode fabrication, design of the surface, and utilization in the sensing.

Devices for real time sensing, wearable electronics, and point of care device applications are currently in higher demand. More reliable and economic methods should be produced to meet these demands. Screen-printed technology is a robust technique in terms of economy, versatility, and ease of use. Many research groups consider screen-printed technology as the best among the other additive manufacturing techniques. Screen-printing electrodes is a technique with simple equipment design, which requires a paste for printing, a meshwork to design the house, and a squeegee to carry out the printing work in an up and down motion. This technique provides immense potential for flexible electronics [22].

16.5 Integration of mediators, pre-anodized screen-printed carbon electrodes

Transfer of the redox equivalents has been carried out by the electrochemical mediators between the active sites of enzymes and the electrodes which triggered the process of bioelectrocatalysis. This process is very helpful in the development of the second-generation biosensors.

There should be ready reduction/oxidation of the redox mediators at the surface of the electrode and redox mediators should interact easily with the biorecognition component. Chemical compounds like osmium or ruthenium complexes, ferrocene derivatives, and viologens are the smaller size compounds frequently used. Horse heart cytochrome c, a small redox protein, has been reported to act like a redox partner in the biosensing applications. It has been reported by Silveira and Almeida [23], that docking two complementary proteins with each other introduces a second level of selectivity to the biosensor and the list of compounds targeted for analysis has been enlarged.

The central biological process is the transfer of electrons between the redox centers of the proteins such as photosynthesis, respiratory chains, metabolic pathways, and also in the regulation of gene expressions where the redox switches are involved [24]. The electrochemistry of enzymes and the proteins not only provides better understanding of the redox reactions but also provides the basis for fabricating the electrochemical biosensors. For example, integrated analytical devices which combine the voltammetric or amperometric tranducers with the biorecognition element displaying the catalytic activity of enzymes [25]. Some enzymes like heme of copper containing proteins, exhibit the direct transfer of electrons on the surface of electrodes [26]. Catalytic activity is not often displayed in the direct electron transfer, it is dependent on the nature of the polypeptide chain which is wrapped around the redox centers of the proteins.

It is very difficult to achieve the electrochemical communication with the electrodes if the cofactors are shielded by the apoprotein. Protein interaction with the transducers can be achieved by modifying the surface of the electrodes thus facilitating electrocatalysis and direct electron transfer. Using the redox mediators for enzyme electrocatalysis can help the easy interaction between the biocatalyst and the electrodes. Mediators have proved to be very helpful for measuring the enzymatic reactions via the electrochemical methods [27].

The most used protein mediator is a small cytochrome namely, horse heart cytochrome c which is a smaller size globular protein having a c type heme. Horse heart cytochrome c protein is involved in the transport of charge in the respiratory protein. Horse heart cytochrome c is a basic protein which interacts very well at the electrode surfaces having negative charge [28]. A well-defined quasi-reversible electrochemistry has been displayed by horse heart cytochrome c on the electrode surfaces like, promoter covered surfaces and polished edge pyrolytic graphite [29]. De Wael [30] has reported a cytochrome c-based biosensor. In this cytochrome c-based biosensor, horse heart cytochrome c has been entrapped by the gelatine hydrogel and cytochrome c peroxidise from *Saccharomyces cerevisae* has been used for delivering a hydrogen peroxide biosensor. Cytochrome c peroxidise from *S. cerevisae* is a monomeric protein involved in the detoxification of hydrogen peroxide [23, 31]. Cytochrome c protein has been employed as an electron shuttle for the bilirubin oxidase from *Myrothecium verrucaria* [32]. Bilirubin oxidase is a multi-copper oxidase having four copper centers. This bilirubin oxidase is found to be involved in the oxidation of bilirubin into the biliverdin with the reduction of dioxygen into the water. Bilirubin oxidase showed the direct electron transfer on the thiol modified gold electrodes [33].

The performance of electrochemical biosensors has been reported to be enhanced or improved by using nanomaterials as electron mediators [34]. The number of electrochemical biosensors which combines phenol oxidase with a variety of materials has been reported, e.g., nanomaterials or polymeric matrices [35, 36], graphene [37], carbon nanotubes [38, 39], metallic nanoparticles, nanostars, metal oxide nanoparticles [40–42], carbon fibers [43], nanostructured thin films [44, 45] and molecular imprinted polymers [46] have proved to be excellent electron mediators.

An interesting replacement of the zero-dimensional nanoparticles is the one-dimensional nanoparticles because of the high surface to volume ratios. Silver nanowires have been reported to be very good mediators due to the presence of properties like excellent thermal properties and higher conductivity [47, 48].

Silver nanowires have been reported for designing a novel catechol electrochemical biosensor. Salvo-Comino *et al* [49] reported the homologous distribution of the enzyme along the silver nanowires as studied by atomic force microscopy and this homologous distribution maximizes the contact surface. There are different nanomaterials reported to modify the electrochemical biosensors as shown in table 16.1.

Transfer of electrons between the electrode and the enzyme is promoted by the larger contact area. Excellent anti-interference characteristics have been shown by the biosensor to the phenols present in wine including quercetin, anillin, catechin, and pyrogallol. Detection of catechol presence in wine samples has been successfully shown by the biosensor. So, the biosensor silver nanowires have been proved to be a stable, sensitive, and rapid voltametric detector of phenols in real time applications [49].

Table 16.1. Different types of nanomaterials used to modify the electrochemical biosensors.

Sl. No.	Analyte	Nano-material used	Detection limit	References
1.	Uranyl	Gold nanoparticles	0.3 μg L^{-1}	[50]
2.	Aflatoxin B1	Au NBPs	0.4 nM	[51]
3.	*Escherichia coli*	Gold nanoparticles	15 CFU mL^{-1}	[52]
4.	Au/CdS QDs/TNTs	Hydrogen peroxide cholesterol	18.73–355.87 μm 0.024–1.2 mM	[53]
5.	AuNP-MoS2-rGO	Carcinoembryonic antigen	0.084 ng mL^{-1}	[54]
6.	Ag/Pd NPs	Ractopamine Clenbuterol Salbutamol	1.52 pg mL^{-1} 1.44 pg mL^{-1} 1.38 pg mL^{-1}	[55]
7.	Au NPs/TiO2	Hydrogen peroxide	5 μm	[56]
8.	Ag NPs	Mucin 1	0.37 fg mL^{-1}	[57]
9.	Pt NFs/PANi	Urea	10 μm	[58]
10.	Cu/rGO-BP	Glucose	11 μm	[59]
11.	Ni/Cu MOF	Glucose	0.51 μm	[60]
12.	MnO2 NFs	Salmonella	19 CFU mL^{-1}	[61]
13.	Fe2O3/NiO/Mn2O3NPs	Folic acid	96.89 pM	[62]
14.	NiO@Au	Lactic acid	11.6 μM	[63]
15.	ZnO-rGO	Dopamine	8.75 pM	[64]
16.	CNT thin-film transistor	DNA	0.88 g L^{-1}	[65]

16.6 Pre-anodized screen-printed carbon electrodes

Like mediator modification, pre-treatment is a different methodology used for improving the electrocatalytic properties of electrodes in screen-printed carbon electrodes. The pre-treatment method mainly focuses on removing the constituents of the organic ink as well as the contaminants which leads to the improvement in the functionalities and the surface roughness of the electrochemical biosensor. Therefore, the two step pre-treatment method, potential cycling method and pre-anodization method have been developed by combining the electrochemical and chemical treatments. Carbon materials have attracted much attention for the development of the screen-printing technology because of its properties like ease of availability, cost effectiveness, and its nontoxic nature [66].

Graphene and carbon nanotubes are the electrocatalysts used primarily for electroanalytical techniques. It has been found that edge plane-like sites enhanced the electrocatalytic nature of the carbon nanotubes. Edge plane pyrolytic graphite electrodes have been found to show the same results as carbon nanotubes [67, 68]. Comparison of the edge plane sites and the oxygen functionalities showed that pre-anodized screen-printed electrodes are highly efficient for electrochemical sensing, enhancing sensitivity as well as selectivity. These electrodes are used for non-enzymatically detecting the analytes which are poorly electroactive.

Characterization of electrodes has been performed via x-ray photoelectron spectroscopy and Raman and scanning electron microscopy has been used to confirm the formation of oxygen functionalities and the edge plane.

Raman spectra showed differences in D band intensities for oxygen plasma treated screen-printed carbon electrodes and screen-printed carbon electrodes, confirming the sites for edge/defect plane. It has been reported that for simultaneous detection of uric acid, ascorbic acid, and dopamine, pre-anodized electrodes are more effective compared with the oxygen plasma treated screen-printed carbon electrode [69].

16.7 Applications

Biosensor applications are wide ranging, aiming to improve the quality of our life. Biosensors are found to be useful in various sectors, e.g., detection of different diseases, environmental monitoring, defense, food safety, pesticide detection, and detection of heavy metals, etc. Detection of biomolecules is one of the main applications of the biosensors in which targets for drugs and indicators of different diseases can be identified [70, 71]. Biosensors are very good platforms for monitoring the quality of food, food safety, traceability and nutritional value [72, 73]. Applications of electrochemical biosensors are discussed in detail in the following sectioon.

16.8 Disposable electrodes in the detection of biomolecules

Biomolecules are present as essential organic compounds in all living systems. These biomolecules play a very important role in maintaining and regulating different metabolic processes in our body. Biomolecules can be classified as proteins, carbohydrates, nucleic acids, lipids, hormones, enzymes etc, playing a vital role in regulating our day-to-day activities. So, the presence of these biomolecules in our body should be monitored from time to time for good health and a sound body. The absence as well as the presence of biomolecules in excess amounts in our body creates many disorders in the body such as diabetes, Parkinson's disease, Alzheimer's disease, osteoporosis, heart attacks, etc. All of these are reducing the average life span of living systems. Therefore, these bioanalytes need to be detected and this can be done with the help of biosensors. Disposable electrodes are reported in the literature, for the analysis of these biomolcules [74].

Pre-anodized screen-printed electrodes are reported to be effective in the analysis of biomolecules such as uric acid, dopamine, ascorbic acid, and the blood samples of human beings, thus producing an electrochemical signal depending on the concentration of different bioanalytes analyzed. Disposable electrodes are found to produce results for uric acid in the blood sample of the human beings in the presence of the ascorbic acid [75]. The detection of creatine with the help of electrodes is found to be very effective [76]. Melamine is a poor electroactive compound but can be detected with the help of electrodes [77]. The role of electrodes has been further extended for detecting glucose oxidase and there is improvement in the direct electron transfer

[78, 79]. Disposable screen-printed electrodes efficiently show lactic acid detection with improved electrocatalytic properties [80].

16.9 Screen-printed electrodes in the detection of food contaminants

We know that healthy and balanced food plays a very important role in maintaining our health. So we need to estimate what contaminants and constituents are present in our food products. To know the quality of the food, contamination and hygiene levels of the food we should monitor products periodically. Low molecular weight toxins have been reported from sea-food, namely domoic acid, breyetoxin, tetrotoxin, and okadaic acid which are known to spoil sea-food. Detection of contaminants in food products has been achieved via immunosensors using disposable electrodes. It has been identified that biosensors detect contaminants with higher accuracy [81, 82].

D-lactate, acetaldehyde, and l-lactate are the analytes in food to be targeted for the analysis via biosensors [83, 84]. Detection of acetaldehyde and lactates in food has using screen-printed electrodes has been reported. Ethanol biosensors are being used for analyzing the level of ethanol in wine and beer [85]. Sometimes the human population is affected by food allergies, which are caused by different allergens. These allergens are proteinaceous in nature and are obtained from eight main groups of foods including, tree nuts, nuts, eggs, wheat, soy, milk, shellfish, and fish [86]. Proteins which are obtained from wheat, milk, fish gelatine, and eggs are used as fining agents during the preparation of wine and these proteins may cause allergic reactions in consumers [87].

Therefore, it is very important to monitor the level of nutrients as well as screening of the food is needed as a priority to obtain good quality food which is secure and safe. Genosensors, antibody-based sensors, and molecularly imprinted polymers are the biosensors used for detecting the allergens in food. Allergens such as Ara h 1, conglutinin, and lysozymes have been detected via antibody-based assays as aptamers [88–90]. Detection of various bacterial strains in different foods has been reported by different researchers as shown in table 16.2.

16.10 Importance of disposable electrodes in pesticide detection

Pesticides are used to keep our crops healthier as well as protecting the crops from various diseases and infestations. Pesticides include repellents, fungicides, herbicides, acaricides, plant growth regulators, and insecticides. Leaching of the pesticides into the ground water through the soil leads to the contamination of water [106]. Pesticides can also enter the water surface by runoff and drift, causing risks to our ecosystem as well as to human health [107]. Pesticide analysis through the conventional techniques includes liquid and gas chromatography coupled with mass spectroscopy [108]. In addition to these techniques, electrochemical biosensors are reported to be successful at detecting pesticides [109, 110]. Electrochemical biosensors are based on the whole cells, enzymes, and the antigen-antibody are reported to the application of pesticide sensing [111].

Table 16.2. Detection of different bacterial strains in various food samples by screen-printed electrodes.

Sl. No.	Bacterial strains	Sensor types	Sample	Analysis time (min)	References
1.	*Salmonella typhimurium*	Immunoassay; Au-coatedMB/SAM/Ab; CdSNP as label; SPCE	Milk	40	[91]
2.	*S. typhimurium*	Immunoassay (sandwich); Ab-coatedMB; Ag measurement; Avidin-SPCE	Green beans, milk, sprouts, and eggs	105	[92]
3.	*S. typhimurium*	Paper-based immunoassay (sandwich);AuNP as label; SPCE	Water	35	[93]
4.	*S. typhimurium and Salmonella aureus* (and *E. coli*)	Antimicrobial petide melittin on MB; SP-IDME (silver)	Apple juice and water	30	[94]
5.	*S. typhimurium* (and *E. coli* O157:H7)	Immunoassay (sandwich); capture biotinylated Ab on stretavidin-MB; GOX as label; SP-IDME (gold)	Chicken carcass (ground beef for *E. coli*)	180	[95]
6.	*Salmonella (no serotype)*	Immunoassay (sandwich); capture Ab on MB; QD (CdTe) dendron as label; BiSPCE	Milk	80	[96]
7.	*Listeria monocytogenes*	Immunoassay (sandwich); Ab capture on MB; Ab detection/urease (as label) modified AuNP; SP-IDE (gold)	Lettuce	150	[97]
8.	*Salmonella pullorum and Salmonella gallinarum*	Immunoassay (sandwich); capture Ab on AuNP-modified MB (SiO2/ Fe3O4); HRP as label; 4-SPCE	Chicken	70	[98]
9.	*Listeria innocua Serovar 6b*	Label-free; Bacteriophage endolysin CBD500 covalent immobilized on SPAuE	Milk	25	[99]
10.	*S. Typhimurium*	Immunoassay; label-free; Ferrocyanide measurement; rG-GO/ SPCE	Orange juice and water	15	[100]
11.	*E. coli*	Immunoassay (sandwich); Ab capture on MB; AuNP as label (catalysing HER); SPCE	Minced beef and water	70	[101]

12.	*E. coli*	Immunoassay (sandwich); Capture Ab on MB;rGO-NR-Au@Pt nanocomposite-detection Ab (measurement of H2O2 reduction); Thionine as mediator; SPCE	Pork and milk	115	[102]
13.	*E. coli*	Immunoassay (sandwich); HRP as label;AuNP/FeDC-SPCE	Milk	35	[103]
14.	*E. coli*	Immunoassay; Ab on AuNP/MB-GOX@PDA; Filtration step; GOX as label;Prussian Blue-modifiedSP-IDME of gold	Ground beef	75 min	[104]
15.	*E. coli*	Immunoassay; Label-free (measurement of Fe(CN) 63-/4); AuNP-SPCE	Milk powder	55 min	[105]

Biosensors based on enzymes are found to be very effective for the sensing of pesticides because of properties such as selectivity, sensitivity, variety of functions, and easy availability. Enzyme-based biosensors are categorized into direct and indirect [112, 113]. Enzymatic biosensors with direct mode measure the concentration of the analyte and the formation of the products during the enzymatic reaction. Enzyme inhibition due to the analyte contact has been monitored by indirect mode biosensors. Acetylcholine esterase is an enzyme which catalyzes the hydrolysis of acetylcholine into acetate and choline in cholinergic neurons [114], this is critical to control the nerve signals in our body.

Detection of organophosphate pesticides is particularly done by using indirect mode biosensors [115]. Organophosphates are found to inhibit the activity of acetylcholine esterase, thus preventing acetylcholine hydrolysis [116]. An acetylcholine esterase based electrochemical biosensor has been developed by Kumar and Sundramoorthy for the detection of methyl parathion pesticide, which is an organophosphate [117].

An electrochemical biosensor has been fabricated by acetylcholine esterase cross linked with a glutaraldehyde glossy carbon electrode modified with carbon nanotubes which has been immobilized on the cross-linked enzyme. Enzymes like, choline oxidase, organophosphorus hydrolase, butyrylcholine esterase, and phosphotriesterase have also been successfully utilized for sensing the organophosphate pesticides [118].

The ability of the biosensor to detect pesticides primarily depends on the immobilization of the enzyme. Sol–gel, cross-linking, and the physical entrapment are the main methods involved in the immobilization. The drawbacks associated with these methods are the denaturation of enzymes under the harsh conditions and the non-renewability of the surface of the electrodes. The use of magnetic materials in the biosensors for sensing pesticides has proved to be an excellent improvement for renewing the surface of electrodes as well as providing a better environment for electrochemical sensing. A hybrid core platform of the biomagnetic glasses containing a silica shell and FeO is used for the immobilization of the acetylcholine esterase. Gold containing ferric oxide is also used to immobilize the acetylcholine esterase, which improves the stability of the matrix. *Escherichia coli* immobilized screen-printed electrodes has been successfully employed to detect methyl parathion [74].

16.11 Environmental sample analysis (determination of pH and dissolved oxygen level in water, estimation of ions in water samples, organic compounds, heavy metal detection)

Water is a basic need of life and needed by every living organism. We know that water occupies 76% of the globe and that only 1% of that water is fresh water. Human activities like deforestation, industrialization, and dumping waste indiscriminately is the main cause of pollution affecting fresh water. This pollution leads to many health problems like tuberculosis, malaria, etc. There are very many changes such as changes in pH, reduced content of dissolved oxygen and very many chemicals that are being introduced into water resources due to dumping waste [74].

16.12 Determination of pH and dissolved oxygen levels in water

Glass electrodes have been used to identify the pH of aquatic systems. These glass electrodes are very unstable and there is a need to calibrate them. A three electrode system is made by a successful printing process which is capable of sensing the pH with improved sensitivity [119, 120]. Electrode application is extended and the use of electrochemical sensing has been reported for hydroxide ions using nickel oxide. The quality of the water sample has been defined by the concentration of dissolved oxygen. Electrochemiluminescence is a method used to detect the concentration of dissolved oxygen in the water samples by electrochemical sensing platform which uses cadmium sulfide electrodes [121].

16.13 Estimation of ions in water samples

Determination of ions such as phosphates and nitrites in water samples is also important because they are toxic in nature. Larger over-potential is needed for the determination of nitrites so the detection of nitrites is very difficult using conventional electrodes. Screen-printed electrodes have extraordinary features which are highly effective for the detection of phosphates and nitrites in water. Manganese oxide screen-printed electrodes are found to be effective for the detection of nitrites [122]. A microelectrode when acting in combination with screen-printed technology produces a novel sensor for detecting nitrites. Effective sensing of nitrites in river water is reported with the help of a screen-printed shallow recessed graphite microelectrode array [123].

Specific ion detection in water samples has been performed using ion selective electrodes, because of the properties of measuring different parameters in colored as well as turbid samples. Phosphate ions are hydrophilic in nature and because of this they are very difficult to detect in water samples. Three different types of electrode are being used to detect the phosphate ions namely, carbon paste, screen-printed, and conventional polyvinyl chloride electrodes [124]. Screen-printed electrodes are among the three electrodes which are functioning well for the detection of phosphates. Ammonium ions detection in river water has been reported to have been performed using a disposable amperometric biosensor which was developed using NADH and glutamate dehydrogenase, 2-oxoglutarate [19].

16.14 Organic compounds

Phenolic compounds and their derivatives are toxic to animals and plants and their presence in the drinking water is very dangerous so the presence of phenolic compounds must be monitored for our safety. Phenolic compounds produce toxic effects on penetrating the surface of plants and the skin of animals [19].

Non-pre-treated screen-printed electrodes have been reported to detect phenolic compounds with direct electron transfer properties [125]. Phenol produces oxidized products which hinder the electrode's surface and make the electrodes inefficient against the analyte thus making detection difficult.

To overcome this problem of oxidized phenol products, surfactants have been used. Determination of bisphenol in sewage and water samples has been carried out using cetyltrimethyl ammonium bromide, a cationic surfactant [125]. There is overlap of the peaks obtained during the detection of phenolic compounds and its derivatives by using non-pre-treated screen-printed electrodes.

The result of the pre-treatment is the removal of organic binders in the screen-printed electrode and the graphite particles are then exposed. Detection of hydroquinone, catechol, and aminophenol in river water samples has been found to produce well separated peaks using pre-treated screen-printed electrodes [19]. Three separate peaks were found showing different isomers of aminophenol in the mixture of the water samples. Various organic compounds contaminating the water were detected using the modified screen-printed electrodes. Three types of dihydroxy benzene derivatives have been detected using a screen-printed electrode coated with gold nanoparticles and multi-walled carbon nanotubes [126]. The use of enzymes such as laccase, polyphenol oxidase enzyme and tyrosinase has been employed to detect phenols in water samples. These enzymes act like catalysts during the reaction where phenols and benzene diols have been oxidized into quinines and other radicals [127]. The process of immobilization is very important while developing the biosensors to prevent the enzymes from leaking and denaturing. The entrapment method is the method used to immobilize the enzymes. Immobilizing agents like, polyvinyl alcohol photopolymer and polyazatidine prepolymer are in use for the immobilization process of enzymes [127, 128]. Another method to immobilize the enzymes is electrodeposition.

Polyphenol oxidase has been reported for electrodeposition using bismuth on the multi-walled carbon nanotube modified screen-printed electrodes [129].

This matrix does not use any binder and is successfully employed to detect the phenols. Detection of 1,2 diaminobenzene has been done using a screen-printed platinum electrode by the deposition and electrochemical polymerization of polyphenol oxidase [130].

Enzymatic biosensors have been reported for the analysis of contaminants in nanomolar to picomolar concentrations effectively. These enzymatic biosensors are found very useful for the real time analysis with ultrasensitivity and quick detection limits. The enzymatic biosensors are the advanced biosensors which are formed by using techniques like novel nanofabrication and nanobioconjugation [131].

16.15 Heavy metal detection

The quality of water has been also defined by the concentration of heavy metals present in the water sample along with the pH, ions, nitrites, organic pollutants, and phosphates etc. A higher concentration of heavy metals in water causes various life-threatening diseases like brain damage, chromosomal mutations, mental illness, cerebral convulsions, etc. The presence of even lower concentration of heavy metals in water proves very harmful to our health so the monitoring of heavy metals in water is extremely important. Heavy metals like cadmium, lead, mercury, and arsenic are neurotoxic in nature so special attention has been given to the detection of these metals [132].

Detection of heavy metals in water has been done by using stripping voltammetric techniques. On the spot analysis of these heavy metals in the environment has been successfully performed by combining screen-printed electrodes with stripping voltammetric techniques. By combining screen-printed electrodes with stripping voltammetric techniques yields accurate, inexpensive, sensitive, and rapid results.

Modified screen-printed electrodes especially with mercury, gold, silver, and bismuthate is highly effective for detecting heavy metals in terms of selectivity and sensitivity [132, 133]. Modified electrodes with bismuth are found to be effective for monitoring the environment because of their eco-friendly nature. Bismuth electrodes undergo hydrolysis in alkaline and neutral mediums easily [134].

Mercury modified electrodes have been found to show the best results over a wider pH range [135]. Mercury showed higher affinity for metals and produces improved sensitivity with lower detection limit. The toxicity of mercury limits its use for the detection of heavy metals. A thin film of mercury has been used for the modification of screen-printed electrodes for the detection of cadmium [136].

Mercury is a heavy metal which is toxic as well as bioaccumulative. Its accumulation in human beings causes various health problems in our respiratory system and kidneys. Gold is a metal which has shown good affinity with mercury [137]. So, gold can be utilizing to detect mercury, in modified or in bare form [138]. Structural changes can occur due to the formation of amalgam. so cleaning is needed to produce the best results. The detection of arsenic without the interference of copper had been reported using citrate capped gold nanoparticles and platinum nanoparticles [139]. The presence of heavy metal can be effectively determined using enzymes.

The effective electrochemical sensing of arsenic has been reported using acetylcholinesterase immobilized on screen-printed electrodes [140]. Bismuth in different forms such as nanoparticles, oxides, and co-deposition of bismuth in *ex situ/in situ* with mercury are found to be better electrocatalysts for the detection of heavy metals simultaneously [74]. Various pollutants like, pesticides, heavy metals, microbial pathogens, endocrine disruptors, biological oxygen demand, dioxins, and monitoring of food can be detected effectively in nanomolar to picomolar concentrations using the nanobiosensors [141–143].

16.16 Conclusions and future perspectives

The use of disposable electrodes is an emerging platform for biosensing applications with outstanding potential. These electrochemical sensors with modified screen-printed electrodes are excellent in the field of monitoring various contaminants and pollutants in our environment. The use of modified screen-printed electrodes is very useful in detecting various contaminants which allow us to live a healthier and safer life. The use of suitable electrocatalysts or modifiers such as, carbonaceous materials, enzymes, and metal nanoparticles are helpful for detecting a number of analytes and allowing us to take a step forward to lead a sophisticated life style. Use of these disposable electrodes and the electrochemical systems play an extraordinary role in the on-site monitoring of various contaminants with properties such as portability, stability, availability, cost effectiveness, selectibility, and sensitivity. Research should be focused with a wider range on the disposable electrodes so that we can effectively utilize the applications of disposable electrodes.

Acknowledgements

The authors are grateful to the Director of the Indian Institute of Information Technology Una, Himachal Pradesh for providing the necessary facility to pursue the present work.

References

[1] Lee Y H and Mutharasan R 2005 *Dept. Chem. Eng., Drexel Univ.* https://globalspec.com/reference/46540/203279/chapter-6-biosensors
[2] Li H, Lui X, Li L, Mu X, Genov R and Mason A J 2016 *Sensors* **17** 74
[3] Zheng H, Ma X, Chen L, Lin Z, Guo L, Qiu B and Chen Z 2013 *Anal. Methods* **5** 5005–9
[4] Wang Z and Dai Z 2015 *Nanoscale.* **7** 6420–31
[5] Krishnan S K, Singh E, Singh P, Meyyappan M and Nalva H S 2019 *RSC Adv.* **9** 8778–881
[6] Bakker E 2004 *Anal. Chem.* **76** 3285–98
[7] Pohanka M and Skladal P 2008 *J. Appl. Biomed.* **6** 57–64
[8] Chen J, Liu B, Gao X and Xu D 2018 *RSC Adv.* **8** 28048–85
[9] Lim S A and Ahmed M U 2016 *RSC Adv.* **6** 24995–5014

[10] Cho I H, Kim D H and Park S 2020 *Biomater. Res.* **24** 6
[11] Wang J 2002 *Acc. Chem. Res.* **35** 811–6
[12] Wang J, Lu J, Farias S A M and Ogorevc B 2000 *Anal. Chem.* **72** 3218–22
[13] Hayat A, Yang C, Rhouati A and Marty J 2013a *Sensors* **13** 15187–208
[14] Metters J P, Kadara R O and Banks C E 2011 *Analyst* **136** 1067–76
[15] Heller A and Feldman B 2008 *Chem. Rev.* **108** 2482–505
[16] Newman J D and Turner A P F 2005 *Biosens. Bioelectron.* **20** 2435–53
[17] Hart J P, Wring S A and TrAC 1997 *Trends Anal. Chem* **16** 89–103
[18] Hayat A, Marty J L and Radi A E 2012a *Electroanalysis* **24** 1446–52
[19] Su W Y, Wang S M and Cheng S H 2011 *J. Electroanal. Chem.* **651** 166–72
[20] Hayat A, Barthelmebs L and Marty J L 2012b *Sens. Actuators B Chem* **171** 810–15
[21] Hayat A, Sassolas A, Marty J L and Radi A 2013b *Talanta* **103** 14–19
[22] Raghav R S, Muthaiyan L, Arockia J J B B, Rajan K S, Swaminathan S, Uma M K and John B B R 2012 *J. Mater. Sci.* **56** 8951–9006
[23] Silveira C M and Almeida M G 2013 *Anal Bioanal Chem* **405** 3619–35
[24] Scheller F W, Wollenberger U, Lei C, Jin W, Ge B, Lehmann C, Lisdat F and Fridman V 2002 *Rev. Mol. Biotechnol.* **82** 411–24
[25] Alvarez-lcaza M and Bilitewski U 1993 *Anal. Chem.* **65** 525A–33A
[26] Ferapontova E and Gorton L 2005 *Bioelectrochemistry* **66** 55–63
[27] Ikeda T and Kano K 2001 *J. Biosci. Bioeng.* **92** 9–18
[28] Moretto L M, Bertoncello P, Vezza F and Ugo P 2005 *Bioelectrochemistry* **66** 29–34
[29] Fedurco M 2000 *Coord. Chem. Rev.* **209** 263–331
[30] De Wael K, Bashir Q, Van Vlierberghe S, Dubruel P, Heering H A and Adriaens A 2012 *Bioelectrochemistry* **83** 15–18
[31] Fishel L A, Villafranca J E, Mauro J M and Kraut J 1987 *Biochemistry* **26** 351–60
[32] Dronov R, Kurth D G, Scheller F W and Lisdat F 2007 *Electroanalysis* **19** 1642–46
[33] Cracknell J A, McNamara T P, Lowe E D and Dalton Trans C F **2011** 6668–75
[34] Lv M, Liu Y, Geng J, Kou X, Xin Z and Yang D 2018 *Biosens. Bioelectron.* **106** 122–8
[35] Forzani E S, Solís V M and Calvo E J 2000 *Anal. Chem.* **72** 5300–7
[36] Apetrei C, Rodríguez-Méndez M and De Saja J A 2011 *Electrochim. Acta* **56** 8919–25
[37] Rocha D P, Dornellas R M, Cardoso R M, Narciso L C, Silva M N, Nossol E, Richter E M and Munoz R A 2018 *Actuators B Chem.* **254** 701–8
[38] Wee Y, Park S, Kwon Y H, Ju Y, Yeon K M and Kim J 2019 *Biosens. Bioelectron.* **132** 279–85
[39] Apetrei I M, Rodriguez-Mendez M L, Apetrei C and de Saja J A 2013 *Sens. Actuat. B Chem.* **177** 138–44
[40] Mattos G J, Moraes J T, Barbosa E C M, Camargo P H, Dekker R F, Barbosa-Dekker A M and Sartori E R 2019 *Bioelectrochemistry* **129** 116–23
[41] Cerrato-Alvarez M, Bernalte E, Bernalte-García M J and Pinilla-Gil E 2019 *Talanta* **193** 93–9
[42] Wang S, Tan Y, Zhao D and Liu G 2008 *Biosens. Bioelectron.* **23** 1781–7
[43] Fu J, Qiao H, Li D, Luo L, Chen K and Wei Q 2014 *Sensors* **14** 3543–56
[44] Salvo-Comino C, González-Gil A, Rodriguez-Valentin J, Garcia-Hernandez C, Martín F, Garcia-Cabezon C and Rodriguez-Mendez M L 2020 *Sensors* **20** 2152
[45] Sun D, Yang D, Wei P, Liu B, Chen Z, Zhang L and Lu J 2020 *ACS Appl. Mater. Interfaces* **12** 41960–8
[46] Salvo-Comino C, Rassas I, Minot S, Bessueille F, Rodriguez-Mendez M L, Errachid A and Jaffrezic-Renault N 2020a *Mater. Sci. Eng.* C **110** 110667

[47] Lee H J, Oh S, Cho K Y, Jeong W L, Lee D S and Park S J 2018 *ACS Appl. Mater. Interfaces* **10** 14124–31
[48] Gao L, Zhu C, Li L, Zhang C W, Liu J, Yu H D and Huang W 2019 *ACS Appl. Mater. Interfaces* **11** 25034–42
[49] Salvo-Comino C, Martin-Pedrosa F, Garcia-Cabezon C and Rodriguez-Mendez M L 2021 *Sensors* **21** 899
[50] Shi S, Wu H, Zhang L, Wang S, Xiong P, Qin Z, Chu M and Liao J 2021 *J. Electroanal. Chem.* **880** 114884
[51] De Castro A C, Alves L M, Siquieroli A C, Madurro J M, Brito-Madurro A G and Microchem J 2020 **155** 104746
[52] Vu Q H, Tran Q H, Vu N P, Anh Y L, Dang T T L, Tonezzer M and Ngyyen T H H 2020 *Mater. Today Commun.* **26** 101726
[53] Khaliq N, Rasheed M A, Khan M, Maqbool M, Ahmad M, Karim S, Nisar A, Schmuki P, Cho S O and Ali G 2021 *Appl. Mater. Interfaces* **13** 3653–68
[54] Jandas P J, Luo J, Prabakaran K, Chen F and Fu Y Q 2020 *Chem. Phys.* **246** 122800
[55] Wang H, Zhang Y, Du B, Ma H, Wu D and Wei Q 2013 *Biosens. Bioelectron.* **49** 14–19
[56] Liu X, Zhang J, Liu S, Zhang Q, Liu X and Wong D K Y 2013 *Anal. Chem.* **85** 4350–6
[57] Wei U P, Zhang Y W and Mao C J 2020 *J. Mater. Chem.* B **8** 2471–5
[58] Jia W, Su L and Lei Y 2011 *Biosens. Bioelectron.* **30** 158–64
[59] Zhu T, Wang X, Chang W, Zhang Y, Maruyama T, Luo L and Zhao X 2021 *Mater. Sci. Eng.* C **120** 111757
[60] Wang B, Luo Y, Gao L, Liu B and Duan G 2021 *Biosens. Bioelectron.* **171** 112736
[61] Xue L, Guo R, Huang F, Qi W, Liu Y, Cai G and Lin J 2021 *Biosens. Bioelectron.* **173** 112800
[62] Alam M M, Rahman M M, Uddin M T, Asiri A M, Uddin J and Islam M A 2020 *Curr. Res. Biotechnol.* **2** 176–86
[63] Maduraiveeran G and Chen A 2021 *Electrochim. Acta* **368** 137612
[64] Verma S, Arya P, Singh A, Kaswan J, Shukla A, Kushwaha H R, Gupta S and Singh S P 2020 *Biosens. Bioelectron.* **165** 112347
[65] Li W, Gao Y, Zhang J, Wang X, Yin F, Li Z and Zhang M 2020 *Nanoscale Adv* **2** 717–23
[66] Verma M L *Nanoscience in Food and Agriculture (Part of Sustainable Agriculture Reviews Book Series* vol 4) (Berlin: Springer Nature) pp 229–45
[67] Banks C E and Compton R G 2006 *Analyst* **131** 15
[68] Banks C E, Crossley A, Salter C, Wilkins S J and Compton R G 2006 *Angew. Chem. Int. Ed.* **45** 2533
[69] Thiyagarajan N, Chang J L, Senthilkumar K and Zen J M 2014 *Electrochem. Commun.* **38** 86
[70] Jolly P, Formisano N and Estrela P 2015 *Chem. Pap* **69** 77–89
[71] Formisano N, Jolly P, Bhalla N, Cromhout M, Flanagan S P and Fogel R 2015 *Sens. Actuators* B **220** 369–75
[72] Sharma T K, Ramanathan R, Rakwal R, Agrawal G K and Bansal V 2015 *Proteomics* **15** 1680–92
[73] Van Dorst B, Mehta J, Bekaert K, Rouah-Martin E, De Coen W and Dubruel P 2010 *Biosens. Bioelectron.* **26** 1178–94
[74] Rajaram R and Mathiyarasu J 2021 *Material Properties and Design* (London: Royal Society of Chemistry) pp 1–26
[75] Tsai T H, Thiagarajan S and Chen S M 2010 *J. Agric. Food Chem.* **58** 4537

[76] Chen J C, Kumar A S, Chung H H, Chien S H, Kuo M C and Zen J M 2006 *Sens. Actuators,* B **115** 473
[77] Liao C W, Chen Y R, Chang J L and Zen J M 2011 *J. Agric. Food Chem.* **59** 9782
[78] Yang T H, Hung C L, Ke J H and Zen J M 2008 *Electrochem. Commun.* **10** 1094
[79] Chiu M H, Wei W C and Zen J M 2011 *Electrochem. Commun.* **13** 605
[80] Sprules S D, Hart J P, Wring S A and Pittson R 1995 *Anal. Chim. Acta* **304** 17
[81] Sirvent M A and Hart A L 2002 *Sens. Actuators,* B **87** 73
[82] Noguer T, Szydlowska D, Marty J and Trojanowicz M 2004 *Pol. J. Chem.* **78** 1679
[83] Avramescu A, Noguer T, Avramescu M and Marty J L 2002 *Anal. Chim. Acta* **458** 203
[84] Avramescu A, Andreescu S, Noguer T, Bala C, Andreescu D and Marty J L 2002 *Anal. Bioanal. Chem.* **374** 25
[85] Azevedo A M, Prazeres D M F, Cabral J M S and Fonseca L P 2005 *Biosens. Bioelectron.* **21** 235
[86] Vasilescu A, Nunes G, Hayat A, Latif U and Marty J L 2016 *Sensors* **16** 1863
[87] Penas E, Lorenzo C D, Uberti F and Restani P 2015 *Molecules* **20** 13144
[88] Cox J C and Ellington A D 2001 *Bioorg. Med. Chem.* **9** 2525
[89] Potty A S R, Kourentzi K, Fang H, Schuck P and Willson R C 2011 *Int. J. Biol. Macromol* **48** 392
[90] Tran D T, Knez K, Janssen K P, Pollet J, Spasicand D and Lammertyn J 2013 *Biosens. Bioelectron.* **43** 245
[91] Freitas M, Viswanathan S, Nouws H P A, Oliveira M B P P and Delerue-Matos C 2014 *Biosens. Bioelectron.* **51** 195–200
[92] Pratiwi F W, Rijiravanich P, Somasundrum M and Surareungchai W 2013 *Analyst* **138** 5011–18
[93] Wonsawat W, Limvongjaroen S, Supromma S, Panphut W, Ruecha N, Ratnarathorn N and Dungchai W 2020 *Analyst* **145** 4637–45
[94] Wilson D, Materón E M, Ibáñez-Redín G, Faria R C, Correa D S and Oliveira O N 2019 *Talanta* **194** 611–618
[95] Xu M, Wang R and Li Y 2016 *Talanta* **148** 200–8
[96] Murasova P, Kovarova A, Kasparova J, Brozkova I, Hamiot A, Pekarkova J, Dupuy B, Drbohlavova J, Bilkova Z and Korecka L 2020 *J. Electroanal. Chem.* **863** 114051
[97] Wang D, Chen Q, Huo H, Bai S, Cai G, Lai W and Lin J 2017 *Food Control* **73** 555–61
[98] Fei J, Dou W, Zhao G and Adv R S C 2015 **5** 74548–56
[99] Tolba M, Ahmed M U, Tlili C, Eichenseher F, Loessner M J and Zourob M A 2012 *Analyst* **137** 5749–56
[100] Mutreja R, Jariyal M, Pathania P, Sharma A, Sahoo D K and Suri C R 2016 *Biosens. Bioelectron.* **85** 707–13
[101] Hassan A R H A A, de la Escosura-Muñiz A and Merkoçi A 2015 *Biosens. Bioelectron.* **67** 511–15
[102] Zhu F, Zhao G and Dou W 2018 *Anal. Biochem.* **559** 34–43
[103] Lin Y H, Chen S H, Chuang Y C, Lu Y C, Shen T Y, Chang C A and Lin C S 2008 *Biosens. Bioelectron.* **23** 1832–7
[104] Xu M, Wang R and Li Y 2016 *Analyst* **141** 5441–9
[105] Huang Y, Wu Z, Zhao G and Dou W 2019 *J. Electron. Mater.* **48** 7960–9
[106] Zambito Marsala R, Capri E, Russo E, Bisagni M, Colla R, Lucini L, Gallo A and Suciu N A 2020 *Sci. Total Environ.* **736** 139730
[107] Sjerps R M A, Kooij P J F, van Loon A and Van Wezel A P 2019 *Chemosphere* **235** 510–18

[108] Díaz-González M, Gutiérrez-Capitán M, Niu P, Baldi A, Jiménez-Jorquera C and Fernández-Sánchez C 2016 *Trends Anal. Chem.* **77** 186–202
[109] Arduini F, Cinti S, Caratelli V, Amendola L, Palleschi G and Moscone D 2019 *Biosens. Bioelectron.* **126** 346–54
[110] Lang Q, Han L, Hou C, Wang F and Liu A 2016 *Talanta* 156–57, 34–41
[111] Pérez-Fernández B, Costa-García A and Muñiz A D I E 2020 *Biosensors* **10** 32
[112] Wang X, Lu X, Chen J and Trends J 2014 *Environ. Anal. Chem* **2** 25–32
[113] Hashemi Goradel N, Mirzaei H, Sahebkar A, Poursadeghiyan M, Masoudifar A, Malekshahi Z V and Negahdari B 2018 *J. Cell. Biochem.* **119** 207–12
[114] Nanda Kumar D, Alex S A, Chandrasekaran N and Mukherjee A 2016 *RSC Adv.* **6** 64769–77
[115] Liu Z, Xia X, Zhou G, Ge L and Li F 2020 *Analyst* **145** 2339–44
[116] Kumar T H V and Sundramoorthy A K 2019 *Anal. Chim. Acta* **1074** 131–41
[117] da Silva W, Ghica M E and Brett C M A 2019 *Sens. Actuators,* B **298** 126862
[118] Kampouris D K, Kadara R O, Jenkinsonand N and Banks C E 2009 *Anal. Methods* **1** 25
[119] Hallam P M, Kampouris D K, Kadara R O, Jenkinsonand N and Banks C E 2010 *Anal. Methods* **2** 1152
[120] Zheng R J, Fang Y M, Qin S F, Song J, Wuand A H and Sun J J 2011 *Sens. Actuators,* B **157** 488
[121] Sljukic B R, Kadara R O and Banks C E 2011 *Anal. Methods* **3** 105
[122] Khairy M, Kadaraand R O and Banks C E 2010 *Anal. Methods* **2** 851
[123] Khaled E, Hassan H N A, Girgis A and Metelka R 2008 *Talanta* **77** 737
[124] Hart J P, Abass A K, Cowell D C and Chappell A 1999 *Electroanalysis* **11** 406
[125] Brugnera M F and Trindade M A G 2010 *Anal. Lett.* **43** 2823
[126] Li D W, Li Y T, Song W and Long Y T 2010 *Anal. Methods* **2** 837
[127] Renedo O D, Alonso-Lomillo M A and Martinez M J A 2007 *Talanta* **73** 202
[128] Fusco M D, Tortolini C, Deriu D and Mazzei F 2010 *Talanta* **81** 235
[129] Merkoci A, Anik U, Cevik S, Cubukcuand M and Guix M 2010 *Electroanalysis* **22** 1429
[130] Akyilmaz E, Kozgus O, Türkmen H and Cetinkaya B 2010 *Bioelectrochemistry* **78** 135
[131] Verma M L 2017 *Environ. Chem. Lett.* **15** 555–60
[132] Ahamed M, Verma S, Kumar A and Siddiqui M K J 2005 *Sci. Total Environ.* **346** 48
[133] Fang H L, Zheng H X, Ou M Y, Meng Q, Fan D H and Wang W 2011 *Sens. Actuators,* B **153** 369
[134] Arduini F, Calvo J Q, Amine A, Palleschiand G and Moscone D 2010 *TrAC, Trends Anal. Chem.* **29** 1295
[135] Mandil A and Amine A 2009 *Anal. Lett.* **42** 1245
[136] Zaouak O, Authier L, Cugnet C, Castetbon A and Potin-Gautier M 2010 *Electroanalysis* **22** 1151
[137] Zhang L, Li D W, Song W, Shi L, Li Y and Long Y T 2010 *IEEE Sens. J.* **10** 1583
[138] Zalups R K 2000 *Pharmacol. Rev.* **52** 113
[139] Gong J, Zhou T, Song D, Zhang L and Hu X 2010 *Anal. Chem.* **82** 567
[140] Bernalte E, Sanchez C M and Gil E P 2011 *Anal. Chim. Acta* **689** 60
[141] Verma M L and Rani V 2021 *Environ. Chem. Lett.* **19** 1657–66
[142] Rani V and Verma M L 2020 *Nanosens. Environ. Appl.* **43** 1–28
[143] Thakur M, Wang B and Verma M L 2022 *Environ. Technol. Innov.* **26** 102371

Chapter 17

The role of carbon materials in sensors

Pallavi Singhal

Sensing, as the name suggests is the detection of something. That can be anything, for example, temperature, pressure, weather, material, enzyme, current etc. The role of sensors in today's world is irreplaceable, this comes from the fact that they are used in almost every industry: biology, health, environment, electronics etc. Smart watches, thermometers, blood pressure monitors, and microphones are few examples of sensors that are routinely used. Development of sensors is a well-known field and sensors developed decades ago are still used today. But the field is developing and currently a lot of research is going on to modify their properties.

Nowadays, much attention has been diverted towards metal ion detection [1–7]. The field is important because of the presence of metal ions everywhere being it water, air, or land. These metal ions can reach the human body easily and if the concentration is more than the permissible limit it may have a toxic effect on the body. There are many state-of-the-art techniques to detect metal ions at trace levels such as atomic absorption spectroscopy, inductively coupled plasma mass spectrometry, voltammetry, etc however, all these techniques call for sample pretreatment, trained personnel, and are high cost [3]. Also, often it is important to perform analysis at the site of interest, as many situations do not afford the luxury of the time required for samples to be collected and sent for analysis as they need rapid action. For example, contamination of a large scale drinking water body, monitoring airborne workplace hazards, etc. Considering these limitations, it is important to find efficient methodologies for *in situ* monitoring and a great deal of work is being carried out in this area [5–7]. In this area optical methods have attracted special attention because of their low cost, visual detection, simplicity, and ease of fabrication [1, 5, 7]. Both colorimetric and luminescence methods have been suggested for detecting a variety of metal ions and the detection limit as well as selectivity can be tuned by changing surface properties, binding group, and material. Particularly, Au nanoparticles (NPs) and semiconductor quantum dots (QDs) are some of the optical materials that have been explored extensively and have shown

doi:10.1088/978-0-7503-5127-0ch17

promising applications in this area [3, 8]. However, their toxicity and high cost calls for other environmentally benign low cost optical sensors.

The other area where sensor development is proceeding at an unprecedented rate is biology [6]. It is well-known that early diagnosis is the key for successful treatment of diseases. Therefore, it is important to develop efficient diagnostic techniques. Biosensors are the devices used for this purpose. They are used to analyze organic/inorganic molecules in living organisms and consist of three components: a detector, which recognizes the target analyte signal; a converter, which converts the detected signal into a useful output; and a signal processor, which analyzes and processes the output signal [6]. The key part in a biosensor is a detector which is usually composed of biological receptors. Deoxyribonucleic acid and enzymes are a few examples of biological receptors, however, the vulnerability and degradation of these receptors are major issues and require proper storage conditions otherwise the sensors may become ineffectiveness. Nowadays, with the advent of nanotechnology it is possible to diagnose diseases more effectively at the molecular and cellular levels. Different materials such as metal NPs [9, 10], semiconductor QDs [11], dye-doped materials [12], have been applied for the sensing and study of biomolecules. However, the toxicity of metal and semiconductor QDs limits their application is biology. Similarly dyes are not environmentally friendly and also degrade fast. Therefore, there is a need to develop environmental friendly, nontoxic, and low cost materials which can be used as biosensors.

In light of this carbon dots (CDs) are attaining particular attention because of their low cost, *in situ* detection and tunability. Since their discovery in 2004 by Xu *et al* [13], a large amount of research has been carried out on their synthesis, photoluminescence properties, and applications. They are used extensively in various applications [14] such as biomedicine [15], catalysis [16], energy [17, 18], photo-electronic applications [19], and sensing [20]. To date a large number of methods have been discovered for CDs synthesis and their properties are monitored extensively. These aspects have been discussed in earlier chapters. In this chapter we will primarily focus on the application of CDs in sensing technology. We have divided the application of CDs as sensors in two categories: Chemo-sensors and biosensors. We will discuss each category in detail and the mechanism of detection for each sensing application.

17.1 CDs as chemo-sensors

Chemical detection by CDs is as extensive area of research and several heavy metals such as Fe^{3+} [21–32], Hg^{2+} [33–41], Zn^{2+} [42, 43], Cd^{2+}, Cu^{2+} [44–48], Au^{3+} [49], Co^{2+} [50], Ni^{2+} [51], Pd^{2+} [52, 53], Pb^{2+} [54], Mn^{2+} [55] etc are being detected using these dots. It has been proposed that the optical characteristics of CDs originate from the photo-excitation of the charge carriers in CQDs. The metal ions interact with these photo-generated charge carriers and based on the interaction fluorescence can be either quenched or enhanced. Several mechanisms such as photo induced electron transfer (PET), aggregation, fluorescence resonance energy transfer (FRET), and surface passivation can lead to the above processes [56]. These

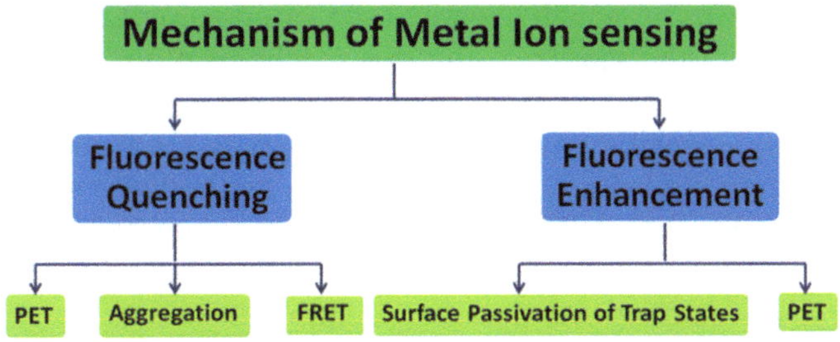

Figure 17.1. Mechanism of metal ion detection by CDs.

processes are depicted in figure 17.1. Largely, the above mechanisms employed for sensing involve interaction between the CDs and target molecules at the intermolecular level.

In fluorescence quenching mechanism, the photo-generated e–h pair recombines non-radiatively. This can occur if either of the photo-generated charge carriers transferred to the metal ion or upon metal ion interaction, the CDs aggregates. Both of these mechanisms quench the fluorescence in CDs thereby leading to their detection. This strategy is particularly designed for the detection of Cu(II) and Fe (III) ions [57, 58]. It has been observed that these ions bind to the carboxyl surface rich CDs and successfully induce effective quenching. The other mechanism which quenches the fluorescence of CDs is FRET. This mechanism occurs when the emission spectra of doner (i.e., CDs) overlap with the absorption spectra of acceptor (i.e., metal ion). The efficiency of this mechanism depends upon the distance between doner and acceptor and decreases with increase in distance. Hg (II) detection via this mechanism has been reported by Cui *et al* based upon CDs-labeled oligodeoxyribonucleotide (ODN) and graphene oxide (GO). The fluorescence of CDs is quenched in the presence of an acceptor, i.e., GO. The addition of Hg (II) ions can induce the recovery of fluorescence as it promotes the release of ODN-CQDs from the GO surface. Hg (II) is known to form a selective TeHg(II) eT duplex in oligonucleotides [59]. The above stated mechanisms detect the metal ions via fluorescence quenching. In many instances metal ion detection via fluorescence enhancement is also being reported [4]. Here, in certain cases metal ions can modify the surface of CDs thereby enhancing their emission yield. The other mechanism by which ion detection via fluorescence enhancement can occur is PET. Here initially the luminescence of CDs is quenched by photoelectron transfer to a bound ion. When a foreign ion is added to the system it binds with the bound ion and luminescence is again regenerated. Fluoride ion sensing by this mechanism is reported by Singhal *et al* [4]. In their study they have synthesized CDs and added Eu^{3+} ions to it. The luminescence gets quenched. Once the fluoride ions are added to this system, the luminescence again regenerated. Metal ion detection via this mechanism has been popularly known as fluorescence *on–off–on* probe. Though many CDs with different functionality have been synthesized to date has been difficult to find suitable CDs which are specific for a particular metal ion and this is one of the major challenges to utilize their properties in a real system.

The other mechanism by which chemical detection using CDs can be realized is chemiluminescence and the sensors using this mechanism are called chemiluminescence chemo-sensors. There are many advantages of this method such as simple instrumentation, high sensitivity, wide linear range, and no interference from the background scattering light. Cu^{2+} detection by this method was reported by Amjadi *et al* [44]. It was observed that $KMnO_4$ generated a stronger chemiluminescence in CDs doped with sulfur and nitrogen. The influence of metal ions in this system was investigated, and it was observed that the chemiluminescence intensity diminished upon exposure to Cu^{2+}. The sensor was also applied for the detection of Cu^{2+} in diluted human plasma and water samples. Detection of organic species is also being carried out using this method [60]. In this area, indomethacin detection by sulfur nitrogen codoped CDs were carried out using various oxidants such as $K_3Fe(CN)_6$, $KMnO_4$, H_2O_2, and $NaIO_4$ and a detection limit of 65 μg L^{-1} was achieved [60].

CDs based electrochemical sensors are also used todays for detection of various chemicals [61–66]. For example, Xi *et al* [62] have developed a hydrogen peroxide sensor using Pd NPs-functionalized N-GQDs@N-doped carbon hollow nanospheres. The sensors showed satisfactory performance in real-time detection of hydrogen peroxide secreted from cancer cells and show an insignificant response to other electroactive species such as cathodic interference from oxygen (O_2) and anodic interferences by dopamine, ascorbic acid, and uric acid. CDs as electrochemical sensors are also used for detection of various organic species [61, 67–71]. CDs particularly GQDs act as multivalent redox species in the electrochemical sensors. Liu *et al* [72] have detected glucose via this mechanism using a uniform three-dimensional graphene nanodots-encaged porous gold electrode and obtained a detection limit of 30 μM.

17.2 CDs as biosensors

CDs have attained particular attention in biosensing because of their non-toxicity and high luminescence yield [73–84]. Based on the detection mechanism; CDs based biosensors are categorized in different categories such as electrochemical biosensors, optical biosensors, piezoelectric biosensors, thermo-electric biosensors etc. In electrochemical biosensors, the target analyte is detected by a bioreceptor which is in direct integration with electrochemical conduction element. CDs are good electron conductors and also have a high surface area that can be used to modify biological receptors. Therefore, CDs are excellent choices for electrochemical biosensor fabrication. In this area Li *et al* [85] constructed graphene quantum dot (GQDs) based electrochemical DNA sensor where pyrolytic graphite electrode was modified with GQDs. Although the performance of the sensor was suitably established the selectivity in complex media, or when there are multiple other sequences of DNA present, is expected to be challenging.

In optical biosensing mechanisms, the luminescence property of CDs is being used. Here the bioreceptor is coupled with CDs and upon analyte interaction a change in emission intensity is recorded. It is important to fabricate CDs with bioreceptor so that a change in emission intensity could be observed. This change in emission occurs either by aggregation of CDs or from electron transfer or FRET mechanism based on the system design. FRET mechanism based biosensor has been designed by Shi *et al* [86] for detecting a gene sequence specific to *Staphylococcus aureus*. They have used graphene QDs (GQDs) and AuNPs for this purpose. For this the GQDs were modified with aminated ssDNA by activating surface carboxylic moieties on the GQDs. The AuNPs were also modified with thiolated DNA probes. It was observed that if the sequence from *S. aureus* was present, it would bridge the probe sequences on the GQDs and that on the AuNPs to bring the two sufficiently close together so that the FRET can be realized. In another study [87], pyrene-functionalized molecular beacons (py-MBs) immobilized GQDs were used for the detection of miRNA. Here molecular beacons were labeled with pyrene and fluorescent dye. The introduction of pyrene to molecular beacons allowed FRET between fluorescent dyes and GQDs. The resulting fluorescent intensity change generated a signal for target miRNA detection with satisfied discrimination abilities in a wide detection range from 0.1 to 200 nM.

17.3 Conclusions and future perspective

In this chapeter we shed a light upon use of CDs as sensors in both chemical and biological detection. Although numerous studies have been carried out using CDs as sensors however, synthetic control, selectivety, and integration in real devices are some of the challenges that need to be resolved. In today's world because of the industrial revolution a lot of metal ions are released into the enviornment. These metal ions reach the human body and pose a serious hazard. Use of CDs based sensors have undoubtedly shown enhanced sensitivity and selectivity for these toxic targets. However, a lot needs to be done. Similarly in biosensing also CDs have shown their applications but detection and selectively in real systems are challenging tasks. CDs are reported to have good interactions with biological cells and organelles. Currently, most sensing techniques realize the detection of target analytes by combining CDs with enzymes, antibodies, and nucleic acids. It is important to develop CDs with some intrinsic properties that could mimic those of biomacromolecules (i.e., enzymes, antibodies, and nucleic acids), thus realizing most detections in a label-free fashion. Despite many advantages; a control over CDs synthesis, origin of their luminesence and mechanism of interaction are still debatable and more work is needed in this direction.

References

[1] Singhal P and Jha S K 2019 A semi quantitative visual probe for fluoride ion sensing in aqueous medium *J. Lumin.* **206** 113–9

[2] Singhal P, Jha S K, Pandey S P and Neogy S 2017 Rapid extraction of uranium from sea water using Fe_3O_4 and humic acid coated Fe_3O_4 nanoparticles *J. Hazard. Mater.* **335** 152–61

[3] Singhal P, Jha S K, Vats B G and Ghosh H N 2017 Electron-transfer-mediated uranium detection using quasi-type II core–shell quantum dots: insight into mechanistic pathways *Langmuir* **33** 8114–22

[4] Singhal P, Vats B G, Jha S K and Green N S 2017 Water-dispersible photoluminescent on–off–on probe for selective detection of fluoride ions *ACS Appl. Mater. Interfaces* **9** 20536–44
[5] Li M, Chen T, Gooding J J and Liu J 2019 Review of carbon and graphene quantum dots for sensing *ACS Sens.* **4** 1732–48
[6] Ji C, Zhou Y, Leblanc R M and Peng Z 2020 Recent developments of carbon dots in biosensing: a review *ACS Sens.* **5** 2724–41
[7] Krishnan S K, Singh E, Singh P, Meyyappan M and Nalwa H S 2019 A review on graphene-based nanocomposites for electrochemical and fluorescent biosensors *RSC Adv.* **9** 8778–881
[8] Saha K, Agasti S S, Kim C, Li X and Rotello V M 2012 Gold nanoparticles in chemical and biological sensing *Chem. Rev.* **112** 2739–79
[9] Zou T, Xing X, Yang Y, Wang Z, Wang Z, Zhao R, Zhang X and Wang Y 2020 Water-soluble ZnO quantum dots modified by (3-aminopropyl)triethoxysilane: the promising fluorescent probe for the selective detection of Cu^{2+} ion in drinking water *J. Alloys Compd.* **825** 153904
[10] Liu W, Zhang X, Wang Z, Wang R, Chen C and Dong C 2019 Nanoparticles assembled CdIn2O4 spheres with high sensing properties towards n-butanol *Nanomaterials.* **9** 1714
[11] Yola M L and Atar N 2019 Development of cardiac troponin-I biosensor based on boron nitride quantum dots including molecularly imprinted polymer *Biosens. Bioelectron.* **126** 418–24
[12] Gubala V, Giovannini G, Kunc F, Monopoli M P and Moore C J 2020 Dye-doped silica nanoparticles: synthesis, surface chemistry and bioapplications *Cancer Nanotechnol.* **11** 1
[13] Xu X, Ray R, Gu Y, Ploehn H J, Gearheart L, Raker K and Scrivens W A 2004 Electrophoretic analysis and purification of fluorescent single-walled carbon nanotube fragments *J. Am. Chem. Soc.* **126** 12736–7
[14] Islam M S, Deng Y, Tong L, Roy A K, Faisal S N, Hassan M, Minett A I and Gomes V G 2017 In-situ direct grafting of graphene quantum dots onto carbon fibre by low temperature chemical synthesis for high performance flexible fabric supercapacitor *Mater. Today Commun.* **10** 112–9
[15] Zhu S *et al* 2011 Strongly green-photoluminescent graphene quantum dots for bioimaging applications *Chem. Commun.* **47** 6858–60
[16] Li H, He X, Kang Z, Huang H, Liu Y, Liu J, Lian S, Tsang C H A, Yang X and Lee S-T 2010 Water-soluble fluorescent carbon quantum dots and photocatalyst design *Angew. Chem. Int. Ed.* **49** 4430–4
[17] Mondal S, Rana U and Malik S 2015 Graphene quantum dot-doped polyaniline nanofiber as high performance supercapacitor electrode materials *Chem. Commun.* **51** 12365–8
[18] Liu W-W, Feng Y-Q, Yan X-B, Chen J-T and Xue Q-J 2013 Superior micro-supercapacitors based on graphene quantum dots *Adv. Funct. Mater.* **23** 4111–22
[19] Shen J, Zhu Y, Yang X, Zong J, Zhang J and Li C 2012 One-pot hydrothermal synthesis of graphene quantum dots surface-passivated by polyethylene glycol and their photoelectric conversion under near-infrared light *New J. Chem.* **36** 97–101
[20] Wang S, Zhang Y, Pang G, Zhang Y and Guo S 2017 Tuning the aggregation/disaggregation behavior of graphene quantum dots by structure-switching aptamer for high-sensitivity fluorescent ochratoxin a sensor *Anal. Chem.* **89** 1704–9
[21] Qu F, Wang S, Liu D and You J 2015 Differentiation of multi-metal ions based on fluorescent dual-emission carbon nanodots *RSC Adv.* **5** 82570–5
[22] Liu M, Xu Y, Niu F, Gooding J J and Liu J 2016 Carbon quantum dots directly generated from electrochemical oxidation of graphite electrodes in alkaline alcohols and the applications for specific ferric ion detection and cell imaging *Analyst* **141** 2657–64

[23] Gong X, Lu W, Paau M C, Hu Q, Wu X, Shuang S, Dong C and Choi M M F 2015 Facile synthesis of nitrogen-doped carbon dots for Fe3+ sensing and cellular imaging *Anal. Chim. Acta* **861** 74–84

[24] Aslandaş A M, Balcı N, Arık M, Şakiroğlu H, Onganer Y and Meral K 2015 Liquid nitrogen-assisted synthesis of fluorescent carbon dots from blueberry and their performance in Fe3+ detection *Appl. Surf. Sci.* **356** 747–52

[25] Patidar R, Rebary B, Bhadu G R and Paul P 2016 Fluorescent carbon nanoparticles as label-free recognizer of Hg^{2+} and Fe^{3+} through effective fluorescence quenching in aqueous media *J. Lumin.* **173** 243–9

[26] Xu J, Zhou Y, Liu S, Dong M and Huang C 2014 Low-cost synthesis of carbon nanodots from natural products used as a fluorescent probe for the detection of ferrum(III) ions in lake water *Anal. Methods* **6** 2086–90

[27] Singh V and Mishra A K 2016 Green and cost-effective fluorescent carbon nanoparticles for the selective and sensitive detection of iron (III) ions in aqueous solution: mechanistic insights and cell line imaging studies *Sensors Actuators* B **227** 467–74

[28] Chen Y, Wu Y, Weng B, Wang B and Li C 2016 Facile synthesis of nitrogen and sulfur co-doped carbon dots and application for Fe(III) ions detection and cell imaging *Sensors Actuators* B **223** 689–96

[29] Zhou M, Zhou Z, Gong A, Zhang Y and Li Q 2015 Synthesis of highly photoluminescent carbon dots via citric acid and Tris for iron(III) ions sensors and bioimaging *Talanta* **143** 107–13

[30] Chandra S, Laha D, Pramanik A, Ray Chowdhuri A, Karmakar P and Sahu S K 2016 Synthesis of highly fluorescent nitrogen and phosphorus doped carbon dots for the detection of Fe^{3+} ions in cancer cells *Luminescence.* **31** 81–7

[31] Shen C, Sun Y, Wang J and Lu Y 2014 Facile route to highly photoluminescent carbon nanodots for ion detection, pH sensors and bioimaging *Nanoscale.* **6** 9139–47

[32] Xu J, Lai T, Feng Z, Weng X and Huang C 2015 Formation of fluorescent carbon nanodots from kitchen wastes and their application for detection of Fe^{3+} *Luminescence.* **30** 420–4

[33] Zhang R and Chen W 2014 Nitrogen-doped carbon quantum dots: facile synthesis and application as a 'turn-off' fluorescent probe for detection of Hg^{2+} ions *Biosens. Bioelectron.* **55** 83–90

[34] Wan X, Li S, Zhuang L and Tang J 2016 l-Tryptophan-capped carbon quantum dots for the sensitive and selective fluorescence detection of mercury ion in aqueous solution *J. Nanopart. Res.* **18** 202

[35] Liu T, Dong J X, Liu S G, Li N, Lin S M, Fan Y Z, Lei J L, Luo H Q and Li N B 2017 Carbon quantum dots prepared with polyethyleneimine as both reducing agent and stabilizer for synthesis of Ag/CQDs composite for Hg^{2+} ions detection *J. Hazard. Mater.* **322** 430–6

[36] Li Z, Ni Y and Kokot S 2015 A new fluorescent nitrogen-doped carbon dot system modified by the fluorophore-labeled ssDNA for the analysis of 6-mercaptopurine and Hg (II) *Biosens. Bioelectron.* **74** 91–7

[37] Patir K and Gogoi S K 2018 Facile synthesis of photoluminescent graphitic carbon nitride quantum dots for Hg^{2+} detection and room temperature phosphorescence *ACS Sust. Chem. Eng.* **6** 1732–43

[38] Gupta A, Chaudhary A, Mehta P, Dwivedi C, Khan S, Verma N C and Nandi C K 2015 Nitrogen-doped, thiol-functionalized carbon dots for ultrasensitive Hg(II) detection *Chem. Commun.* **51** 10750–3

[39] Qin X, Lu W, Asiri A M, Al-Youbi A O and Sun X 2013 Microwave-assisted rapid green synthesis of photoluminescent carbon nanodots from flour and their applications for sensitive and selective detection of mercury(II) ions *Sensors Actuators* B **184** 156–62

[40] Gui R, Jin H, Wang Z, Zhang F, Xia J, Yang M, Bi S and Xia Y 2015 Room-temperature phosphorescence logic gates developed from nucleic acid functionalized carbon dots and graphene oxide *Nanoscale.* **7** 8289–93

[41] Xu X-Y and Yan B 2016 Fabrication and application of a ratiometric and colorimetric fluorescent probe for Hg^{2+} based on dual-emissive metal–organic framework hybrids with carbon dots and Eu^{3+} *J. Mater. Chem.* C **4** 1543–9

[42] Zhang Z, Shi Y, Pan Y, Cheng X, Zhang L, Chen J, Li M-J and Yi C 2014 Quinoline derivative-functionalized carbon dots as a fluorescent nanosensor for sensing and intracellular imaging of Zn^{2+} *J. Mater. Chem.* B **2** 5020–7

[43] Zhang Z, Sun W and Wu P 2015 Highly photoluminescent carbon dots derived from egg white: facile and green synthesis, photoluminescence properties, and multiple applications *ACS Sustain. Chem. Eng.* **3** 1412–8

[44] Amjadi M, Manzoori J L, Hallaj T and Azizi N 2017 Sulfur and nitrogen co-doped carbon quantum dots as the chemiluminescence probe for detection of Cu^{2+} ions *J. Lumin.* **182** 246–51

[45] Hu C, Yu C, Li M, Wang X, Yang J, Zhao Z, Eychmuller A, Sun Y-P and Qiu J 2014 Chemically tailoring coal to fluorescent carbon dots with tuned size and their capacity for Cu (II) detection *Small.* **10** 4926–33

[46] Yang S, Sun J, Li X, Zhou W, Wang Z, He P, Ding G, Xie X, Kang Z and Jiang M 2014 Large-scale fabrication of heavy doped carbon quantum dots with tunable-photoluminescence and sensitive fluorescence detection *J. Mater. Chem.* A **2** 8660–7

[47] Vedamalai M, Periasamy A P, Wang C-W, Tseng Y-T, Ho L-C, Shih C-C and Chang H-T 2014 Carbon nanodots prepared from o-phenylenediamine for sensing of Cu^{2+} ions in cells *Nanoscale.* **6** 13119–25

[48] Ahmad K, Gogoi S K, Begum R, Sk M P, Paul A and Chattopadhyay A 2017 An interactive quantum dot and carbon dot conjugate for pH-sensitive and ratiometric Cu^{2+} sensing *ChemPhysChem* **18** 610–6

[49] Gu J, Hu D, Wang W, Zhang Q, Meng Z, Jia X and Xi K 2015 Carbon dot cluster as an efficient 'off–on' fluorescent probe to detect Au(III) and glutathione *Biosens. Bioelectron.* **68** 27–33

[50] Shi J, Lu C, Yan D and Ma L 2013 High selectivity sensing of cobalt in HepG2 cells based on necklace model microenvironment-modulated carbon dot-improved chemiluminescence in Fenton-like system *Biosens. Bioelectron.* **45** 58–64

[51] Tan X W, Romainor A N B, Chin S F and Ng S M 2014 Carbon dots production via pyrolysis of sago waste as potential probe for metal ions sensing *J. Anal. Appl. Pyrolysis* **105** 157–65

[52] Sharma V, Saini A K and Mobin S M 2016 Multicolour fluorescent carbon nanoparticle probes for live cell imaging and dual palladium and mercury sensors *J. Mater. Chem.* B **4** 2466–76

[53] Xiong C, Liang W, Wang H, Zheng Y, Zhuo Y, Chai Y and Yuan R 2016 *In situ* electropolymerization of nitrogen doped carbon dots and their application in an electrochemiluminescence biosensor for the detection of intracellular lead ions *Chem. Commun.* **52** 5589–92

[54] Qian Z S, Shan X Y, Chai L J, Chen J R and Feng H 2015 A fluorescent nanosensor based on graphene quantum dots–aptamer probe and graphene oxide platform for detection of lead (II) ion *Biosens. Bioelectron.* **68** 225–31
[55] Gogoi N, Barooah M, Majumdar G and Chowdhury D 2015 Carbon dots rooted agarose hydrogel hybrid platform for optical detection and separation of heavy metal ions *ACS Appl. Mater. Interfaces* **7** 3058–67
[56] Devi P, Rajput P, Thakur A, Kim K-H and Kumar P 2019 Recent advances in carbon quantum dot-based sensing of heavy metals in water *TrAC, Trends Anal. Chem.* **114** 171–95
[57] Salinas-Castillo A *et al* 2013 Carbon dots for copper detection with down and upconversion fluorescent properties as excitation sources *Chem. Commun.* **49** 1103–5
[58] Qu K, Wang J, Ren J and Qu X 2013 Carbon dots prepared by hydrothermal treatment of dopamine as an effective fluorescent sensing platform for the label-free detection of Iron(III) ions and dopamine *Chem.—A Eur. J.* **19** 7243–9
[59] Cui X, Zhu L, Wu J, Hou Y, Wang P, Wang Z and Yang M 2015 A fluorescent biosensor based on carbon dots-labeled oligodeoxyribonucleotide and graphene oxide for mercury (II) detection *Biosens. Bioelectron.* **63** 506–12
[60] Hallaj T, Amjadi M, Manzoori J L and Azizi N 2017 A novel chemiluminescence sensor for the determination of indomethacin based on sulfur and nitrogen co-doped carbon quantum dot–KMnO4 reaction *Luminescence.* **32** 1174–9
[61] Nguyen H V *et al* 2016 Electrochemical sensing of etoposide using carbon quantum dot modified glassy carbon electrode *Analyst* **141** 2665–75
[62] Xi J, Xie C, Zhang Y, Wang L, Xiao J, Duan X, Ren J, Xiao F and Wang S 2016 Pd nanoparticles decorated n-doped graphene quantum dots@N-doped carbon hollow nanospheres with high electrochemical sensing performance in cancer detection *ACS Appl. Mater. Interfaces* **8** 22563–73
[63] Ju J and Chen W 2015 *In situ* growth of surfactant-free gold nanoparticles on nitrogen-doped graphene quantum dots for electrochemical detection of hydrogen peroxide in biological environments *Anal. Chem.* **87** 1903–10
[64] Xu Q, Yuan H, Dong X, Zhang Y, Asif M, Dong Z, He W, Ren J, Sun Y and Xiao F 2018 Dual nanoenzyme modified microelectrode based on carbon fiber coated with AuPd alloy nanoparticles decorated graphene quantum dots assembly for electrochemical detection in clinic cancer samples *Biosens. Bioelectron.* **107** 153–62
[65] Ran X, Qu Q, Qian X, Xie W, Li S, Li L and Yang L 2018 Water-soluble pillar are ne functionalized nitrogen-doped carbon quantum dots with excellent supramolecular recognition capability and superior electrochemical sensing performance towards TNT *Sensors Actuators* B **257** 362–71
[66] Chen D, Zhuang X, Zhai J, Zheng Y, Lu H and Chen L 2018 Preparation of highly sensitive Pt nanoparticles-carbon quantum dots/ionic liquid functionalized graphene oxide nanocomposites and application for H2O2 detection *Sensors Actuators* B **255** 1500–6
[67] Yang S, Liang J, Luo S, Liu C and Tang Y 2013 Supersensitive detection of chlorinated phenols by multiple amplification electrochemiluminescence sensing based on carbon quantum dots/graphene *Anal. Chem.* **85** 7720–5
[68] Rao H, Ge H, Wang X, Zhang Z, Liu X, Yang Y, Liu Y, Liu W, Zou P and Wang Y 2017 Colorimetric and fluorometric detection of protamine by using a dual-mode probe consisting of carbon quantum dots and gold nanoparticles *Microchim. Acta* **184** 3017–25

[69] Hasanzadeh M, Hashemzadeh N, Shadjou N, Eivazi-Ziaei J, Khoubnasabjafari M and Jouyban A 2016 Sensing of doxorubicin hydrochloride using graphene quantum dot modified glassy carbon electrode *J. Mol. Liq.* **221** 354–7

[70] Arvand M and Hemmati S 2017 Magnetic nanoparticles embedded with graphene quantum dots and multiwalled carbon nanotubes as a sensing platform for electrochemical detection of progesterone *Sensors Actuators* B **238** 346–56

[71] Arvand M, Abbasnejad S and Ghodsi N 2016 Graphene quantum dots decorated with Fe3O4 nanoparticles/functionalized multiwalled carbon nanotubes as a new sensing platform for electrochemical determination of l-DOPA in agricultural products *Anal. Methods* **8** 5861–8

[72] Wang J, Zhu H, Xu Y, Yang W, Liu A, Shan F, Cao M and Liu J 2015 Graphene nanodots encaged 3-D gold substrate as enzyme loading platform for the fabrication of high performance biosensors *Sensors Actuators* B **220** 1186–95

[73] Chenthattil R and Manjunatha J G 2018 Sensitive electrochemical analysis of resorcinol using polymer modified carbon paste electrode: a cyclic voltammetric study *Anal. Bioanal. Electrochem.* **10** 488–98

[74] Hareesha N and Manjunatha J G 2020 Elevated and rapid voltammetric sensing of riboflavin at poly(helianthin dye) blended carbon paste electrode with heterogeneous rate constant elucidation *J. Iran. Chem. Soc.* **17** 1507–19

[75] Hareesha N and Manjunatha J G 2020 Fast and enhanced electrochemical sensing of dopamine at cost-effective poly(DL-phenylalanine) based graphite electrode *J. Electroanal. Chem.* **878** 114533

[76] Hareesha N and Manjunatha J G 2020 A simple and low-cost poly (DL-phenylalanine) modified carbon sensor for the improved electrochemical analysis of riboflavin *J. Sci.: Adv. Mater. Devices.* **5** 502–11

[77] Hareesha N, Manjunatha J G, Raril C and Tigari G 2019 Sensitive and selective electrochemical resolution of tyrosine with ascorbic acid through the development of electropolymerized alizarin sodium sulfonate modified carbon nanotube paste electrodes *ChemistrySelect.* **4** 4559–67

[78] Hareesha N, Manjunatha J G G, Raril C and Tigari G 2019 Design of novel surfactant modified carbon nanotube pasteelectrochemical sensor for the sensitive investigation of tyrosineas a pharmaceutical drug *Adv. Pharm. Bull.* **9** 132–7

[79] Manjunatha J G 2020 A surfactant enhanced graphene paste electrode as an effective electrochemical sensor for the sensitive and simultaneous determination of catechol and resorcinol *Chem. Data Coll.* **25** 100331

[80] Manjunatha J G, Swamy K and Awitdrus A 2013 Electrochemical studies of dopamine, ascorbic acid and their simultaneous determination at a poly (rosaniline) modified carbon paste electrode: a cyclic voltammetric study *Anal. Bioanal. Electrochem.* **5** 426–38

[81] Manjunatha J G, Swamy K, Awitdrus A and Mamatha G 2013 Simultaneous determination of ascorbic acid, dopamine and uric acid at poly (aniline blue) modified carbon paste electrode: a cyclic voltammetric study *Int. J. Pharm. Pharm. Sci.* **5** 355–61

[82] Manjunatha J G, Swamy K, Mamatha G, Gilbert O and Sherigara B 2011 Poly(maleic acid) modified carbon paste electrode for simultaneous detection of dopamine in the presence of uric acid: a cyclic voltammetric study *Anal. Bioanal. Electrochem.* **3** 146–59

[83] Tigari G and Manjunatha J G 2020 A surfactant enhanced novel pencil graphite and carbon nanotube composite paste material as an effective electrochemical sensor for determination of riboflavin *J. Sci.: Adv. Mater. Devices.* **5** 56–64
[84] Tigari G and Manjunatha J G 2019 Electrochemical preparation of poly(arginine)-modified carbon nanotube paste electrode and its application for the determination of pyridoxine in the presence of riboflavin: an electroanalytical approach *J. Anal. Test.* **3** 331–40
[85] Zhao J, Chen G, Zhu L and Li G 2011 Graphene quantum dots-based platform for the fabrication of electrochemical biosensors *Electrochem. Commun.* **13** 31–3
[86] Shi J, Chan C, Pang Y, Ye W, Tian F, Lyu J, Zhang Y and Yang M 2015 A fluorescence resonance energy transfer (FRET) biosensor based on graphene quantum dots (GQDs) and gold nanoparticles (AuNPs) for the detection of mecA gene sequence of *Staphylococcus aureus Biosens. Bioelectron.* **67** 595–600
[87] Zhang H *et al* 2015 Universal fluorescence biosensor platform based on graphene quantum dots and pyrene-functionalized molecular beacons for detection of MicroRNAs *ACS Appl. Mater. Interfaces* **7** 16152–6

Chapter 18

Prospects of carbon based sensors for hormones detection

Didem Nur Unal, Ozge Selcuk, Cem Erkmen, Sevinc Kurbanoglu and Bengi Uslu

Carbon is a unique element with allotropes such as graphite, fullerenes, and diamond, thanks to its different hybridization abilities. Along with nanotechnological developments, carbon nanomaterials are the materials that are frequently encountered with many reviews, books, and research studies in electrode modification in electrochemical sensors. They are widely used as electrode modifiers due to their high electrical conductivity, chemical stability, large surface area, antifouling resistance, and great adsorption abilities against various analytes. Electrochemical sensors are one of the important scientific developments in recent years that continue to develop thanks to their features such as fast response times for analyzing various analytes, low-cost production, easy use, and directing chip production by integrating into miniature systems. Hormones are known as special chemicals that are secreted from endocrine cells in the body through glands, diffusion, circulation, and affect the activities of target cells. Their analyses are important as they are biologically important substances that regulate the functioning of organs in the human body. In this chapter, types of hormones, their functions in the body, and carbon-based nanomaterials, including fullerenes, graphene, carbon nanotubes, carbon nanowires, carbon fibers, carbon dots, are discussed. Moreover, electrode modification with carbon nanoparticles and applications of carbon-based electrochemical sensors in hormone analysis is also summarized.

18.1 Introduction

For bodily functions to work correctly, the various parts and organs of the body must communicate with each other to maintain homeostasis. This communication is also necessary for the organism to respond appropriately to internal and external environment changes. The two central systems responsible for controlling and

doi:10.1088/978-0-7503-5127-0ch18

communicating with the body are the neurological and endocrine systems. The endocrine organs secrete hormones that circulate throughout the body and control several responses, including fertility, growth, and development. The nervous system typically facilitates the transmission of information between different parts of the body in a matter of seconds. Hormonal communication, based on the production and release of hormones from various glands and their transport through the bloodstream, is better suited to situations that require more extensive and prolonged regulatory action. The two communication systems, therefore, complement each other [1].

18.1.1 Hormones

Hormones are special chemicals that are secreted from endocrine cells in the body through glands, diffusion, circulation, and affect the activities of target cells. Hormones produced by humans, animals, and plants have important roles in functioning living organisms and regulating many physiological conditions. Hormones are produced and secreted in the hypothalamus, pituitary gland, thyroid gland, pineal gland, pancreas, adrenal gland, ovaries, and testes. The hormone must be transported from the cell in which it is produced to the tissue (target tissue) in which it is to act. Hormones are usually named according to the tissue in which they are first found or their main effects. However, it is now known that the same hormone is produced in different tissues. It is important to understand the functions of these important chemicals, and that is why they are divided into various groups [1] (figure 18.1).

18.1.1.1 Hypothalamus

The hypothalamus is an organ in the brain and secretes some hormones. There are hormones called neurotransmitters that enable communication between nerves in regions of the brain other than the hypothalamus [2, 3]. Its primary function is to keep the body in metabolic balance. Therefore, it regulates blood pressure, body temperature, fluid–electrolyte balance, and body weight. The function of the

Figure 18.1. Classification of hormones.

hormones secreted by the hypothalamus is to ensure the secretion of hormones from the pituitary gland [3–5].

These hormones are called regulatory hormones or factors [6]:

- Gonadotropin-releasing hormone (GnRH) secretes the hormones FSH and LH from the pituitary gland.
- Growth hormone-releasing hormone (GHRH) secretes growth hormones from the pituitary gland.
- TSH-releasing hormone (TRH) secretes the hormone TSH from the pituitary gland.
- Corticotropin-releasing hormone (CRH) releases the hormone ACTH (also called corticotropin) from the pituitary gland.
- Dopamine inhibits prolactin secretion from the pituitary gland.
- Somatostatin inhibits the secretion of growth hormone and TSH hormone from the pituitary gland. Somatostatin is also secreted by the pancreas, intestinal membrane (mucosa), and thyroid parafollicular C cells. In addition to growth hormone, it prevents the secretion of many hormones such as insulin, glucagon, gastrin, and secretin.

18.1.1.2 Pituitary gland

The pituitary gland consists of two parts, the anterior pituitary, and the posterior pituitary. Thanks to the hormones secreted from the anterior pituitary, other glands in our body work and enable them to make hormones. In other words, the pituitary gland controls all the glands in the body like an orchestra conductor [2, 3].

The hormones secreted from the anterior pituitary are [2]:

- Follicle-stimulating hormone (FSH): stimulates the growth of follicle cells and the release of estrogen from follicle cells during the menstrual cycle in females.
- Luteinizing hormone (LH): is a gonadotropic hormone that stimulates ovulation. The target organ is the interstitial cells in the testes in males, which secrete testosterone, and in females, the ovaries. LH secretion is controlled by a gonadotropin-releasing hormone, which is regulated in feedback by hormones such as progesterone, estrogen, and testosterone.
- Prolactin: has two functions in women; with estrogen, it stimulates the development of mammary ducts in the mammary glands during pregnancy and stimulates milk production in the mammary tissue after childbirth.
- Growth hormone: this does not have a specific target organ but stimulates growth by affecting all body parts. Its most apparent effect is accelerating growth by increasing tissue mass (protein synthesis) and stimulating cell division in children and adolescents and providing continuity of the epiphyseal plates by acting directly on the epiphyseal plates of the long bones.
- Adrenocorticotropic hormone (ACTH): increases the production and secretion of steroid hormones called glucocorticoids from the adrenal gland. ACTH secretion is regulated by the corticotropin-releasing hormone secreted from the hypothalamus. Stress, ADH, and other hormones regulate

corticotropin-releasing hormone secretion by feedback mechanisms, increasing the secretion of ACTH.
- Thyroid-stimulating hormone (TSH): stimulates the secretion and synthesis of thyroid gland hormones, and secretion is controlled by a thyrotropin-releasing hormone produced in the hypothalamus.

There are two hormones secreted from the posterior pituitary [4]:
- Anti-diuretic hormone (ADH) or vasopressin: the main task is to contribute to the regulation of body fluid balance; its target organ is the kidneys. It increases water reabsorption from the kidneys and reduces water excretion through urine. In cases where the amount of blood fluid decreases, such as dehydration and coma, ADH secretion increases.
- Oxytocin: a hormone that stimulates the contraction of the uterine muscle during childbirth and the secretion of milk after birth.

18.1.1.3 Pineal gland
The hormone melatonin is secreted by a gland in the brain called the pineal gland. Melatonin allows the body to adjust to the differences between day and night. Melatonin is produced in the pineal gland by the conversion of the amino acid tryptophan to serotonin, which in turn is converted to melatonin. Melatonin acts under the control of the suprachiasmatic nucleus in the hypothalamus [7, 8].

18.1.1.4 Thyroid gland
The thyroid gland is an organ located in the front part of our neck. The job of the thyroid gland is to produce, store, and release thyroid hormones into the blood when needed, thus regulating our metabolism. The thyroid gland secretes the hormones T3 and T4 [5].

18.1.1.5 Parathyroid gland
Parathyroid hormone is secreted by the parathyroid gland. The secretion of the parathyroid hormone depends on the level of calcium in the blood. When the calcium level in the blood is low, parathyroid hormone is secreted. This hormone increases the level of calcium in the blood by acting directly on the kidneys and bones and indirectly on the intestines. When the blood calcium level is high, less parathyroid hormone is secreted [9].

The kidney also plays an important role in regulating the level of calcium in the blood. Parathyroid hormone contributes to the activation of vitamin D in the kidney. Parathyroid hormones affect the reabsorption of calcium, which is filtered by the kidney. Parathyroid hormones affect the bones and ensure the excretion of calcium and phosphorus from the bones. Parathyroid hormone also increases calcium absorption from the intestine by vitamin D. Parathyroid hormone increases calcium absorption through the kidneys and increases phosphate excretion through the urine. Skeletal, intestinal, renal, parathyroid hormones, and vitamin D play an important role in keeping calcium balanced in the body, especially in the blood [10].

18.1.1.6 Adrenal glands

The adrenal glands have an outer part called the 'cortex' and an inner part called the 'medulla.' The hormones adrenaline and noradrenaline are produced in the inner part, the medulla. Hormones such as cortisol, aldosterone, DHEA, and testosterone are produced in the outer part, the cortex [11].

The adrenal gland uses cholesterol from the blood to make these hormones. The uptake of cholesterol by the adrenal gland and the production of the hormones are stimulated by the hormone ACTH secreted by the pituitary gland [12].

18.1.1.7 Testes

The testes not only produce sperm but also secrete the male hormone testosterone. Spermatogenesis and testosterone secretion are controlled by the hormones FSH and LH secreted by the pituitary gland.

Cells in the testes called Leydig cells produce testosterone. The hypothalamus in the brain tells the pituitary gland how much testosterone to produce. For this purpose, the hypothalamus secretes the hormone GnRH, which goes to the pituitary gland. The GnRH hormone stimulates the secretion of the hormones FSH and LH from the pituitary gland. The hormone LH secreted by the pituitary gland increases the secretion of testosterone from the Leydig cells, while the hormone FSH secreted by the pituitary gland ensures sperm production in the seminiferous tubules. The secreted testosterone reduces LH secretion from the pituitary gland [3, 13].

18.1.1.8 Ovaries

The main hormones secreted by the ovary are estrogen, progesterone, and androgen, all of which are made from cholesterol. For these hormones to be secreted by the ovary, adequate and regular secretion of the hormones FSH and LH from the pituitary gland is required [6, 14, 15].

Hormones circulate in the body at low concentrations, and therefore, it is very important to develop analytical instruments for hormone analysis that allow analysis with high sensitivity and low detection limits. Thanks to the numerous advantages they provide, carbon-based electrochemical sensors have become very popular today for the analysis of these substances, thanks to the possibility of easy, low cost, sensitive, and low detection limits. In addition, considering that hormone levels are important in the diagnosis of disease, the combination of sciences such as health and electroanalytical chemistry is very important [16–18].

18.1.2 Carbon-based electrochemical sensors

Electrochemical sensors are one of the important scientific developments in recent years that continue to develop thanks to their features such as fast response times for analyzing various analytes, low-cost production, easy use, and directing chip production by integrating into miniature systems. Electrochemical sensors basically consist of a converter system in which various electroanalytical techniques such as voltammetric [19, 20], impedimetric [21, 22], and amperometric [23] are used, and a working electrode that provides electrochemical signal transmission and can be

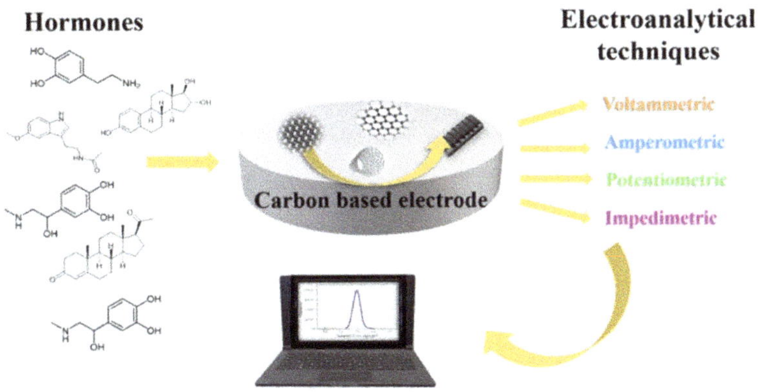

Scheme 18.1. Electrochemical carbon-based sensors for hormone detection.

developed specifically for the analyte. The meeting of electrochemical sensors with nanometer-sized materials and systems reveals the definition of nanosensor; with nanosensors using electrodes modified with different nanomaterials, the possibility of making more selective, sensitive, and faster analyzes of various pharmaceutical, biological and chemical compounds increases.

Carbon is a unique element with allotropes such as graphite, fullerenes, and diamond, thanks to its different hybridization abilities. Along with nanotechnological developments, carbon nanomaterials are the materials that are frequently encountered with many reviews, books, and research studies in electrode modification in electrochemical sensors. They are widely used as electrode modifiers due to their high electrical conductivity, chemical stability, large surface area, antifouling resistance, and great adsorption abilities against various analytes [24].

Since carbon can form different surface bonds, it has a complex surface compared to other materials. The importance of surface events in electrochemistry increases the importance of carbon electrodes. They are frequently used in electroanalytical applications due to their low cost and the possibility of working in a wide potential range [25]. Carbon can bond with many functional groups on its surface to place many analytes on the electrode surface. Carbon electrodes modified with different materials such as biological molecules, carbon, and metal nanomaterials have made the electrochemical analysis of many hormones possible (scheme 18.1). Carbon electrodes that are frequently used in electrochemical hormone analysis are glassy carbon electrodes (GCE) [26], carbon paste electrodes (CPE) [27], screen-printed carbon electrodes (SPE) [17, 28], and pencil graphite electrodes (PGE).

18.2 Carbon-based nanomaterials

Since carbon is one of the versatile elements with different hybridization abilities such as sp, $sp^{2,}$ and sp^3, it is used in many fields. The chains of different lengths and electronic properties create important electronic, optical, mechanical, and chemical carbon-based nanomaterials [29]. There are different types of carbon nanomaterials such as fullerenes [30], carbon dots [31], carbon nanotubes [32], carbon nanofibers

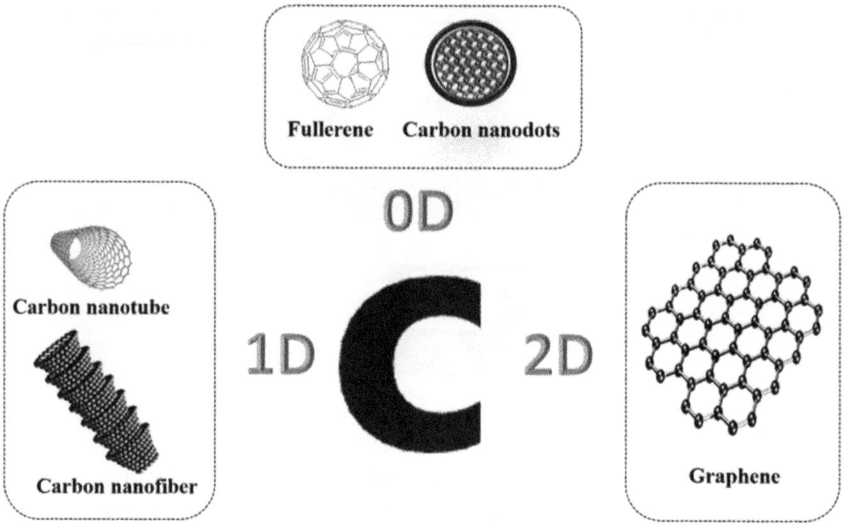

Figure 18.2. 0D, 1D, and 2D carbon-based nanomaterials.

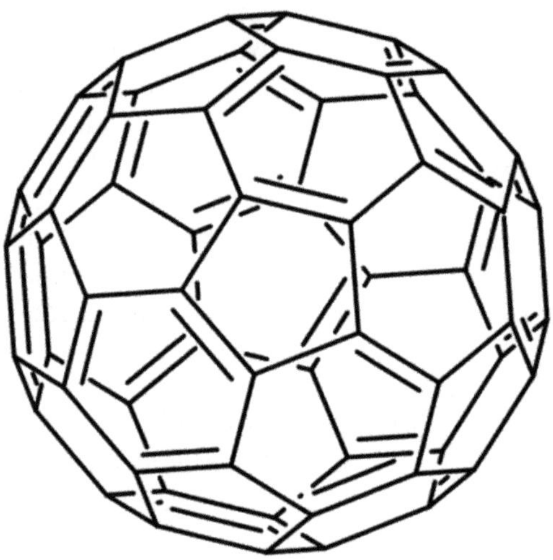

Figure 18.3. Structure of fullerenes.

[33], graphene [34], 0-dimensional (0D), 1-dimensional (1D), 2-dimensional (2D) (figure 18.2).

18.2.1 Fullerenes

Fullerenes or C60 is a class of zero-dimensional allotropes of carbon consisting of 60 carbon atoms. It has attracted attention in different fields since it was discovered in 1985 [30]. The hollow sphere or hexagonal structure formed by carbon atoms is shown in figure 18.3. The combination of five or six rings of fullerenes with a high double bond character gives fullerenes the ability to undergo addition reactions that

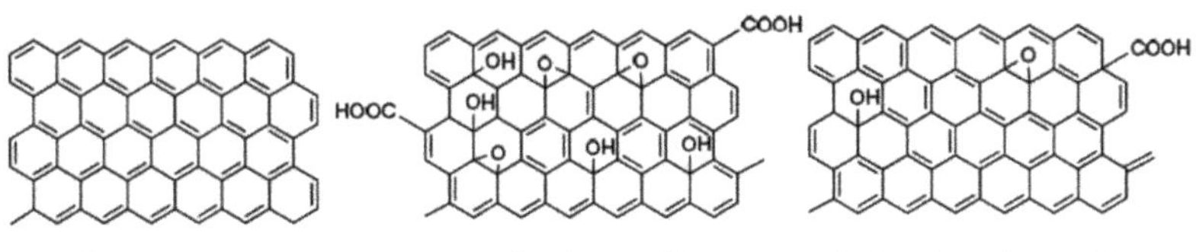

Figure 18.4. Structure of graphene, graphene oxide, and reduced graphene oxide. Reproduced with permission from [42].

give rise to numerous mono and multiple adducts. These addition reactions can compete with electrocatalytic reactions. Electron affinities, ionization potentials make fullerenes important materials for electrocatalytic applications [35].

18.2.2 Graphene

Graphene (Gr) is a 2D nanomaterial with a thickness of 0.35–1.6 nm, formed as a result of the honeycomb-like arrangement of carbon atoms connected by sp^2 bonds. It was first introduced in 2004 by a group of researchers from the University of Manchester [34]. Due to its excellent electrical, thermal, and optical properties, it has been given great scientific focus by modern scientists. Unlike CNTs, it is a cheaper and non-toxic material. Gr (figure 18.4) is used as an excellent conductive surface with a very large surface area to deposit catalytic nanoparticles (NPs) as well as an electrochemical sensor. It is synthesized from graphite. Compared to graphite, graphene provides uniform and more electroactive fields and density, which decreases on potentials and increases surface area (even higher than SWCNTs) to immobilize biomolecules [36].

The graphene family of materials includes several layers of the graphene sheet, GO, and rGO sheet. –OH, –COOH, –CO, and epoxy groups increase the interlayer distance resulting in a hydrophilic layer. Therefore, GO is an oxygenated graphene sheet. Partial reduction of GO results in the reduced graphene-like layer called reduced graphene oxide (rGO) (figure 18.4). Compared to graphite, graphene facilitates the immobilization of biomolecules by increasing the surface area more [37].

It has similar strength and stiffness to CNTs except for its high Young's modulus, good electrical and thermal conductivity. Besides these advanced features; problems such as thickness, poor dispersion, and re-stacking can limit its use in applications [38]. Graphene can be synthesized by mechanical exfoliation (scotch tape method) [39], chemical vapor deposition (CVD) [40] and the widely used Hummer's method [41].

18.2.3 Carbon nanotubes

The discovery of CNTs dates back to the 1960s, with the identification of graphite whiskers in the form of wrapped graphite sheets by Bacon [43]. It was first discovered in 1991 during Iijima's fullerene exploration [32]. Structurally, they are composed of hexagonal carbon bonds of sp^2 hybridized carbon atoms arranged into hollow, cylindrical nanostructures. They have high aspect ratios, large surface areas, ultra-small diameters, and non-planar structures and have unique electrochemical and thermal properties. CNTs can be classified as single-walled or multiwalled. They are divided into two according to the number of graphene layers in their structures. These;

- Single-walled CNTs (SWCNTs) and their outer diameter range from 0.4 nm to 2 nm.
- The outer diameter of multiwalled CNTs (MWCNTs) ranges from 2 to 100 nm, depending on the synthesis conditions [44, 45].

The general production methods of CNTs: arc discharge [46], CVD [47], and laser ablation [48] CNTs have received significant attention in various applications. Due to their unique and extraordinary structural, electrical, mechanical, thermal, and optical properties, they have different applications in different fields, including electrochemical sensors [49]. Large-scale production of MWCNTs is relatively easy compared to SWCNTs, which has increased the former's popularity for biomedical applications. Although MWCNTs are more stable than SWCNTs, they are more inert and less soluble in aqueous media. CNTs tend to form bundles or crystalline ropes and become entangled due to their high aspect ratio and strong van der Waals (more specifically, π–π stack) interactions [36, 50].

Besides its unique properties, it has dispersion problems in aqueous solutions. To solve this problem, CNTs are functionalized by covalent [51] and non-covalent [52] methods. In the covalent process of CNTs, nitric acid or a mixture of concentrated nitric acid and sulfuric acid is used as the oxidizing agent for their oxidation. Chemical modification of nanotubes by oxidation is followed by subsequent esterification or amidization of carboxyl groups. Non-covalent methods ensure that the structural and electrical properties of CNTs are preserved. Non-covalent approaches are based on the interaction of the hydrophobic part of the adsorbed molecules with the sidewalls of the nanotubes through van der Waals, p–p, CH–p, and other interactions [52]. The hydrophilic part of the molecules gives solubility in an aqueous solution. The advantage of non-covalent methods is the addition of various functional groups to the surface of CNTs while keeping the p-system of graphene sheets intact. As a result, the physical properties and chemical structure of CNT do not change by non-covalent methods, unlike in the case of covalent chemical bonding. However, these methods shorten the length of CNTs [53].

18.2.4 Carbon nanowires

Nanowires are wires about 1 nanometer in size. The simple structure of nanowires provides them with easy electron transfer properties [54]. In this way, it allows its use in the field of electrochemical sensors [55, 56].

18.2.5 Carbon fibers

Carbon nanofibers (CNFs) are cylindrical wire-shaped materials formed by stacking graphene sheets in different arrangements. Their strength and electrical conductivity are similar to CNTs, with lengths up to 10 μm and diameters between 10 and 500 nm [57].

They have more edge regions than CNTs due to the different arrangements of graphene sheets in their structure. These edge regions enable electron transfer with electroactive species in the analyte in the electrochemical analysis [58, 59]. CNFs are advantageous in electrochemical nanosensor applications due to their excellent electrical conductivity, biocompatibility, large surface area, and ease of fabrication. They are also ideal materials specifically for use with fast voltammetric techniques such as fast-scan cyclic voltammetry. In addition to these advantages, it is necessary to prevent contamination of the carbon surface with epoxy during the preparation stage. Carbon fibers are used in microelectrodes because of their excellent electrical conductivity, strength, and durability [55, 60, 61]. They can be produced using synthesis methods such as arc discharge [62], laser ablation [63], chemical vapor deposition (CVD) [64], plasma enhanced chemical vapor deposition (PECVD) [65], and electrospinning [66]. In addition, they can be made functional according to the sensor mechanism to be designed [67].

18.2.6 Carbon dots

Carbon is composed of carbon dots (CDs), carbon quantum dots (CQDs), graphene quantum dots (GQDs), carbon nanodots (CNDs), which have different forms according to their structural properties, surface groups, carbon core structures is the general name of the material class [68]. 0D GQDs consist of a single or several layers of graphene and are under 10 nm in size [68]. These materials offer several superior qualities such as chemical inertness, fluorescence activity, photostability, tunable low cytotoxicity, high surface area, luminescence emission, and excellent biocompatibility. They are synthesized by two methods, top-down and bottom-up approaches [69]. GQD's features such as electron and quantum trapping, low toxicity, and biocompatibility have made their use widespread in the field of electrochemical sensors as well as in many other fields [70]. GQDs dissolve well in organic solvents such as dimethylformamide (DMF), tetrahydrofuran (THF), acetone, dimethyl sulfoxide (DMSO), and ethanol. Surface functionalities are produced from carbon-containing materials such as GQDs, fullerene, glucose, graphite, graphene oxide (GO), CNTs, and carbon fibers (CFs) used as precursors [71].

CQDs have a crystal lattice with chemical groups on their surface and a quantum confinement effect (QCE). They have a crystal lattice with surface chemical groups. In GQDs, surface groups are added to interlayer defects or edges that contribute to different QCE and edge properties. It is possible to adjust the photoluminescence wavelength provided from CQDs by size regulation of CQDs. The height of GQDs is usually less than a graphene sheet with a lateral dimension of less than ~100 nm [71].

18.3 Electrode modification with carbon nanoparticles

As explained in the previous section, carbon nanomaterials (CNM) are used to design electrochemical sensor platforms with high sensitivity and selectivity due to their excellent electrical conductivity and superior properties. CNMs are modified to various carbon electrodes such as screen-printed electrodes (SPE) [20, 72, 73], glassy carbon electrodes (GCE) [26, 74], carbon paste electrodes (CPE) [75, 76], pencil graphite electrodes (PGE) [77], or metallic electrodes [78] such as Au, Pt, and nanosensor platforms are prepared. Various modification methods have been developed for carbon-based sensors prepared with CNMs.

Current strategies for electrode modification with CNMs include physical adsorption [79], electrodeposition [80] chemical bonding [81] electropolymerization [82] by adding to graphite powder in CPEs [83]. Modification with drop-casting is one of the easiest and fastest ways to produce carbon-based modified electrodes. CNM suspensions can be dropped in small volumes on the bare electrode to form stable surfaces. Due to strong π–π interactions with CNMs on the electrode surface, modification processes are easily performed. In addition to these advantages, there may be stability problems.

Using electrodes modified with CNMs (CPE, SPE, GCE, etc) has the following advantages:

(1) Fast electron transfer between CNMs and target molecules in chemical sensors to achieve better peak separation between different analytes.
(2) The larger electroactive surface area of CNMs enhances the electrochemical reaction and therefore has better sensitivity.
(3) CNMs provide a stable carrier for the immobilization of biomolecules. Therefore, successful immobilization and further functionalization of CNMs on the electrode surface effectively improve electrode performance [24, 38, 49].

18.4 Applications of carbon-based electrochemical sensors in hormone analysis

Hormones are crucial compounds for human health, which are secreted directly into the blood and carried to parts of the body such as organs or tissues to perform their physiological functions. There are many types of hormones that act on different aspects of bodily functions and processes. Reliable and sensitive analytical methods are, therefore, needed for the determination of hormones. Consequently, the development of sensitive analytical methods to monitor hormones in the different matrices is fundamentally important. Recent selected electrochemical methods based on carbon-based materials and carbon-based electrodes for hormone analysis are tabulated in table 18.1. This table presents the basic parameters such as the used electroanalytical method, the modification architecture and electrodes, measurement medium, linear range, and LOD.

As a carbon material, graphene and its derivatives have been preferred as a modification material with several advantages. Easy preparation, cost-effectiveness, and high sensitivity to biomolecules are some of these advantages. Moreover, homogeneous dispersion of different materials in the mixture can be obtained due to the hydrophilicity and thermal stability features of GO. In their study, Zhao *et al* developed AuNPs/5-Amino-2-mercaptobenzimidazole (AMBI)/rGO composite sensor that exhibits electrocatalytic activity for the determination of progesterone. The

Table 18.1. Applications of carbon-based electrodes and carbon nanomaterial modified sensors in hormone analysis.

Hormone	Method	Electrode	Medium	Linear range	Limit of deetection	Applications	References
Adrenaline	DPV	MIP/rGO/GCE	0.1 M PBS (pH 7.2)	0.015–40 μM	3 nM	Urine and ampoule	[102]
Adrenaline	CV	Poly (Alanine)/CPE	0.2 M PBS (pH 7.0)	10–60 μM	28.0 nM	NS	[103]
Adrenaline	DPV	PPMAM/CNTPE	0.2 M PBS (pH 7.0)	10–110 μM	4.5 $\times 10^{-8}$ M	Human serum	[27]
Adrenaline	SWV	NiO/CNTs/CPE	0.1 M PBS (pH 8.0)	0.08–900.0 μM	0.01 μM	Urine and ampoule	[104]
Adrenaline	DPV	EDDPT/GO/CPE	0.1 M PBS (pH 7.0)	1.5–600.0 μM	0.65 μM	Human serum	[105]
Cholecalciferol	SWV	LN-NO/GCE	0.5 M NaOH	0–2.6 $\times 10^{-5}$ M	8.3 $\times 10^{-7}$ M	River water	[106]
Chorionic gonadotropin hormone	Open circuit potential	Pt-labeled-hCG Ab/ Ag-hCG/BSA/ PtNPs/SPE	50 mM PBS (pH 6.0)	0–10 ng mL^{-1}	0.28 ng mL^{-1}	Human urine	[107]
Chorionic gonadotropin hormone	DPV	MWCNT/CS-GA/ hCG/MIP/GCE	PBS (pH 7.4)	0.5 pg mL^{-1}– 250 ng mL^{-1}	0.035 pg mL^{-1}	Human serum	[108]
Chorionic gonadotropin hormone	SWV	CNOs/AuNPs/PEG/ GCE	10 mM PBS (pH 7.4)	0.1 fg mL^{-1}– 1 ng mL^{-1}	0.1 fg mL^{-1}	Human urine	[108]
Cortisol	Chronoamperometry	BSA/Ab-Cortisol/ erGO/SPE	100 mM PBS (pH 7.4)	0–200 ng mL^{-1}	0.1 ng mL^{-1}	NS	[73]
Cortisol	CV	MIP/PANI- PMSAN/SPE	500 mM KCl, 20 mM $K_4Fe(CN)_6$ and 20 mM $K_3Fe(CN)_6$ solution (pH = 6.0)	1–1000 ag mL^{-1}	2 ag mL^{-1}	Urine	[93]
17β-estradiol				0.001–1000 fg mL^{-1}	9 ag mL^{-1}		
Progesterone				0.001–10 fg mL^{-1}	2.5 ag mL^{-1}		
Testosterone				1–1000 ag mL^{-1}	10 ag mL^{-1}		
Dienestrol	SWV	AgNPs/f-MWCNT/ GCE	PBS (pH 3.0)	5.0 $\times 10^{-8}$–6.3 $\times 10^{-7}$ M	8.4 nM	Bovine urine	[109]
Dopamine	SDLSV	MnO_2/ N-GO/ GCE	0.1 M PBS (pH 7.0)	1.0 $\times 10^{-7}$–10^{-4} M	3.4 $\times 10^{-8}$ M	Human serum	[110]
Dopamine	Amperometry	Fe_2O_3/N-rGO/ GCE	PBS (pH 7.5)	0.5 $\times 10^{-6-3.4}$ $\times 10^{-4}$ M	0.49 $\times 10^{-6}$ M	Injection solution	[111]
Dopamine	LSV	α-Fe_2O_3/rGO/GCE	0.1 M PBS (pH 6.0)	1.0$\times 10^{-8}$–9.0$\times 10^{-4}$ M	3.26 $\times 10^{-9}$ M	Human serum	[112]
Epinephrine	DPV	Au/PILs/PPyNTs/ GCE	0.05 M PBS (pH 7.4)	35–960 μM	298.9 nM	NS	[113]

Epinephrine	DPV	CS/MNP/SPE	0.1 M PBS (pH 7.0)	0.1–0.8 μM	14 nM	Urine and in cerebrospinal synthetic fluids	[114]
Epinephrine	DPV	Pt-NiO/MWCNTs/GCE	0.1 M PBS (pH 7.0)	0.5–300 μM	0.035 μM	Human serum, urine and injection solution	[115]
Epinephrine	DPV	ZnO-Pt/CNTs/GCE	(pH 7.0)	0.5–250.0 μM	0.5 μM	Human serum and urine	[116]
Epinephrine	DPV	CoAl-OOH/rGO nanosheets/GCE	0.1 M PBS (pH 7.0)	0.1–25 μM	0.023 μM	Human serum, urine and injection solution	[117]
Epinephrine	DPV	GNR/GCE	0.1 M PBS (pH 7.0)	6.4×10^{-6}–1.0×10^{-4} M	2.1×10^{-6} M	Injection solution	[118]
Epinephrine	DPV	SDS-WO3 NPs/GCE	0.1 M PBS (pH 7.0)	0.009–1000 μM	1.8 nM	Human serum	[119]
Epinephrine	DPV	mpg-C_3N_4/PANI/CdO/GCE	0.1 M PBS (pH 7.4)	0.05–80 μM100–1000 μM	0.011 μM	Human serum	[100]
Epinephrine	Amperometry	Bi_2O_3@rGO/SPE	0.05 M PBS (pH 7.0)	0.01–550.25 μM	2.14 nM	Human serum and rat serum	[120]
Epinephrine Melatonine	DPV	$ZnFe_2O_4$/CPE	0.1 M PBS (pH 5.0)	5–100 μM 6.5–145 μM	0.7 μM 3 μM	Ampule and tablet	[121]
17α-ethinylestradiol 17β-estradiol	SWV	SPE	0.3 M BRBS (pH 10.0)	~1.7 mg L^{-1}	1365 μg L^{-1} 0.953 μg L^{-1}	Human urine and tablet	[122]
17α-ethinylestradiol 17β-estradiol		SPCNTE		~2.0 mg L^{-1}	1112 μg L^{-1} 1191 μg L^{-1}		
Diethylstilbestrol	SWV	CeO_2 NPs/CPE	0.2 M PBS (pH 7.5)	10–100 nM 100–1200 nM	0.8 nM	Underground and lake water	[123]
17β-estradiol				10–100 nM 100–1200 nM	1.3 nM		
Diethylstilbestrol 17β-estradiol	DPV	Fe_3O_4-NC/GCE	0.04 M BRBS (pH 8.69)	0.01–12 μM 0.01–20 μM	4.6 nM 4.9 nM	Toner	[124]
17β-Estradiol	DPAdSV	MMIP/Magneto CPE	0.1 M PBS (pH 7.0)	6×10^{-8}–175×10^{-6} M	2×10^{-8} M	River water and raw milk	[125]

(*Continued*)

Table 18.1. (*Continued*)

Hormone	Method	Electrode	Medium	Linear range	Limit of deetection	Applications	References
17β-Estradiol	DPV	AF1-ADA/ON1/AF2-Au/LSGE	PBS (pH 7.4)	1.0×10^{-13}–1.0×10^{-9} M	63.1 fM	Milk	[126]
17β-Estradiol	SWV	Fe_3O_4-MIP/SPE	0.1 M PBS (pH 8.0)	0.05–10 μM	20 nM	River water	[127]
17β-Estradiol	LSV	CDs-PANI/GCE	PBS (pH 1.0)	0.001–100 μm	43 nm	Drinking water, tap water, river water and human serum	[128]
17β-Estradiol	DPV	RGO/CuTthP/GCE	0.1 M PBS (pH 7.0)	0.10–1.0 μM	5.3 nM	River water	[129]
17β-Estradiol	DPV	MWCNT/THI/AuNPs/SPCE	0.1 M PBS (pH 7.4)	0.01–100 ng mL^{-1}	10 pg mL^{-1}	Clinical serum	[130]
17β-Estradiol	SWV	AgNPs:ErGO/PG	0.1 M PBS (pH 7.4)	0.001–175 μM	0.046 nM	Human serum and urine	[131]
17β-Estradiol	DPV	MIP–MWCNT–AuNP/SPE	0.05 M PBS (pH 7.0)	1.0 × 10–15–1.0 × 10–6 M	2.5 × 10–16 M	Fish serum	[132]
17β-Estradiol	Amperometry	CB/GCE	0.1 M PBS (pH 6.2)	0.15×10^{-6}–3.5 × 10–6 M	9.2×10^{-8} M	Tablets	[74]
17β-Estradiol	LSV	N-TiO_2-TP/GCE	0.2 M PBS (pH 7.0)	9.9×10^{-6}–1.4×10^{-5} M	1.7×10^{-7} M	Tap water and synthetic urine	[133]
17β-Estradiol	DPV	CuPc-P6LC-Nafion/SPE_f	0.1 M PBS (pH 5.0)	8.0×10^{-8}–7.3×10^{-6} M	5.0×10^{-9} M	River water and human urine	[134]
17β-Estradiol Progesterone	DPV	GQDs-PSSA/GO/GCE	0.1 M CBS (pH 6.0)	1–100 nM 0.1–6.0 μM	0.23 nM 0.31 nM	Human serum and tablet	[135]
Estratriol	DPV	rGO–SbNPs/GCE	0.1 M PBS (pH 7.0)	0.20–1.4 μM	0.5 nM	Natural water	[136]
Estratriol	DPV	RGO/AgNWs/AgNPs/SPCE	0.01 M PBS (pH 3.5)	1–90 μM	0.58 μM	Synthetic urine	[137]
Estratriol	Amperometry	MWCNT/SPE	0.5 M ABS (pH 5.0)	1.0–1000 μM	0.53 μM	Estriol powder	[138]
Estratriol	DPV	CNB-AgNP/GCE	0.1 M PBS (pH 7.0)	0.2–3 μM	0.5 μM	Creek water	[139]
Follicle-Stimulating Hormone	DPV	anti-FSH/BSA/FSH rGO/MWCNTs/Thi/AuNPs/SPE	0.1 M PBS (pH 7.4)	1–250 mIU mL^{-1}	0.05 mIU mL^{-1}	Human serum	[140]
Ghrelin	DPV	BSA/anti-GHRL/Strep/PEI/CS/HfO_2-Pr_6O_{11}/SPE	0.1 M PBS (pH 7.4)	0.01 pg mL^{-1}–50 ng mL^{-1}	0.006 pg mL^{-1}	Human serum	[141]
Ghrelin Peptide YY	DPV	rGO/SPE	0.1 M PBS (pH 7.4)	10^{-3}–100 ng mL^{-1} 10^{-4}–10 ng mL^{-1}	1.0 pg mL^{-1} 0.02 pg mL^{-1}	Human serum and saliva	[142]
Insulin	DPV	Ppy@Gr/PGE	0.04 M BRBS (pH 11.0)	0.225–1.235 μM	8.65nM	Human serum	[143]

Insulin	DPV	Au-AgNPs/GQDs/GCE	0.1 M PBS (pH 7.0)	10–120 μM	1.1 nM	Insulin glargine injection	[144]
Insulin	Chronoamperometry	NiNPs/CNTs/CFME	0.1 M NaOH	2.0–20.0 μM	270 nM	NS	[108]
Leptin	DPV	Au/Ce_3NbO_7/CeO_2 Nanospheres/SPE	0.1 M PBS (pH 7.4)	0.5–12,000 ± 500 pg mL^{-1}	0.138 pg mL^{-1}	ELISA Human plasma	[94]
Melatonine	SWV	Gr-Av inked paper electrode	0.5 M McIlvaine BS (pH 7.0)	10.0–100.0 μM	0.49 μM	Human urine, human serum, and capsules	[145]
Melatonine	SWV	GPT/WPE	0.1 M PBS (pH 7.4)	0.80–100 μM	32.5 nM	Tablet, saliva, and urine	[146]
Melatonine Serotonine	DPV	CNTs-graphene/SPCE	50 mM PBS (pH 7.6)	0.005–3 mM 0.001–2.5 mM	1.1 μM 0.4 μM	Tablet and capsule	[147]
Parathyroid hormone	SWV	MB-APBA-HRP-Ab-PE-PTH/SPE	0.1 M PBS (pH 7.0)	10–300 pg mL^{-1}	11.56 pg mL^{-1}	Human serum	[148]
Parathyroid hormone	CV EIS	PTH/FBS/Ab_1-MWCNT-AuNP/SPCE	0.1 M PBS (pH 7.0)	1–10^5 pg mL^{-1}	0.092 pg mL^{-1} 0.033 pg mL^{-1}	Human serum	[149]
Parathyroid hormone	DPV SWV	Ag-PTH/ab-HRP/AuNPs/MWCNT/SPE	0.1 M PBS (pH 7.4)	1–300 pg mL^{-1}	0.886 pg mL^{-1} 0.065 pg mL^{-1}	Serum	[20]
Progesterone	Photoelectrochemistry	CDs-GO/GCE	10 mM PBS (pH 7.4)	0.5–180 nM	0.17 nM	Human serum	[150]
Progesterone	Amperometry	biotin-Ab/blocking reagent/Ag/Thi-GO/GCE	0.01 M PBS (pH 7.4)	0.0002–0.2 ng mL^{-1}	6.3×10^{-5} ng mL^{-1}	Milk	[151]
Progesterone	Amperometry	Mn(III)-SB/AuNPs/GCE	0.1 M NaOH	35–800 μM	11.4 nM	Cow's milk	[152]
Progesterone	Amperometry	WO_3 NB nanoballs@multi-layered GR sheets/GCE	0.05 M PBS (pH 7.0)	0.025–1792.5 μM	4.28 nM	Human serum	[153]
Progesterone	Amperometry	HRP-P4-(P4)-anti-P4-Protein-G-MBs/SPE	0.05 M PBS (pH 6.0)	0.02–100 ng mL^{-1}	5 pg mL^{-1}	Saliva	[154]
Progesterone	DPV	GQDs–NiO-AuNFs/f-MWCNTs/SPE	0.1 M PBS (pH 7.4)	0.01–1000 nM	1.86 pM	Human serum, injection, and tablet	[155]

(Continued)

Table 18.1. (*Continued*)

Hormone	Method	Electrode	Medium	Linear range	Limit of deetection	Applications	References
Progesterone	SWV	GO-IMZ/GCE	0.1 M NaOH	0.22–14.0 µM	68 nM	Tablet	[156]
Progesterone	DPV	PEDOT/ZrO_2-NPs/GCE	0.1 M CBS (pH 4)	1–100 nM100–6 × 10^3 nM	0.32 nM	Human serum and injection	[157]
Progesterone	DPV	Fe_3O_4@GQD/f-MWCNTs/GCE	0.1 M PBS (pH 7.0)	0.01–0.5 µM0.5–3.0 µM	2.18 nM	Human blood serum and ampoule	[158]
Progesterone	DPV	Tin-nanorod/GCE	0.2 M NaOH with 0.05 mM of CTAB	40–600 µM	0.12 µM	Pharmaceutical product	[159]
Progesterone	SWV	AuNPs/AMBI/rGO/SPE	0.2 M PBS(pH 6.0)	0.9×10^{-9}–27×10^{-6} M	0.28 nM	Calf serum and milk	[28]
Prolactin	DPV	BSA/anti-PRL/EDC/NHS/MPA/Bi_2S_3/PPy/SPE	0.1 M PBS (pH 7.4)	1–250 ng mL^{-1}	0.130 ng mL^{-1}	Human and mouse serum	[160]
Prolactin	DPV	AP-Strept-Biotin-DNA/AuNPs/dAb-PRL-cAb/CS/AuNPs/SWCNTs/Graphene/GCE	0.01 M PBS (pH 7.0)	50–3,200 pg mL^{-1}	47 pg mL^{-1}	Spiked serum	[161]
Prolactin	DPV	HRP-dAb-PRL-cAb/ AuNPs/IL@CPE	0.1 M PBS (pH 7.0)	25.0–2,000.0 mIU L^{-1}	12.5 mIU L^{-1}	Human serum	[162]
Prolactin	Amperometry	PdPtNDs@NH_2-MIL-53 (Fe)-dAb-PRL-cAb/AuNPs@NH_2GS/GCE	0.01 M PBS (pH 7.4)	0.005–500.0 ng mL^{-1}	1.15 pg mL^{-1}	Human serum	[163]
Prolactin	Amperometry	HRP-dAb-PRL-b-cAb-NA-MBs/SPE	0.05 M PBS (pH 6.0)	1.1–100 ng mL^{-1}	0.32 ng mL^{-1}	Saliva and breast milk	[164]

Testosterone	DPV	rGO/GCE	Borate buffer (pH 5.4) with 2 mM CTAB	2.0–210.0 nM	0.1 nM	Human serum and urine	[165]
Testosterone	EIS	MIP/GO/GCE	ABS (pH 4.5)	1 fM – 1 µm	0.4 fM	Human serum	[166]
Thyroid-stimulating hormone	DPV	HRP-Ab/TSH/Ab/TGA/AuNPs/IL-CPE	0.1 M PBS (pH 7.0)	0.2–90.0 ng mL^{-1}	0.1 ± 0.02 ng mL^{-1}	Clinical serum	[83]
Thyroid-stimulating hormone	EIS	anti-TSH/AuNPs/SPE	0.01 M PBS (pH 7.4)	0.001–150 µIU mL^{-1}	0.001 µIU mL^{-1}	NS	[167]
Thyroid-stimulating hormone	SWV	gly/ab-TSH/azo/GCE	0.1 M PBS (pH 8.0)	0.2–20.0 µIU mL^{-1}	0.04 µIU mL^{-1}	Human serum	[168]
Thyronamine	SWV	MIP/SPCE	0.1 M PBS (pH 7.4)	0–10 µM	0.081 µM	NS	[169]
Thyroxin	DPV	(Fe_3O_4/GO–AuNPs@MIP)/CPE	0.1 M HCl	4×10^{-8}–7×10^{-7} M	3×10^{-9} M	Human urine and serum	[75]

Abbreviations: Ab: antibody, Ag: antigen, AgNPs: silver nanoparticles, AgNWs: silver nanowires, AMBI: 5-amino-2-mercaptobenzimidazole, AuNPs: gold nanoparticles, Bi_2S_3:bismuth sulfide, BSA: bovine serum albumin, CB: carbon black, CD: carbon dots, Ce_3NbO_7/CeO_2:cerium niobate/cerium oxide hollow nanospheres, CFME: carbon fiber microelectrode, CNB: carbon black nanoballs, CNO: carbon nano-onions, CNT: carbon nanotube CNTPE: carbon nanotube paste electrode, CPE: carbon paste electrode, CS:chitosan, CTAB: cetyltrimethylammonium bromide, CuPc: copper phthalocyanine, CuTthP: Cu(II)-meso-tetra(thien-2-yl)porphyrin, EDDPT:2-(5-Ethyl-2,4-dihydroxyphenyl)–5,7-dimethyl-4H-pyrido[2,3-d][1, 3]thiazine-4-one, ErGO: electrochemically reduced graphene oxide, Fe_3O_4-NC: Fe_3O_4-doped nanoporous carbon, GA: glutaraldehyde, GCE: glassy carbon electrode, gly:glycine, GNRs: graphene nanoribbons, GO: graphene oxide, GO-IMZ: imidazole-functionalized graphene oxide, GPT: conductive ink composed of nail polish and graphite, GQDs: graphene quantum dots, Gr-AV: graphite-automotive varnish, HfO_2-Pr_6O_{11}:hafnium oxide-praseodymium oxide nanoflakes, HRP: horseradish peroxidase, LN-NO: perovskite-type $LaNiO_3$-nanoxide, LSGE: laser-scribed graphene electrode, MBs: magnetic microbeads, MPA: 3-mercaptopropionic acid, mpg: mesoporous polymeric graphitic, MMIP: magnetic molecularly imprinted polymer, MWCNT: multiwalled carbon nanotubes, N-: nitrogen-doped, NA-MBs: magnetic microparticles functionalized with neutravidin, NS: Not stated N-TiO_2: nitrogen-doped titanium dioxide, NiNP: nickel nanoparticle, P6LC: printex 6L carbon, PANI: poly(aniline), PEDOT: poly(3,4-ethylenedioxythiophene), PEI: poly(ethyleneimine), PEG: polyethylene glycol, PG: pyrolytic graphite, PILs: poly (ionic liquids), PMSAN: poly(m-aminobenzenesulfonic acid), PPMAM: poly(phenylmethanoic acid), PPy: poly(pyrrole), PPyNTs: poly(pyrrole) nanotubes, rGO:reduced graphene oxide, SB: Schiff base, SbNPs:antimony nanoparticles, SDLSV:second-derivative linear sweep voltammetry, SPE: screen-printed electrode, SPCNTE: screen-printed carbon nanotube electrode, SPE_f: functionalized screen-printed electrode, Strep: streptavidin, TiO_2: titanium dioxide, THI: thionine, TP: thin film of tapioca, WO_3 NBs: tungsten trioxide nanoballs, WPE: waterproof paper substrate electrode, $ZnFe_2O_4$: zinc ferrite nanoparticles, ZrO_2-NPs: zirconium oxide nanoparticles.

AMBI is a component of vitamin B12. It has biologically active and presents a remarkable catalytic effect. In this study, to obtain GO suspension, the GO was dispersed in ethanol, and then this suspension was added to AMBI in a 1:1 (v/v) ratio. AuNPs were prepared by $HAuCl_4$ in boiling water to obtain a nanogold solution. After that, AMBI/GO was added to a nanogold solution. When sodium hydroxide and sodium borohydride was added, the mixture was constantly stirred to obtain rGO. Finally, prepared composite by centrifugation dispersed in ultrapure water. To prepare the modified electrode, 10 μL of AuNP/AMBI/rGO suspension was dropped on to the SPCE surface and dried at 40 °C for further use. CV results suggested that the bare SPCE has a well-defined oxidation peak at 131.87 μA. Also, the peak current increase of 14.30% relative to bare SPCE when SPCE modified with GO/AMBI. As a result of EIS, the Rct value of AMBI/GO was found 66.4 Ω, compared to the bare SPCE Rct value of 178.9 Ω. So, it can be said that AMBI/GO composite improved the rate of electron transfer and the electroactive surface area. Under optimal experimental conditions, for the first time, the developed AuNP/AMBI/rGO sensor exhibited a linear range from 0.9×10^{-9} to 27×10^{-6} mol L^{-1}, the LOD of 0.28×10^{-9} mol L^{-1}. Consequently, the developed sensor can be a potential platform for further electrochemical sensing applications for different compounds [28].

Glassy carbon is obtained as a result of the carbonization of phenol/formaldehyde polymers or polyacrylonitrile under pressure between 1000 °C and 3000 °C [84]. GCEs are preferred in electrochemical analyses due to their robustness, non-reactivity, wide potential range, and good electrical properties. In addition, the smooth and uniform structure of their surfaces is one of the reasons for their preference compared to CPE. The use of GCEs as modified or naked in the analysis of various hormones is encountered. It is widely used for the electrochemical detection of serotonin. Serotonin exhibits a well-defined but somewhat broad electro-oxidation peak at ~400 mV at the bare glassy carbon electrode [85, 86].

In 2015, a glassy carbon electrode (GCE, diameter 3 mm) was functionalized using gold nanoparticles-graphene nanocomposite for the detection of the peptide hormone prolactin from human serum samples [87]. First, the graphene solution was prepared by dissolving 1 mg graphene in 1 mL chitosan solution (0.5 wt%), followed by 2 h sonication to obtain a black homogeneous solution. From this solution, 10 μL was dropped onto the cleaned GCE, and the modified electrode was immersed in 3 mM $HAuCl_4$ solution. The electrodeposition of gold was achieved between 0.2 to 0.6 V by cyclic voltammetry. Then, 10 μL of 67 μg mL^{-1} Ab1, (prolactin antigen, monoclonal prolactin antibodies) was further dropped onto the AuNPs/graphene/chitosan modified electrode. This electrode was kept at 4 °C overnight. For further step, AuNPs/graphene/chitosan modified electrode was immersed in 0.5% (wt) BSA for 30 min this prepared immunosensor was incubated in different concentrations of prolactin solution and biotin-labeled monoclonal prolactin antibodies (biotinylated Ab2 solution) for 40 min, followed by washing steps with PBS buffer. 10 μl of 2.0 μg mL^{-1} streptavidin-alkaline phosphatase solution was dropped onto AuNPs/graphene/chitosan modified electrode and incubated for 30 min at 37 °C. To follow the interaction, DPV measurements were conducted from 0.1 to 0.5 V in 5.0 ml diethanolamine solution containing 1.2 mg mL^{-1} α-naphthyl phosphate. A sandwich-type immune complex was formed on the sensing interface. The electrochemical cell consisted of a three-electrode system; the prepared AuNPs-graphene/Ab1/BSA/PRL/Ab2 modified GCE as WE, Ag/AgCl as a RE, and a Pt wire as a CE. The proposed method showed linearity in the logarithm of PRL in the concentration range from 0.1 to 50 ng mL^{-1}, and the detection limit is estimated to be 38.9 pg mL^{-1} (S/N=3). As potential interfering proteins such as BSA, angiotensionII (AngII), and tumor necrosis factor-α (TNF-α) with the concentration of 100 ng mL^{-1} were investigated, and the magnitude of the recorded current is negligible compared to the current arising from the designed immunosensors. Reproducibility of the suggested immunosensor was also proved to be reliable with different Prolactin concentrations of 100 pg mL^{-1} and 1 ng mL^{-1} with five repetitions using the same immunosensors. The prepared AuNPs-graphene/Ab1/BSA/PRL/Ab2 modified GCE immunosensor was applied for the determination of prolactin in serum samples gained from the First Affiliated Hospital of Chongqing Medical University. The results showed that the immunosensor has the capacity to measure prolactin in complex matrices with a recovery percentage from 98.2% to 100.7%.

SPEs are produced with screen printing technology and have been used in electrochemical sensors since the 1990s [88, 89]. The screen printing technique is required to produce SPEs; it consists of a patterned stencil or a screened mesh of a specific size and shape on a paste-like substrate with suitable ink formulations [90]. The surface material used is usually ceramic or plastic [91]. SPE has tremendous technology that includes a working electrode (WE), a pseudo-reference electrode (RE), and a counter electrode (CE) required for electrochemical analysis [92]. Since SPEs are frequently used in the analysis of hormones such as serotonin and estrogen, there are examples of reviews about this in recent years [17, 86].

Undoubtedly, there are many advantages that provide electrochemical methods different from other analytical methods. While a single measurement can be taken with chromatographic or spectrophotometric methods, electrochemical systems enable multichannel measurements for simultaneous and no time-consuming analysis. In 2019, Lee *et al* used a multichannel electrochemical system combined with conductive molecularly imprinted polymers (MIP) for ultrasensitive detection and determination of cortisol, progesterone, testosterone, and 17β-estradiol in human urine samples using SPEs. In this study, these steroid hormones were used as templates, while aniline and metanilic acid were used to generate molecular imprints. Modified electrodes were prepared by electropolymerization on the surface of the screen-printed carbon electrode. This MIP-based sensor, prepared by electrochemical synthesis, offered advantages such as low cost and fast preparation. Under optimal conditions, cyclic voltammetry (CV) was used in the presence of 20 mM potassium ferricyanide ($K_3[Fe(CN)_6]$), 20 mM potassium ferrocyanide ($K_4[Fe(CN)_6]$) for the determination of all analytes. The linear relationships between concentrations and peak currents were found as 1–1000 ag mL^{-1}, 0.001–1000 fg mL^{-1}, 0.001–10 fg mL^{-1}, and 1–1000 ag mL^{-1} for cortisol, estradiol, progesterone, and testosterone, respectively. Moreover, the limit of detection values were of ag mL^{-1} level for all compounds using screen-printed carbon electrodes [93].

Nowadays, Ce- and Nb-based materials have been of enormous interest to many researchers. They are suitable as modification materials for biosensing applications owing to their enormous properties such as high corrosion resistance, low toxicity, and biocompatibility. In 2021, Liu *et al* used Ce_3NbO_7/CeO_2 modified SPE for the sensitive detection of leptin. When Ce_3NbO_7/CeO_2 was modified on the SPE surface, these nanomaterials exhibited a large surface area, higher catalytic activity, porous structure, and more active sites towards the leptin. Firstly, the hydrothermal method was used to synthesize Ce_3NbO_7/CeO_2 hollow spheres. This prepared Ce_3NbO_7/CeO_2 was dispersed in double distilled water by sonochemical method. Then, 6 μl of Ce_3NbO_7/CeO_2 was dropped on the surface of SPE. In the second step, AuNPs were electrodeposited on the Ce_3NbO_7/CeO_2/SPE using 10 mM $HAuCl_4$ solution. For this purpose, 20 deposition cycles were applied in a potential range between −0.4 and −0.9 V at 50 mV s^{-1} scan rate. After Au/ Ce_3NbO_7/CeO_2/SPE was rinsed with double distilled water, the electrode was incubated in 3-mercatopropionic acid (10 mM) solution for 4 h. After MPA/Au/Ce_3NbO_7/CeO_2/SPE was rinsed with buffer solution, 6 μl 1-ethyl-3-(3-dimethylaminopropyl) carbodiimide (EDC)/N-Hydroxysuccinimide (NHS) solution was dropped surface of MPA/Au/ Ce_3NbO_7/CeO_2/SPE and dried at room temperature. Next, to fabricate LEP/BSA/anti-LEP/EDC/NHS/MPA/Au/Ce_3NbO_7/CeO_2/SPE as a working electrode, EDC/NHS/MPA/Au/ Ce_3NbO_7/CeO_2/SPE was incubated in LEP antibody (Anti-LEP), BSA and LEP antigen, respectively. Under optimum experimental conditions, a good linear range from 0.5 pg mL^{-1} to 12 000 ± 500 pg mL^{-1} with a LOD of 0.138 pg mL^{-1} was obtained by developed immunosensor. Moreover, leptin was successfully determined in plasma samples [94].

Among electrode transducers, the carbon paste electrode (CPE) exhibits several advantages, such as high sensitivity, wide potential window, and low background current. CPEs were invented in the late 1950s [95] as an alternative to the dripping mercury electrode. Graphite powder and binders such as Nujol (mineral oil), paraffin oil, silicone oil are used in making a conventional CPE. It takes its name from the paste-like substance produced from this mix. The capacity of electron transfer greatly affects the ratio of binder and graphite powder forming the paste [96–98]. In recent years, to fabricate carbon ionic liquid (ILs) electrodes, ILs have become an attractive material as a non-conductive organic binder. For example, Beitollahi *et al* fabricated an ionic liquid carbon paste electrode (CILE) using a mix of graphite powder, ionic liquid BPPF6, and paraffin oil to determine the determination of prolactin. After homogeneous carbon paste was obtained, an appropriate amount of the carbon paste was filled to the end of a glass tube. Then, gold nanoparticles were electrodeposited onto CILE by CV. To fabricate immunosensor, nano Au/CILE was immersed in thioglycolic acid (TGA) aqueous solution and removed the residue of TGA by rinsed several times with water. After that, prolactin/antibody was covalently bonded to TGA/ nano Au/CILE electrode. Before each measurement, the electrode was incubated using various prolactin concentrations and excessive horseradish peroxidase-labeled prolactin antibody. The developed immunosensor was stored at 4 °C when not in use. The experimental conditions such as pH of supporting electrolyte, time of electrodeposition, and incubation temperature were optimized. The fabricated immunosensor provided a linear concentration range between 25.0 and 2000.0 mIU L^{-1} with a detection limit 12.5 mIU L^{-1} in human serum samples [99].

In another study, Moreira *et al* [76] suggested a platform based on ionic liquid-supported magnetite nanoparticles using a carbon paste electrode for the electroanalytical determination of 17β-estradiol and estriol. The authors used transmission electron microscopy, scanning electron microscopy, and zeta potential techniques for the characterization of the nanoparticles, and using the square-wave voltammetric technique, detection of 17β-estradiol and estriol were achieved in B-R buffer pH 12.0. The linear range was obtained between 0.10 and 1.0 μmol L^{-1}, for 17β-estradiol, and between1.0 to 10.0 μmol L^{-1}, for estriol. The detection limits were calculated as 50.0 nmol L^{-1} and 300.0 nmol L^{-1} for 17β-estradiol and estriol, respectively. As a real sample application, these potential endocrine disruptors in pharmaceutical and cream formulations were used with acceptable recovery results; the study was completed.

Carbon nitride (C_3N_4), as an N-doped carbon-based material, is an attractive material owing to remarkable features, unique 2D structure, low toxicity to humans, environmentally friendly, and higher chemical and thermal stability. Although C_3N_4 exhibits a large bandgap, C_3N_4 is combined with various materials such as conducting polymers and metals or metal oxides to adjust the bandgap. Among different polymers, polyaniline (PANI) is an excellent conductive material and host for C_3N_4. When PANI and C_3N_4 are combined for using sensor modification, obtained composite improves the limitations of the weak mechanical strength of PANI and the low conductivity of C_3N_4. In their study, Bonyadi *et al* developed mesoporous polymeric graphitic-C_3N_4/PANI/cadmium oxide nanoparticles (mpg-C_3N_4PANI/CdO) nanocomposite for the determination of epinephrine in human serum samples. In this study, after the GCE surface was cleaned mechanically, mpg-C_3N_4, PANI, and CdO nanoparticles were electrodeposited to the previous surfaces, respectively. Electrochemical characterization results suggested that the surface of CdO/PANI/mpg-C_3N_4/GCE exhibited a faster electron transfer rate than the other electrodes. Moreover, unique features such as large electroactive surface area, excellent conductivity, and mesoporosity were obtained by CdO/PANI/mpg-C_3N_4. Under optimal conditions, the developed sensor exhibited two linear ranges in the range of 0.05–80 μM and 100–1000 μM for epinephrine, and the LOD was of 0.011 μM [100]. Among other carbon nanomaterials, graphene quantum dots (GQD) provide remarkable conductivity, higher electroactive surface area, low toxicity, good biocompatibility, and cost–effective design sensors. Similarly, multi-walled carbon nanotubes (MWCNTs) have enjoyed widespread attention owing to their unique electronic and nano-structural properties, good stability, and higher surface area. Nowadays, many researchers have used magnetite (Fe_3O_4) nano-particles to improve the sensitivity of sensors due to easy preparation, low toxicity, and high adsorption ability. However, Fe_3O_4 nanoparticles may not be very stable over time because they are very likely to aggregate. For this reason, Fe_3O_4

nanoparticles may combine with GQD to change the surface of Fe_3O_4 nanoparticles during or after the synthesis process. In 2017, Arvand *et al* fabricated a 3D hybrid platform that performs synergic and electrocatalytic characteristics taking into account the features of Fe_3O_4, GQDs and MWCNTs mentioned above. The developed Fe_3O_4@GQD/f-MWCNTs/GCE sensor was used for the determination of progesterone in human serum and pharmaceutical samples. In this study, $FeCl_3{\cdot}6H_2O$ was added to GQDs solution at room temperature under vigorous stirring. Then the $FeCl_2{\cdot}4H_2O$ was added slowly to the solution containing Fe^{3+}/GQDs, which was vigorously stirred. Finally, the pH of Fe_3O_4@GQDs solution was adjusted by ammonia solution. In order to fabricate the modified sensor, f-MWCNTs: Fe_3O_4@GQD (3:1) were added in N,N-dimethylformamide (DMF) and ultrasonicated to form a stable black suspension. Finally, 10 μl f-MWCNTs: Fe_3O_4@GQD suspension was drop casted on the GCE surface and dried at room temperature. After analytical conditions were optimized, the linear concentration range was found in the wide range of 0.01–0.5 and 0.5–3.0 μM with 2.18 nM detection limit [28].

Li *et al* [101] developed an electrochemical immunosensor for the detection of anti-Mullerian hormone from five rabbit tumor model samples, using the conventional immunohistochemical staining method and proposed immunosensor based on graphene oxide nanosheets, poly(3,4-ethylenedioxythiophene) (PEDOT) nanorods, and Au nanocomposite. First, the PEDOT nanorods were grown on graphene nanosheets. In this procedure, $FeCl_3$ (1 ml, 1 M) was added into aqueous graphene oxide dispersion (1 ml, 0.5 mg mL^{-1}), and sonicated for 10 min After that 3,4-ethylenedioxythiophene solution in $CHCl_3$ (25 mg mL^{-1}) was introduced to the reaction medium. The solution was heated at 50 °C under static conditions and left to react for 24 h, then the upper layer mixture was centrifuged and washed with distilled water and ethanol. 8 ml of the PEDOT/GO suspension was mixed with $HAuCl_4{\cdot}3H_2O$ (1 ml, 5 mM) and 0.1 M $NaBH_4$ aqueous solution. Finally, AuNPs/PEDOT/GR was prepared and stored at 4 °C before use. The surface of the GCE was modified with the synthesized AuNPs/PEDOT/GR and incubated in 20 μg mL^{-1} anti-AMH antibody solution (Ab/AuNPs/PEDOT/GR/GCE). This was followed by the addition of 5 μl, 1 wt% BSA, and incubation in different concentrations of anti-Mullerian hormone for 50 min DPV method was used to follow the interaction and x-ray diffraction, Raman, and IR methods to characterize the synthesized nanocomposite. The Ab/AuNPs/PEDOT/GR/GCE was employed for the determination of anti-Mullerian hormone, obtaining linearity related to the logarithm of the anti-Mullerian hormone concentration range from 0.0004 to 40 ng mL^{-1}, with a calculated LOD of 0.21 pg mL^{-1}. The authors suggested that the designed immunosensor had lower detection limits than the enzyme-linked immunosorbent assays in the literature. The real sample applications were also performed using five rabbit tumor model samples and reported with low RSD% values, which clearly indicate the reliability of the proposed method. Moreover, the selectivity studies showed a very good tolerance when other potential interference molecules were added to the medium.

18.5 Conclusion

Carbon, as one of the basic elements of the periodic table, has an outstanding contribution to the development of carbon-based nanomaterials. In recent years, apart from the known synthesis methods, the synthesis of carbon-based materials found in nature with the green synthesis method has attracted the attention of scientists. It is predicted that the use of these unique nanomaterials in electrochemical sensors will increase day by day with the development of new synthesis and easy modification methods. The important electrocatalytic activities, biocompatible structures, and stability provided by carbon-based sensors have made their use widespread in the analysis of different analytes. It is hoped that it will bring important developments in the field of health, especially with the development of these sensors, which enable the determination of low amounts of biologically important substances that regulate the functioning of organs from the human body, such as hormones.

References

[1] Katsuyama T and Moulton V R 2021 Hormones *Systemic Lupus Erythematosus* 2nd edn, ed G C Tsokos (New York: Academic) pp 105–12

[2] Phelps C 2007 The hypothalamus and its releasing factors and hormones *xPharm: The Comprehensive Pharmacology Reference* ed S J Enna and D B Bylund (New York: Elsevier) pp 1–6

[3] Emerald M 2016 Pituitary gland: pituitary hormones *Encyclopedia of Food and Health* ed B Caballero, P M Finglas and F Toldrá (Oxford: Academic) pp 392–400

[4] Zallone A and Colaianni G 2020 Posterior pituitary hormones and bone *Encyclopedia of Bone Biology* ed M Zaidi (Oxford: Academic) pp 519–26

[5] Liyanarachchi K D and Debono M 2017 Physiology of the pituitary, thyroid, parathyroid and adrenal glands *Surg.* **35** 542–55

[6] Malcomson R D G and Nagy A 2015 The endocrine system *Keeling's Fetal and Neonatal Pathology* (Cham: Springer) pp 671–702

[7] Pevet P, Klosen P and Felder-Schmittbuhl M P 2017 The hormone melatonin: animal studies *Best Pract. Res. Clin. Endocrinol. Metab.* **31** 547–59

[8] Hardeland R, Cardinali D P, Srinivasan V, Spence D W, Brown G M and Pandi-Perumal S R 2011 Melatonin-A pleiotropic, orchestrating regulator molecule *Prog. Neurobiol.* **93** 350–84

[9] Izquierdo-Lahuerta A 2021 The parathyroid hormone-related protein/parathyroid hormone 1 receptor axis in adipose tissue *Biomolecules* **11** 1570

[10] Crooks M and Bendall S 2021 Parathyroid hormone and vitamin D: from bench to bedside *Orthop. Trauma* **35** 267–73

[11] Gallagher T F, Yoshida K, Roffwarg H D, Fukushima D K, Weitzman E D and Hellman L 1973 ACTH and cortisol secretory patterns in man *J. Clin. Endocrinol. Metab.* **36** 1058–68

[12] Boonen E and Van Den Berghe G 2014 Cortisol metabolism in critical illness: Implications for clinical care *Curr. Opin. Endocrinol. Diabetes Obes.* **21** 185–92

[13] Bernard D J and Brûlé E 2020 Anterior pituitary: glycoprotein hormones from gonadotrope (FSH and LH) and thyrotrope (TSH) cells *Hormonal Signaling in Biology and Medicine* ed G Litwack (New York: Academic) ch 7 pp 119–44

[14] Neave N 2007 Hormones and the endocrine system *Hormones and Behaviour: A Psychological Approach* (Cambridge: Cambridge University Press) pp 22–47

[15] Lu X, Zhu Y and Zhu T 2017 Endocrine system *Adverse Effects of Engineered Nanomaterials* (Amsterdam: Elsevier) pp 339–55
[16] Zhang J and Gopinath S C B 2020 Quantification of cortisol for the medical diagnosis of multiple pregnancy-related diseases *3 Biotech.* **10** 35
[17] Musa A M, Kiely J, Luxton R and Honeychurch K C 2021 Trends in analytical chemistry recent progress in screen-printed electrochemical sensors and biosensors for the detection of estrogens *Trends Anal. Chem.* **139** 116254
[18] Khoshnevisan K, Honarvarfard E, Torabi F, Maleki H, Baharifar H, Faridbod F, Larijani B and Khorramizadeh M R 2020 Electrochemical detection of serotonin: a new approach *Clin. Chim. Acta* **501** 112–19
[19] Antoniazzi C, Lima C A D, Marangoni R and Spinelli A 2018 Voltammetric determination of 17β-estradiol in human urine and buttermilk samples using a simple copper (II) oxide-modified carbon paste electrode *J. Solid State Electrochem.* **22** 1373–83
[20] Chen G, Liu C and Wu W 2021 Electrochemical immunosensor for serum parathyroid hormone using voltammetric techniques and a portable simulator *Anal. Chim. Acta* **1143** 84–92
[21] Özcan H M, Yildiz K, Çakar C, Aydin T, Asav E, Sağiroğlu A and Sezgintürk M K 2015 Ultrasensitive impedimetric biosensor fabricated by a new immobilisation technique for parathyroid hormone *Appl. Biochem. Biotechnol.* **176** 1251–62
[22] Senturk H, Eksin E, Işık Ö, İlaslan Z, Mısırlı F and Erdem A 2021 Impedimetric aptasensor for lysozyme detection based on carbon nanofibres enriched screen-printed electrodes *Electrochim. Acta* **377** 138078
[23] Erkmen C, Demir Y, Kurbanoglu S and Uslu B 2021 Multi-purpose electrochemical tyrosinase nanobiosensor based on poly (3,4 ethylenedioxythiophene) nanoparticles decorated graphene quantum dots: applications to hormone drugs analyses and inhibition studies *Sens. Actuators, B Chem.* **343** 130164
[24] Asadian E, Ghalkhani M and Shahrokhian S 2019 Electrochemical sensing based on carbon nanoparticles: a review *Sens. ActuatorsB. Chem.* **293** 183–209
[25] Yang N, Swain G M and Jiang X 2016 Nanocarbon electrochemistry and electroanalysis: current status and future perspectives *Electroanalysis* **28** 27–34
[26] Gomes E S, Leite F R F, Ferraz B R L and Mour H A J L 2019 Voltammetric sensor based on cobalt-poly (methionine)-modified glassy carbon electrode for determination of estriol hormone in pharmaceuticals and urine *J. Pharm. Anal.* **9** 347–57
[27] Charithra M M, Manjunatha J G, Sreeharsha N, Mohammed S and Asdaq B 2021 Polymerized carbon nanotube paste electrode as a sensing material for the detection of adrenaline with folic acid *Monatsh. Chem.* **152** 411–20
[28] Zhao X, Zheng L, Yan Y, Cao R and Zhang J 2021 An electrocatalytic active AuNPs/5-amino-2-mercaptobenzimidazole/rGO/SPCE composite electrode for ultrasensitive detection of progesterone *J. Electroanal. Chem.* **882** 115023
[29] Kurbanoglu S, Uslu B and Ozkan S A 2017 Carbon-based nanostructures for electrochemical analysis of oral medicines *Micro and Nano Technologies* ed E Andronescu and A Grumezescu (Amsterdam: Elsevier) ch 28 pp 885–938
[30] Kroto H W, Heath J R, O'Brien S C, Curl R F and Smalley R E 1985 C60: Buckminsterfullerene *Nature* **318** 162–3
[31] Xu X, Ray R, Gu Y, Ploehn H J, Gearheart L, Raker K and Scrivens W A 2004 Electrophoretic analysis and purification of fluorescent single-walled carbon nanotube fragments *J. Am. Chem. Soc.* **126** 12736–7

[32] Iijima S 1991 Helical microtubules of graphitic carbon *Nature* **354** 56–8
[33] Chen Y, Guo L P, Johnson D J and Prince R H 1998 Plasma-induced low-temperature growth of graphitic nanofibers on nickel substrates *J. Cryst. Growth* **193** 342–6
[34] Katsnelson M I 2007 Graphene: carbon in two dimensions *Mater. Today* **10** 20–7
[35] Sherigara B S, Kutner W and Souza D 2003 Electrocatalytic properties and sensor applications of fullerenes and carbon nanotubes *Electroanalysis* **15** 753–72
[36] Erol O, Uyan I, Hatip M, Yilmaz C, Tekinay A B and Guler M O 2018 Recent advances in bioactive 1D and 2D carbon nanomaterials for biomedical applications *Nanomed. Nanotechnol. Biol. Med.* **14** 2433–54
[37] Jastrzębska A M, Kurtycz P and Olszyna A R 2012 Recent advances in graphene family materials toxicity investigations *J. Nanopart. Res. Interdiscip. Forum Nanoscale Sci. Technol.* **14** 1320
[38] Ananda Murthy H C, Gebremedhn Kelele K, Ravikumar C R, Nagaswarupa H P, Tadesse A and Desalegn T 2021 Graphene-supported nanomaterials as electrochemical sensors: a mini review *Results Chem.* **3** 100131
[39] Liu Q, Shi J, Zeng L, Wang T, Cai Y and Jiang G 2011 Evaluation of graphene as an advantageous adsorbent for solid-phase extraction with chlorophenols as model analytes *J. Chromatogr.* A **1218** 197–204
[40] Reina A, Jia X, Ho J, Nezich D, Son H, Bulovic V, Dresselhaus M S and Kong J 2009 Large area, few-layer graphene films on arbitrary substrates by chemical vapor deposition *Nano Lett.* **9** 30–5
[41] Hummers W S and Offeman R E 1958 Preparation of graphitic oxide *J. Am. Chem. Soc.* **80** 1339
[42] Evtugyn G, Porfireva A, Shamagsumova R and Hianik T 2020 Advances in electrochemical aptasensors based on carbon nanomaterials *Chemosensors* **8** 96
[43] Bacon R 1960 Growth, structure, and properties of graphite whiskers *J. Appl. Phys.* **31** 283–90
[44] Zhang D-B, Akatyeva E and Dumitrică T 2011 Bending ultrathin graphene at the margins of continuum mechanics *Phys. Rev. Lett.* **106** 255503
[45] Odom T W, Huang J-L, Kim P and Lieber C M 2000 Structure and electronic properties of carbon nanotubes *J. Chem. Phys.* **104** 2794–809
[46] Ando Y and Iijima S 1993 Preparation of carbon nanotubes by arc-discharge evaporation *Jpn. J. Appl. Phys.* **32** L107–9
[47] Che G, Lakshmi B B, Martin C R, Fisher E R and Ruoff R S 1998 Chemical vapor deposition based synthesis of carbon nanotubes and nanofibers using a template method *Chem. Mater.* **10** 260–7
[48] Guo T, Nikolaev P, Thess A, Colbert D T and Smalley R E 1995 Catalytic growth of single-walled manotubes by laser vaporization *Chem. Phys. Lett.* **243** 49–54
[49] Kaur Billing B 2021 Carbon nanotubes and its potential application in sensing *ChemistrySelect* **6** 9571–90
[50] Liu Z, Tabakman S, Welsher K and Dai H 2009 Carbon nanotubes in biology and medicine: *in vitro* and *in vivo* detection, imaging and drug delivery *Nano Res.* **2** 85–120
[51] Balasubramanian K and Burghard M 2005 Chemically functionalized carbon nanotubes *Small* **1** 180–92
[52] Kocharova N, Ääritalo T, Leiro J, Kankare J and Lukkari J 2007 Aqueous dispersion, surface thiolation, and direct self-assembly of carbon nanotubes on gold *Langmuir* **23** 3363–71

[53] Gupta S, Murthy C N and Prabha C R 2018 Recent advances in carbon nanotube based electrochemical biosensors *Int. J. Biol. Macromol.* **108** 687–703
[54] Tang Y H, Wang N, Zhang Y F, Lee C S, Bello I and Lee S T 1999 Synthesis and characterization of amorphous carbon nanowires *Appl. Phys. Lett.* **75** 2921–3
[55] Kurbanoglu S and Ozkan S A 2018 Electrochemical carbon based nanosensors: a promising tool in pharmaceutical and biomedical analysis *J. Pharm. Biomed. Anal.* **147** 439–57
[56] Qin Z, Li Z J and Yang B C 2012 Synthesis of carbon nanowires as electrochemical electrode materials *Mater. Lett.* **69** 55–8
[57] Endo M, Kim Y A, Hayashi T, Fukai Y, Oshida K, Terrones M, Yanagisawa T, Higaki S and Dresselhaus M S 2002 Structural characterization of cup-stacked-type nanofibers with an entirely hollow core *Appl. Phys. Lett.* **80** 1267–9
[58] Hatzimarinaki M, Vamvakaki V and Chaniotakis N 2009 Spectro-electrochemical studies of acetylcholinesterase in carbon nanofiber-bioinspired silica nanocomposites for biosensor development *J. Mater. Chem.* **19** 428–33
[59] Vamvakaki V and Chaniotakis N A 2007 Carbon nanostructures as transducers in biosensors *Sens. Actuators B Chem.* **126** 193–7
[60] Andrews R, Jacques D, Rao A M, Rantell T, Derbyshire F, Chen Y, Chen J and Haddon R C 1999 Nanotube composite carbon fibers *Appl. Phys. Lett.* **75** 1329–31
[61] Guzmán A, Agüí L, Pedrero M, Yáñez-Sedeño P and Pingarrón J M 2002 Carbon fiber cylindrical microelectrode-based detector for the determination of antithyroid drugs *Talanta* **56** 577–84
[62] Iijima S and Ichihashi T 1993 Single-shell carbon nanotubes of 1-nm diameter *Nature* **363** 603–5
[63] Andreas T *et al* 1996 Crystalline ropes of metallic carbon nanotubes *Science* **273** 483–7
[64] Singh C, Shaffer M S P and Windle A H 2003 Production of controlled architectures of aligned carbon nanotubes by an injection chemical vapour deposition method *Carbon N. Y.* **41** 359–68
[65] Saidin M A R, Ismail A F, Sanip S M, Goh P S, Aziz M and Tanemura M 2012 Controlled growth of carbon nanofibers using plasma enhanced chemical vapor deposition: Effect of catalyst thickness and gas ratio *Thin Solid Films* **520** 2575–81
[66] Siddiqui S, Arumugam P U, Chen H, Li J and Meyyappan M 2010 Characterization of carbon nanofiber electrode arrays using electrochemical impedance spectroscopy: effect of scaling down electrode size *ACS Nano* **4** 955–61
[67] Kour R, Arya S, Young S, Gupta V and Bandhoria P 2020 Review—recent advances in carbon nanomaterials as electrochemical biosensors *J. Electrochem. Soc.* **167** 037555
[68] Hu C, Li M, Qiu J and Sun Y-P 2019 Design and fabrication of carbon dots for energy conversion and storage *Chem. Soc. Rev.* **48** 2315–37
[69] Sun Y-P *et al* 2006 Quantum-sized carbon dots for bright and colorful photoluminescence *J. Am. Chem. Soc.* **128** 7756–7
[70] Ding X, Niu Y, Zhang G, Xu Y and Li J 2020 Electrochemistry in carbon-based quantum dots *Chem.—An Asian J.* **15** 1214–24
[71] Kumar Y R, Deshmukh K, Sadasivuni K K and Pasha S K K 2020 Graphene quantum dot based materials for sensing, bio-imaging and energy storage applications: a review *RSC Adv.* **10** 23861–98
[72] Dai M, Haselwood B, Vogt B D and La Belle J T 2013 Amperometric sensing of norepinephrine at picomolar concentrations using screen printed, high surface area mesoporous carbon *Anal. Chim. Acta* **788** 32–8

[73] Tuteja S K, Ormsby C and Neethirajan S 2018 Noninvasive label-free detection of cortisol and lactate using graphene embedded screen-printed electrode *Nano-Micro Lett.* **10** 41

[74] Smajdor J, Piech R, Martyna Ł and Paczosa-bator B 2018 Glassy carbon electrode modi fi ed with carbon black for sensitive estradiol determination by means of voltammetry and fl ow injection analysis with amperometric detection *Anal. Biochem.* **544** 7–12

[75] Mardani L, Vardini M T, Es'Haghi M and Ghorbani Kalhor E 2020 Design and construction of a carbon paste electrode modified with molecularly imprinted polymer-grafted nanocomposites for the determination of thyroxin in biological samples *Anal. Methods* **12** 333–44

[76] Moreira F, Santana E R and Spinelli A 2020 Ionic liquid-supported magnetite nanoparticles as electrode modifier materials for estrogens sensing *Sci. Rep.* **10** 1955

[77] Ghanbary E, Asiabani Z, Hosseini N, Hossein S and Kaki S 2020 The development of a new modified graphite pencil electrode for quantitative detection of Gibberellic acid (GA3) herbal hormone *Microchem. J.* **157** 105005

[78] Bibani M, Ktari N, Fourati N, Zerrouki C and Kalfat R 2021 A better understanding of diethylstilbestrol electro-oxidation: towards the design of an electrochemical *Electroanalysis* **33** 1014–23

[79] Baghayeri M, Nodehi M, Veisi H, Tehrani M B and Maleki B 2019 The role of pramipexole functionalized MWCNTs to the fabrication of Pd nanoparticles modified GCE for electrochemical detection of dopamine *J. Pharm. Sci.* **27** 593–603

[80] Cai R, Rao W, Zhang Z, Long F and Yin Y 2014 An imprinted electrochemical sensor for bisphenol A determination based on electrodeposition of a graphene and Ag nanoparticle modified carbon electrode *Anal. Methods* **6** 1590–7

[81] Zehani N, Fortgang P, Saddek Lachgar M, Baraket A, Arab M, Dzyadevych S V, Kherrat R and Jaffrezic-Renault N 2015 Highly sensitive electrochemical biosensor for bisphenol A detection based on a diazonium-functionalized boron-doped diamond electrode modified with a multiwalled carbon nanotube-tyrosinase hybrid film *Biosens. Bioelectron.* **74** 830–5

[82] Xin X, Sun S, Wang M and Jia R 2020 Sensitive detection of estriol with an electrochemical sensor based on core–shell N-MWCNT/GONR-imprinted electrode *Ionics* **26** 2633–41

[83] Beitollahi H, Ivari S G and Torkzadeh-Mahani M 2018 Application of antibody–nanogold–ionic liquid–carbon paste electrode for sensitive electrochemical immunoassay of thyroid-stimulating hormone *Biosens. Bioelectron.* **110** 97–102

[84] Zittel H E and Miller F J 1995 A glassy-carbon electrode for voltammetry *Anal. Chem.* **2** 200–3

[85] Goyal R N, Oyama M, Gupta V K, Singh S P and Sharma R A 2008 Sensors for 5-hydroxytryptamine and 5-hydroxyindole acetic acid based on nanomaterial modified electrodes *Sens. Actuators B Chem.* **134** 816–21

[86] Sharma S, Singh N, Tomar V and Chandra R 2018 A review on electrochemical detection of serotonin based on surface modified electrodes *Biosens. Bioelectron.* **107** 76–93

[87] Sun X, Jiang Z, Wang H and Zhao H 2015 Highly sensitive detection of peptide hormone prolactin using gold nanoparticles-graphene nanocomposite modified electrode *Int. J. Electrochem. Sci.* **10** 9714–24

[88] Pereira P A R, Gomes N O, Machado S A S and Oliveira Junior O N 2019 Simultaneous, ultrasensitive detection of hydroquinone, paracetamol and estradiol for quality control of tap water with a simple electrochemical method *J. Electroanal. Chem.* **848** 113317–9

[89] Wring S A, Hart J P, Bracey L and Birch B J 1990 Development of screen-printed carbon electrodes, chemically modified with cobalt phthalocyanine, for electrochemical sensor applications *Anal. Chim. Acta* **231** 203–12
[90] Mohamed H M 2016 Screen-printed disposable electrodes: Pharmaceutical applications and recent developments *TrAC, Trends Anal. Chem.* **82** 1–11
[91] Fletcher S 2015 Screen-printed carbon electrodes *Electrochemistry of Carbon Electrodes* pp 425–44
[92] Niu X, Lan M, Zhao H, Chen C, Li Y, Zhu X, Li Y and Zhu X 2013 Review: electrochemical stripping analysis of trace heavy metals using screen-printed electrodes *Anal. Lett.* **2719** 2479–502
[93] Lee M, Thomas J L, Liu W, Zhang Z, Liu B, Yang C and Lin H 2019 A multichannel system integrating molecularly imprinted conductive polymers for ultrasensitive voltammetric determination of four steroid hormones in urine *Microchim. Acta* **186** 695
[94] Liu X *et al* 2021 Template-free synthesis of mesoporous Ce_3NbO_7/CeO_2 hollow nanospheres for label-free electrochemical immunosensing of leptin *Sens. Actuators B Chem.* **341** 130005
[95] Adams R 1958 Carbon paste electrodes *Anal. Chem.* **30** 1576
[96] Tajik S, Beitollahi H, Nejad G, Safaei M, Van Le Q, Varma R S, Jang H W and Shokouhimeh M 2020 Developments and applications of nanomaterial-based carbon paste electrodes *RSC Adv.* **10** 21561–81
[97] Kalcher K 1990 Chemically modified carbon paste electrodes in voltammetric analysis *Electroanalysis* **2** 419–33
[98] Švancara I, Vytřas K, Barek J and Zima J 2001 Carbon paste electrodes in modern electroanalysis *Crit. Rev. Anal. Chem.* **31** 311–45
[99] Beitollahi H, Nekooei S and Torkzadeh-Mahani M 2018 Amperometric immunosensor for prolactin hormone measurement using antibodies loaded on a nano-Au monolayer modified ionic liquid carbon paste electrode *Talanta* **188** 701–7
[100] Bonyadi S, Ghanbari K and Ghiasi M 2020 All-electrochemical synthesis of a three-dimensional mesoporous polymeric g-C3N4/PANI/CdO nanocomposite and its application as a novel sensor for the simultaneous determination of epinephrine, paracetamol, mefenamic acid, and ciprofloxacin *New J. Chem.* **44** 3412–24
[101] Li L, Yu Y, Chen P and Liu W 2017 A label-free electrochemical immunosensor for anti-mullerian hormone detection on graphene-Au nanocomposite *Int. J. Electrochem. Sci.* **12** 7911–20
[102] Zaidi S A 2018 Utilization of an environmentally-friendly monomer for an efficient and sustainable adrenaline imprinted electrochemical sensor using graphene *Electrochim. Acta* **274** 370–7
[103] Harisha K V, Swamy B E K, Jayadevappa H and Ganesh P S 2016 Electrochemical sensor for the determination of adrenaline in presence of acetaminophen at poly(alanine) modified carbon paste electrode: a cyclic voltammetric study *J. Biosens. Bioelectron.* **7** 1–6
[104] Kumar Gupta V 2017 Electrochemical determination of adrenaline using voltammetric sensor employing NiO/CNTs based carbon paste electrode *Int. J. Electrochem. Sci.* **8** 248–57
[105] Tezerjani M D, Benvidi A, Dehghani Firouzabadi A, Mazloum-Ardakani M and Akbari A 2017 Epinephrine electrochemical sensor based on a carbon paste electrode modified with hydroquinone derivative and graphene oxide nanosheets: Simultaneous determination of epinephrine, acetaminophen and dopamine *Measurement* **101** 183–9

[106] Raymundo-Pereira P A, Lima A R F and Machado S A S 2016 A nanostructured label-free platform based on an ultrathin film for ultrasensitive detection of a secosteroid hormone *RSC Adv.* **6** 34458–67

[107] Charoenkitamorn K, Tue P T, Kawai K, Chailapakul O and Takamura Y 2018 Electrochemical immunoassay using open circuit potential detection labeled by platinum nanoparticles *Sensors* **18** 444

[108] Lu L, Liang L, Xie Y, Tang K, Wan Z and Chen S 2018 A nickel nanoparticle/carbon nanotube-modified carbon fiber microelectrode for sensitive insulin detection *J. Solid State Electrochem.* **22** 825–33

[109] Filho M M S L, Correa A A, Silva F D C, Carvalho F A O, Mascaro L H and Oliveira T M B F 2019 A glassy carbon electrode modified with silver nanoparticles and functionalized multiwalled carbon nanotubes for voltammetric determination of the illicit growth promoter dienestrol in animal urine *Microchim Acta* **186** 525

[110] Li Q, Xia Y, Wan X, Yang S, Cai Z, Ye Y and Li G 2020 Morphology-dependent MnO2/nitrogen-doped graphene nanocomposites for simultaneous detection of trace dopamine and uric acid *Mater. Sci. Eng.* C **109** 110615

[111] Yang Z, Zheng X and Zheng J 2017 A facile one-step synthesis of Fe2O3/nitrogen-doped reduced graphene oxide nanocomposite for enhanced electrochemical determination of dopamine *J. Alloys Compd.* **709** 581–87

[112] Liu J, Sun L, Li G, Hu J and He Q 2021 Ultrasensitive detection of dopamine via electrochemical route on spindle-like α-Fe2O3 Mesocrystals/rGO modified GCE *Mater. Res. Bull.* **133** 111050

[113] Mao H, Zhang H, Jiang W, Liang J, Sun Y, Zhang Y, Wu Q, Zhang G and Song X M 2017 Poly(ionic liquid) functionalized polypyrrole nanotubes supported gold nanoparticles: An efficient electrochemical sensor to detect epinephrine *Mater. Sci. Eng.* C **75** 495–502

[114] Mattioli I A, Cervini P and Cavalheiro É T G 2020 Screen-printed disposable electrodes using graphite-polyurethane composites modified with magnetite and chitosan-coated magnetite nanoparticles for voltammetric epinephrine sensing: a comparative study *Microchim. Acta* **187** 318

[115] Dehdashti A and Babaei A 2020 Designing and characterization of a novel sensing platform based on Pt doped NiO/MWCNTs nanocomposite for enhanced electrochemical determination of epinephrine and tramadol simultaneously *J. Electroanal. Chem.* **862** 113949

[116] Samadzadeh A, Sheikhshoaie I and Karimi-Maleh H 2019 Simultaneous determination of epinephrine and tyrosine using a glassy carbon electrode amplified with ZnO-Pt/CNTs nanocomposite *Curr. Anal. Chem.* **15** 166–71

[117] Wu C, Li J, Liu X, Zhang H, Li R, Wang G, Wang Z, Li Q and Shangguan E 2021 Simultaneous voltammetric determination of epinephrine and acetaminophen using a highly sensitive CoAl-OOH/reduced graphene oxide sensor in pharmaceutical samples and biological fluids *Mater. Sci. Eng.* C **119** 111557

[118] Sainz R *et al* 2020 Chemically synthesized chevron-like graphene nanoribbons for electrochemical sensors development: determination of epinephrine *Sci. Rep.* **10** 14614

[119] Anithaa A C, Asokan K and Sekar C 2017 Voltammetric determination of epinephrine and xanthine based on sodium dodecyl sulphate assisted tungsten trioxide nanoparticles *Electrochim. Acta* **237** 44–53

[120] Manavalan S, Rajaji U, Chen S M, Govindasamy M, Selvin S S P, Chen T W, Ali M A, Al-Hemaid F M A and Elshikh M S 2019 Sonochemical synthesis of bismuth(III) oxide

decorated reduced graphene oxide nanocomposite for detection of hormone (epinephrine) in human and rat serum *Ultrason. Sonochem.* **51** 103–10

[121] Tavakkoli N, Soltani N, Shahdost-fard F, Ramezani M, Salavati H and Jalali M R 2018 Simultaneous voltammetric sensing of acetaminophen, epinephrine and melatonin using a carbon paste electrode modified with zinc ferrite nanoparticles *Microchim. Acta* **185** 479

[122] Triviño J J, Gómez M, Valenzuela J, Vera A and Arancibia V 2019 Determination of a natural (17B-estradiol) and a synthetic (17A-ethinylestradiol) hormones in pharmaceutical formulations and urine by adsorptive stripping voltammetry *Sens. Actuators, B Chem.* **297** 126728

[123] Souza M B, Santos J S, Pontes M S, Nunes L R, Oliveira I P, Lopez Ayme A J, Santiago E F, Grillo R, Fiorucci A R and Arruda G J 2022 CeO_2 nanostructured electrochemical sensor for the simultaneous recognition of diethylstilbestrol and 17β-estradiol hormones *Sci. Total Environ.* **805** 150348

[124] Chen X, Shi Z, Hu Y, Xiao X and Li G 2018 A novel electrochemical sensor based on Fe_3O_4-doped nanoporous carbon for simultaneous determination of diethylstilbestrol and 17β-estradiol in toner *Talanta* **188** 81–90

[125] Nunes da Silva D, Leijoto de Oliveira H, Borges K B and Pereira A C 2021 Sensitive determination of 17β-estradiol using a magneto sensor based on magnetic molecularly imprinted polymer *Electroanalysis* **33** 506–14

[126] Chang Z, Zhu B, Liu J, Zhu X, Xu M and Travas-Sejdic J 2021 Electrochemical aptasensor for 17β-estradiol using disposable laser scribed graphene electrodes *Biosens. Bioelectron.* **185** 113247

[127] Lahcen A A, Baleg A A, Baker P, Iwuoha E and Amine A 2017 Synthesis and electrochemical characterization of nanostructured magnetic molecularly imprinted polymers for 17-β-Estradiol determination *Sens. Actuators B Chem.* **241** 698–705

[128] Supchocksoonthorn P, Alvior Sinoy M C, de Luna M D G and Paoprasert P 2021 Facile fabrication of 17β-estradiol electrochemical sensor using polyaniline/carbon dot-coated glassy carbon electrode with synergistically enhanced electrochemical stability *Talanta* **235** 122782

[129] Moraes F C, Rossi B, Donatoni M C, De Oliveira K T and Pereira E C 2015 Sensitive determination of 17β-estradiol in river water using a graphene based electrochemical sensor *Anal. Chim. Acta* **881** 37–43

[130] Wang Y, Luo J, Liu J, Li X, Kong Z, Jin H and Cai X 2018 Electrochemical integrated paper-based immunosensor modified with multiwalled carbon nanotubes nanocomposites for point-of-care testing of 17β-estradiol *Biosens. Bioelectron.* **107** 47–53

[131] Raj M and Goyal R N 2019 Silver nanoparticles and electrochemically reduced graphene oxide nanocomposite based biosensor for determining the effect of caffeine on Estradiol release in women of child-bearing age *Sens. Actuators B Chem.* **284** 759–67

[132] Futra D, Heng L Y, Jaapar M Z, Ulianas A, Saeedfar K and Ling T L 2016 A novel electrochemical sensor for 17β-estradiol from molecularly imprinted polymeric microspheres and multiwalled carbon nanotubes grafted with gold nanoparticles *Anal. Methods* **8** 1381–9

[133] Mauruto de Oliveira G C, de Palma E P, Kunita M H, Antigo Medeiros R, de Matos R, Francisco K R and Janegitz B C 2017 Tapioca biofilm containing nitrogen-doped titanium dioxide nanoparticles for electrochemical detection of 17-β estradiol *Electroanalysis* **29** 2638–45

[134] Wong A, Martin A, Luiz E, Fatibello-filho O, Del M and Taboada P 2019 Voltammetric determination of 17β-estradiol in different matrices using a screen-printed sensor modified with CuPc, Printex 6L carbon and nafion film *Microchem. J.* **147** 365–73

[135] Arvand M and Hemmati S 2017 Analytical methodology for the electrocatalytic determination of estradiol and progesterone based on graphene quantum dots and poly(sulfosalicylic acid) co-modified electrode *Talanta* **174** 243–55

[136] Cesarino I, Cincotto F H and Machado S A S 2015 A synergistic combination of reduced graphene oxide and antimony nanoparticles for estriol hormone detection *Sens. ActuatorsB. Chem.* **210** 453–9

[137] Zhao Q, Faraj Y, Liu L, Wang W, Xie R, Liu Z and Ju X 2020 Simultaneous determination of dopamine, uric acid and estriol in maternal urine samples based on the synergetic effect of reduced graphene oxide, silver nanowires and silver nanoparticles in their ternary 3D nanocomposite *Microchem. J.* **158** 105185

[138] Ochiai L M, Agustini D, Figueiredo-filho L C S, Banks C E, Marcolino-junior L H and Bergamini M F 2017 Chemical electroanalytical thread-device for estriol determination using screen-printed carbon electrodes modified with carbon nanotubes *Sens. ActuatorsB. Chem.* **241** 978–84

[139] Raymundo-pereira P A *et al* 2017 Sensitive detection of estriol hormone in creek water using a sensor platform based on carbon black and silver nanoparticles *Talanta* **174** 652–9

[140] Fan Y, Guo Y, Shi S and Ma J 2021 An electrochemical immunosensor based on reduced graphene oxide/multiwalled carbon nanotubes/thionine/gold nanoparticle nanocomposites for the sensitive testing of follicle-stimulating hormone *Anal. Methods* **13** 3821–8

[141] Li X, Lin L-Y, Wang K-Y, Li J, Feng L, Song L, Liu X, He J-H, Sakthivel R and Chung R-J 2021 Streptavidin-functionalized-polyethyleneimine/chitosan/HfO2-Pr6O11 nanocomposite using label-free electrochemical immunosensor for detecting the hunger hormone ghrelin *Compos. Part B Eng.* **224** 109231

[142] Martínez-García G, Agüí L, Yáñez-Sedeño P and Pingarrón J M 2016 Multiplexed electrochemical immunosensing of obesity-related hormones at grafted graphene-modified electrodes *Electrochim. Acta* **202** 209–15

[143] Ebrahimiasl S, Fathi E and Ahmad M 2018 Electrochemical detection of insulin in blood serum using Ppy/GF nanocomposite modified pencil graphite electrode *Nanomed. Res. J.* **3** 219–28

[144] Wang J 2021 Au–Ag nanoparticles-graphene quantum dots as sensor for highly sensitive electrochemical determination of insulin level in pharmaceutical samples *Int. J. Electrochem. Sci.* **16** 211016

[145] Freitas R C, Orzari L O, Ferreira L M C, Paixão T R L C, Coltro W K T, Vicentini F C and Janegitz B C 2021 Electrochemical determination of melatonin using disposable self-adhesive inked paper electrode *J. Electroanal. Chem.* **897** 115550

[146] Camargo J R, Andreotti I A A, Kalinke C, Henrique J M, Bonacin J A and Janegitz B C 2020 Talanta waterproof paper as a new substrate to construct a disposable sensor for the electrochemical determination of paracetamol and melatonin *Talanta* **208** 120458

[147] Gomez F J V, Martín A, Silva M F and Escarpa A 2015 Screen-printed electrodes modified with carbon nanotubes or graphene for simultaneous determination of melatonin and serotonin *Microchim. Acta* **182** 1925–31

[148] Malla P, Liao H-P, Liu C-H and Wu W-C 2021 Electrochemical immunoassay for serum parathyroid hormone using screen-printed carbon electrode and magnetic beads *J. Electroanal. Chem.* **895** 115463

[149] Malla P, Chen G-C, Liao H-P, Liu C-H and Wu W-C 2021 Label-free parathyroid hormone immunosensor using nanocomposite modified carbon electrode *J. Electroanal. Chem.* **880** 114917

[150] Zhu Y, Xu Z, Gao J, Ji W and Zhang J 2020 An antibody-aptamer sandwich cathodic photoelectrochemical biosensor for the detection of progesterone *Biosens. Bioelectron.* **160** 112210

[151] Dong X-X, Yuan L-P, Liu Y-X, Wu M-F, Liu B, Sun Y-M, Shen Y-D and Xu Z-L 2017 Development of a progesterone immunosensor based on thionine-graphene oxide composites platforms: improvement by biotin-streptavidin-amplified system *Talanta* **170** 502–8

[152] Shamsipur M, Pashabadi A, Taherpour A, Bahrami K and Sharghi H 2017 Manganese mediated oxidation of progesterone in alkaline medium: mechanism study and quantitative determination *Electrochim. Acta* **225** 292–302

[153] Govindasamy M, Subramanian B, Wang S-F, Chinnapaiyan S, Jothi Ramalingam R and Al-lohedan H A 2019 Ultrasound-assisted synthesis of tungsten trioxide entrapped with graphene nanosheets for developing nanomolar electrochemical (hormone) sensor and enhanced sensitivity of the catalytic performance *Ultrason. Sonochem.* **56** 134–42

[154] Serafín V, Martínez-García G, Aznar-Poveda J, Lopez-Pastor J A, Garcia-Sanchez A J, Garcia-Haro J, Campuzano S, Yáñez-Sedeño P and Pingarrón J M 2019 Determination of progesterone in saliva using an electrochemical immunosensor and a COTS-based portable potentiostat *Anal. Chim. Acta* **1049** 65–73

[155] Samie H A and Arvand M 2020 Label-free electrochemical aptasensor for progesterone detection in biological fluids *Bioelectrochemistry* **133** 107489

[156] Gevaerd A, Blaskievicz S F, Zarbin A J G, Orth E S, Bergamini M F and Marcolino-Junior L H 2018 Nonenzymatic electrochemical sensor based on imidazole-functionalized graphene oxide for progesterone detection *Biosens. Bioelectron.* **112** 108–13

[157] Arvand M, Elyan S and Ardaki M S 2019 Facile one-pot electrochemical synthesis of zirconium oxide decorated poly(3,4-ethylenedioxythiophene) nanocomposite for the electrocatalytic oxidation and detection of progesterone *Sens. Actuators B Chem.* **281** 157–67

[158] Arvand M and Hemmati S 2017 Magnetic nanoparticles embedded with graphene quantum dots and multiwalled carbon nanotubes as a sensing platform for electrochemical detection of progesterone *Sens. Actuators B Chem.* **238** 346–56

[159] Das A and Sangaranarayanan M V 2018 A sensitive electrochemical detection of progesterone using tin-nanorods modified glassy carbon electrodes: voltammetric and computational studies *Sens. Actuators B Chem.* **256** 775–89

[160] Sakthivel R, Lin L-Y, Lee T-H, Liu X, He J-H and Chung R-J 2022 Disposable and cost-effective label-free electrochemical immunosensor for prolactin based on bismuth sulfide nanorods with polypyrrole *Bioelectrochemistry* **143** 107948

[161] Li S, Yan Y, Zhong L, Liu P, Sang Y, Cheng W and Ding S 2015 Electrochemical sandwich immunoassay for the peptide hormone prolactin using an electrode modified with graphene, single walled carbon nanotubes and antibody-coated gold nanoparticles *Microchim. Acta* **182** 1917–24

[162] Beitollahi H, Nekooei S and Torkzadeh-mahani M 2018 Talanta amperometric immunosensor for prolactin hormone measurement using antibodies loaded on a nano-Au monolayer modi fi ed ionic liquid carbon paste electrode *Talanta* **188** 701–7

[163] Zhang F, Huang F, Gong W, Tian F, Wu H, Ding S, Li S and Luo R 2021 Multi-branched PdPt nanodendrites decorated amino-rich Fe-based metal-organic framework as signal

amplifier for ultrasensitive electrochemical detection of prolactin *J. Electroanal. Chem.* **882** 115032

[164] Arévalo B, Serafín V, Campuzano S, Yáñez-Sedeño P and Pingarrón J M 2021 Electrochemical immunosensor for the determination of prolactin in saliva and breast milk *Microchem. J.* **169** 106589

[165] Heidarimoghadam R, Akhavan O, Ghaderi E, Hashemi E, Mortazavi S S and Farmany A 2016 Graphene oxide for rapid determination of testosterone in the presence of cetyltrimethylammonium bromide in urine and blood plasma of athletes *Mater. Sci. Eng.* C **61** 246–50

[166] Liu W, Ma Y, Sun G, Wang S, Deng J and Wei H 2017 Biosensors and bioelectronics molecularly imprinted polymers on graphene oxide surface for EIS sensing of testosterone *Biosens. Bioelectron.* **92** 305–12

[167] Saxena R and Srivastava S 2019 A sensitive and one-step quantification of thyroid stimulating hormone using nanobiosensor *Mater. Today Proc.* **18** 1351–57

[168] Smaniotto A, Mezalira D Z, Zapp E, Gallardo H and Vieira I C 2017 Electrochemical immunosensor based on an azo compound for thyroid-stimulating hormone detection *Microchem. J.* **133** 510–7

[169] Pacheco J G, Rebelo P, Cagide F, Gonçalves L M, Borges F, Rodrigues J A and Delerue-Matos C 2019 Electrochemical sensing of the thyroid hormone thyronamine (T0AM) via molecular imprinted polymers (MIPs) *Talanta* **194** 689–96

www.ingramcontent.com/pod-product-compliance
Ingram Content Group UK Ltd.
Pitfield, Milton Keynes, MK11 3LW, UK
UKHW051042120926
8556IPUK00030B/800

9 780750 351256